Porsche 911 SC

Service Manual

Coupe, Targa and Cabriolet

1978, 1979, 1980, 1981, 1982, 1983

BentleyPublishers®
.com

BENTLEY PUBLISHERS® | Automotive Reference™

Bentley Publishers, a division of Robert Bentley, Inc.
1734 Massachusetts Avenue
Cambridge, MA 02138 USA
800-423-4595 / 617-547-4170

Information that makes the difference®

BentleyPublishers®.com

Technical contact information

We welcome your feedback. Please submit corrections and additions to our Porsche technical discussion forum at:

`http://www.BentleyPublishers.com`

Updates and corrections

We will evaluate submissions and post appropriate editorial changes online as updates or tech discussion. Appropriate updates and corrections will be added to the book in future printings. Check for updates and corrections for this book before beginning work on your vehicle. See the following web address for additional information:

`http://www.BentleyPublishers.com/updates/`

WARNING—Important Safety Notice

Do not use this manual unless you are familiar with automotive repair procedures and safe workshop practices. This manual illustrates the workshop procedures required for some maintenance and service work. It is not a substitute for full and up-to-date information from the vehicle manufacturer or for proper training as an automotive technician. Note that it is not possible for us to anticipate all of the ways or conditions under which vehicles may be serviced or to provide cautions as to all of the possible hazards that may result.

We have endeavored to ensure the accuracy of the information in this manual. Please note, however, that considering the vast quantity and the complexity of the service information involved, we cannot warrant the accuracy or completeness of the information contained in this manual.

FOR THESE REASONS, NEITHER THE PUBLISHER NOR THE AUTHOR MAKES ANY WARRANTIES, EXPRESS OR IMPLIED, THAT THE INFORMATION IN THIS BOOK IS FREE OF ERRORS OR OMISSIONS, AND WE EXPRESSLY DISCLAIM THE IMPLIED WARRANTIES OF MERCHANTABILITY AND OF FITNESS FOR A PARTICULAR PURPOSE, EVEN IF THE PUBLISHER OR AUTHOR HAVE BEEN ADVISED OF A PARTICULAR PURPOSE, AND EVEN IF A PARTICULAR PURPOSE IS INDICATED IN THE MANUAL. THE PUBLISHER AND AUTHOR ALSO DISCLAIM ALL LIABILITY FOR DIRECT, INDIRECT, INCIDENTAL OR CONSEQUENTIAL DAMAGES THAT RESULT FROM ANY USE OF THE EXAMPLES, INSTRUCTIONS OR OTHER INFORMATION IN THIS BOOK. IN NO EVENT SHALL OUR LIABILITY WHETHER IN TORT, CONTRACT OR OTHERWISE EXCEED THE COST OF THIS MANUAL.

Before attempting any work on your Porsche, read **Warnings and Cautions** (pages v, vi) and any **WARNING** or **CAUTION** that accompanies a procedure in this manual. Review the **WARNINGS** and **CAUTIONS** each time you prepare to work on your Porsche.

Your common sense and good judgment are crucial to safe and successful service work. Read procedures through before starting them. Think about whether the condition of your car, your level of mechanical skill, or your level of reading comprehension might result in or contribute in some way to an occurrence which might cause you injury, damage your car, or result in an unsafe repair. If you have doubts for these or other reasons about your ability to perform safe repair work on your car, have the work done at an authorized Porsche dealer or other qualified shop.

Part numbers listed in this manual are for identification purposes only, not for ordering. Always check with your authorized Porsche dealer to verify part numbers and availability before beginning service work that may require new parts.

Special tools required to perform certain service operations are identified in the manual and are recommended for use. Use of tools other than those recommended in this manual may be detrimental to the car's safe operation as well as the safety of the person servicing the car.

The vehicle manufacturer will continue to issue service information updates and parts retrofits after the editorial closing of this manual. Some of these updates and retrofits will apply to procedures and specifications in this manual. We regret that we cannot supply updates to purchasers of this manual.

ISBN 978-0-8376-1705-3 Job code: P983-10, editorial closing 10/00.

Library of Congress Cataloging-in-Publication Data for 2000 edition

Porsche 911 SC : service manual : Coupe, Targa, and Cabriolet : 1978, 1979, 1980, 1981, 1982, 1983.
p. cm.
Includes index.
ISBN 978-0-8376-0290-4 (pbk. : alk. paper)
1. Porsche 911 automobile--Maintenance and repair--Handbooks, manuals, etc.

TL215.P75 P633 2000
629.28'722--dc21

00-064196

Manufactured in the United States of America.

202510R003

Foreword

For the Porsche owner with mechanical skills, and for independent service professionals, this manual includes maintenance and repair procedures the Porsche 911 SC is likely to need during its service life, including many procedures and specifications that were available in an authorized Porsche dealer service department as the time this manual was prepared. The Porsche owner with no intention of working on his or her car will find that owning and referring to this manual will make it possible to be better informed and to more knowledgeably discuss repairs with a professional automotive technician.

For those intending to do maintenance and repair on the 911 SC, it is essential that safety equipment be used and safety precautions observed when working on your vehicle. A minimum safety equipment list includes hand protection, eye protection, and a fire extinguisher. A selection of good quality hand tools is also needed. This includes a torque wrench to ensure that fasteners are tightened in accordance with specifications. In some cases, the text refers to special tools that are recommended or required to accomplish adjustments or repairs. These tools are often identified by their Porsche special tool number and illustrated. Additional information on safety and tools is in **010 General**.

Disclaimer

We have endeavored to ensure the accuracy of the information in this manual. When the vast array of data presented in the manual is taken into account, however, no claim to infallibility can be made. We therefore cannot be responsible for the result of any errors that may have crept into the text. Please also read the **WARNING–Important Safety Notice** on the copyright page at the beginning of this manual.

A thorough prereading of each procedure, the **WARNINGS** and **CAUTIONS** found on pages v, vi, and any **WARNING** or **CAUTION** that accompanies the procedure is essential. Reading a procedure before beginning work helps you determine in advance the need for specific skills, identify hazards, prepare for appropriate capture and handling of hazardous materials and the need for particular tools, safety equipment, and replacement parts such as gaskets.

Bentley Publishers encourages comments from the readers of this manual with regard to errors, and suggestions for improvement of our product. These communications have been and will be carefully considered in the preparation of this and other manuals. If you identify inconsistencies in the manual, you may have found an error. Please contact the publisher and we will endeavor to post applicable corrections on our website. Review updates and corrections that we have posted before beginning work. Please see the following web address:

`http://www.BentleyPublishers.com/updates/`

Porsche continues to issue service information and parts retrofits after the editorial closing of this manual. Some of this updated information may apply to procedures and specifications in this book. For the latest information, please see the following web address:

`https://techinfo.porsche.com`

Porsche offers extensive warranties, especially on components of the fuel delivery and emission control systems. Therefore, before deciding to repair a Porsche that may be covered wholly or in part by any warranties issued by Porsche Car North America, Inc., consult your authorized Porsche dealer. You may find that the dealer can make the repair either free or at minimum cost. Regardless of its age, or whether it is under warranty, your Porsche is both an easy car to service and an easy car to get serviced. So if at any time a repair is needed that you feel is too difficult to do yourself, a trained Porsche technician is ready to do the job for you.

Bentley Publishers

Please read these warnings and cautions before proceeding with maintenance and repair work.

WARNINGS

- See also **Cautions** on page vi

- Some repairs may be beyond your capability. If you lack the skills, tools and equipment, or a suitable workplace for any procedure described in this manual, we suggest you leave such repairs to an authorized Porsche dealer service department or other qualified shop.

- Do not re-use any fasteners that are worn or deformed in normal use. Many fasteners are designed to be used only once and become unreliable and may fail when used a second time. This includes, but is not limited to, nuts, bolts, washers, self-locking nuts or bolts, circlips and cotter pins. Always replace these fasteners with new parts.

- Never work under a lifted car unless it is solidly supported on stands designed for the purpose. Do not support a car on cinder blocks, hollow tiles or other props that may crumble under continuous load. Never work under a car that is supported solely by a jack. Never work under the car while the engine is running.

- If you are going to work under a car on the ground, make sure that the ground is level. Block the wheels to keep the car from rolling. Disconnect the battery negative (–) terminal (ground strap) to prevent others from starting the car while you are under it.

- Never run the engine unless the work area is well ventilated. Carbon monoxide kills.

- Finger rings, bracelets and other jewelry should be removed so that they cannot cause electrical shorts, get caught in running machinery, or be crushed by heavy parts.

- Tie long hair behind your head. Do not wear a necktie, a scarf, loose clothing, or a necklace when you work near machine tools or running engines. If your hair, clothing, or jewelry were to get caught in the machinery, severe injury could result.

- Do not attempt to work on your car if you do not feel well. You increase the danger of injury to yourself and others if you are tired, upset or have taken medication or any other substance that may keep you from being fully alert.

- Illuminate your work area adequately but safely. Use a portable safety light for working inside or under the car. Make sure the bulb is enclosed by a wire cage. The hot filament of an accidentally broken bulb can ignite spilled fuel or oil.

- Catch draining fuel, oil, or brake fluid in suitable containers. Do not use food or beverage containers that might mislead someone into drinking from them. Store flammable fluids away from fire hazards. Wipe up spills at once, but do not store the oily rags, which can ignite and burn spontaneously.

- Always observe good workshop practices. Wear goggles when you operate machine tools or work with battery acid. Gloves or other protective clothing should be worn whenever the job requires working with harmful substances.

- Greases, lubricants and other automotive chemicals contain toxic substances, many of which are absorbed directly through the skin. Read the manufacturer's instructions and warnings carefully. Use hand and eye protection. Avoid direct skin contact

- Disconnect the battery negative (–) terminal (Ground strap) whenever you work on the fuel system or the electrical system. Do not smoke or work near heaters or other fire hazards. Keep an approved fire extinguisher handy.

- Friction materials (such as brake pads or shoes or clutch discs) contain asbestos fibers or other friction materials. Do not create dust by grinding, sanding, or by cleaning with compressed air. Avoid breathing dust. Breathing any friction material dust can lead to serious diseases and may result in death.

- Batteries give off explosive hydrogen gas during charging. Keep sparks, lighted matches and open flame away from the top of the battery. If hydrogen gas escaping from the cap vents is ignited, it will ignite gas trapped in the cells and cause the battery to explode.

- Connect and disconnect battery cables, jumper cables or a battery charger only with the ignition switched off, to prevent sparks. Do not disconnect the battery while the engine is running.

- Do not quick-charge the battery (for boost starting) for longer than one minute. Wait at least one minute before boosting the battery a second time.

- Do not allow battery charging voltage to exceed 16.5 volts. If the battery begins producing gas or boiling violently, reduce the charging rate. Boosting a sulfated battery at a high charging rate can cause an explosion.

- The air conditioning system is filled with chemical refrigerant, which is hazardous. The A/C system should be serviced only by trained technicians using approved refrigerant recovery/recycling equipment, trained in related safety precautions, and familiar with regulations governing the discharging and disposal of automotive chemical refrigerants.

- Do not expose any part of the A/C system to high temperatures such as open flame. Excessive heat will increase system pressure and may cause the system to burst.

- Some aerosol tire inflators are highly flammable. Be extremely cautious when repairing a tire that may have been inflated using an aerosol tire inflator. Keep sparks, open flame or other sources of ignition away from the tire repair area. Inflate and deflate the tire at least four times before breaking the bead from the rim. Completely remove the tire from the rim before attempting any repair.

- The ignition system produces high voltages that can be fatal. Avoid contact with exposed terminals and use extreme care when working on a car with the engine running or the ignition switched on.

- Place jack stands only at locations specified by the manufacturer. The vehicle lifting jack supplied with the vehicle is intended for tire changes only. A heavy duty floor jack should be used to lift vehicle before installing jack stands. See **010 General**.

- Battery acid (electrolyte) can cause severe burns. Flush contact area with water, seek medical attention.

- Aerosol cleaners and solvents may contain hazardous or deadly vapors and are highly flammable. Use only in a well ventilated area. Do not use on hot surfaces (engines, brakes, etc.).

continued on next page

Please read these warnings and cautions before proceeding with maintenance and repair work.

CAUTIONS—

- See also **Warnings** on page v

- If you lack the skills, tools and equipment, or a suitable workshop for any procedure described in this manual, we suggest you leave such repairs to an authorized Porsche dealer or other qualified shop.

- Porsche is constantly improving its cars and sometimes these changes, both in parts and specifications, are made applicable to earlier models. Therefore, part numbers listed in this manual are for reference only. Always check with your authorized Porsche dealer parts department for the latest information.

- Before starting a job, make certain that you have all the necessary tools and parts on hand. Read all the instructions thoroughly, do not attempt shortcuts. Use tools appropriate to the work and use only replacement parts meeting Porsche specifications. Makeshift tools, parts and procedures will not make good repairs.

- Use pneumatic and electric tools only to loosen threaded parts and fasteners. Never use these tools to tighten fasteners, especially on light alloy parts. Always use a torque wrench to tighten fasteners to the tightening torque specification listed.

- Be mindful of the environment and ecology. Before you drain the crankcase, find out the proper way to dispose of the oil. Do not pour oil onto the ground, down a drain, or into a stream, pond or lake. Dispose of in accordance with Federal, State and Local laws.

- Always make sure ignition is off before disconnecting battery.

- Label battery cables before disconnecting. On some models, battery cables are not color coded.

- If a normal or rapid charger is used to charge battery, the battery must be disconnected and removed from the vehicle in order to avoid damaging paint and upholstery.

- Do not quick-charge the battery (for boost starting) for longer than one minute. Wait at least one minute before boosting the battery a second time.

- Connect and disconnect a battery charger only with the battery charger switched off.

- Sealed or "maintenance free" batteries should be slow-charged only, at an amperage rate that is approximately 10% of the battery's ampere-hour (Ah) rating.

- Do not allow battery charging voltage to exceed 16.5 volts. If the battery begins producing gas or boiling violently, reduce the charging rate. Boosting a sulfated battery at a high charging rate can cause an explosion.

Vehicle Identification Number (VIN), decoding

Some of the information in this book applies only to cars of a particular model year or range of years. For example, 1978 refers to the 1978 model year but does not necessarily match the calendar year in which the car was manufactured or sold. To be sure of the model year of a particular car, check the Vehicle Identification Number (VIN) on the car.

The VIN is a unique sequence of numbers and letters assigned by Porsche to identify each individual car. When decoded, the VIN tells the country and year of manufacture; make, model and serial number; assembly plant and even some equipment specifications.

- Up to 1979: Porsche used 10-digit VIN.
- 1980: VIN consisted of 10 characters (letters and numbers).
- 1981 and later: 17-character VIN (letters and numbers) standardized.

VIN information for 1978 - 1979 911 SC

VIN for the 1978 - 1979 911 SC is located on an ID plate under the front hood on the right side.

Sample VIN:	9 1 1	8	2	0	0 0 0 1
position	1 2 3	4	5	6	7 8 9 10

VIN position	Description	Decoding information	
1 - 3	Model	911	911 SC
		930	Turbo (930)
4	Model year	8	1978
		9	1979
5	Engine type	2	911 SC (US)
		3	911 SC (RoW, Japan)
		7	Turbo (930) (RoW, Japan)
		8	Turbo (930) (US)
6	Body type	0	Coupe
		1	Targa, Cabriolet
7 - 10	Serial number		Sequential production number specific to vehicle

NOTE —
RoW indicates rest of world.

VIN information for 1980 911 SC

VIN for the 1980 911 SC is located on an ID plate under the front hood on the right side.

Sample VIN:	9 1	A	O	1	4	0 0 2 1
position	1 2	3	4	5	6	7 8 9 10

VIN position	Description	Decoding information	
1 - 2	Model	91	911 SC
		93	Turbo (930)
3	Model year	A	1980
4	Assembly plant	O	Stuttgart, Germany
5	Model supplement	1	911 SC
		0	Turbo (930)
6	Engine type	3	911 SC (RoW)
		7	911 SC (US)
		7	Turbo (930) (RoW)
7 - 10	Serial number		Sequential production number specific to vehicle

NOTE —
RoW indicates rest of world.

Vehicle Identification Number (VIN), decoding *(cont.)*

VIN information for 1981 - 1983 911 SC

VIN for the 1981 - 1983 911 SC is located on an ID plate under the front hood on the right side.

The 10th character is the model year code. The letters I, O, Q and U are not used for model year designation. Examples: B for 1981, C for 1982, D for 1983 etc.

Sample VIN: W P O A A 0 9 1 X B S 1 2 0 0 0 1
position 1 2 3 4 5 6 7 8 9 10 11 12 13 14 15 16 17

VIN position	Description	Decoding information	
1	Manufacturing country	W	Germany
2	Manufacturer	P	Dr. Ing. h.c. F. Porsche AG
3	Vehicle type	O	Passenger vehicle
4 - 6	Body code		US code or ZZZ for RoW
7 - 8	Model type	91	First two digits of Porsche model (1981 - 1983 911 SC = 91)
9			
10	Model year	B C D	1981 1982 1983
11	Assembly plant	S	Stuttgart, Germany
12	Model type	1	Last digit of Porsche model (1981 - 1983 911 SC = 1)
13	Body type	0 2 4 5 6 7	Coupe, Japan and RoW Coupe, US Targa, Japan and RoW Cabriolet, Japan and RoW Targa, US (Japan, 1981 only) Cabriolet, US
14 - 17	Serial number		Sequential production number specific to vehicle

NOTE —
RoW indicates rest of world.

010 General

TABLES

GENERAL

This section of the manual is intended to help the beginner get started. To begin with there is a discussion on **How to Use This Manual**. Tips on mechanic's skills and workshop practices that can be found under **Getting Started.** The basic tools needed to do most of the procedures in this manual are found under **Tools**. The section ends with a quick reference guide to **Emergencies**, including basic troubleshooting information.

> ***WARNING —***
> *Do not use this manual unless you are familiar with basic automotive repair procedures and safe workshop practices. This manual illustrates the workshop procedures required for most service work; it is not a substitute for full and up-to-date information from the vehicle manufacturer or for proper training as an automotive technician. Note that it is not possible for us to anticipate all of the ways or conditions under which vehicles may be serviced or to provide cautions as to all of the possible hazards that may result.*

> ***WARNING —***
> *Your common sense and good judgment are crucial to safe and successful service work. Read procedures through before starting them. Think about whether the condition of your car, your level of mechanical skill, or your level of reading comprehension might result in or contribute in some way to an occurrence that might cause you injury, damage your car, or result in an unsafe repair. If you have doubts for these or other reasons about your ability to perform safe repair work on your car, have the work done at an authorized Porsche dealer or other qualified shop.*

HOW TO USE THIS MANUAL

The manual is divided into ten sections:

0 General and Maintenance
1 Engine
2 Engine Management, Exhaust and Engine Electrical
3 Transmission
4 Suspension, Brakes and Steering
5 Body–Assembly
6 Body–Components and Accessories
7 Body–Trim, Seats
8 Heating, Ventilation, Air Conditioning
9 Electrical System

0 General and Maintenance covers the recommended maintenance schedules and service procedures. It also includes a section covering basic instructions, tips and helpful hints for the do-it-yourselfer.

The next nine sections are repair based and are further broken down into three digit repair groups. Each major section begins with a **General** repair group, e.g. **100 Engine–General.** These "00" (double zero) groups are mostly descriptive in nature, covering topics such as theory of operation and troubleshooting. The remainder of the repair groups contain repair information. The last major section contains detailed electrical system information, wiring diagram schematics, and electrical component locations.

A master listing of the ten major sections and the corresponding individual repair groups can be found on the inside front cover.

Each repair group begins with a table of contents listing the major subject headings within the group. Page numbers throughout the manual are organized according to the repair group system. For example, you can expect to find repair information on steering (Repair Group **480**) beginning on page 480-1. A comprehensive index can be found at the back of the manual.

Warnings, cautions and notes

Throughout this manual are many passages with the headings **WARNING**, **CAUTION** or **NOTE**. These important messages have different meanings.

> ***WARNING —***
> *The text under this heading warns of unsafe practices that are very likely to cause injury, either by direct threat to the person(s) performing the work or by increased risk of accident or mechanical failure while driving.*

> ***CAUTION —***
> *A caution calls attention to important precautions to be observed during the repair work that will help prevent accidentally damaging the car or its parts.*

NOTE —
A note contains additional information or tips intended to clarify the repair procedure or aid in doing a better job or completing it more easily.

Always read each **WARNING**, **CAUTION** and **NOTE** at the front of the manual and when they appear in repair procedures. These messages are important. Read them before you begin any maintenance or repair job.

GETTING STARTED

Most of the necessary maintenance and minor repair that an automobile will need can be done with ordinary tools, even by owners with little or no experience in car repair. Below is some important information on how to work safely, a discussion of what tools will be needed and how to use them.

Safety

Although an automobile presents many hazards, common sense and good equipment can help ensure safety. Many accidents happen because of carelessness. Pay attention and stick to these few important safety rules.

WARNING —
- *Never run the engine in the work area unless it is well-ventilated. The exhaust should be vented to the outside. Carbon monoxide (CO) in the exhaust kills.*
- *Remove all neckties, scarfs, loose clothing, or jewelry when working near a running engine or power tools. Tuck in shirts. Tie long hair and secure it under a cap. Severe injury can result from these things being caught in rotating parts.*
- *Remove rings, watches, and bracelets. Aside from the dangers of moving parts, metallic jewelry conducts electricity and may cause shorts, sparks, burns, or damage to the electrical system when accidentally contacting the battery or other electrical terminals.*
- *Disconnect the battery negative (–) cable whenever working on or near the fuel system or anything electrically powered.*
- *Never work under a lifted car unless it is solidly supported on jack stands. Do not support a car on cinder blocks, bricks, or other objects that may shift or crumble under continuous load. Never work under a car that is supported only by a jack.*
- *Keep sparks, lighted matches, and any open flame away from the top of the battery. Hydrogen gas emitted by the battery is highly flammable. Any nearby source of ignition may cause the battery to explode.*
- *Never lay tools or parts in the engine compartment or on top of the battery. They may fall into confined spaces and be difficult to retrieve, become caught in belts or other rotating parts when the engine is started, or cause electrical shorts and damage to the electrical system.*
- *Illuminate the work area adequately and safely. A fluorescent type light is best because it gives off less heat. If using a light with a normal incandescent bulb, use rough service bulbs to avoid breakage. The hot filament of an accidentally broken bulb can ignite spilled fuel or oil.*
- *The fuel system is designed to retain pressure even when the ignition is off. When working with the fuel system, loosen the fuel lines slowly to allow the residual pressure to dissipate gradually. Take precautions to avoid spraying fuel.*
- *Fuel is highly flammable. When working around fuel, do not smoke or work near heaters or other fire hazards. Keep an approved fire extinguisher handy.*

Lifting the car

For repairs that require raising the car, the specified jacking points should be used to raise the car safely and avoid damage. There are two side jacking points to temporarily lift the vehicle for tire changing. The jack supplied with the car by Porsche can only be used at the vehicle side points. When using a floor jack or automotive lift, the four jacking point closer to the wheels should be used.

> ***WARNING —***
> *Watch the jack closely. Make sure it stays stable and does not shift or tilt. As the car is raised, the car may roll slightly and the jack may shift.*

Raising car safely using vehicle jack

– Park car on flat, level surface.

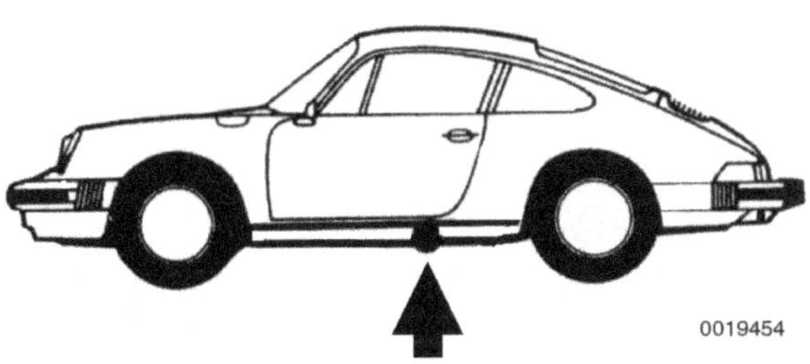

0019454

< Remove plastic plug from jack receptacle in rocker panel (**arrow**). Place jack fully into square receptacle in rocker panel. Make sure jack is resting on flat, solid ground. Use a board or other support to provide a firm surface for the jack, if necessary.

– Raise car slowly with the jack while constantly checking position of jack and car.

– Once car is raised, block wheel on opposite side of vehicle, to prevent car from unexpectedly rolling.

> ***WARNING —***
> - *Do not rely on the transmission or the emergency brake to keep the car from rolling. They are not a substitute for positively blocking the opposite wheel.*
> - *Never work under a car that is supported only by a jack. Use jack stands that are properly designed to support the car. See* **Tools**.

Working under car safely

Lifting or raising the vehicle for repairs requires a hydraulic jack or lifting equipment. In addition to the instructions below, follow the manufacturer's directions for the use of the lifting equipment.

– Disconnect negative (–) battery cable.

0019454

< Position jack or lift arms beneath specified lifting points (**arrows**). Make sure jack is resting on flat, solid ground.

> ***WARNING —***
> *When raising the car using a floor jack or a hydraulic lift, carefully position the jack pad to prevent damaging the car body. A suitable liner (wood, rubber, etc.) should be placed between the jack and the car to prevent body damage.*

– Raise car slowly while constantly checking position of lifting equipment and car.

– Use at least two jack stands to support the car. A jack is a temporary lifting device and should not be used alone to support the car while you are under it. Use jack stands designed for the purpose of supporting a car. For more information on jack stands, see **Tools** below.

> ***WARNING —***
> *Do not use wood, concrete blocks, or bricks to support a car. Wood may split. Blocks or bricks, while strong, are not designed for that kind of load, and may break or collapse.*

– Place jack stands on firm, solid surface. If necessary, use a flat board or similar solid object to provide a firm footing.

– Lower car slowly until its weight is fully supported by jack stands. Watch to make sure that the jack stands do not tip or lean as the car settles on them. Raise the jack again for additional support.

– Observe all jacking precautions when raising car to remove jack stands.

ADVICE FOR THE BEGINNER

The tips that follow are general advice to help the do-it-yourself Porsche owner perform repairs and maintenance tasks more easily and more professionally.

Planning ahead

Most of the repairs and maintenance tasks described in this manual can be successfully completed by anyone with basic tools and abilities. To prevent getting in too deep, know what the whole job requires before starting. Read the procedure thoroughly from beginning to end to know just what to expect and what parts will have to be replaced.

Cleanliness

Keeping things organized, neat, and clean is essential to doing a good job.

0019513

◄ When working in the engine compartment, a fender cover will protect the finish from scratches and other damage. Make sure the car is relatively clean so that dirt under the cover does not scratch the finish. Most custom covers have storage pockets for holding removed fasteners and parts and a special pad (**arrow**) for the A/C compressor. Removing the A/C compressor and laying it aside without disconnecting the refrigerant lines is an easy procedure, and is required for certain procedures.

Any repair job will be less troublesome if the parts are clean. For cleaning old parts, there are many solvents and parts cleaners commercially available.

For cleaning parts prior to assembly, commercially available aerosol parts cleaner or brake cleaner are handy to use, and the cleaner will evaporate completely.

> ***WARNING —***
> *Virtually all solvents used for cleaning parts are highly flammable and hazardous to health, especially in aerosol form. Use with extreme care. Do not smoke. Do not use these products indoors or near any source of heat, sparks or flame.*

Let any solvent or cleaning product dry completely. Low-pressure, dry compressed air is helpful if available. Also, use only lint-free rags for cleaning and drying.

> ***WARNING —***
> *Avoid getting tools or clothing near the battery. Battery electrolyte is a corrosive acid. Be careful with brake fluid, as it can damage the car's paint. Finally, keep rubber parts such as hoses and belts free from oil or gasoline, as they will cause the material to soften and fail prematurely.*

Non-reusable fasteners

Many fasteners used on the cars covered by this manual must be replaced with new ones once they are removed. These include but are not limited to: bolts, nuts (self-locking, nylock, etc.), cotter pins, studs, roll pins, clips and washers. Genuine Porsche parts should be the only replacement parts used for this purpose.

Some bolts are designed to stretch during assembly and are permanently altered, rendering them unreliable once removed. These are known as torque-to-yield fasteners. Always replace fasteners where instructed to do so. Failure to replace these fasteners could cause vehicle damage and personal injury. See an authorized Porsche dealer for applications and ordering information.

Tightening fasteners

When tightening the bolts or nuts that attach a component, it is always good practice to tighten the bolts gradually and evenly to avoid misalignment or over-stressing any one portion of the component. For components sealed with gaskets, this method helps to ensure that the gasket will seal properly and completely.

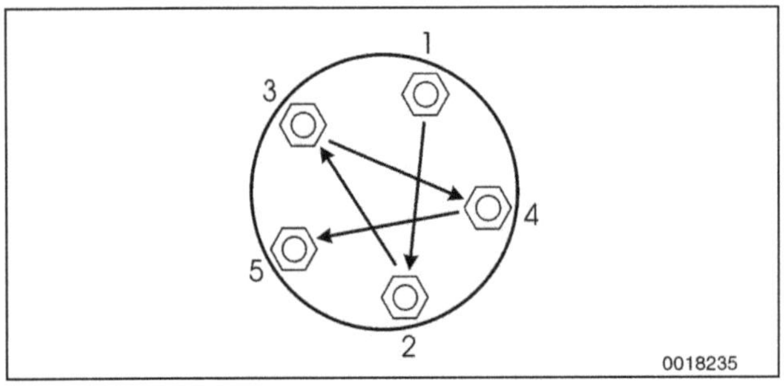

Where there are several fasteners, tighten them in a sequence alternating between opposite sides of the component. Repeat the sequence in stages until all the bolts are evenly tightened to the proper specification.

For some repairs a specific tightening sequence is necessary, or a particular order of assembly is required. Such special conditions are noted in the text, and the necessary sequence is described or illustrated. Where no specific torque is listed, **Table a** can be used as a general guide for tightening fasteners.

> ***WARNING —***
> **Table a** *is a general reference only. The values listed in the table are not intended to be used as a substitute for torques specifically called out in the text throughout this manual.*

NOTE —
- *Metric bolt classes or grades are marked on the bolt head.*
- *Do not confuse wrench size with bolt diameter size. For a listing of the common wrenches used on various bolt diameters, see* **Basic tool requirements**.

Table a. General bolt tightening torques in Nm (max. permissible)

Bolt diameter	Bolt Class (according to DIN 267)					
	5.6	**5.8**	**6.8**	**8.8**	**10.9**	**12.9**
M5	2.5	3.5	4.5	6	8	10
M6	4.5	6	7.5	10	14	17
M8	11	15	18	24	34	40
M10	23	30	36	47	66	79
M12	39	52	62	82	115	140
M14	62	82	98	130	180	220
M16	94	126	150	200	280	340
M18	130	174	210	280	390	470

Gaskets and seals

Gaskets should not be reused. Once a gasket has been used, it is no longer capable of making as good a seal as when new, and is much more likely to leak. For this reason, always plan to use new gaskets for reassembly. Some gaskets are directional. Make sure that these are installed correctly. This same logic applies to any part used for sealing, including rubber O-rings and copper sealing washers.

In places where a shaft must pass through a housing, flexible lip seals are used to keep the lubricating oil or grease from leaking out past the rotating shaft. Seals should never be reused once they have been removed. When removing a seal, be careful not to scratch or otherwise damage the metal surfaces.

When installing a new seal, it a good idea to coat the seal with oil to aid installation. Some seals are directional and special installation instructions apply. Make sure a seal is installed with the lip facing the correct way. Normally the lip faces the inside. Note the installation direction of the old seal before removing it.

Electrical testing

Many electrical problems can be understood and solved with only a little fundamental knowledge of how electrical circuits function.

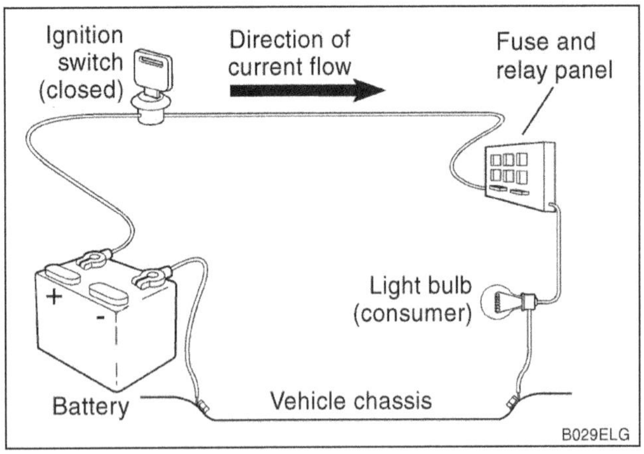

Electric current only flows in a complete circuit. To operate, every electrical device in the car requires a complete circuit including a voltage source and a path to ground. The positive (+) side of the battery is the original voltage source, and ground is any return path to the negative (–) side of the battery, whether through the wiring harness or the car body. Except for portions of the charging system, all electrical current in the car is direct current (DC) and flows from positive (+) to negative (–).

◄ Switches are used to turn components on or off by completing or interrupting the circuit. A switch is "open" when the circuit is interrupted, and "closed" when the circuit is completed. See **900 Electrical System–General** for electrical troubleshooting.

Wire repairs

Repairs to a wiring harness require special care to make the repair permanent. The wire ends must be clean. If frayed or otherwise damaged, cut off the end. If the wire is too short, splice in a new piece of wire of the same size and make two connections.

Use connectors designed for the application. Crimped-on or soldered-on connectors are best. Crimp connectors and special crimping pliers are widely available. If soldering, use needlenose pliers to hold the wire near the solder joint and create a "heat dam". This keeps the heat and the solder from traveling up the wire. Always use a solder made specifically for electrical work (rosin core).

NOTE —

Twisting wires together to make a repair is not recommended. Corrosion and vibration will eventually spoil the connection and may lead to irreparable damage to sensitive electronic components.

Insulate the finished connection. Electronics stores can supply heat-shrinkable insulating tubing that can be placed onto the wire before connecting, slid over the finished joint, and shrunk to a tight fit with a heat gun or hair dryer. The next best alternative is electrical tape. Make sure the wire is clean and free of solder flux or other contamination. Wrap the joint tightly to seal out moisture. See **900 Electrical System–General** for more information.

BUYING PARTS

Many of the maintenance and repair tasks in this manual call for the installation of new parts, or the use of new gaskets and other materials when reinstalling parts. Most often, the parts that will be needed should be on hand before beginning the job. Read the introductory text and the complete procedure to determine which parts will be needed.

NOTE —
For some bigger jobs, partial disassembly and inspection are required to determine a complete parts list. Read the procedure carefully and, if necessary, make other arrangements to get the necessary parts while your car is disassembled.

Genuine Porsche parts

Genuine Porsche replacement parts from an authorized Porsche dealer are designed and manufactured to the same high standards as the original parts. They will be the correct material, manufactured to the same specifications, and guaranteed to fit and work as intended by the engineers who designed the car. Some genuine Porsche parts have a limited warranty.

Porsche is constantly updating and improving their cars, often making improvements during a given model year. Porsche may recommend a newer, improved part as a replacement, and your authorized dealer's parts department will know about it and provide it. The Porsche parts organization is best equipped to deal with any Porsche parts needs.

Non-returnable parts

Some parts cannot be returned for credit, even if they are the wrong parts for the car. The best example is electrical parts, which are almost universally considered non-returnable because they are so easily damaged internally.

Buy electrical parts carefully, and be as sure as possible that a replacement is needed, especially for expensive parts such as electronic control modules. It may be wise to let an authorized Porsche dealer or other qualified shop confirm your diagnosis before replacing an expensive part that cannot be returned.

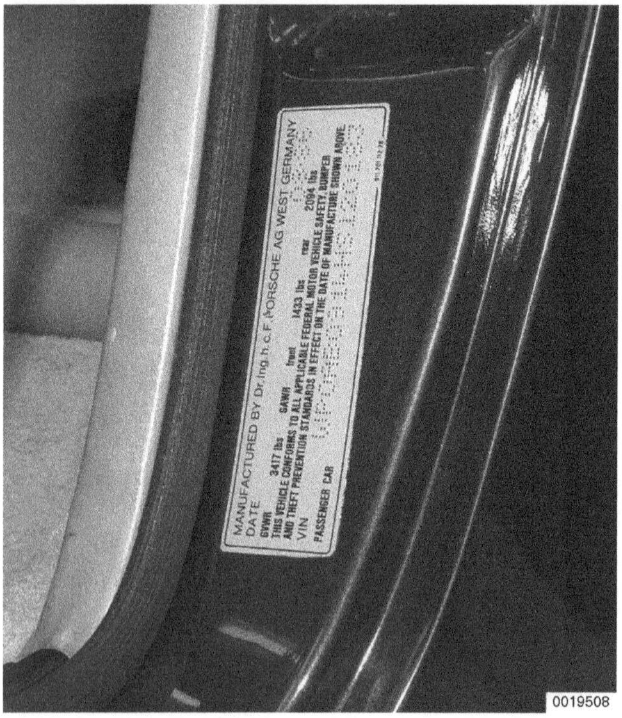

Information you need to know

Model. When ordering parts it is important that you know the correct model designation for your car. Models covered in this manual are 911 SC coupe, targa and cabriolet configurations.

Model year. This is not necessarily the same as date of manufacture or date of sale. A 1981 model may have been manufactured in late 1980, and perhaps not sold until early 1981. It is still a 1981 model. Model years covered by this manual are 1978 to 1983.

◀ **Date of manufacture**. This information is helpful when ordering replacement parts or determining if any of the warranty recalls are applicable to your car. The label on the driver's door pillar (**arrow**) will specify the month and year that the car was built.

◀ **Vehicle Identification Number (VIN)**. This is a combination of letters and numbers that identify the particular car. The VIN appears on the state registration document, and on the car itself. One location is in on the left windshield pillar, another is on the luggage compartment lid (**arrow**). The sticker on the luggage compartment lid contains the following information:

1. Vehicle Identification Number
2. Vehicle code
3. Engine and transmission code
4. Paint and interior code
5. Option Codes

A duplicate of this label is also contained in the vehicle's original paper work (Warranty and Maintenance card).

0019512

Engine number. The engine number is stamped on the left crankcase half, below the fan housing (**arrow**). 911 cars covered in this manual are powered by a 6-cylinder engine. For information on engine codes and engine applications, see **100 Engine–General**.

Engine numbers

- 1978-1979 (930/04) . . (49 states) 6280001, 6290001-on
- 1978-1979 (930/06) . . (California) 6580001, 6590001-on
- 1980 (930/07) 6400001-on
- 1981 (930/16) 6410001-on
- 1982 (930/16) 64C0001-on
- 1983 (930/16) 64D0001-on

0019510

Paint code. The paint code is on a sticker on the right fender inside the luggage compartment (**arrow**).

Transmission. The transmission type with its identifying code may be important when buying clutch parts, seals, gaskets, and other transmission-related parts.

Transmission

- 1978 - 1983
  with cable-operated clutch . . . . 915, 5-speed manual

1978 - 1981

The transmission number has seven digits, e.g.: **7190012**.

Digit 1	Transmission for 6-cylinder engine (**7**)
2	5-speed (**1**)
3	Model year (**9** = 1979)
Digits 4 - 8	Serial number (**0072**)

1982 - 1983

The transmission number has eight digits, e.g.: **74 C 3495**

Digit 1	Transmission for 6-cylinder engine (**7**)
2	5-speed (**4**)
3	Model year (**C** = 1982)
Digits 4 - 8	Serial number (**3495**)

For more information, see **300 Transmission and Clutch–General**.

SERVICE

Porsche dealers are uniquely qualified to provide service for Porsche cars. Their authorized relationship with the large Porsche service organization means that they are constantly receiving new tools and equipment, together with the latest and most accurate repair information.

The Porsche dealer's service technicians are highly trained, very capable, and intensely committed to supporting the Porsche product. They share the owner's interest in Porsche value, performance, and reliability. On the other hand, there are many independent shops that specialize in Porsche service and are capable of doing high quality repair work. Checking with other Porsche owners for recommendations on service facilities is a good way to learn of reputable Porsche shops in your area.

TOOLS

Most maintenance can be accomplished with a small selection of the right tools. Tools range in quality from inexpensive junk, which may break at first use, to very expensive and well-made tools for the professional.

Many reputable tool manufacturers offer good quality, moderately priced tools with a lifetime guarantee. These are your best buy. They cost a little more, but they are good quality tools that will do what is expected of them.

Basic tool requirements

The basic hand tools described below can be used to accomplish most of the simple maintenance and repair tasks.

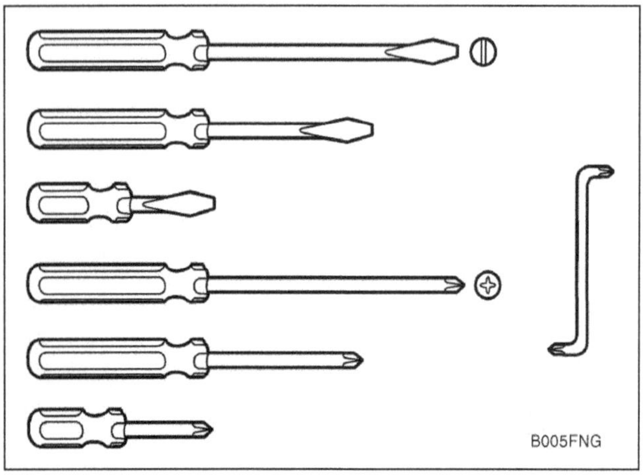

The common flat-blade **screwdriver** and the Phillips type will handle almost all screws used on Porsches. Two or three different sizes of each type will be required, since a screwdriver of the wrong size will damage the screw head.

A complete set of screwdrivers should also include Torx® type screwdrivers.

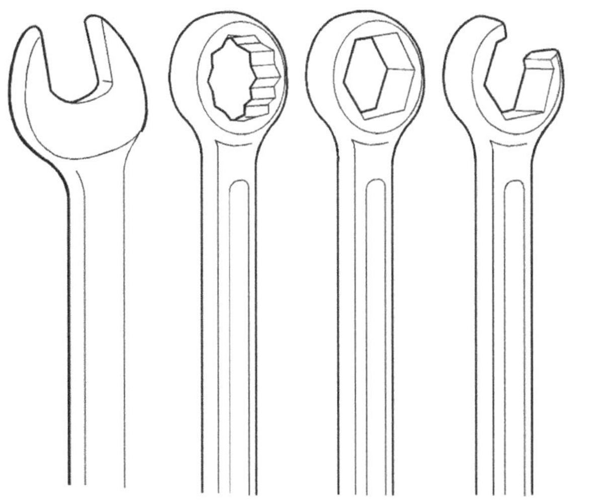
B002FNG

Wrenches come in different styles for different uses. The basic open-end wrench is the most widely used, but grips on only two sides. It can spread apart and slip off more easily. The box-end wrench has better grip on all six sides of a nut or bolt.

A 12-point box-end can loosen a nut or bolt where there is less room for movement, while a 6-point box-end provides better grip. For hex fasteners on fluid lines, like brake lines and fuel lines, a flare-nut wrench offers the advantages of a box-end wrench with a slot that allows it to fit over the line.

The combination wrench is the most universal. It has one open-end and one box-end. 10 mm and 13 mm wrenches are the most common sizes needed. A more complete set of wrenches would include 6 mm through 19 mm sizes.

Sockets can be combined with extensions and universal joints (swivels) to reach fasteners more easily than hand wrenches. Sockets come with different drive size connections. The most common drive sizes are ¼ in., 3/8 in., and ½ in.

Sockets come in 6-point and 12-point styles. For use with a ratchet, the 6-point offers a better grip on tight nuts and bolts. 6 mm to 19 mm sockets are the most needed sizes. Below is a list of typical bolt diameters and the corresponding wrench sizes.

Fastener diameters and wrench sizes

- M5 . . . . 8 mm
- M6 . . . . 10 mm
- M8 . . . . 12 mm, 13 mm
- M10 . . . . 14 mm, 15 mm, 17 mm
- M12 . . . . 19 mm
- M14 . . . . 22 mm

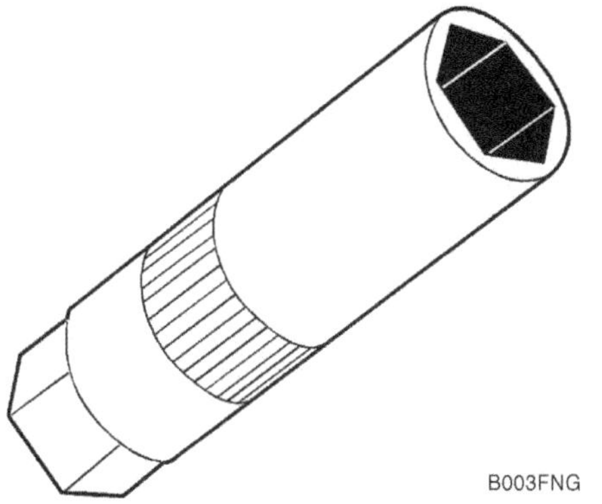
B003FNG

A special 13/16 in. **spark plug socket** is the correct size, is deep enough to accommodate a spark plug's length, and includes a rubber insert to both protect the spark plug from damage and grip it for easier removal.

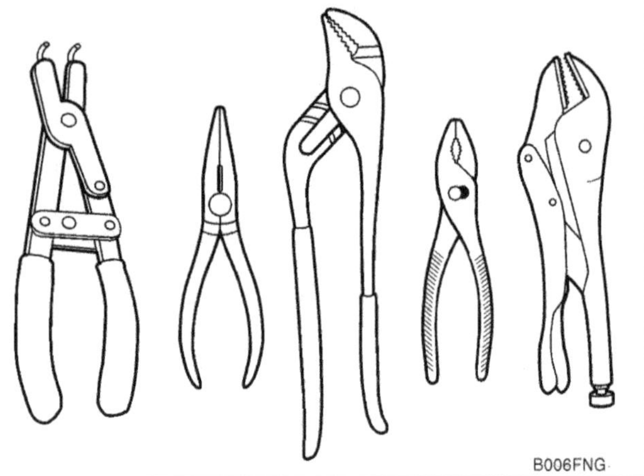

B006FNG

Most **Pliers** are used for holding irregular objects, bending, or crimping. Some have special applications.

A needlenose plier is used for gripping small and poorly accessible objects, and is useful for wiring and other electrical work. A locking plier such as the Vise-Grip® is useful because of its tight grip.

Snap-ring and circlip pliers with special tipped jaws are used to remove and install snap-rings or circlips. A Channellock® or water pump plier has adjustable jaws that can be quickly changed to match the size of the object being held to give greater leverage.

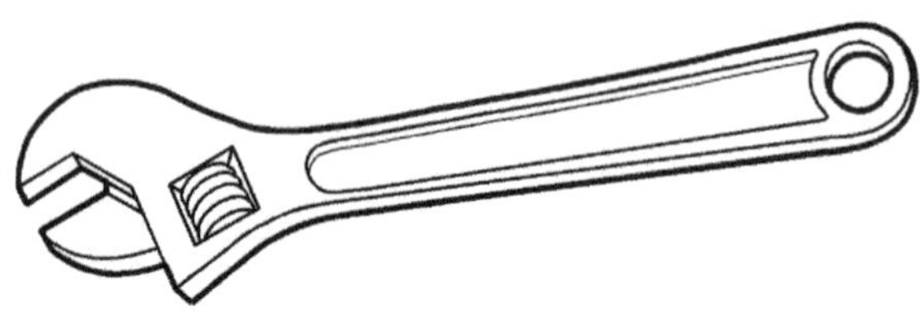

B008FNG

An **adjustable wrench** can be a useful addition to a small tool kit. It can substitute in a pinch, if two wrenches of the same size are needed to remove a nut and bolt. Use extra care with adjustable wrenches, as they tend to loosen, slip and damage fasteners.

Compared to a wrench of the correct size, an adjustable wrench is always second best. They should only be used when the correct size wrench is not available. Choose one of average size range, about 6 to 8 inches in length.

Jack stands

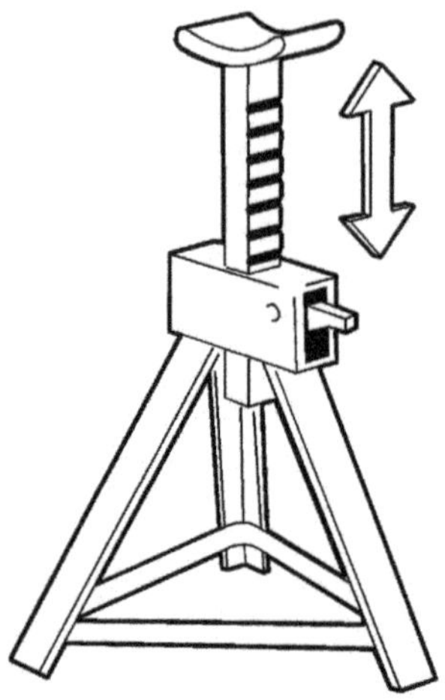

B021FNG

Strong jack stands are extremely important for any work that is done under the car. Use only jack stands that are designed for the purpose. Blocks of wood, concrete, bricks, etc. are not safe or suitable substitutes.

> ***WARNING —***
> *A jack should be used only to raise the vehicle and should not be used to support the car for a long period. Always use jack stands to support a raised vehicle.*

Jack stands are available in several styles. The best ones are made of heavy material for strength, have a wide base for stability, and are equipped to positively lock in their raised positions. Get the best ones available.

Oil change equipment

Changing engine oil requires a 15 mm socket or box-end wrench to loosen and tighten the drain plugs and a drain pan (at least 12 qt. capacity). A wide, low drain pan will fit more easily under the car. An oil filter removal wrench is also required. Use a funnel to pour the new oil into the engine.

Engine oil service is covered in **030 Maintenance**.

Torque wrench

A torque wrench is used to precisely tighten threaded fasteners to a predetermined value. Some of the repair procedures in this manual include Porsche-specified torque values in Newton-meters (Nm) and the equivalent values in pound-feet, more commonly written as foot-pounds (ft-lb).

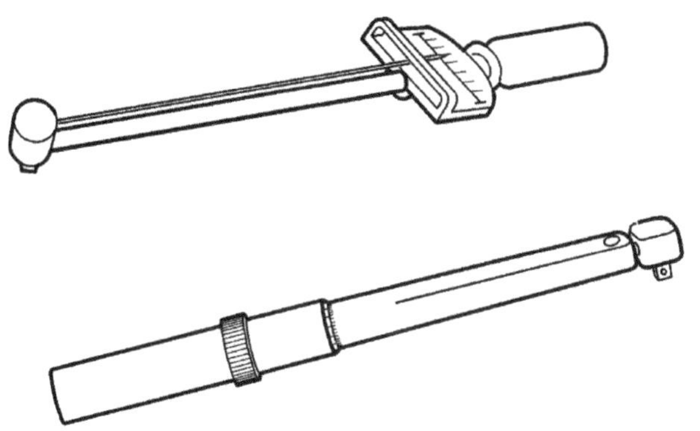

B018FNG

Several types of torque wrenches are available. They all do the same job, but offer different convenience features at different prices. The most convenient ones have a built-in ratchet, and can be preset to indicate when a specific torque value has been reached. Follow the wrench manufacturer's directions for use to achieve the greatest accuracy.

A torque wrench with a range up to about 250 Nm (185 ft-lb) has adequate capacity for most of the repairs covered in this manual. For recommended torque values of 10 Nm or below, the English system equivalent is given in inch-pounds (in-lb). These small values may be most easily reached using a torque wrench calibrated in inch-pounds. To convert inch-pounds to foot-pounds, divide by 12.

Feeler gauges

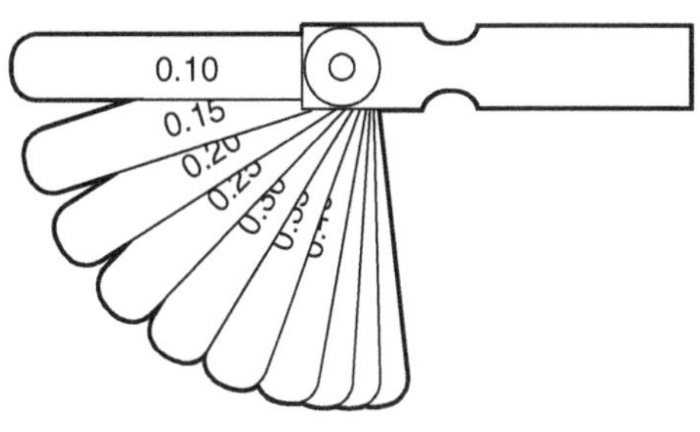

B020FNG

Feeler gauges are thin metal strips of precise thickness, used to measure small clearances such as valve clearance. They are normally available as a set, covering a range of sizes.

Digital multimeter

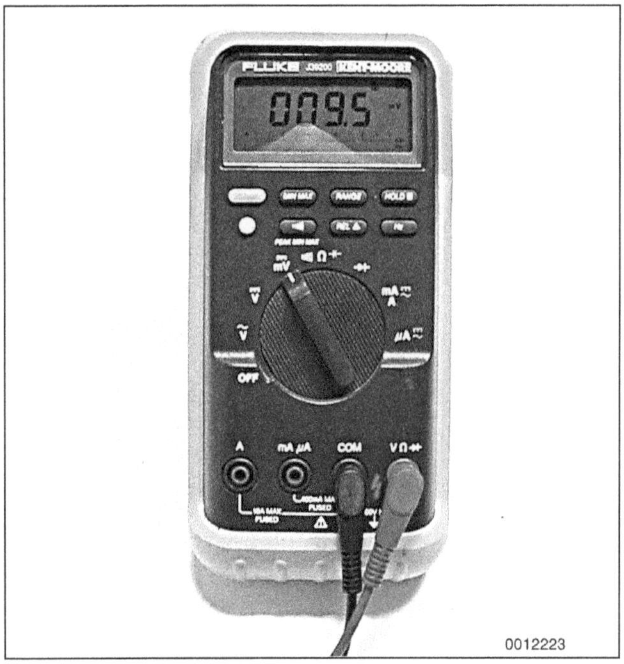

0012223

Many of the electrical tests in this manual call for the measurement of resistance (ohms) or voltage values. For safe and accurate tests of ignition, fuel injection, and emission control systems, the multimeter should be digital, with high (at least 10,000 ohms) input impedance. Some meters have automotive functions such as dwell and pulse width that are useful for troubleshooting ignition and fuel injection problems.

> ***CAUTION —***
> *The CIS fuel injection and CDI ignition systems, and other electronic systems, may be damaged by the high current draw of a test light with a normal incandescent bulb. Always use a high impedance digital multimeter or an LED test light for all electrical testing.*

Porsche special tools

Some of the repairs covered in this manual call for the use of Porsche special tools. Many of the special tools are simply the best thing to use to do the job correctly. In these cases, the tool is identified with a part number and an ordering number, where available.

This, however, does not automatically mean that the job is too complicated or out of reach of the novice. For example an affordable and very useful special tool is a hose clamping tool. This tool is helpful to clamp off fluid lines for a neater job.

0019041

There are some jobs for which expensive special tools are essential, and not a cost-effective purchase for one-time repair by the do-it-yourself owner. An example is the rear wheel bearing press. This tool is more expensive than having the job done by a professional who already owns the tool. This manual includes these jobs for the benefit of those with the necessary experience and access to tools.

Most of the Porsche special tools can be purchased through an authorized Porsche dealer. As an alternative, some special tools mentioned may be purchased from the following tool manufacturers and/or distributors:

- **Baum Tools Unlimited, Inc.**
  P.O. Box 586, Sarasota, FL 3477-5867
  (800) 848-6657
  (941) 927-1414
  http://www.baumtools.com
- **Assenmacher Specialty Tools**
  6440 Odell Place, Boulder, CO 80301
  (303) 530-2424
  http://www.asttool.com
- **Schley Products Inc.**
  5350 E. Hunter Ave., Anaheim Hills, CA 92807
  (714) 693-7666
  http://www.sptool.com
- **Zelenda Automotive, Inc.**
  66-02 Austin Street, Forest Hills, NY 11375
  (718) 896-2288
  http://www.zelenda.com

NOTE —
When ordering special tools from the dealer, have the tool ordering number on hand. This number normally has more digits in it than the tool number itself. When ordering tools directly from Baum Tools Unlimited, Inc., either the tool number or the ordering number will be sufficient.

EMERGENCIES

Changing a tire

Stop the car on as flat a surface as possible, in a place where you can be easily seen by other drivers. Turn on the emergency flashers, and set out flares or emergency markers well behind the car. Chock the wheel diagonally opposite to the one being changed. Firmly set the parking brake. Passengers should get out of the car and stand away from the road.

WARNING —
If a tire goes flat while driving, pull well off the road. Changing a tire on a busy street or highway is very dangerous. If necessary, drive a short distance on the flat tire to get to a safe place. It is much better to ruin a tire or rim than to risk being hit.

0019504

- Take collapsible spare tire, jack and tool kit from luggage compartment. The jack and tool kit are located ahead of spare tire.

- Loosen wheel bolts while car is on ground, but leave them a little snug.

◀ Remove plastic plug from jack receptacle in rocker panel (**arrow**). Place jack fully into square receptacle in rocker panel. Make sure jack is resting on flat, solid ground. Use board or other support to provide a firm surface for jack, if necessary.

– Raise car just until wheel is fully off ground. Remove wheel nuts and wheel. Place removed road wheel in plastic bag provided and store in luggage compartment.

– Install spare wheel. Install wheel nuts and tighten them by hand.

– Remove electric air compressor from tool kit and connect it to cigarette lighter socket. Inflate tire to specified pressure. Check tire pressure using a tire gauge.

Tire inflation pressure
- Collapsible spare . . . . 2.2 bar (32 psi)

– Lower car to ground and fully tighten nuts in a crisscross pattern. Torque wheel nuts.

Tightening torque
- Wheel to wheel hub . . . . 130 Nm (96 ft-lb)

WARNING —
Do not drive more than 50 mph with the collapsible spare tire. Adhere to a moderate a driving style. This tire is for emergency use and short distances only. Remount the full size road wheel as soon as possible.

NOTE —
When the air is removed to from the collapsible spare tire, it will return to its original shape after cooling down for several hours.

Car will not start

If the engine turns over slowly or not at all, especially on cold mornings, the battery may not be sufficiently charged. Jump-starting the battery from another car may help.

WARNING —
Push starting (or bump starting) a car is not recommended.

NOTE —
Be sure to read the cautions under **Jump starting car** *prior to boosting a low battery. Failure to follow the cautions may result in damage to the electronic components in the car.*

If the engine is turning over at normal speed with the starter motor, the battery and starter are fine. Check to make sure that there is fuel in the tank. Do not rely on the fuel gauge because it may be faulty. Instead, remove the gas filler cap and rock the car. If there is gas in the tank, you should hear a sloshing sound at the filler neck. If there is plenty of fuel in the tank, see **100 Engine–General** for in-depth diagnostics and troubleshooting procedures.

Jump starting car

Cars with discharged or dead batteries can be jump-started from another car. When jump-starting the engine, always note the following warnings.

> ***WARNING —***
> - *Battery acid (electrolyte) can cause severe burns, and will damage the car and clothing. If electrolyte is spilled, wash the surface with large quantities of water. If it gets into eyes, flush them with water for several minutes and call a doctor.*
> - *Batteries produce explosive and noxious gasses. Keep sparks and flames away. Do not smoke near batteries.*
> - *Do not jump-start the engine if you suspect that the battery is frozen. Trapped gas may explode. Allow the battery to thaw first.*
> - *Do not quick-charge the battery (for boost starting) for longer than one minute, and do not exceed 15 volts at the battery with the boosting cables attached. Wait at least one minute before boosting the battery a second time.*

– Place cars close together, but do not allow them to touch each other. Turn off engine of car with good battery.

0019485

◄ Open luggage compartment lid and locate battery (**arrow**) on left hand side of luggage compartment.

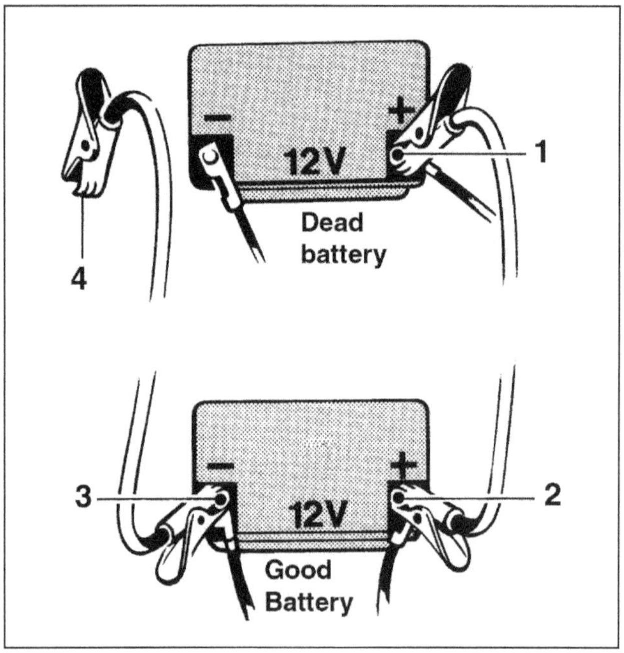

- Connect one end of positive (+) cable to positive (+) post of good battery (**1**). Connect other end of positive (+) cable to positive (+) post of dead battery (**2**).

- Connect one end of negative (–) cable to negative (–) battery post of good battery (**3**). Connect opposite end of negative (–) cable to engine block of car with dead battery (**4**).

- Start car with good battery and run engine at about 2,000 rpm, then start car with dead battery.

- With engine at idle, carefully disconnect jumper cables, starting with negative cable on engine block.

NOTE —
The engine should be run for at least an hour to recharge the battery.

Alternator warning light on

The alternator warning light monitors the alternator, the V-belt, and the engine cooling fan. If the light flickers on or comes on continuously, the alternator V-belt is most likely loose or broken. The belt should be tensioned or replaced as described **030 Maintenance**.

CAUTION —
Do not drive car with broken V-belt. Overheating of engine will result.

Low oil pressure

If the red engine oil pressure warning light illuminates or the oil pressure gauge drops to below normal, immediately stop the car and turn off the engine. Have the vehicle flat-bedded or towed to an authorized Porsche service center and have the engine oil pressure checked.

CAUTION —
Do not drive car with oil pressure warning light on until the oil pressure system has been diagnosed and repaired as necessary.

NOTE —
Detailed information on checking the oil pressure is covered in **170 Engine Lubrication**.

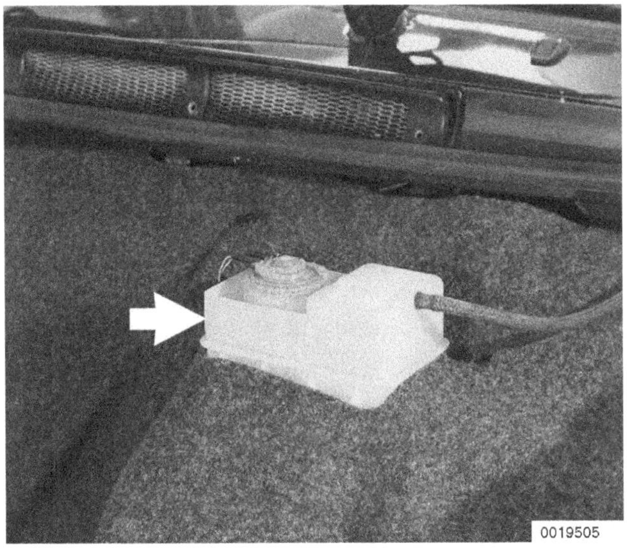

0019505

Brake fluid level

The brake warning light may be an indicator of brake fluid loss. Problems with the brake system should be checked and repaired immediately. Check the brake fluid level at the reservoir (**arrow**) in the luggage compartment. See **470 Brakes–Hydraulic** for more information.

Dim lights

Headlights and instrument panel illumination that are dim or gradually getting dimmer generally indicate a problem with the battery or charging system. The battery charge indicator light may come on as the lights are dimming. In either case, the engine and accessories are running off the battery alone, and will soon discharge it altogether.

If possible, do not stop the engine unless you have the capability to jump start it. There may not be enough power in the starting system to restart the engine. Instead, turn off as many electrical consumers as possible. This will allow the car to be driven farther before you lose all battery power.

With the engine and ignition off, check for loose or corroded battery cables or wires at the battery or the alternator. Disconnecting, cleaning, and reinstalling corroded wires and connectors may solve the problem. Also check drive belt tension as described in **030 Maintenance**.

Towing

The cars covered by this manual should be towed with a tow truck using wheel lift or flat bed equipment. Do not tow the car on all four wheels except for very short distances to move it to a safe place.

Tow hooks are located under the front bumper.

When towing the car from the front with the rear wheels on the ground, a maximum distance of 20 miles is acceptable. If the car needs to be towed further, have the rear wheels placed on dollies.

> ***CAUTION—***
> *Do not tow with sling-type equipment. The front spoilers and bumper covers may sustain damage.*

Spare parts kit

Carrying a basic set of spare parts can prevent a minor breakdown from turning into a major annoyance. Many of the following items won't allow you to do major repair work on the car, but they will help in the event of the failure of something that can disable the car or compromise its safety.

Spare parts kit

- Basic contents:
  Fuel pump relay (relay is located main fuse/relay panel.)
  Alternator, air pump and A/C drive belts (V-belts)
  Engine oil (one or two quarts)
  Fuse assortment (5A, 8A, 16A, 25A)
- Additional contents:
  Exterior lighting bulbs (headlight, brakelight, turn signal, and taillight)
  Wiper blades
  Brake fluid (new unopened bottle, DOT 3 or higher specification)

030 Maintenance

TABLES

Special Tools

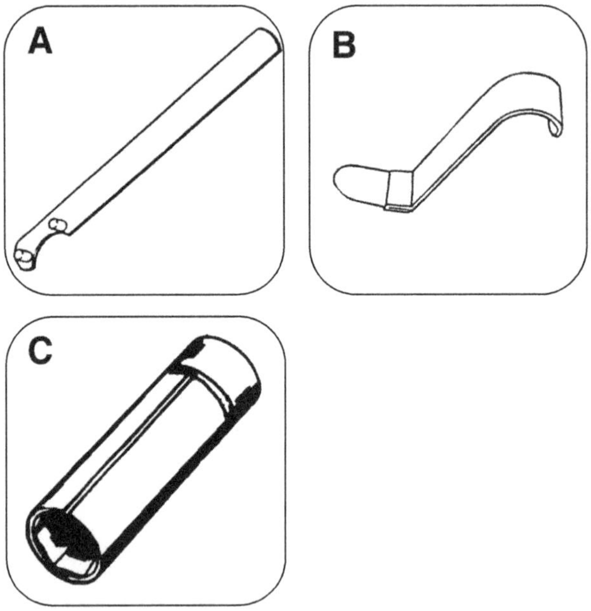

Special tools

- **A** P 208 . . . . Generator pulley spanner (Source: Baum Tools Unlimited)
- **B** P 213 b. . . . . Valve adjusting feeler gauge (Porsche ordering #000 721 213 20)
- **C** Baum 5045. . . . . Oxygen sensor slotted socket (Source: Baum Tools Unlimited)

GENERAL

The information given in this repair group includes the routine checks and maintenance steps recommended by Porsche to ensure long and reliable vehicle operation.

NOTE —

Porsche is constantly updating their recommended maintenance procedures and requirements. The information contained here is as accurate as possible at the time of publication. If there is any doubt about what procedures apply to a specific model or model year, or what intervals should be followed, remember that an authorized Porsche dealer has the latest information on factory-recommended maintenance.

Special tools

Some of the procedures in this section require the use of special tools.

MAINTENANCE TABLES

Table a through **Table d** on the following pages list the routine maintenance tasks specified by Porsche. The intervals of these tasks are determined by elapsed vehicle mileage and/or time.

Except where noted, the maintenance items listed apply to all models and model years covered by this manual. The columns on the right side of each table give quick-reference information about the job. The bold text refers to the repair group where additional repair information can be found.

Table a. Oil service – every 7,500 miles

Maintenance item	Tools required	New parts required	Warm engine required	Dealer service recommended
Engine compartment maintenance				
Change oil and oil filter	∗	∗	∗	

Table b. Maintenance schedule – every 15,000 miles

Maintenance item	Tools required	New parts required	Warm engine required	Dealer service recommended
Engine compartment maintenance				
Engine: Visually inspect for leaks.				
Crankcase ventilation: Inspect for leaks and check hose connections.				
Fuel system: Inspect for leaks and check hose connections.				
Intake air: Check hoses, lines and connections.				
Change oil and oil filter.	*	*	*	
Check and adjust valve clearance.	*	*		
Throttle cable: lubricate.				
Under car maintenance				
Manual transmission: Inspect for leaks. Check oil level and add as necessary. **340 Manual Transmission–Controls and Case**.	*	*	*	
Fuel system: Inspect for leaks and check hose connections.				
Fuel tank, fuel lines: Inspect all connections for leaks.				
Brake system: Check brake pedal play and correct if necessary. Visually inspect brake hoses and lines for leaks, visible damage, corrosion. Visually inspect brake pads and rotors for wear. Check brake fluid level and top up as necessary.	*			
Clutch: Check clutch cable for play.				
Control arms: Check mounting bolts tightness.	*			
Exhaust system: Visually inspect for leaks and damage. Check exhaust system mounting bolts for tightness. **260 Exhaust System**				
Steering: Check steering rack and steering linkages. Check steering rack mounting bolts for tightness. Inspect bellows for damage. Check tie rod ends for play and inspect dust covers for damage.	*			
Drive axles: check boots for leaks or damage.				
Check parking brake operation, adjust cable if necessary. **460 Brakes–Mechanical**	*			
Front and rear suspension: Check for visible damage. Check ball joints for wear and inspect ball joint dust covers for damage. **401 Front Suspension**				*
Front wheel bearings: Check for play and, if necessary, correct.	*			
Inspect wheels and tires, including spare. Check tire pressure and condition.	*			
Body and Interior Maintenance				

Continued on next page

Table b. Maintenance schedule – every 15,000 miles (continued)

Maintenance item	Tools required	New parts required	Warm engine required	Dealer service recommended
Hinges, latches: Lubricate door and hood hinges and hood latch. Check door stop function.				
Locks: Check central door lock operation. Check engine lid lock and luggage compartment lock and safety catch.				
Seals for door, hood and roof: Inspect and replace as necessary.				
Check headlight and fog light aiming and adjust as necessary.	*			*
Vehicle lighting: Check operation of headlights, parking lights, back-up lights, license plate lights, interior lights, glove compartment light, instrument panel lights, turn signals, emergency flashers, stop lights, headlight flasher and dimmer switch.				
Horn: Check operation. Optional electrical equipment: Check operation.				
Check function of air conditioning. Check refrigerant charge at sight glass, where applicable. Tighten A/C compressor mounting bolts. **870 Air Conditioner**.	*			
Check function and condition of seatbelts.				
Check windshield washer fluid level and concentration, add as necessary. Check operation of washer system. Check condition of wiper blades. **920 Windshield Wipers and Washers**		*		
Battery: Check fluid level and if necessary, top up.				
Road Test				
Check braking, clutch operation, cruise control, steering, HVAC, and instruments.				

Table c. Maintenance schedule – additional items every 30,000 miles

Maintenance item	Tools required	New parts required	Warm engine required	Dealer service recommended
Engine compartment maintenance				
Change oil and oil filter	*	*	*	
Check and adjust valve clearance.	*	*		
V-belt: Check condition and tension.				
Spark plugs: Replace (at least every two years)		*		
Air filter: Replace filter element		*		
Oxygen sensor: Replace (1980 - 1983 models).	*	*		

Table d. Maintenance schedule – time/mileage

Maintenance item	every 24 months	every 48 months	every 60,000 miles	New parts required	Dealer service recommended
Manual transmission: change gear oil.			✻	✻	
Fuel filter: Replace.			✻	✻	
Brake fluid: Flush and replace brake fluid. **460 Brakes–Hydraulic**	✻			✻	

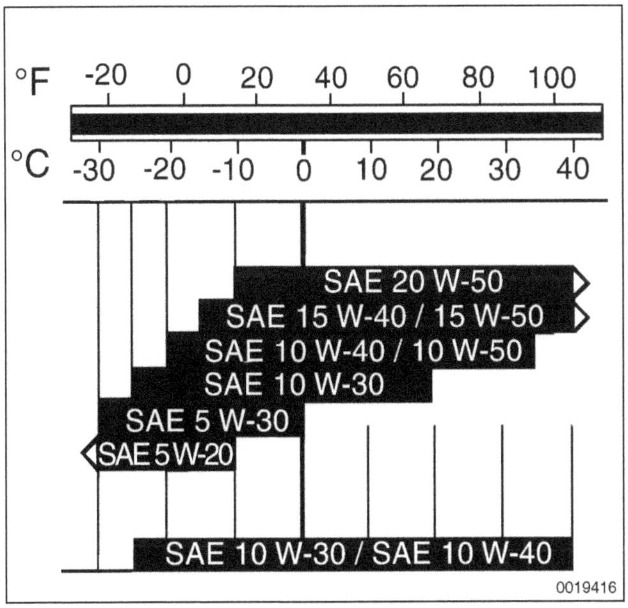

FLUID AND LUBRICANT SPECIFICATIONS

◄ Engine oil viscosity (SAE grade) vs. operating temperature range shown for the Porsche engine covered in this manual.

The fluids and lubricants specified by Porsche are listed in **Table e**.

> ***WARNING —***
> *The use of fluids that do not meet Porsche's specifications may impair performance and reliability, and may void warranty coverage.*

Table e. Fluids and lubricants

Fluid	Approximate capacity	Specification
Engine oil (with filter change)		
Total dry fill	13 L (13.7 US qt.)	API service SH or SJ
Oil and filter change	10.5 L (11 US qt.)	API service SH or SJ
Manual transmission (drain and fill)		
915 type	3.0 L (3.2 US qt.)	Hypoid gear oil SAE 90 (GL-5)
Brake fluid		
All models	0.42 pt.	SAE DOT 3 or DOT 4

NOTE —
Synthetic manual transmission fluid is not approved by the manufacturer. Consult with an authorized Porsche Dealer for recommendations on synthetic transmission fluid use in the 915 transmission.

ENGINE OIL SERVICE

If the car is used primarily for short trips in slow moving traffic, or routinely operated aggressively, the oil should be changed at shorter intervals (3,000 mile intervals).

Engine oil and filter, changing

> ***WARNING —***
> - *When changing oil, pull engine drain plug away from drain hole quickly to avoid being burned by hot oil. It will run out quickly when the plug is removed.*
> - *Use gloves to protect your hands.*

– Run engine for a few minutes to warm oil. Turn engine off.

– With car on level ground, place drain pan under crankcase.

NOTE —
The car will not need to be raised if a shallow drain pan is used.

◀ Through production date 3/83: Loosen M6 nuts (**arrows**) and remove crankcase sump cover. Let oil drain into pan.

NOTE —
If there is a drain plug in the center of the crankcase sump cover (original production cover), use the center plug to drain oil.

◀ From production date 3/83: Loosen and remove crankcase drain plug (**arrow**) on left side of crankcase. Let oil drain into pan.

◀ Loosen and remove oil supply tank drain plug (**arrow**).Let oil drain into pan.

Remove oil filter. Lubricate oil filter seal and install new filter hand tight. Tighten filter an additional ½ turn.

> ***CAUTION —***
> *Overtightening the oil filter will make the next change much more difficult, and may deform the gasket, causing leaks.*

– When oil flow has diminished to an occasional drip, reinstall drain plug (if applicable) with new metal sealing washer and tighten. Reinstall oil sump cover (if applicable) with new gaskets.

Tightening torques

- Oil drain plug to crankcase (from 3/83). . 70 Nm (51 ft-lb)
- Oil drain plug to sump cover (to 3/83). . . 42 Nm (31 ft-lb)
- Sump cover to crankcase (to 3/83)(M6) . . 10 Nm (7 ft-lb)

– Refill oil supply tank with oil. Oil specifications are found previously under **Fluid and Lubricant Specifications**.

– Start engine and allow to reach operating temperature. With engine running at idle, check engine oil level. Check for leaks at drain plug and oil filter.

> ***CAUTION —***
> *Dispose of used oil properly. Use tight-sealing containers and mark them clearly. Check with the place of purchase about disposal.*

ENGINE COMPARTMENT MAINTENANCE

Air filter element, replacing

The specified maintenance intervals for replacing the air filter element are based on normal use. If the car is operated primarily in dusty conditions, the air filter should be serviced more frequently. A dirty air filter starves the engine for air, reducing power output.

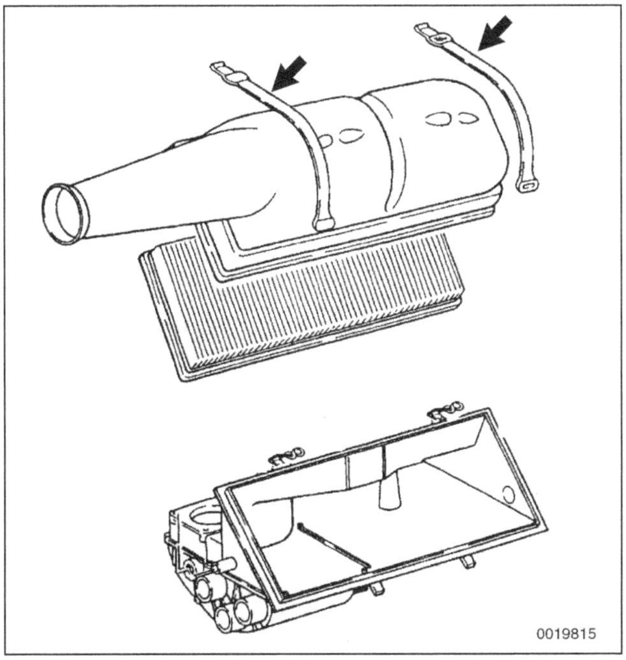

The air filter housing is fastened to the intake manifold housing with two rubber retaining straps (**arrows**).

- To replace air filter element, unfasten both straps from clips at top of housing.

– Separate air filter housing and remove filter element.

– Wipe clean inside of air filter housing using a lint-free cloth

– Install new filter, making sure that the metal side of the filter element faces the intake manifold side.

– Assemble air filter housing and refasten rubber straps.

Spark plugs, replacing

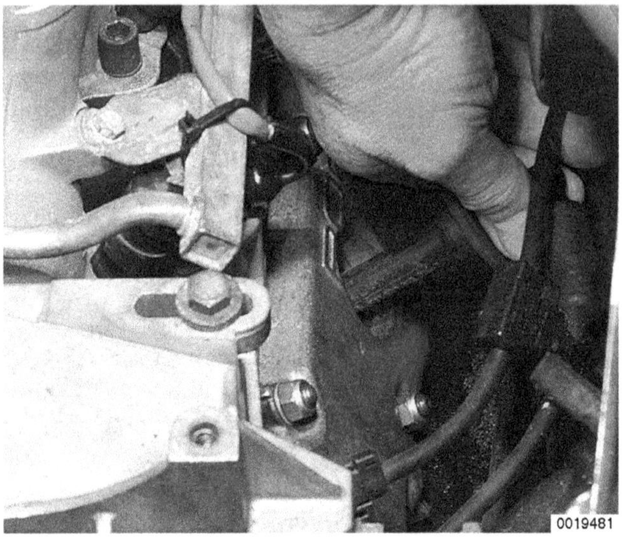
0019481

Carefully remove spark plug wire by pulling straight up on boot. Blow away any dirt from around the base of plug to prevent it from entering engine when plug is removed.

> ***CAUTION —***
> *Spark plugs should be replaced one at a time so that the spark plug wires do not get mixed up. If all of the wires need to be removed together, label each wire so that they can be reinstalled to the correct spark plug.*

– Use 13/16 in. spark plug socket to remove and reinstall spark plugs.

- Use a spark plug gap gauge to check gap, if applicable. If necessary, bend outer electrode slightly to adjust gap to meet specification.
- Do not bend or file center electrode.
- Thread plugs into cylinder head by hand to prevent cross-threading.
- Torque spark plugs.

Tightening torque
- Spark plug to cylinder head . . . . 30 Nm (22 ft-lb)

NOTE —
Abnormal spark plug condition or engine damage could also result from spark plugs of the wrong specification. Check replacement plugs carefully and follow the spark plug manufacturer's recommendations.

Table f. Spark plug application

Model	Spark plug (Bosch)	Electrode gap
911 SC	WR7DC (copper) WR7DP (platinum)	0.7 - 0.8 mm (0.028 - 0.032 in.)

Distributor cap/rotor/spark plug wires, replacing

The distributor cap, rotor, and spark plug wires deliver high-voltage spark to the spark plugs. They are subject to insulation breakdown, corrosion fouling, and electrode wear and damage. The components should be inspected and replaced as necessary at the intervals listed under **Table b** or **Table c** to ensure maximum engine efficiency. Guidelines for visual inspection and testing, and instructions for replacement are found in **280 Ignition System**.

Fuel filter, replacing

The fuel filter is located in the left rear of the engine compartment.

– Disconnect negative (–) battery cable.

0019591

◄ Loosen filter clamping bracket. Counterhold and loosen fittings at top and bottom **(arrows)** of filter. Soak up spilled fuel.

- Install filter in correct direction and tighten fittings while counterholding fuel filter.

NOTE —
The direction of fuel flow is indicated on the filter.

WARNING —
Fuel will be spilled when the filter is removed. Do not smoke or work near heaters or other fire hazards. Keep a fire extinguisher handy.

CAUTION —
Clean thoroughly around the filter connections before removing. Make sure no dirt gets into the fuel lines.

Valve clearance, adjusting

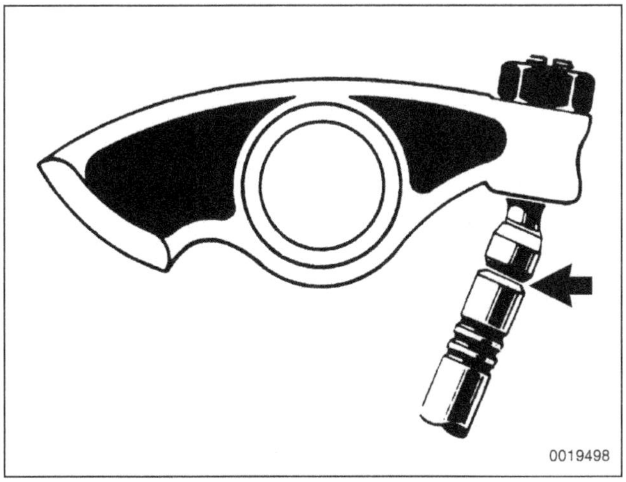

0019498

◄ The valve clearance is checked and adjusted with the engine cold (ambient temperature of approximately 20°C/68°F).

– Drain oil from engine crankcase.

WARNING —
Do not remove the spark plugs to make the engine turn easier. Carbon deposits that have built up on the plug can fall between the valve and seat with the plug is removed. This can result in inaccurate valve adjustment.

– Raise rear of car and support safely on jack stands.

WARNING —
Make sure the car is stable and well supported at all times. Use a professional automotive lift or jack stands designed for the purpose. A floor jack is not adequate support.

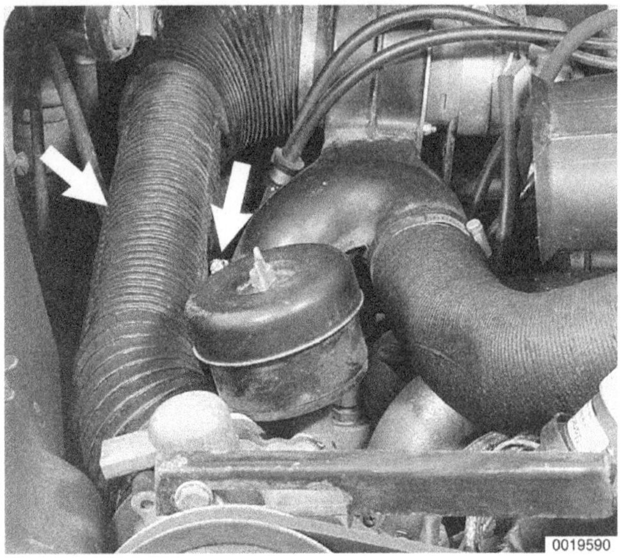

◄ Remove both heater hoses (**arrows**) on left side of engine.

– Remove air filter cover and filter element.

– Disconnect A/C compressor electrical harness connector.

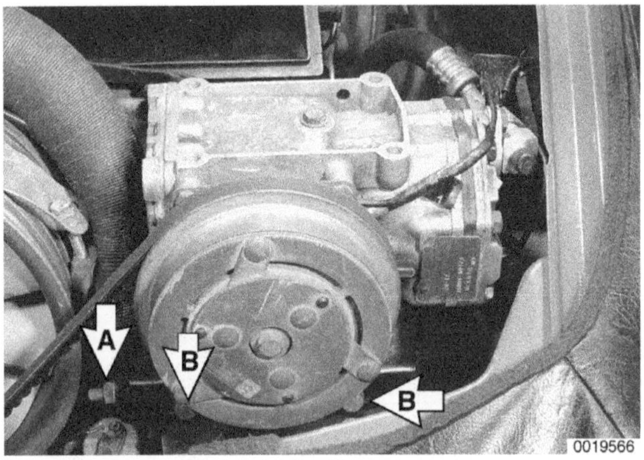

◄ Back off A/C belt adjusting fastener (**A**) and detach belt.

- Remove A/C compressor mounting bolts (**B**) and remove compressor. (One mounting bolt in back not shown.)
- Set compressor aside without disconnecting A/C lines.

– Unplug all spark plug connectors and set aside.

– Remove distributor cap from distributor. Do not remove spark plug wires from cap.

– Remove upper and lower valve covers.

◄ Using a 23 mm socket on fan pulley nut, rotate fan in a clockwise direction until distributor rotor is pointing to cylinder 1 mark (**arrow**) on distributor body.

- **Z1** mark on crankshaft pulley must also align with mark on lower fan housing. If all marks align, engine is at cylinder 1 top dead center (TDC).

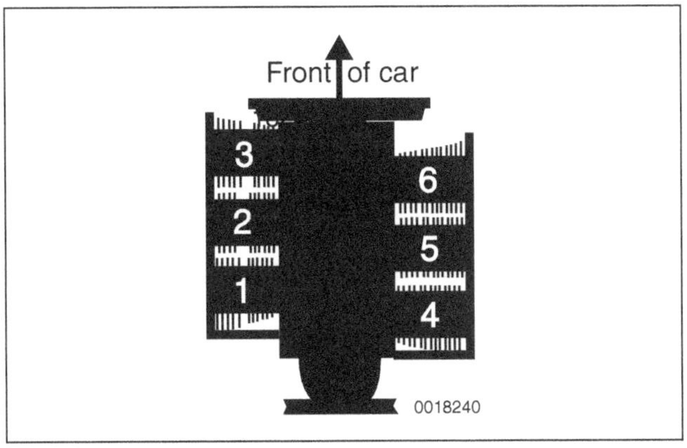

The valve adjusting sequence is 1–6–2–4–3–5

Valve clearance

- Adjusting value . . . . 0.10 mm (0.004 in.)

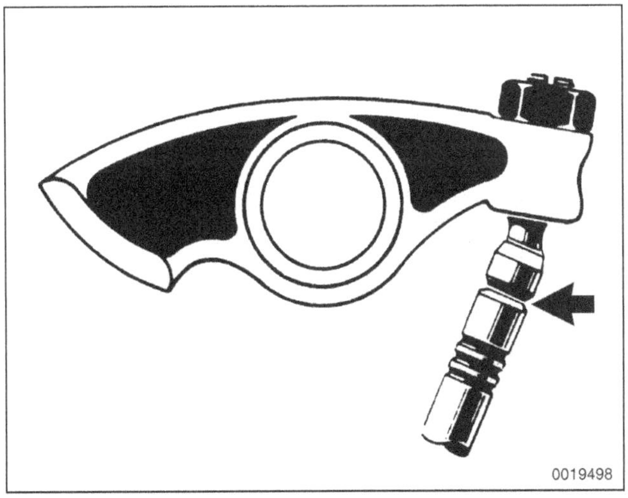

Check clearance of cylinder 1 intake valve by sliding a feeler gauge (special tool P 213 b) between rocker arm adjusting screw and the valve stem (**arrow**).

- If feeler gauge slides in easily, valve clearance is too big. If gauge will not slide in at all, clearance is too small.
- Repeat procedure for exhaust valve.

If necessary, loosen valve adjuster lock nut and turn adjuster with screwdriver until a light resistance is felt at feeler gauge.

– While holding adjuster in position, tighten adjuster locking nut.

– Recheck adjustment after tightening lock nut.

Turn crankshaft pulley 120° (clockwise direction). Adjust valves for cylinder 6 using above procedure.

NOTE—

*The crankshaft is marked in 120 degree intervals starting with **Z1**. With the crankshaft at TDC, the **Z1** mark will be at 12 o-clock, the next mark will be at 4 o'clock and the last mark will be at 8 o'clock.*

– Turn crankshaft clockwise 120° and adjust valves for cylinder 2.

– Turn crankshaft clockwise 120° and adjust valves for cylinder 4.

– Turn crankshaft clockwise 120° and adjust valves for cylinder 3.

– Turn crankshaft clockwise 120° and adjust valves for cylinder 5.

– Clean valve cover surfaces on camshaft housing to remove any gasket material that may be stuck to the surface.

– Clean all gasket material and oil from valve covers.

– Install valve covers with new gaskets, washers and locknuts.

Tightening torques

- Oil drain plug to crankcase (from 3/83). . 70 Nm (51 ft-lb)
- Oil drain plug to sump cover (to 3/83). . . 42 Nm (31 ft-lb)
- Sump cover to crankcase (to 3/83)(M6) . . 10 Nm (7 ft-lb)
- Valve cover to camshaft case (M8). . . . . . . 8 Nm (6 ft-lb)

– Top up engine oil, run engine and check for leaks.

Battery maintenance

The battery is located in the luggage compartment, on the left side. When servicing the battery, always disconnect the negative (–) cable first, and connect it last.

– Simple maintenance of the battery and its terminal connections will ensure maximum starting performance, especially in colder temperatures. For a more detailed discussion of the battery and charging system, see **270 Battery, Starter, Alternator**.

NOTE—

A secure battery hold-down is important in order to prevent vibrations and road shock from damaging the battery internally.

Battery cable clamps should be tight. The terminals, the cable clamps, and the battery case should be clean and free of white deposits. Even a thin layer of dust containing conductive acid salts can cause battery discharge.

To remove battery corrosion, begin by disconnecting the cables. Disconnect the negative (–) cable first.

Clean the terminal posts and the cable clamps with a wire brush. Clean the main chassis ground terminal next to the battery.

Corrosion can be washed away with a baking soda and water solution that will neutralize the acid. Apply the solution carefully as it will also neutralize the acid inside the battery. Lightly coat the outside of the terminals, hold down screw, and clamps with petroleum jelly, grease, or a commercial battery terminal corrosion inhibitor.

> ***WARNING —***
> - *Battery acid is caustic and dangerous. Take care to keep it from contacting eyes, skin, or clothing. Wear eye protection. Extinguish all smoking materials and do not work near any open flames.*
> - *Hydrogen gas given off by the battery during charging is explosive. Do not smoke. Keep open flames away from the top of the battery, and prevent electrical sparks by turning off the battery charger before connecting or disconnecting it.*

> ***CAUTION —***
> *To avoid causing battery acid damage to the interior, remove battery from car before using a battery charger.*

Battery electrolyte should be maintained at the correct level just above the battery plates and their separators. The correct level is approximately 5 mm (¼ in.) above the top of battery plates or to the top of the indicator marks (if applicable). The battery plates and the indicator marks can be seen once the filler caps are removed. If the electrolyte level is low, replenish it by adding distilled water only.

Throttle housing and linkage, lubricating

The linkage at the throttle housing should be lubricated at the normal maintenance intervals. Use a general purpose oil on the joints and bearings of the linkage. Use a multipurpose lube on the bearing points of the throttle plate.

V-belts, checking and replacing

Replacement of V-belts every four years is recommended. When belts are replaced, store the old set in the luggage compartment for emergency use.

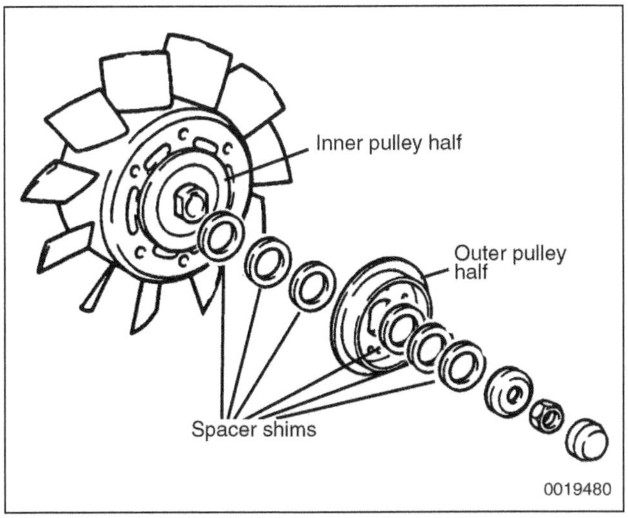

The V-belt is tensioned through a split pulley and shim arrangement. Six shims are installed at the factory and arranged on either side of the pulley depending on the adjustment required to correctly tension the belt. All six shims should remain on the alternator shaft.

Removing shims from the between the pulleys halves increases belt tension. Placing additional shims between the pulley halves decreases the tension.

A special spanner wrench is needed to hold the belt pulley stationary while loosening the large nut on the alternator shaft.

Inspect belts with the engine off. Twist the belt to inspect its sidewalls and bottom. Belt structural damage, glazed or shiny sidewalls caused by a loose belt, or layer separation caused by oil contamination are all reasons to replace a belt.

– Remove A/C compressor belt. See **870 Air Conditioner**.

0019530

◄ While holding alternator belt pulley with special tool P 208 spanner wrench (**arrow**), loosen and remove pulley mounting nut and separate pulley. Note arrangement of shims on either side of pulley.

– Remove belt from crankshaft pulley.

– Install new belt and reinstall shims and pulley. Install shims, washer, and nut. Tighten nut, making sure pulley seats properly.

Tightening torque
- Alternator pulley to alternator . . . . 40 Nm (30 ft-lb)

– Check belt tension by applying thumb pressure at belt midpoint. The belt should deflect approximately 5 mm (0.2 in.).
  - If adjustment is necessary, reposition shims. Run engine for a few minutes (at least 1,500 rpm), then recheck belt tension.

Oxygen sensor, replacing (1980 - 1983 models)

The oxygen sensor monitors engine combustion efficiency by measuring the oxygen content of the exhaust gasses. The output signal from the sensor is used to precisely control fuel delivery. Any problems with the oxygen sensor will directly affect exhaust emissions and the way the engine runs. For more information on testing the oxygen sensor system, see **240 Fuel injection**.

Replacement of the oxygen sensor at the specified interval ensures that the engine and emission control system will continue to operate as designed.

0018764

◄ To remove sensor (**arrow**) from catalyst:
- Detach oxygen sensor harness connector in engine compartment.
- Remove sensor from catalyst.
- Lightly coat the threads of new sensor with anti-seize compound before installing.

NOTE —
A special deep 22 mm socket for replacing the oxygen sensor is available from most automotive parts stores or from Baum Tools Unlimited. The socket has a groove cut down one side to allow the sensor to be installed without damaging the wire harness.

Tightening torque
- Oxygen sensor to catalytic converter. . . . . 50 - 60 Nm (36 - 43 ft-lb)

CHASSIS AND UNDER-CAR MAINTENANCE

Tires, checking inflation pressure

Correct tire inflation pressures are important to handling and stability, fuel economy, and tire wear. Tire pressures change with temperature. Pressures should be checked often during seasonal temperature changes. Correct tire pressure information can be found on an information label (**arrow**) on the rear engine compartment apron and in the owner's manual.

> ***WARNING —***
> *Do not inflate any tire to a higher pressure than the tire's maximum inflation pressure listed on the sidewall. Use care when adding air to warm tires. Warm tire pressures can increase as much as 4 psi (0.3 bar) over their cold pressures.*

Tires, rotating

The tires must not be rotated because of front-to-rear size difference.

Wheels, aligning

Porsche recommends checking front and rear wheel alignment once a year and whenever new tires are installed. See **440 Wheels, Tires, Alignment** for a more detailed discussion of alignment requirements and specifications.

Front wheel bearings, checking play

Routine maintenance of the tapered front wheel bearings includes checking for play and adjusting as necessary.

– To check wheel bearing play, first loosen front wheel lug nuts and then raise front of car and support safely using jack stands.

> ***WARNING —***
> *Make sure the car is stable and well supported at all times. Use a professional automotive lift or jack stands designed for the purpose. A floor jack is not adequate support.*

– Remove front wheel and pry off grease cap from wheel hub.

Check that wheel bearing thrust washer can just barely be moved back and forth with a light push on the tip of a screwdriver (do not pry or twist screwdriver).

– If necessary, adjust wheel bearing play as described in **401 Front Suspension**.

– Install grease cap. Reinstall road wheel.

Tightening Torque

- Wheel to wheel hub . . . . 130 Nm (96 ft-lb)

Brake system, inspecting

Routine maintenance of the brake system includes maintaining an adequate level of brake fluid in the reservoir, checking brake pads for wear, checking hand brake function, and inspecting the system for fluid leaks or other damage.

Brake hoses and lines, inspecting

Check that all brake hoses are correctly routed to avoid chafing or kinking. Inspect unions and brake calipers for signs of fluid leaks. Inspect rigid lines for corrosion, dents, or other damage. Inspect flexible hoses for cracking. Replace faulty hoses or lines as described in **460 Brakes–Mechanical**.

CAUTION —

- *Brake fluid is poisonous. Do not siphon brake fluid by mouth. Wear safety glasses and rubber gloves when working with brake fluid or brake pads to prevent contamination of cuts.*
- *Brake fluid will damage painted surfaces. Immediately clean brake fluid spilled on painted surfaces thoroughly using water.*
- *Incorrect installation or overtightening hoses, lines, and unions may cause chafing or leakage. This can lead to partial or complete brake system failure.*

Brake pad/rotor wear, checking

< Check brake pad thickness (**A**) through opening in caliper. See **460 Brakes–Mechanical** for brake pad replacement.

Brake Pad Minimum Thickness (Dimension A)
- Front and rear disc brake pads....... 2.0 mm (0.08 in.)

– Check brake rotor for minimum thickness using a micrometer. Check rotor for deep scoring or uneven surface wear. Replace rotors in pairs if any faults are found. See **460 Brakes–Mechanical** for brake rotor replacement.

Brake Rotor Minimum Thickness (after machining)
- Front .... 19.1 mm (0.752 in.)
- Rear .... 18.5 mm (0.728 in.)

NOTE —
For parking brake lining wear, see **Parking brake, checking** *below.*

WARNING —
Friction materials such as brake linings may contain asbestos fibers. Do not create dust by grinding, sanding, or cleaning the pads with compressed air. Avoid breathing asbestos dust, as it may result in serious diseases and cancer, or in death.

Brake fluid, replacing

Porsche strictly recommends replacing the brake fluid once every two years. This will help protect against corrosion and the effects of moisture in the fluid. See **470 Brakes–Hydraulic** for brake fluid flushing procedures.

Parking brake, checking

The parking brake may require periodic adjustment depending on use. Adjust the parking brake if the brake lever can be pulled up more than 4-5 clicks on the ratchet mechanism. Check that the cable moves freely. A complete description of the parking brake and parking brake adjustment can be found in **460 Brakes–Mechanical**.

NOTE —
The parking brake may lose some of its effectiveness over time if it is not used frequently. This is due to corrosion build-up on the parking brake drum. To remove corrosion, apply the parking brake just until it begins to grip, then pull the lever up one more stop (click). Drive the car approximately 400 meters (1,300 ft.) and release the brake. To recheck the adjustment of the parking brake see **460 Brakes–Mechanical**.

Exhaust system, inspecting

Exhaust system life varies widely according to driving habits and environmental conditions. If short-distance driving predominates, the moisture and condensation in the system will not fully dry out. This will lead to early corrosion damage and more frequent replacement.

Check for restrictions due to dents or kinks. Check for weakness or perforation due to rust. Check that exhaust system fasteners are tight. Check to see that all support bands are in place and properly supporting the system. Exhaust system replacement is covered in **260 Exhaust System**.

Manual transmission service

Manual transmission service consists of inspecting for leaks, checking and changing the gear oil, checking the clutch and clutch cable. See **340 Manual Transmission–Controls and Case** and **301 Clutch**.

Manual transmission oil, checking and replacing

Porsche recommends checking the manual transmission oil level and changing the oil at specified intervals. Check and fill the transmission with the car on a level surface. Transmission oil level checking and oil replacement procedures are covered in **340 Manual Transmission–Controls and Case**.

Front suspension and steering linkages, inspecting

Inspection of the front suspension and steering includes a check of all moving parts for wear and excessive play. Also inspect the small rubber seals and boots for cracks or tears that could allow the entry of dirt, water, and other contaminants. See **401 Front Suspension**.

Rear suspension, inspecting

Rear suspension inspection consists of checking for damaged or worn suspension components, leaky shock absorbers, and checking the rear drive axle rubber boots for damage. See **420 Rear Suspension, Axle Shafts**.

Fuel tank and fuel lines, inspecting

– Inspect the fuel tank, fuel lines, and fuel system for damage or leaks.
  - Check for fuel leaks in the engine compartment or fuel odors in the passenger compartment.
  - Check for a damaged fuel tank.
  - Check for any evaporative emissions hoses that may have become disconnected, checking carefully at the charcoal canister and evaporative emissions purge valve (where applicable).
  - See **240 Fuel Injection** and **201 Fuel Supply** for component locations and additional information.

> ***WARNING —***
> *When checking for fuel leaks, the engine must be cold. A hot exhaust manifold or exhaust system could cause the fuel to ignite or explode causing serious personal injury. Ventilate the work area and clean up spilled fuel immediately.*

BODY AND INTERIOR MAINTENANCE

Windshield wiper blades, replacing

Normally, replacing the wiper blades twice a year (before and after the cold season) works well. See **920 Windshield Wipers and Washers** for more information about wiper system troubleshooting and wiper blade replacement.

NOTE —
When replacing the wiper blades, check the tightness of the wiper arm nuts. Loose wiper arms can lead to damaged wiper linkages. If found to be loose, align wiper arms and tighten the nuts to the torque listed in **920 Windshield Wipers and Washers**.

Body lubrication

The body and door hinges, the luggage compartment and engine lid latch, and the door check rods can be lubricated with a medium weight engine oil. Lubricate the seat runners with multipurpose grease.

Lubricate the sunroof guide rails with silicone spray. If door weatherstrips are sticking, lubricate them with silicone spray or talcum powder. Do not apply any oil to rubber parts.

The door locks and lock cylinders can be lubricated with an oil that contains graphite.

Seatbelts

Dirt and other abrasive particles will damage seatbelt webbing. If it is necessary to clean seatbelts, use a mild soap solution. Bleach and other strong cleaning agents may weaken the belt webbing.

> ***WARNING —***
> *Do not clean the seatbelt webbing using dry cleaning or other chemicals. Allow wet belts to dry before allowing them to retract.*

The condition of the seatbelt webbing and the function of the retractor mechanisms should be inspected.

100 Engine–General

TABLES

GENERAL

2.4/2.7 CIS Engine

0019651

The 1978 - 1983 Porsche SC models are fitted with a 3.0 liter, air cooled, horizontally-opposed engine. General engine specifications are shown in **Table a**. Note that there were three slightly different variations of the engine.

The 3.0 liter SC engine uses Continuous Injection System (CIS) for fuel management and Capacitor Discharge Ignition (CDI) for ignition management. See **200 Engine Management–General** for more detailed information on system operation.

Table a. Engine specifications

Year	Engine code	No. of cylinders	Displacement liters (cu. in.)	Compression ratio	Horsepower Kw @ rpm (Hp @ rpm)	Torque Nm @ rpm (lb-ft @ rpm)
1978 - 1979	930/04 930/06	6	2994 (182.7)	8.5:1	128 @ 5500 (172 @ 5500)	265 @ 4200 (189 @ 4,200)
1980	930/07	6	2994 (182.7)	8.5:1	128 @ 5500 (172 @ 5500)	265 @ 4200 (189 @ 4,200)
1981 - 1983	930/16	6	2994 (182.7)	9.3:1	132 @ 5500 (180 @ 5500)	265 @ 4200 (189 @ 4,200)

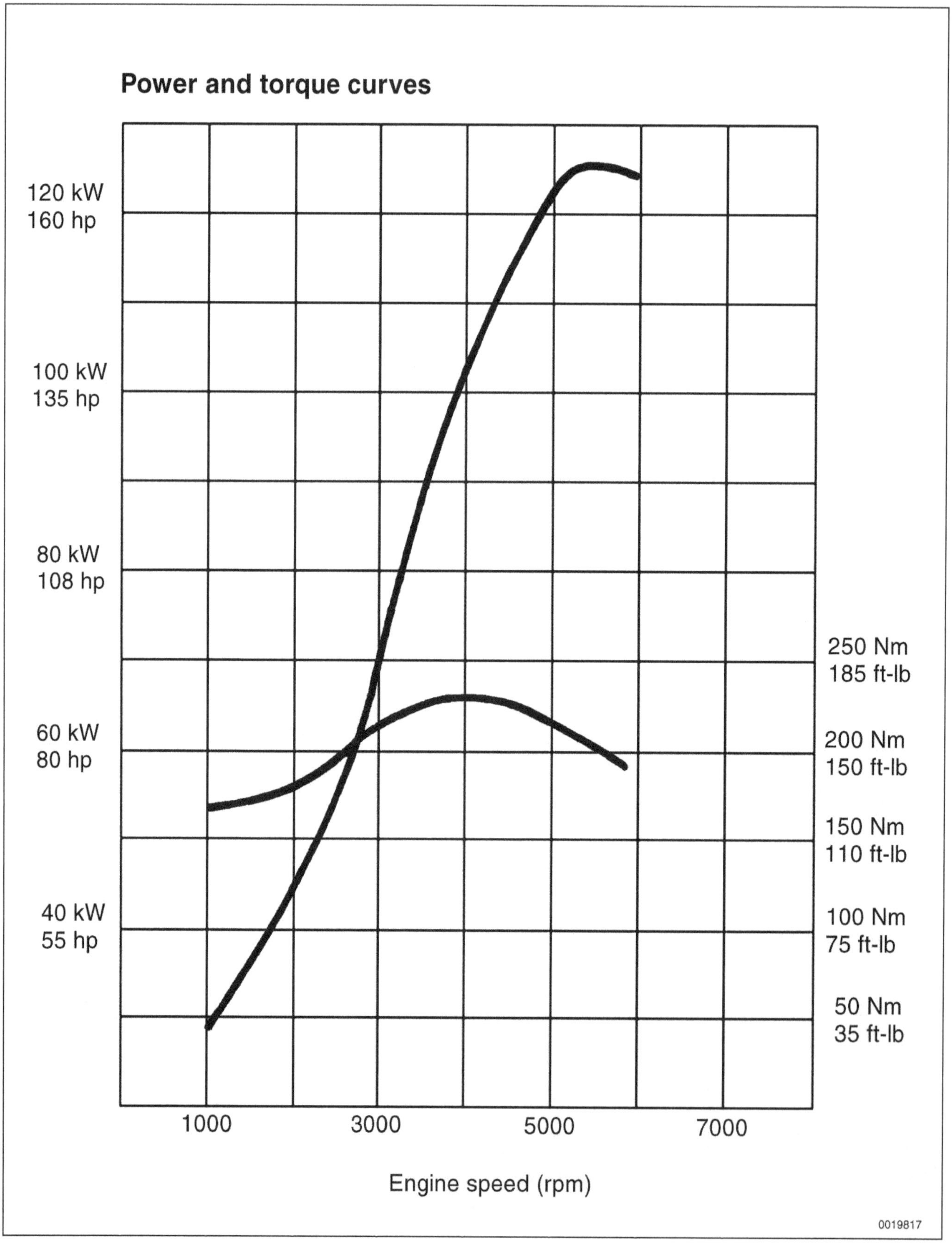
Power and torque curves
120 kW
160 hp
100 kW
135 hp
80 kW
108 hp
60 kW
80 hp
40 kW
55 hp
250 Nm
185 ft-lb
200 Nm
150 ft-lb
150 Nm
110 ft-lb
100 Nm
75 ft-lb
50 Nm
35 ft-lb
1000
3000
5000
7000
Engine speed (rpm)
0019817

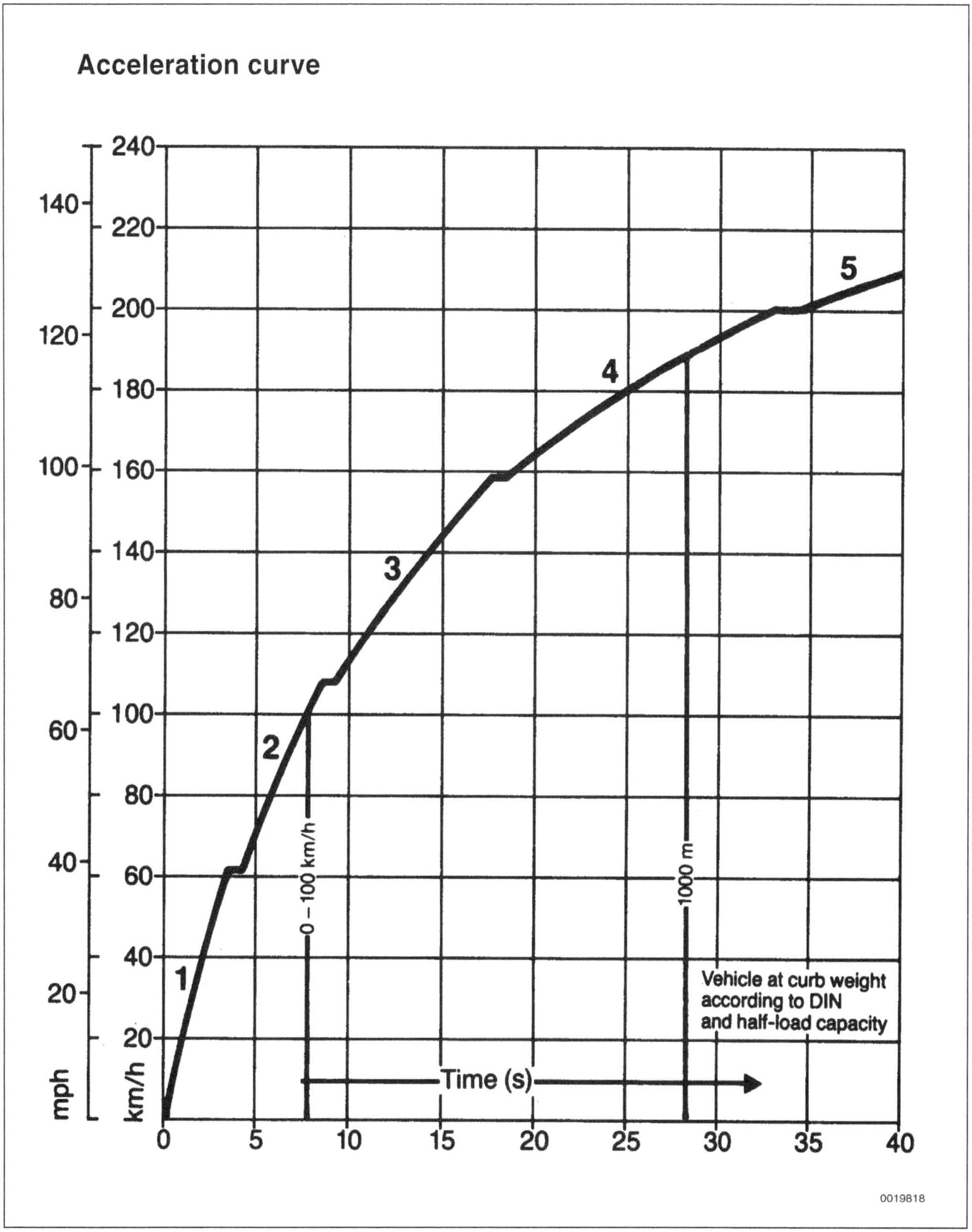
Acceleration curve
mph
km/h
140
120
100
80
60
40
20
240
220
200
180
160
140
120
100
80
60
40
20
1
2
3
4
5
0 – 100 km/h
1000 m
Time (s)
0
5
10
15
20
25
30
35
40
Vehicle at curb weight
according to DIN
and half-load capacity
0019818

911S engine cross section

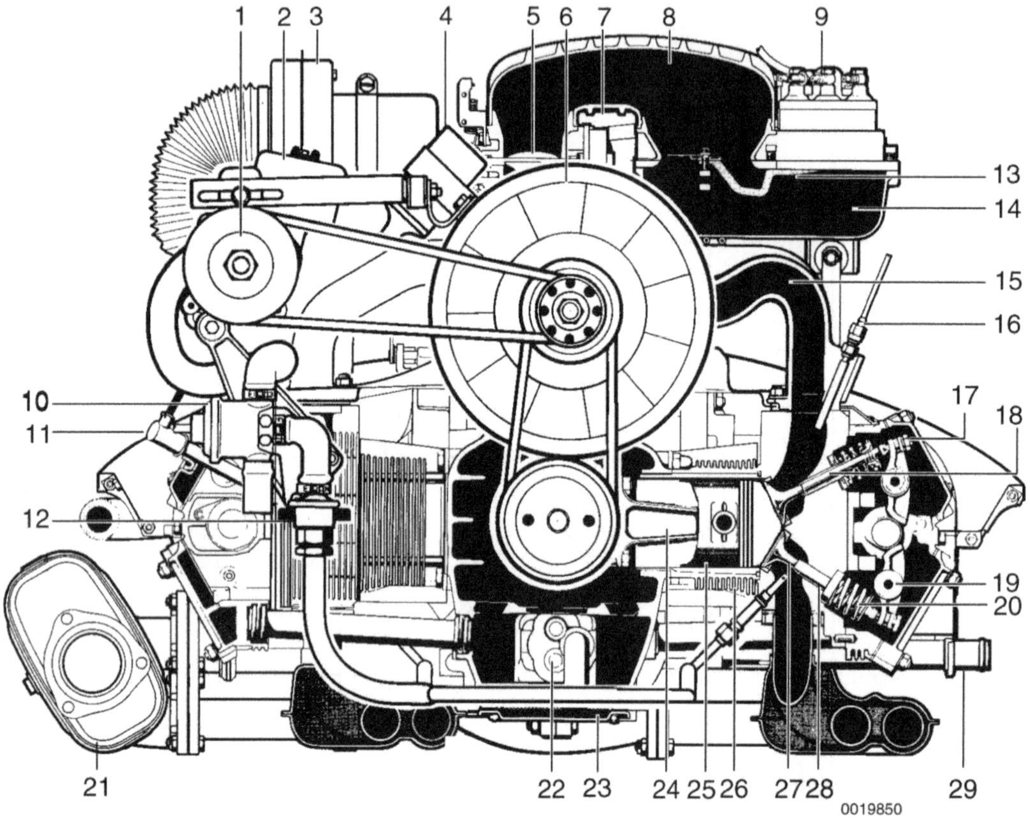

1. Air pump
2. Air pump filter
3. Heater blower
4. Ignition coil
5. Throttle valve
6. Engine cooling fan
7. Control pressure regulator
8. Air intake boot
9. Fuel distributor
10. Air pump diverter valve
11. Spark plug wire end
12. Air pump one-way check valve
13. Air sensor plate
14. Intake housing
15. Intake runner
16. Fuel injector
17. Valve adjuster
18. Intake valve
19. Camshaft
20. Valve spring
21. Catalytic converter
22. Oil pump
23. Oil drain/screen plate
24. Connecting rod
25. Piston
26. Cylinder
27. Exhaust valve
28. Valve guide
29. Oil tank connection

Engine case (crankcase)

The two-piece engine case is cast aluminum and is bolted together along the centerline of the engine around the crankshaft. The crankcase does not use an oil sump. Instead a dry-sump lubrication system is employed. The crankcase oil drain plug is located in the oil screen cover in the bottom of the crankcase.

Crankshaft

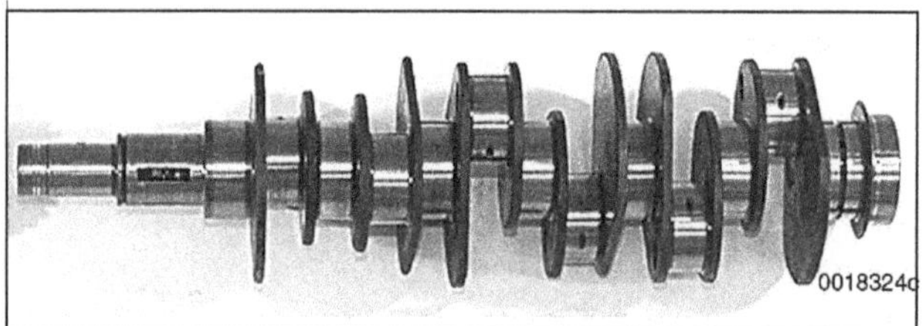

The crankshaft rotates in replaceable split-shell main bearings. Oilways drilled into the crankshaft provide bearing lubrication. Flexible lip seals are installed at both ends of the crankshaft. The crankshaft has a stroke of 70.4 mm (2.77 in.).

0018725

Pistons

The forged pistons are of the three-ring type with two upper compression rings and a lower one-piece oil scraper ring. The pistons have a nominal diameter of 95 mm (3.74 in.). Full-floating piston pins are retained with circlips.

0019521

Connecting rods

The connecting rods use replaceable split-shell bearings at the crankshaft end and solid bushings at the piston pin end. The bolts are stretch type bolts and should always be replaced during engine assembly. The fasteners are tightened using the torque angle tightening method.

0019581

Cylinders

IIndividual aluminum cylinders are Alusil® (KS) or Nikasil® (Mahle) process treated for long life.

– A sealing ring is used between the cylinder head and the cylinder.

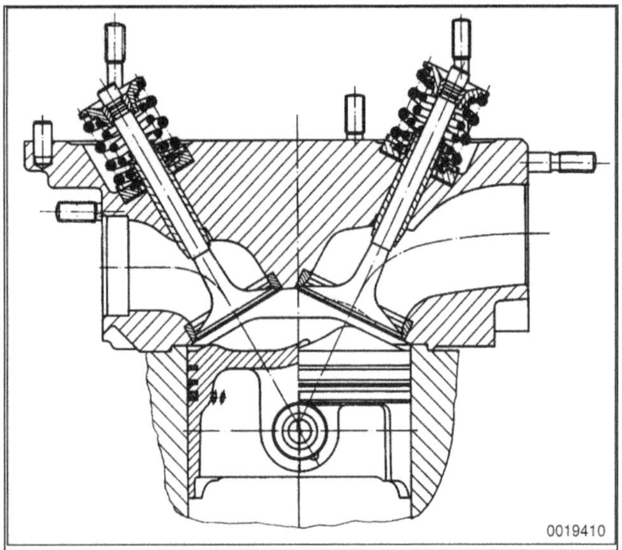

0019410

Cylinder head and valvetrain

The aluminum cylinder heads use chain-driven overhead camshafts and two valves per cylinder. The cylinder heads employ a crossflow design for greater power and efficiency. Intake air enters each combustion chamber from one side while exhaust gasses exit from the other. Oilways in the camshaft housings provide lubrication for the camshaft(s) and valvetrain. Oil is returned to the crankcase via six oil return tubes. The valve heads have a diameter of 49 mm (1.929 in.) on the intake side and 41.5 mm (1.634 in.) on the exhaust side.

Timing chains are kept under constant tension by one of two types of tensioner

0019637

The original-equipment design is a spring-loaded tensioner with hydraulic vibration damper. Oil is fed to this type of tensioner by means of splash.

0019810

An oil-pressure-fed (Carrera type) tensioner can be installed as a retrofit update. See **152 Camshaft Timing Chains** for update information and procedures.

Tensioner are shipped in collapsed position. Remove lock pin (**arrow**)after installation.

Engine lubrication system components

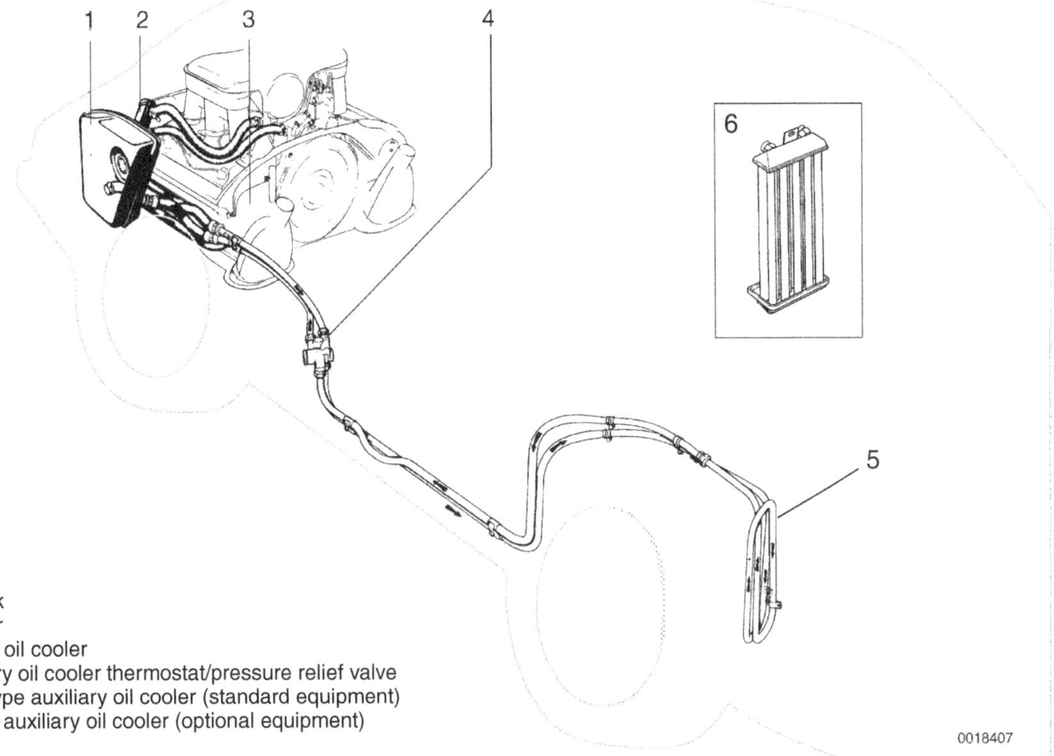

1. Oil tank
2. Oil filler
3. Engine oil cooler
4. Auxiliary oil cooler thermostat/pressure relief valve
5. Tube-type auxiliary oil cooler (standard equipment)
6. Finned auxiliary oil cooler (optional equipment)

Lubrication system

The unique 13 liter (13.5 US quart) lubrication system is a dry sump design. A tandem oil pump (return/delivery) supplies pressurized oil to engine and also pumps the oil back to the oil tank. The pump is gear driven directly off the intermediate shaft.

The engine oil is cooled via two oil coolers. The primary oil cooler is mounted to the side of the crankcase. Flow through it is controlled via a thermostat that prevents oil from circulating until the oil has reached a specified temperature.

An auxiliary oil cooler is mounted ahead of the right front tire in the fender well. An auxiliary oil cooler thermostat/pressure relief valve also controls flow into this cooler. See **170 Lubrication System** for additional information.

Engine cooling

Engine cooling is provided by a V-belt driven fan (**arrow**), which also drives the alternator. Ducting provides control of the air over the engine.

MECHANICAL TROUBLESHOOTING

This troubleshooting section applies to problems affecting the basic engine assembly–the crankcase, cylinder heads, valve train and engine internal moving parts.

Only a few basic functions are required of the engine. The crankcase, cylinder heads and their moving parts must fit together properly, operate smoothly, seal well enough to create and maintain compression, and keep pistons, valve train, and ignition properly timed. The problems discussed in this troubleshooting section are those that affect one or more of these functions.

Troubleshooting specifically for the lubrication system can be found in **170 Lubrication System**. Troubleshooting for starting and running problems can be found in **200 Engine Management–General**.

Basic troubleshooting principles

Begin troubleshooting with careful observation, keeping in mind the following questions:

How has the symptom developed? A symptom that develops quickly is probably caused by a problem that can be corrected by simple maintenance or repair. A symptom that has developed gradually over time, especially after fifty or sixty thousand miles, is more likely an indication of general wear and the need for more comprehensive overhaul work.

Is the symptom engine-speed dependent? A noise that is caused by an engine mechanical problem will be dependent mainly on engine speed, with similar symptoms regardless of changes in car speed. Noises that repeatedly occur only in a certain rpm range suggest a vibration problem. Noises that change with car speed are more likely due to drivetrain or running gear problems.

Is the symptom load dependent? Forces at work inside a running engine vary as the demand for power varies. Symptoms that are more severe during hard acceleration indicate certain kinds of problems. Symptoms that are more apparent at no load or high vacuum (example: coasting at high rpm) point to other problems. Note that higher engine loads also affect the fuel and ignition systems, which may be responsible for high-load performance problems.

Is the symptom temperature dependent? Does it only occur when the engine is cold? Does it change as the engine warms up? How? Metal parts expand and contract with changes in temperature. Clearances change. Oil viscosity and cooling system pressure change. In troubleshooting symptoms that change as the engine gets warm, look for an engine characteristic that changes with temperature.

Noise

In order to run smoothly under harsh conditions, the internal engine parts are made to precise dimensions, assembled with precision clearances, and lubricated by a pressurized oiling system.

Most **unidentified engine noises** result from clearances that have become too large due to worn parts, lack of adequate lubrication, or both. The importance of lubrication cannot be over-emphasized. For best results, troubleshooting engine noises should only be done when the oil and filter have been recently changed and the oil level is correct.

High-pitched metallic tapping noises are caused by relatively small, lightweight parts. These noises are most likely an indication of excessive clearances in the valve train. Valve train noise accompanied by burning oil (blue-gray smoke in the exhaust), particularly at start-up or when decelerating from high rpm, is an indication of worn valve guides that can only be remedied by reconditioning or replacement of the cylinder head(s).

In a high-mileage engine, a **light metallic rattle or chatter** under acceleration, accompanied by increased oil consumption and smoking, may indicate severely worn or broken piston rings. Since this diagnosis means overhaul or replacement of the engine, the problem should be further investigated with a compression or cylinder leak-down test. A vacuum gauge is also helpful when diagnosing engine mechanical faults. These tests are described later in this section.

Deep metallic knocking sounds are caused by excessive clearances between heavier components. Closer analysis of the noise will often help identify the problem. Piston slap, caused by excessive piston skirt to cylinder wall clearance, is worse when the engine is cold and may be accompanied by increased oil consumption and reduced compression due to accelerated piston ring wear. A double knock, most pronounced at idle or low load, is due to excessive clearance at the piston pin and upper connecting rod bushing.

Deep hollow knock from crankshaft bearings become worse as the engine warms up. A noise that is very pronounced under load, perhaps louder during the transition from acceleration to coasting, is most likely caused by a damaged connecting rod bearing. Crankshaft main bearings produce a lower, dull knock. An intermittent knock, that may be most apparent when depressing or releasing the clutch, indicates excessive crankshaft end play. These problems seldom occur as isolated failures. They are almost always an indication of poor overall engine condition and can only be properly corrected by complete engine overhaul or replacement.

Rumbling or groaning from the engine compartment may not indicate engine problems at all, but rather a worn bearing or bushing in an engine-driven accessory. These include the alternator, the engine cooling fan and the A/C compressor. To check these accessories, run the engine briefly with the drive belt(s) disconnected and see if the noise has stopped. Once the drive belt is removed from a component, turning the pulley and shaft by hand may also reveal a bad bearing or bushing. A properly functioning accessory should turn smoothly.

Fluid leaks

Fluid leaking from and around the engine is most likely either oil or fuel. Look for wet spots on the engine to help pinpoint the source.

Fuel leaks are usually obvious by the smell of gasoline. Repair fuel leaks immediately.

The most likely sources of engine **oil leaks** are the oil filter gasket, the crankcase oil seals, the valve cover gaskets, or the oil screen pan gaskets. Because the crankcase is under slight vacuum when the engine is running, some oil leaks may not be apparent until the engine is turned off. Other sources of oil leaks include:

- breather housing
- engine mounted oil thermostat
- oil pressure switch
- oil return pipes
- oil tank breather hose
- temperature switch (1981 - 83)

When investigating fluid leaks, **brake fluid** and **transmission oil** should be also be considered as potential agents.

Smoking

Smoke in the exhaust indicates something getting into the combustion chamber and being burned. The color of the smoke identifies the contaminant.

Blue-gray smoke is from oil. Oil smoke, probably accompanied by increased oil consumption and oil residue on spark plugs, indicates that engine oil is getting past piston rings, valve guides, the cylinder head gasket, or some combination of the three. Use a compression test for diagnosis as described later.

In an older engine, compression pressures that are even but below specifications suggest worn piston rings and cylinder walls, and the need to overhaul or replace the engine. If smoking is most obvious under high engine vacuum, such as while coasting at high rpm, and compression pressures are within specifications, leaking valve guide oil seals or worn valve guides are a probable cause.

Black smoke is caused by the engine getting too much fuel. Driveability troubleshooting is covered later in this section and also in **200 Engine Management–General**.

Excessive oil consumption

Some oil consumption is normal and indicates healthy flow and distribution of the vital lubricant in the engine. The oil level must be checked, and occasionally corrected, between oil changes. Aside from leaks, increased oil consumption will usually be accompanied by some smoking, however slight, and the causes of excessive oil consumption are the same as those for oil smoke in the exhaust. As with smoking symptoms, gradual increases are caused by worn piston rings and/or valve guides. Sudden high oil consumption suggests broken rings. See **Smoking** above for more troubleshooting information.

Poor fuel consumption and low power

Poor fuel consumption and low power can suggest problems with the fuel or ignition systems, particularly on a low-mileage engine. On an engine with high mileage, low compression may be the cause.

Normal wear of the valves, piston rings, and cylinder walls decreases their ability to make a good seal. The engine becomes less efficient and has to work harder, using more fuel to produce the same amount of power. Engine condition can be evaluated with a compression test as described later.

Engine not running

An engine problem that affects ignition or valve timing may prevent the engine from starting or running. The camshaft timing chains and sprockets are responsible for correctly timing the actions of the valves and the ignition system. A worn camshaft timing chain may jump teeth, throwing off all the engine's timing functions, and still appear to be perfectly normal. Begin by doing a compression test as described later.

To check camshaft timing, see **152 Camshaft Timing Chains**. Other troubleshooting information for an engine that fails to start can be found in **200 Engine Management–General**.

Table b lists symptoms of engine problems, their probable causes, and suggested corrective actions. A boldface number in the corrective action column indicates the repair group in the manual where the applicable test and repair procedures can be found.

When troubleshooting an engine that fails to start or runs poorly, first check its mechanical condition—particularly if the engine has high mileage.

Table b. Engine mechanical troubleshooting

Symptom	Probable cause	Corrective action
Pinging or rattling noise under load, uphill or accelerating, especially from low speeds. Indicates detonation or pre-ignition	a. Fuel does not meet manufacturer's octane requirements	a. Switch to higher octane fuel.
	b. Overheating	b. Check engine oil cooler system. Repair group **170**
	c. Spark plugs damaged or wrong heat range	c. Replace spark plugs. Repair group **280**
	d. Air-fuel mixture too lean	d. Adjust mixture. Repair group **240**
Screeching or squealing noise under load. Goes away when coasting. Indicates slipping V-belt	a. Loose, worn, or damaged V-belt(s)	a. Inspect, replace, or tighten belt(s). Repair group **030**
	b. Excessive belt loads due to failed engine-driven component	c. Locate and replace failed component. Repair group **270** or **870**
Growling or rumbling, varies with engine rpm. Bad bearing or bushing in an engine-driven accessory	a. Alternator worn	a. Repair group **270**
	b. Air conditioning compressor faulty	b. Replace compressor. Repair group **870**
Light metallic tapping noise, varies directly with engine speed. Oil warning light **not** illuminated	a. Low oil pressure and defective warning light circuit	a. Check oil pressure. Repair group **170**
	b. Valve clearances out of adjustment	b. Adjust valve clearance. Repair group **030**
	c. Defective rocker arm(s) or rocker arm shaft(s)	c. Check rocker arms/shafts and replace as required. Repair group **151**
Light metallic knock, varies directly with engine speed. Oil warning light blinking or fully illuminated (may be most noticeable on hard stops or cornering) Indicates lack of sufficient oil supply	a. Low oil level	a. Check and correct oil level. Repair group **030**
	b. Restricted (dirty) oil filter	b. Change engine oil and filter. Repair group **030**
	c. Insufficient oil pressure	c. Check oil pressure. Repair group **170**

Table b. Engine mechanical troubleshooting

Symptom	Probable cause	Corrective action
Blue-gray exhaust smoke, oily spark plugs. Indicates oil burning in combustion chamber	a. Contaminated or gasoline-diluted engine oil	a. Change engine oil and filter. Repair group **030**
	b. Leaking valve stem oil seals	b. Replace valve stem oil seals. Requires engine disassembly and cylinder head removal to repair.
	c. Worn valve guides	c. Overhaul or replace cylinder head. Repair groups **102** and **150**
	d. Worn, broken, or incorrectly installed pistons or piston rings	d. Overhaul or replace engine. Repair groups **102** and **131**
Black exhaust smoke	a. Rich air/fuel mixture	a. Adjust mixture. Repair group **240**
Engine runs badly, pops and backfires	a. Spark plug wires installed incorrectly	a. Install wires correctly. Repair group **280**
	b. Incorrect valve timing	c. Check camshaft timing chains. Repair group **152**
Engine will not start or run. Starter operates, engine turns over at normal speed	a. Failed ignition system or fuel system	a. Repair group **200**
	b. Broken timing chain	b. Check camshaft rotation as engine turns over. Repair group **152**
	c. Jumped timing chain	c. Check camshaft timing. Repair group **152**

Compression test

A test of compression pressures in the individual cylinders will tell a lot about the condition of the engine without the need disassemble the engine. The test is relatively simple. It requires compression tester, spark plug wrench, and a few other hand tools.

To obtain accurate results, the battery and starter must be capable of turning the engine at normal cranking speed. The area around the spark plugs should be clean, to avoid getting debris inside the engine when the spark plugs are removed. Because engine temperature may affect compression, the most accurate results are obtained when the engine is at normal operating temperature.

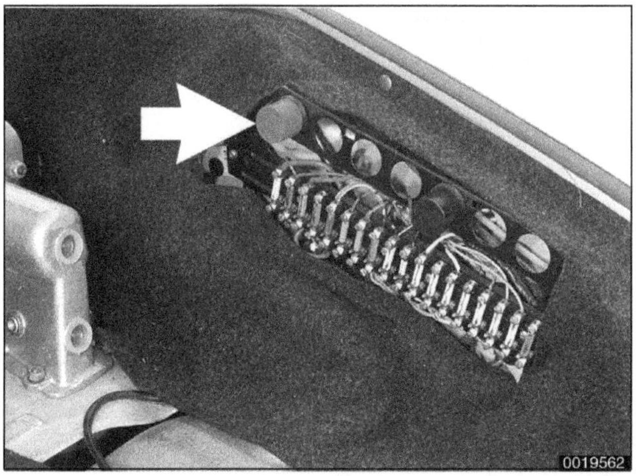

◄ Disable fuel pump by removing fuel pump relay (**arrow**), located in luggage compartment.

Disable ignition system by unplugging capacitor discharge control module harness connector (**arrow**) in engine compartment to left of engine.

> ***WARNING —***
> *The ignition system produces high voltages that can be fatal. Avoid contact with exposed terminals and use caution when working on a car with the ignition switched on or the engine running.*

- Remove heater blower elbow on left side of engine and air filter cover and filter element on right side.
- Remove and label spark plug wires.
- Remove and number spark plugs.

NOTE —
Check the spark plugs for oil deposits that may indicate poor cylinder sealing, then set them aside in order. Used spark plugs should be reinstalled in the same cylinder from which they were removed.

- Thread compression tester into cylinder 1 spark plug hole tight enough to seal hole.
- With parking brake set, transmission in neutral, and throttle held wide open, crank engine with starter. Crank engine for 4 to 5 full revolutions. Record value.

NOTE —
The compression gauge reading should increase with each compression stroke and reach near its maximum reading in 3-4 strokes. Look for consistent and even readings across cylinders, usually in the range of 110-150 psi with no more than a 10% difference between cylinders.

- Remove gauge from cylinder 1 and release pressure from gauge.
- Repeat test for other cylinders. Record gauge reading for each cylinder.
- Compare readings from each cylinder.

NOTE —
Removing the spark plugs may cause built-up carbon on the plug to fall onto the exhaust valve seat. If this happens, the exhaust valve may not seat fully, resulting in a low compression reading at that cylinder. If you suspect a partially open valve, reinstall the plug and run the engine at 2,500 rpm for a minute or so. Remove the plug and repeat the test.

- Reinstall spark plugs and spark plug wires in their original locations. The remainder of installation is reverse of removal.

Tightening torque
- Spark plug to cylinder head . . . . 25 Nm (18 ft-lb)

Low compression is evidence of poorly sealed combustion chambers. Generally, compression pressures that are relatively even but low indicate worn piston rings and/or cylinder walls. Erratic values tend to indicate valve leakage or a broken cylinder head stud. Dramatic differences, between cylinders are often the sign of a failed head gasket, burned valve, or broken piston ring. There are other tests that can further isolate the problem.

Wet compression test

To further help analyze the source of poor compression, repeat the compression test, this time with about a teaspoon of oil squirted into each cylinder. The oil will temporarily help seal between the piston rings and the cylinder wall, practically eliminating leakage past the rings for a short time. If this test yields higher compression reading than the "dry" compression test, there is probably leakage between the piston rings and cylinder walls, due either to wear or to broken piston rings. Little or no change in compression reading indicates other leakage, probably from the valves.

Cylinder leak-down test

The most conclusive diagnosis of low compression symptoms requires a leak-down test. Using a special tester and a supply of compressed air, each cylinder is pressurized. The rate at which the air leaks out of the cylinder, as well as where the air leaks out, can accurately pinpoint the magnitude and location of the leakage. Any engine compression diagnosis that will require major disassembly should first be confirmed by the more accurate leak-down test. Because the test requires special equipment and experience, it may be desirable to have it done by a Porsche dealer or other qualified repair shop.

Vacuum gauge test

A vacuum gauge can be a useful tool to diagnose a variety of engine problems. Care must be taken in interpreting the readings and the movements of the gauge needle. In many instances, the readings may indicate several problems and further testing may be required to isolate the exact problem. The vacuum gauge should be connected to a vacuum source on the intake manifold and the engine should be at operating temperature.

The engine vacuum gauge measures manifold vacuum. Manifold vacuum varies with different engine operating conditions and also with different engine problems. The manner in which the vacuum reading varies from the normal reading can indicate the type of engine problem. **Table c** lists vacuum gauge readings, their probable causes, and corrective actions.

Table c. Vacuum gauge readings

Vacuum gauge reading	Probable cause	Corrective action
High and steady reading (15-22 in./Hg.)	a. Normal performance. Engine in good condition	a. No corrective action required
Low and steady reading(10-15 in./Hg.)	b. Incorrect ignition timing (retarded)	b. Check ignition timing. Repair group **280**
	c. Incorrect valve timing	c. Inspect camshaft timing. Repair group **152**
	d. Low compression	d. Test compression as described above
	e. Throttle valve sticking	e. Remove intake air boot and check throttle movement. Repair group **240**
	f. Leaking intake manifold gasket or fuel injector seals	f. Inspect intake manifold gasket. Remove injectors and replace O-rings. Repair groups **200** and **240**
Very low but steady reading at idle (below 10 in./Hg.) Engine idles very rough and stalls	a. Large intake air leak	a. Visually inspect for faulty gaskets, hoses, or connections. Check the oil filler cap seal.
Gauge reading drift/float at idle	a. Minor intake air leak	a. Visually inspect for intake air leaks
	b. Air/fuel mixture incorrect (rich)	b. Check air/fuel mixture. Repair group **240**
Gauge reading fairly steady at idle but vibrates rapidly when engine speed is increased	a. Worn (weak) valve springs	a. Replace valve springs Repair group **150**
	b. Ignition miss	b. Test ignition system. Repair group **280**
	c.	c.
Gauge reading steady at idle but drops regularly	a. Sticking or faulty valve or incorrect valve clearance. (Needle drops when faulty valve operates)	a. Adjust valve clearance. Repair group **030**. If valve clearance is correct, inspect valves. Repair group **102**
	b. One or more spark plugs not firing	b. Replace faulty spark plugs or spark plug wires. Repair group **280**
	c.	c.
Gauge reading vibrates rapidly at idle but steadies when engine speed is increased	a. Worn valve guides	a. Check valve guides and repair or replace as necessary. Repair group **150**
Gauge reading gradually drops to 0 in./Hg at idle	a. Plugged or restricted exhaust system	a. Check exhaust system for restrictions, especially check for a plugged catalytic converter. Repair group **260**

101 Engine Removal and Installation

Special Tools

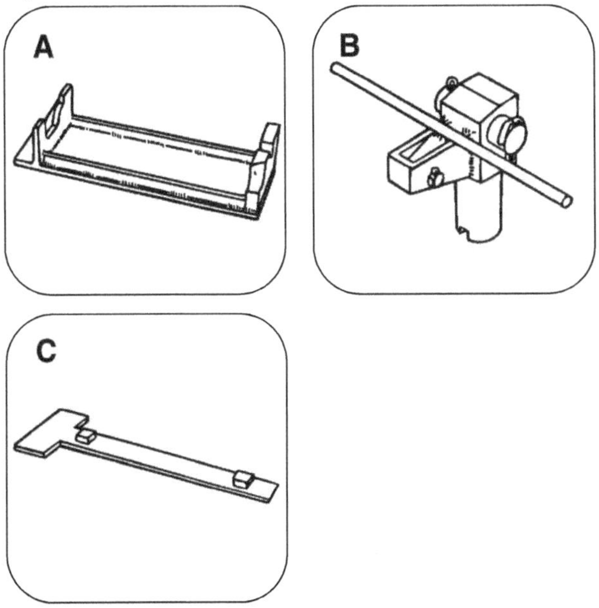

Special tool

A 9111 . . . . Engine fitting plate
(Porsche ordering #000 721 911 10)
B 9111/1 . . . . Adapter
(Porsche ordering #000 721 911 11)
C 9111/2 . . . . Support plate
(Porsche ordering #000 721 911 12)

GENERAL

This repair group covers engine removal and installation. The complete engine/transmission assembly is removed as a single unit through the bottom of the engine bay.

Removal and installation of the engine/transmission assembly is best accomplished by raising the car on an automotive lift. Support the engine/transmission from below using a heavy duty rolling cart. The complete engine with transmission weighs approximately 272 Kg (600 lbs.).

Special tools

For an alternate removal procedure, some special jack adaptors are recommended to support the engine/transmission assembly. The jack adaptors allow the rear of the engine to pivot down and the engine/transmission assembly to be pulled out from the rear of the vehicle.

WARNING —
Due to risk of personal injury, be sure the engine is cold before beginning the removal procedure.

CAUTION —
- *Cover rear fender surfaces and painted areas surrounding engine bay before beginning removal procedure.*
- *As an aid to installation, label all components, wires, and hoses before removing them.*
- *Do not reuse gaskets, O-rings or seals during reassembly.*

ENGINE AND TRANSMISSION, REMOVAL AND INSTALLATION

Engine/transmission, removing

– Disconnect negative (–) battery cable and cover battery terminal to keep cable from accidentally contacting battery terminal.

> ***WARNING —***
> *Avoid getting tools or clothing near the battery. Battery electrolyte is a corrosive acid. Always wear eye protection when working on or near the battery.*

– Raise vehicle on lift to a comfortable working height to allow access to top of engine.

> ***WARNING —***
> *Be sure vehicle is stable and well supported. Removal of the engine/transmission assembly will upset the balance of the vehicle. Use tall jack stands to support the vehicle at all four corners.*

– Disconnect A/C harness connector.

< Back off A/C belt adjusting fastener (**A**) and detach belt.

- Remove A/C compressor mounting bolts (**B**) and remove compressor. (One mounting bolt in back not shown (C).)
- Set compressor aside without disconnecting A/C lines.

> ***WARNING —***
> *A/C refrigerant hoses and components are under high pressure. Do not loosen, disconnect or damage A/C refrigerant lines or fittings.*

– Remove air filter cover and element from air filter housing.

< Disconnect breather hoses (**arrows**) at right side of engine compartment.

◄ Remove air pump belt adjusting support (**A**) and belt (**B**).

– Disconnect hoses from air pump.

– Remove air pump drive pulley.

– Remove air pump mounting bolt and remove pump.

◄ Remove heater hose (**arrow**) between heater blower fan on top of engine and left heat exchanger.

◄ Remove cover from electric panel in left side of engine compartment.

- Detach 14-pin harness connector from panel (**A**).
- Detach voltage regulator electrical harness connector, if applicable (**B**).
- Remove harness connector from CDI module (**C**).

< Disconnect hoses in left side of engine compartment. Counterhold fuel fittings when loosening.

- Disconnect fuel supply line at top and bottom of fuel filter (**arrows**).
- Disconnect fuel return line from rear of left side of engine.

> ***WARNING —***
> *Fuel will be discharged. Wrap a shop towel around fuel line fittings when loosening. Do not smoke or work around heaters or other fire hazards. Keep an approved fire extinguisher handy.*

– Disconnect power brake vacuum hose.

< Disconnect cruise control cable from vacuum module at rear left of engine compartment:

- Remove screws holding cable end flange to vacuum diaphragm. Pull out end flange.
- Squeeze plastic retainer clip and push forward to release cable.

< Working inside passenger compartment, remove floor mats at rear of center tunnel.

NOTE —
Slide driver's seat forward to gain access to rear center tunnel.

- Remove tunnel access cover.
- Disconnect vehicle speed sensor wires at connector (**A**). Push wires through rear bulkhead.
- Pull back rubber boot on shift linkage. Loosen 4 mm allen set screw (**B**) and pull shift rod out of shift coupler.

– Raise vehicle to gain access to bottom of engine

◀ Disconnect accelerator pull rod ball-joint end (**arrow**) from throttle relay lever on left side of transmission.

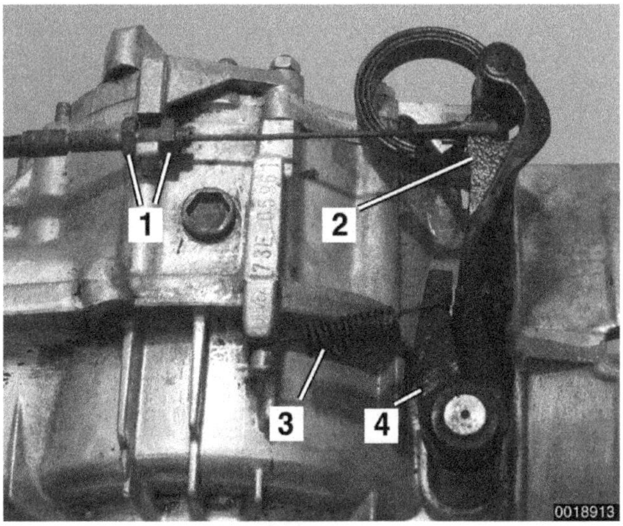

◀ Remove clutch release lever:

- Loosen clutch cable mounting nuts and remove cable from boss(**1**) and then disconnect cable from release lever (**2**).
- Disconnect spring (**3**) on positioning lever, remove circlip (**4**) from lever, and slide lever off splined end of clutch cross-shaft.

◀ Use large screwdriver to press clutch release lever forward. Remove release lever from clutch cross-shaft.

> ***WARNING —***
> *Once the clutch release lever passes a certain point, the auxiliary spring will force the release lever forward. Keep fingers and hands clear when pressing lever forward.*

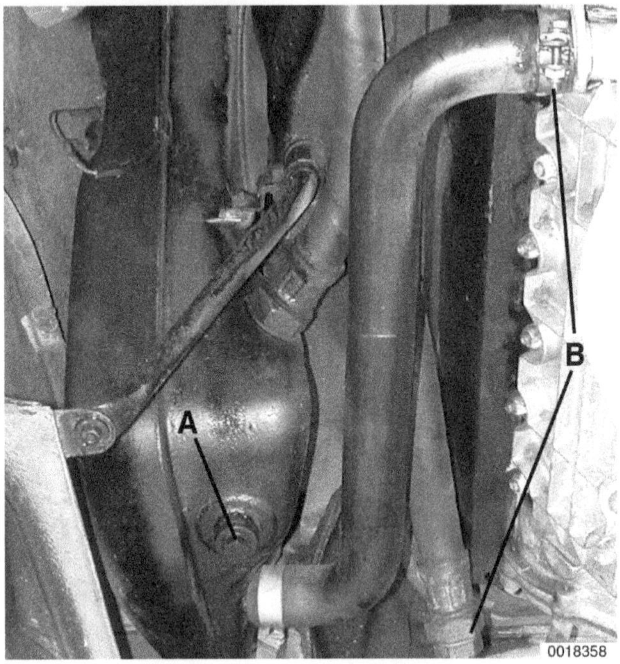

– Drain crankcase oil.

◄ Remove drain plug (**A**) and drain oil from oil tank. Disconnect oil lines (**B**). Counterhold line fitting when loosening.

◄ Remove rear stabilizer bar.

- Unbolt stabilizer bracket and stabilizer link bolts (**arrows**) at left and right sides and lower bar out of car (left side shown).

– Disconnect wiring from starter motor.

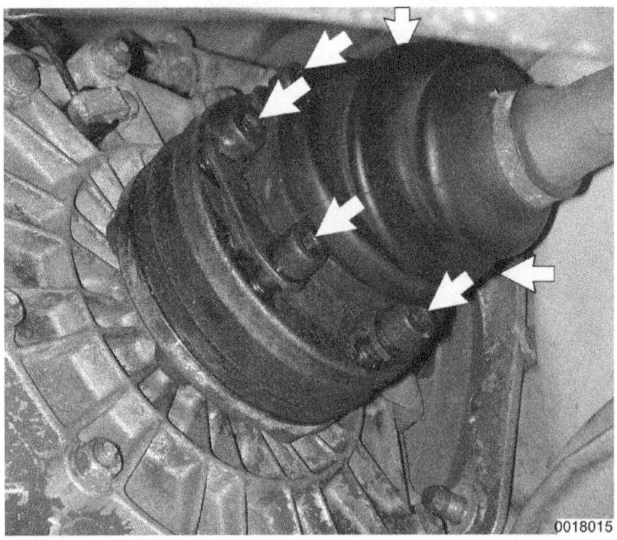

◄ Loosen inner constant velocity (CV) joint bolts (**arrows**) from differential drive flanges (left side shown). Suspend shafts from body with mechanic's wire.

NOTE —

Cover CV joint with a plastic bag and tape bag to CV joint shaft to keep dirt out.

– Remove left and right heater hoses from heat exchanger connector.

– Lower vehicle until weight of engine/transmission assembly engine is supported by hydraulic support or equivalent.

NOTE —

If using a floor jack, raise it as much as possible to allow for maximum travel when it is lowered with the engine.

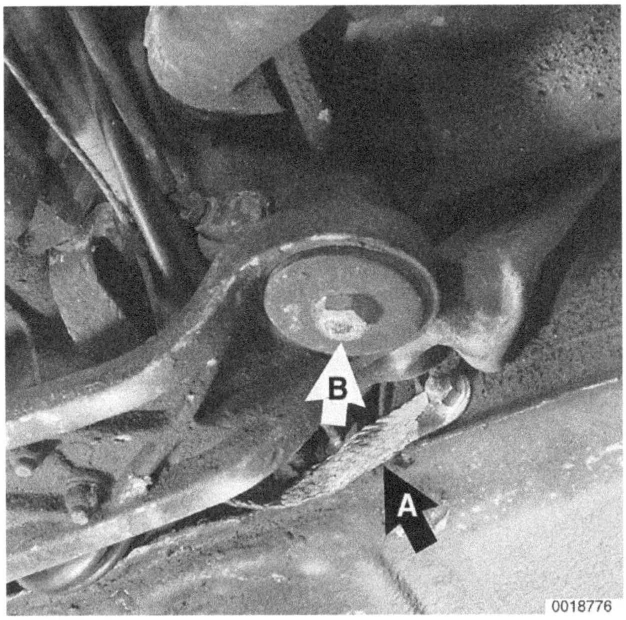

0018776

◀ Disconnect ground strap (**A**) between transmission and body.

- While supporting engine/transmission assembly, carefully remove two transmission mount bolts (**B**). Right side shown.

> ***WARNING —***
> *Check often to make sure that the car remains stable and well-supported while the engine is being removed.*

0019585

◀ Remove 2 engine support crossmember bolts (**arrow**) (right side shown).

◀ Carefully lift car (or lower engine/transmission assembly) while allowing engine to tilt backward as assembly is removed from rear of car.

NOTE —

If separating the engine from the transmission, see **Engine/transmission, separating and rejoining.**

Engine/transmission, installing

◀ Place engine/transmission assembly under engine/transmission bay. Lower car or raise engine/transmission.

– Tilt engine back and guide shift rod into center tunnel hole.

– With engine 3 in. from its mounts connect heat exchanger hoses.

> ***CAUTION —***
>
> *Connecting heater hoses after engine is bolted in may result in damaged hoses.*

– Carefully raise engine/transmission assembly up to its installed position.

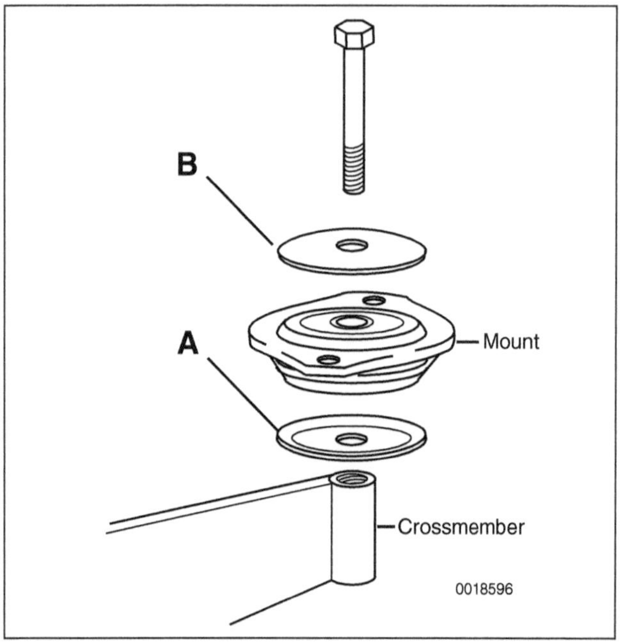

◀ Place washer between engine mount and crossmember with washer concave facing up (**A**).

- Insert engine mounting bolt with concave washer facing down (**B**). Press bolt down through engine mount and washer into crossmember.
- Start engine mounting bolts into crossmember but do not tighten them.

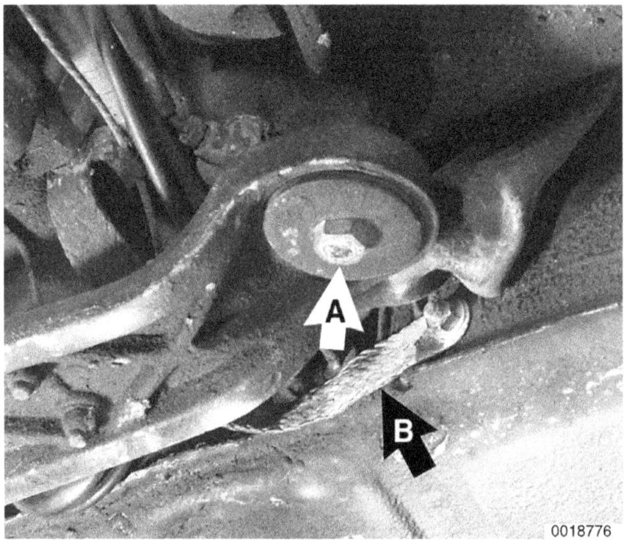

◀ Install transmission mount bolts and washers (**A**). Arrange washers as per engine mount (shown above).

– Reattach transmission to body ground strap (**B**).

– Tighten engine and transmission mounting bolts.

Tightening torques
- Crossmember to transmission .... 25 Nm (18 ft-lb)
- Crossmember to crossmember mount .. 80 Nm (58 ft-lb)
- Ground strap nut to body stud .... 18 Nm (14 ft-lb)

◀ Install clutch release levers and clutch cable on bottom of transmission.

NOTE —
See **301 Clutch** *for detailed instructions on clutch adjustments.*

◀ Install right and left CV joints and tighten bolts (**arrows**).

Tightening torque
- CV joint to transmission output flange
  Hexagon type bolts..............39 - 46 (29 - 34 ft-lb)
  Expansion type bolts.............26 - 30 (19 - 22 ft-lb)

– Connect starter power feed cable.

< Reinstall stabilizer bar and tighten bolts (**arrows**).

Tightening torques

- Stabilizer to body (M8) . . . . 25 Nm (18 ft-lb)
- Stabilizer to trailing arm link (M12) . . . . . 90 Nm (66 ft-lb)

– Install new sealing rings on both oil drain plugs and install.

Tightening torques

- Oil drain plug to crankcase (from 3/83). . 70 Nm (51 ft-lb)
- Oil drain plug to sump cover (to 3/83). . . 42 Nm (31 ft-lb)
- Sump cover to crankcase (to 3/83)(M6) . . 10 Nm (7 ft-lb)

< Reconnect oil tank S-hose and return hose (**arrows**). Counterhold return hose fitting (**A**) while tightening.

– Connect throttle linkage to relay lever on side of transmission.

< Working inside passenger compartment, reconnect vehicle speed sensor harness connectors (**A**) through hole in center tunnel.

- Connect shift rod to coupler and tighten allen set screw (**B**).
- Install access cover and floor mats.

Tightening torque

- Shift rod allen set screw to coupler. . . . . . . 8 Nm (6 ft-lb)

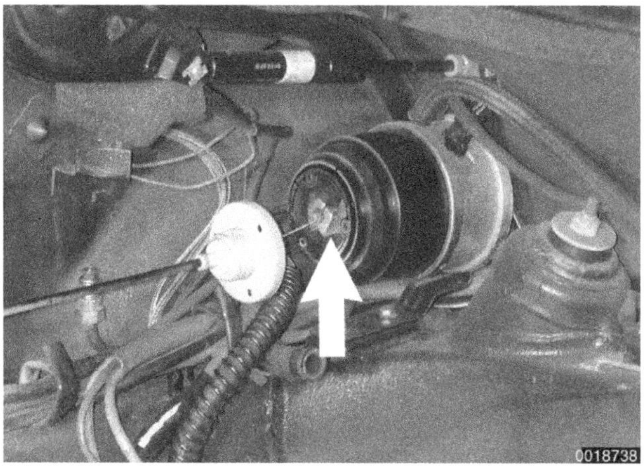

◄ Attach cruise control cable to diaphragm housing. Pull back on locking plastic clip until it snaps into metal retainer (**arrow**).

- Mount cruise control cable flange to diaphragm housing.

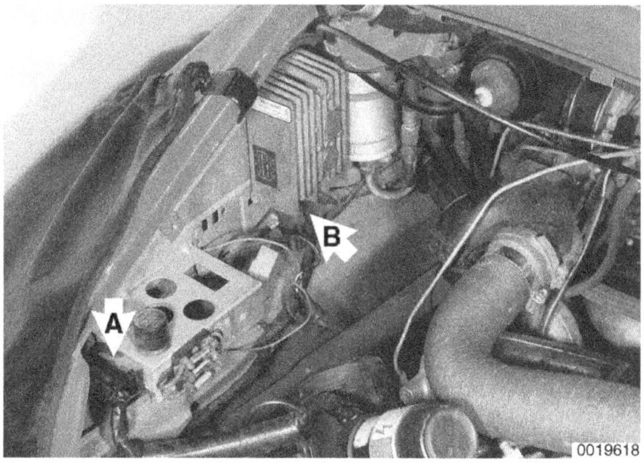

◄ Reconnect left side engine electrical harness connectors:

- Multi pin connector in left rear fuse/relay panel.(arrow A)
- Ignition module connector.(arrow B)

– Install rear fuse/relay panel cover.

– Remainder of installation is reverse of removal.

- Reinstall A/C compressor and adjust drive belt.
- Replace oil filter.
- Add 8 quarts of engine oil.
- Reconnect battery ground terminal.

Tightening torque
- A/C compressor to mount . . . . .25 Nm (18 ft lb)

– Start engine with a helper checking for leaks or any other problems.

– With engine running and at normal operating temperature add 2.5 quarts of engine oil.

Engine/transmission, separating and rejoining

– Detach wiring harness connectors (**arrows**) from transmission.

◄ Remove transmission to engine mounting nuts (**arrows**) and slide transmission off engine.

– Installation is reverse of removal.

NOTE —
As the transmission slides onto the engine studs, turn the clutch cross-shaft so that the clutch release fork hooks into the release bearing.

Engine mount assembly SC coupe

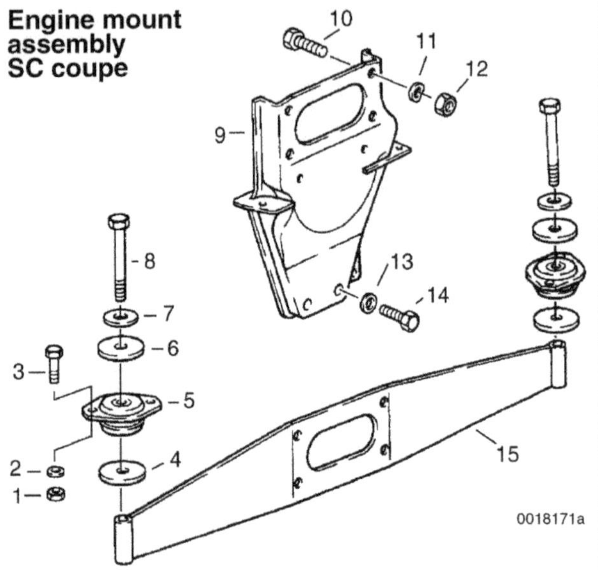

1. Hex nut M8
2. Lockwasher
3. Hex bolt M8x35
4. Washer
5. Bearing
6. Washer
7. Washer
8. Hex bolt M12x1.5x95
9. Engine mount carrier
10. Hex bolt M10x18
11. Spring washer
12. Hex nut M10
13. Washer
14. Hex bolt M10
15. Crossmember

Engine mount assembly SC cabriolet

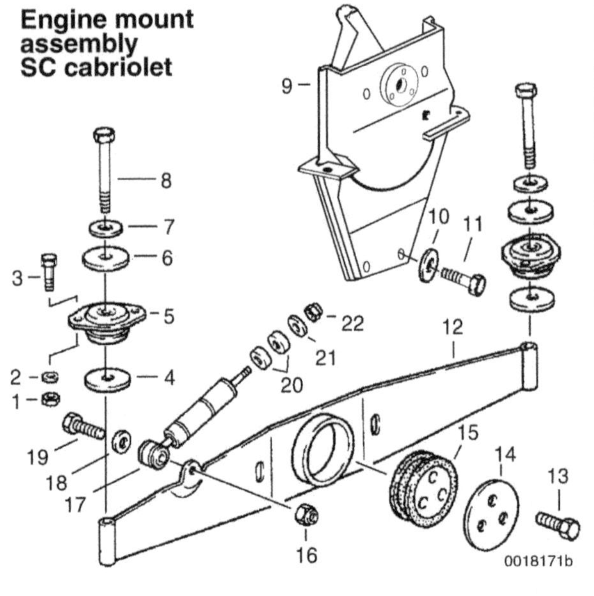

1. Hex nut M8
2. Lockwasher
3. Hex bolt M8x35
4. Washer
5. Bearing
6. Washer
7. Washer
8. Hex bolt M12x1.5x95
9. Engine mount carrier
10. Washer
11. Hex bolt M10
12. Crossmember
13. Hex bolt M10x35
14. End plate
15. Torsion bearing
16. Locknut M10
17. Vibration damper
18. Washer
19. Hex bolt M10x45
20. Spacer rings
21. Washer
22. Locknut M6

ENGINE/TRANSMISSION MOUNTS

The engine or transmission crossmember rubber mounts can be damaged or weakened due to age, oil contamination or accident. This damage can cause excessive noise or vibration in the passenger compartment.

Engine mount, removing and installing

< Refer to illustrations for component locations.

Removal of the rear engine mounting crossmember and mount bracket is covered in **102 Engine Disassembly and Assembly**.

Tightening torques

- Engine crossmember mount to body . . . 25 Nm (18 ft-lb)
- Engine crossmember to crossmember mount. . . . . 80 Nm (58 ft-lb)
- Engine mount carrier to crankcase. . . . . 40 Nm (30 ft-lb)
- Engine mount crossmember to carrier . . . . 40 Nm (30 ft-lb)

Transmission crossmember

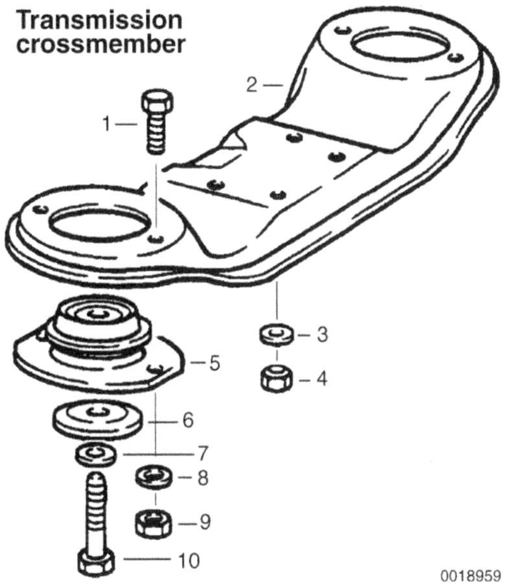

1. Hex bolt M8x16
2. Crossmember
3. Washer
4. Locknut
5. Mount
6. Concave washer
7. Lockwasher
8. Lockwasher
9. Hex nut M8
10. Hex bolt

Transmission mount, replacing

◄ Rubber transmission crossmember mounts in 1978 - 1983 SC models can be replaced without removing crossmember from transmission.

– Raise vehicle on lift and support safely.

> ***WARNING —***
> *Be sure vehicle is stable and well supported. Removal of the engine/transmission assembly will upset the balance of the vehicle. Use tall jack stands to support the vehicle at all four corners.*

– Support transmission while removing transmission crossmember mount long bolts.

– Lower transmission just enough to reach over crossmember and remove bolts holding mount to crossmember.

– Installation is reverse of removal.

Tightening torques

- Transmission crossmember mount to body . . . . 80 Nm (58 ft-lb)
- Transmission crossmember to crossmember mount. . . . . 80 Nm (56 ft-lb)

102 Engine Disassembly and Assembly

Special Tools

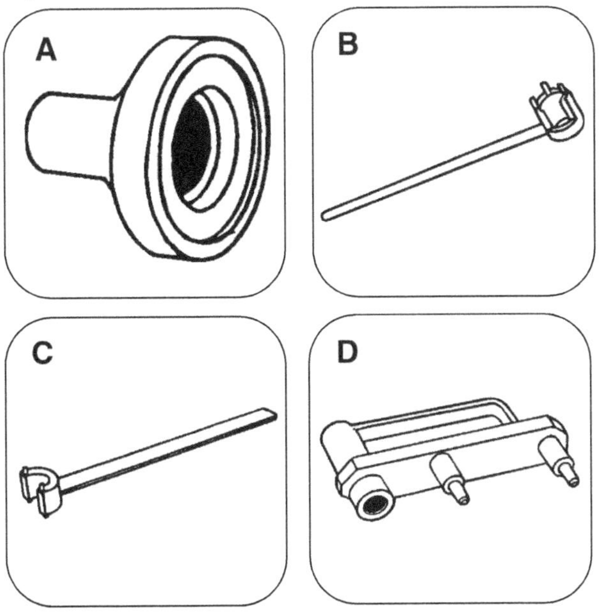

Special tool

A 9126. . . . . Flywheel end oil seal installer (Porsche ordering #000 721 912 60)

B 9191. . . . Camshaft sprocket counterholding wrench (1982 on) (Porsche ordering #000 721 919 10)

C 9236. . . . . .Pulley spanner wrench (Porsche ordering #000 721 923 60)

D P 238 a. . . . . Flywheel locking tool (Porsche ordering #000 721 238 10)

GENERAL

This repair group covers engine disassembly and reassembly. Engine removal procedures, including separating the transmission from the engine, are covered in **101 Engine Removal and Installation**.

Special tools and equipment

Many of the procedures in this repair group require special tools. Be sure to read the procedures through before beginning work on engine disassembly.

If you lack the skills and tools, or a suitable workplace for servicing the engine, we suggest you leave these repairs to an authorized Porsche dealer or other qualified shop. Completely reconditioned engines are available from qualified Porsche repair facilities.

Though you may not have the skills and knowledge for doing the actual internal repairs, it is possible to save some of the expense of professional repair by removing and installing the engine yourself, using the procedures described in **101 Engine Removal and Installation**. It is important to realize, however, that a partially disassembled engine may be a problem for a mechanic. We strongly advise against taking the engine apart to begin any repairs that cannot be properly finished.

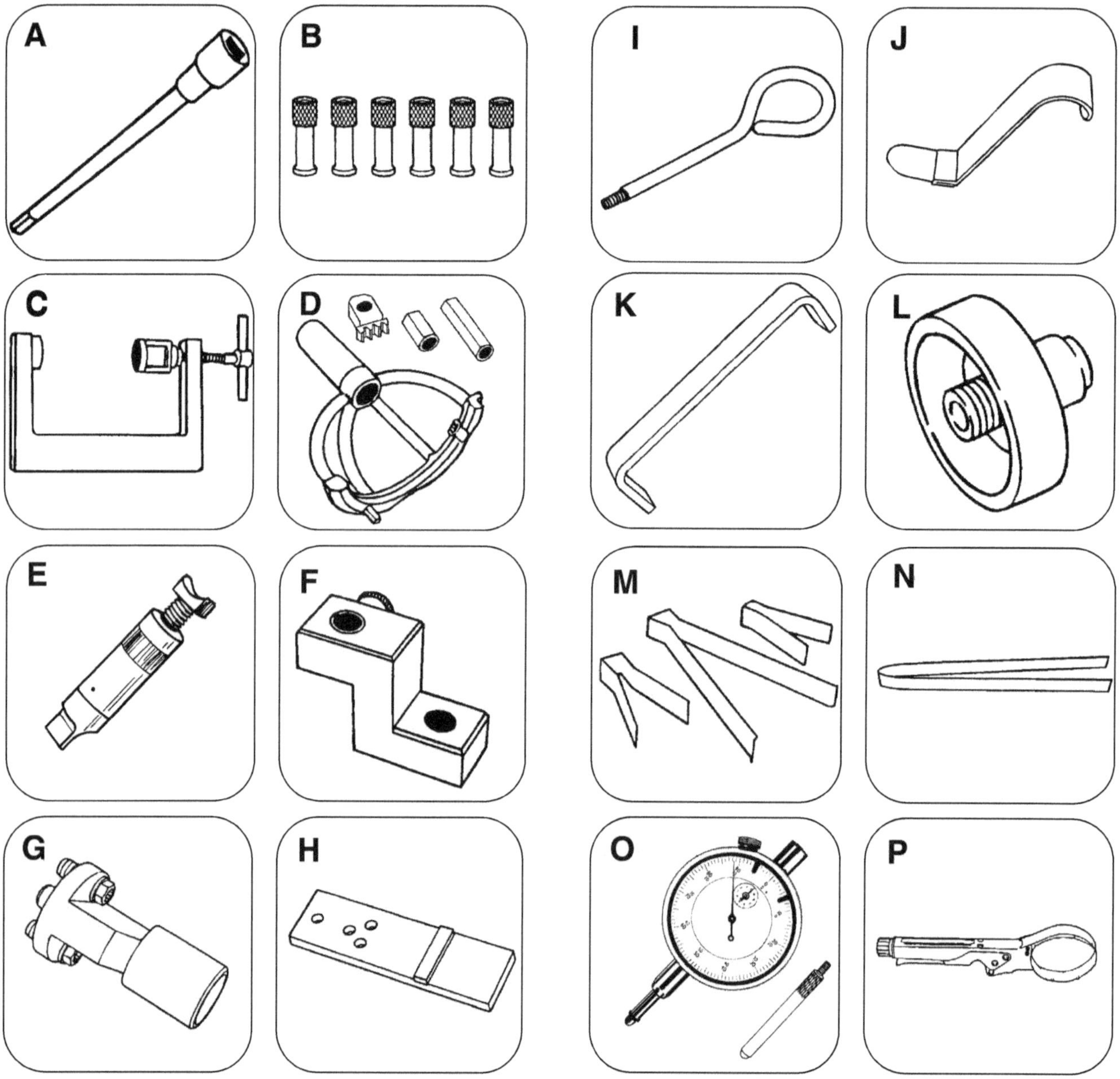

Special tool

A P 119 . . . . . 10 mm allen wrench for cylinder head nut removal (Porsche ordering #000 721 119 00)
B P 140 . . . . Nuts for securing cylinders during engine assembly (Porsche ordering #000 721 140 00)
C P 200 a. . . . . Valve spring compressor (Porsche ordering #000 721 140 00)
alternate: Baum 1020 (source: Baum Tools Unlimited)
D P 201/201 b . . . . Four-arm engine mount and adapters (Porsche ordering #000 721 201 00/20)
E P 204 . . . . .Timing chain idler brace (source: Baum Tools Unlimited)
F P 207 . . . . Dial indicator mounting bracket (Porsche ordering #000 721 207 00)
G P 209 . . . . Crankshaft mounting bracket (source: Baum Tools Unlimited)
H P 209 a. . . . . Crankshaft mounting bracket (Porsche ordering #000 721 209 10)
I P 212 . . . . Camshaft sprocket lockpin tool (Porsche ordering #000 721 212 00)
J P 213 b. . . . . Valve adjusting feeler gauge (Porsche ordering #000 721 213 20)
K P 214 . . . . Spring-loaded tensioner lock (1978 - 1981 models) (Porsche ordering #000 721 214 00)
L P 216 a. . . . . Pulley end oil seal installer (Porsche ordering #000 721 216 10)
M P 221 . Left connecting rods supports during engine assembly (Porsche ordering #000 721 221 00)
N P 222 . . . . . Left timing chain support during engine assembly (Porsche ordering #000 721 222 00)
O US 1008a . . . . Piston ring compressor (source: Baum Tools Unlimited)
P 20A, 21 . . . Dial indicator and extension for setting valve timing (source: Baum Tools Unlimited)

Engine repairs

Some repair operations do not require engine removal and disassembly. Some of the more common engine-installed operations are as follows:

- Camshaft timing adjustment and timing chain and tensioner replacement. See **152 Camshaft Timing Chains**.
- Camshafts and rocker arm replacement. See **151 Camshafts and Camshaft Cases**.
- Leaky engine oil return tube replacement. See **170 Engine Lubrication**.

The repair information given here requires that the engine be removed from the car. The disassembly procedure should be carried out on an appropriate engine stand in a clean environment.

Reconditioning procedures and specifications are not covered here. Once the engine is disassembled to the component level, refer to the appropriate repair group for applicable reconditioning information.

- **130 Crankcase and Crankshaft**
- **131 Pistons and Cylinders**
- **150 Cylinder Heads**
- **151 Camshafts and Camshaft cases**
- **152 Camshaft Timing Chains**

> ***CAUTION—***
> - *The pulley end of the engine is referred to as the* ***front****. The flywheel end of the engine is referred to as the* ***rear****.*
> - *If the engine is disassembled because of bearing failure, a thorough flushing of the lubrication system must be carried out prior to reassembly. Refer to* **130 Crankcase and Crankshaft** *and* **170 Engine Lubrication**.

Component upgrades

The Porsche 911 engine was produced for the U. S. market from 1965 to 1989. During that production run, several large scale redesigns and many gradual improvements and modifications were incorporated into the cars. Aftermarket suppliers also provided (and continue to provide) redesigned parts and components.

This manual only covers original factory designed parts or upgrades shipped to the United States. If an engine being disassembled contains parts, components or equipment not found in the factory parts lists, we recommend you consult with a qualified Porsche repair shop or dealer for further information.

Troubleshooting

Before disassembling the engine for reconditioning, make as many diagnostic tests as possible. Be sure that the fuel system, including the air filter, and the ignition system, including the spark plugs and spark plug wires, are in good condition and that there are no blockages in the engine air intake.

Troubleshooting specifically for the lubrication system can be found in **170 Lubrication System**. Troubleshooting for general starting and running problems can be found in **200 Engine Management–General**.

Make the following tests to assess the condition of the engine while it is still assembled. For details of how to perform each test, as well as other diagnostic information, see the **Mechanical Troubleshooting** section of **100 Engine–General**.

Compression test. This is the simplest and most widespread of diagnostic tests for engine condition. It is possible to get a rough idea of the condition of the valves and piston rings using this test.

Wet compression test. The results of the dry compression test can be refined using this test.

Leak-down test. This test uses compressed air pumped into each cylinder through the spark plug hole to pinpoint compression leaks in that cylinder.

Vacuum test. Perform this test if the engine is still running. Manifold vacuum varies with different engine problems. See **Table c** in **100 Engine–General**.

ENGINE, DISASSEMBLY

The disassembly procedure is divided into subsections as follows:

- **Engine peripherals, removing**
- **Camshaft timing chains and camshafts, removing**
- **Camshaft cases and cylinder heads, removing**
- **Pistons and cylinders, removing**
- **Crankcase, disassembling**
- **Crankshaft, disassembling**

NOTE —
Be sure engine oil is thoroughly drained before starting the disassembly.

Crankcase assembly

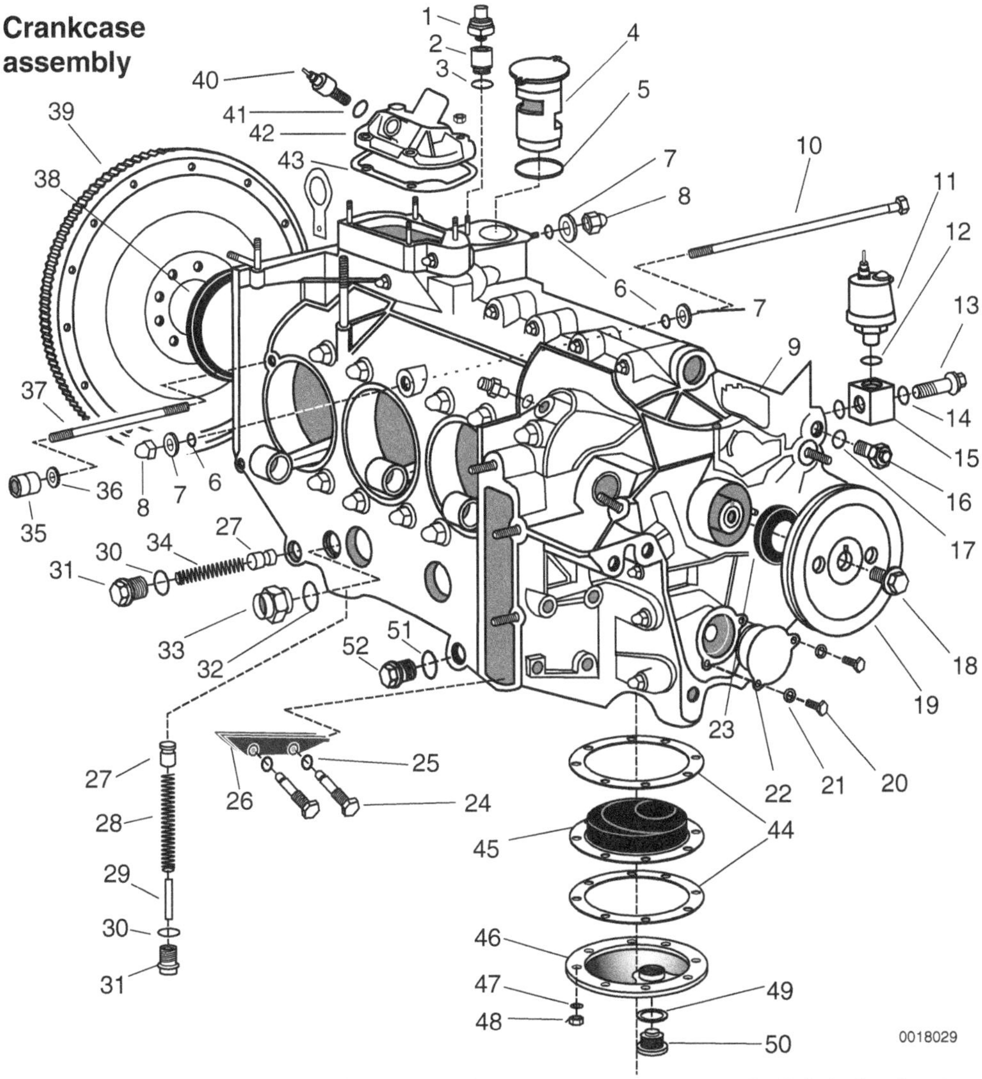

1. Oil pressure switch
   -tighten to 20 Nm (15 ft-lb)
2. Switch adapter
   -tighten to 35 Nm (26 ft-lb)
3. Sealing ring, (18 x 24mm)
4. Oil thermostat
5. Thermostat sealing O-ring (39.2 x 3 mm)
6. Case bolt sealing ring
7. Case bolt beveled washer
8. Case bolt cap nut
   -tighten to 35 Nm (26 ft-lb)
9. Crankcase
10. Case bolt
11. Oil pressure sensor
    -tighten to 35 Nm (26 ft-lb)
12. Oil pressure sensor sealing ring
    (8 x 24 mm)
13. Oil pressure sensor adapter bolt
    -tighten to 35 Nm (26 ft-lb)
14. Adapter bolt sealing ring (12 x 16 mm)
15. Oil pressure sensor adapter housing
16. Oil temperature sensor
    -tighten to 25 Nm (18 ft-lb)
17. Temperature sensor sealing ring
18. Front pulley bolt
    -tighten to 80 Nm (59 ft-lb)
19. Front pulley
20. Intermediate shaft cover bolt
21. Intermediate shaft cover bolt washer
22. Intermediate shaft cover
23. Crankshaft oil seal, front
24. Chain guide rail bolt
    -torque to 25 Nm (18 ft-lb)
25. Guide rail bolt sealing ring (10 x 16 mm)
26. Chain guide rail
27. Oil pressure relief valve piston
28. Oil pressure relief valve spring
29. Oil pressure relief valve sleeve
30. Sealing ring (18 x 24 mm)
31. Plug
    -tighten to 60 Nm (44 ft-lb)
32. Line adapter sealing ring (22 x 27mm)
33. Oil line adapter
    -tighten to 120 Nm (89 ft-lb)
34. Oil pressure safety valve spring
35. Cylinder head allen nut
    -tighten to 32 Nm (24 ft-lb)
36. Cylinder head washer
37. Cylinder head stud
38. Crankshaft oil seal, rear
39. Flywheel
40. Oxygen sensor system
    temperature switch (35°C)
    -tighten to 25 Nm (18 ft-lb)
41. Temperature sensor sealing ring
42. Breather cover
43. Breather cover gasket
44. Oil sump cover gasket (up to 3/83 prod.)
45. Oil sump screen (up to 3/83 prod.)
46. Oil sump cover (up to 3/83 prod.)
47. Washer (M6)
48. Nut (M6)
    -tighten to 10 Nm (7 ft-lb)
49. Oil drain plug sealing ring
50. Oil drain plug (up to 3/83 prod.)
    -tighten to 42 Nm (32 ft-lb)
51. Oil drain plug sealing ring
52. Oil drain plug (from 3/83 prod.)
    -tighten to 70 Nm (51 ft-lb)

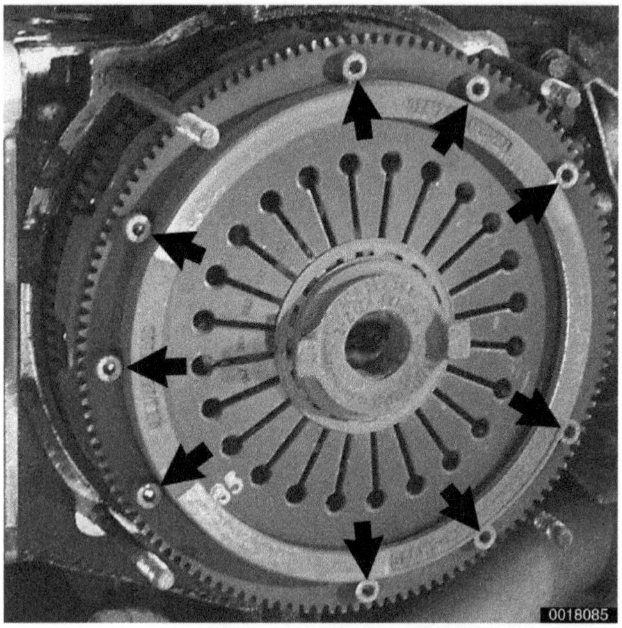

Engine peripherals, removing

Remove clutch bolts (**arrows**). Remove clutch from flywheel.

Hold flywheel stationary with Porsche special tool P 238a (remove inner pin from tool) or equivalent locking device and loosen flywheel bolts (**arrows**). Remove flywheel.

- Remove metal shrouding from flywheel end of engine.
- Mount engine on engine stand.

> ***WARNING —***
> *A 4-arm or 5-arm engine mount and adapters (Porsche special tool P 201 and P 201 b) should be used to mount the engine. 3-arm and "generic" engine mounts are dangerous and may damage the engine block.*

- Remove metal shrouding from pulley end of engine.
- Remove engine support crossmember.
- Rotate engine on stand so exhaust system is on top. Lock engine stand in this position.

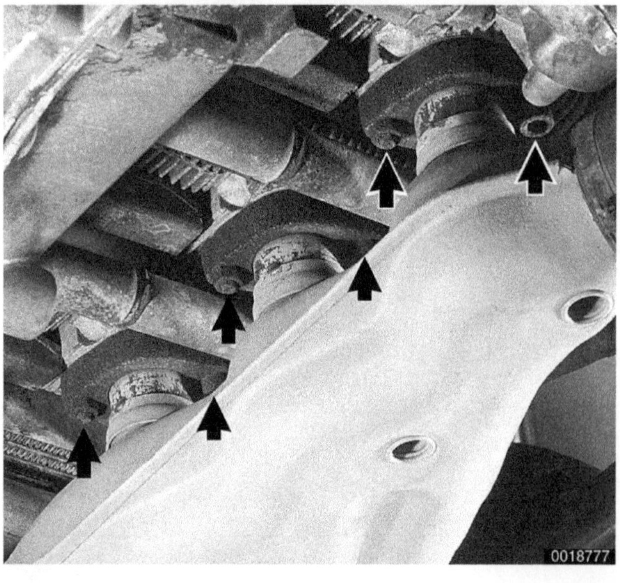

Remove heat exchanger mounting nuts (**arrows**). Right side shown.

- Remove muffler retaining straps.
- Remove heat exchanger crossover tube.
- Remove exhaust system as a complete unit. See **260 Exhaust System**.
- Remove muffler support bracket.
- Remove engine oil screen pan from bottom of crankcase.
- Rotate engine so that intake side is up. Lock engine stand in this position.

– Remove A/C compressor mounting bracket and engine support bracket bolts.

NOTE —
There is a blind mounting nut on reverse side of engine support bracket.

∢ Remove intake runner fasteners (**arrows**) from cylinder heads (qty: 12).

- Left side runner 1 shown. Other runners similar.

∢ Disconnect electrical harness connectors on engine:

- Oil pressure gauge sender (**A**)
- Oil temperature sender (**B**)
- Oil pressure (warning light) switch (**C**)

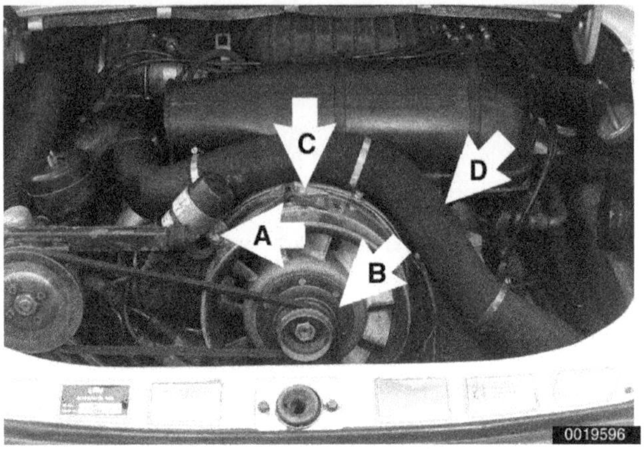

◄ Remove cooling fan/alternator housing:

- Remove ignition coil (**A**).
- Remove fan/alternator drive belt (**B**). See **030 Maintenance**.
- Remove allen bolt (**C**) from fan housing strap.
- Remove heater hose (**D**).
- Remove bolts from engine cooling shroud.
- Pull alternator out partially and unbolt alternator ground strap from top of engine crankcase.

NOTE —

Once the drive belt is removed, reinstall alternator belt pulley and shims and tighten nut fingertight.

– Disconnect spark plug connectors. Unclip and remove distributor cap with wires.

– Remove left and right metal shrouding.

◄ To remove throttle linkage at flywheel end of crankcase:

- Remove intake manifold support nut (**A**)
- Pry off throttle linkage from bellcrank (**B**).
- Remove bellcrank spring (**C**).

– Carefully lift off complete fuel system, intake manifolds and alternator with cooling shroud.

◄ Remove oil cooler mounting nuts from top and bottom (**arrows**) of cooler.

– Slide oil cooler off mounting studs.

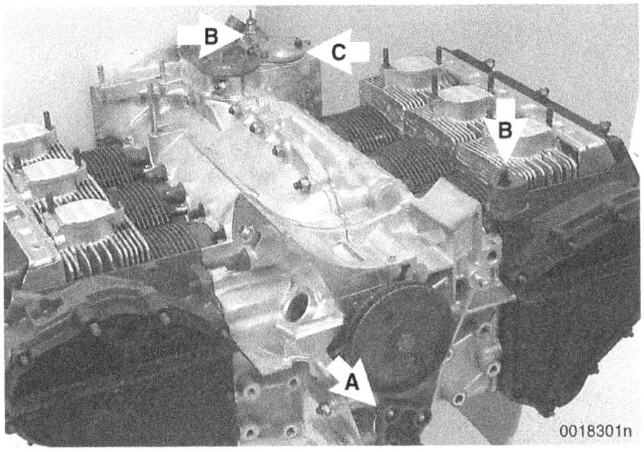

Remove following parts from engine:

- Intermediate shaft end cover (**A**)
- Oil pressure sensors (**B**)
- Oil thermostat (**C**)

Camshaft timing chains and camshafts, removing

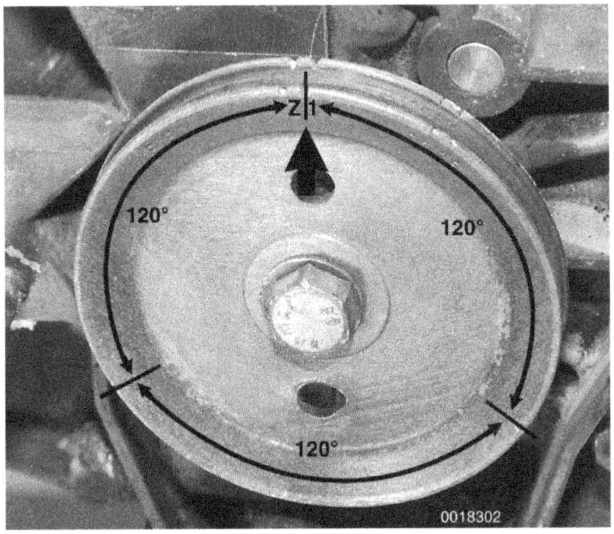

Set crankshaft to **Z1** (TDC) cylinder 1 compression stroke (**arrow**).

– Remove intake and exhaust valve covers on both sides of engine.

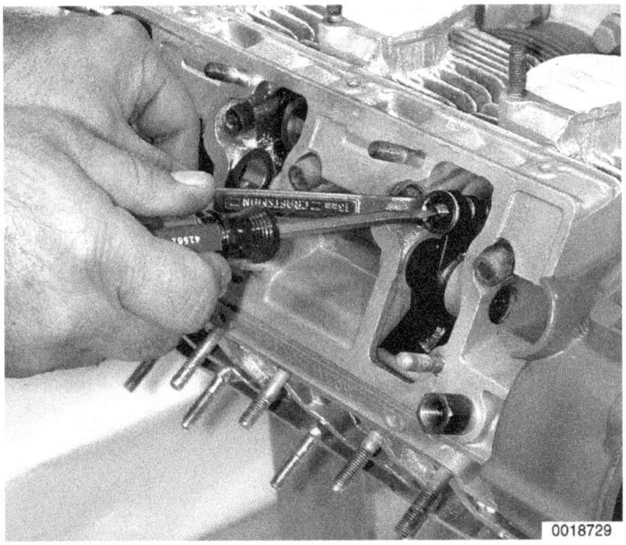

Loosen all valve adjuster lock nuts and fully back off adjusting screws for all valves.

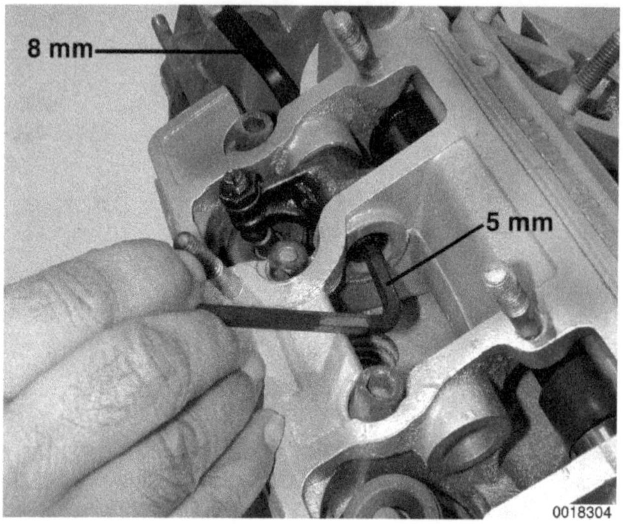

Working at cylinder 1, counterhold rocker shaft nut with 8 mm allen wrench, then loosen opposite end with 5 mm allen wrench.

NOTE —
Mark location of each rocker arm. Do not interchange used parts.

Gently tap rocker shaft out and remove rocker arm.

- Turn crankshaft 120° in normal direction of rotation. Remove cylinder 6 rocker arms using above procedure.
- Following firing order, turn crankshaft 120° and remove rocker arms for next cylinder. Repeat until all rocker arms are removed.

NOTE —
The firing order is 1 - 6 - 2 - 4 - 3 - 5.

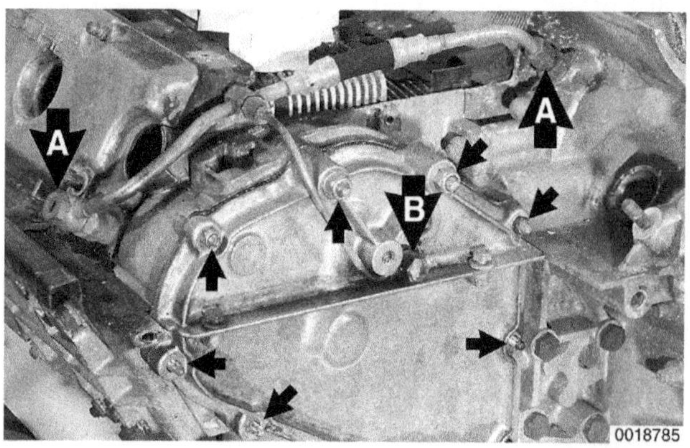

If equipped with Carrera type timing chain tensioners: Remove left camshaft and timing chain tensioner oil feed line pressure fittings (**A** and **B**).

- Remove left chain housing cover fasteners (**small arrows**), and remove cover.

◄ Loosen left camshaft sprocket:

- On 1978 - 1981 models: Use Porsche special tools P 202 and P 203 or equivalent to counterhold camshaft and loosen sprocket mounting nut.
- On 1982 - 1983 models: Use Porsche special tool 9191 or equivalent to counterhold sprocket while loosening sprocket bolt.

◄ If equipped with spring-loaded chain tensioner:

- Compress spring-loaded chain tensioner and lock down using Porsche special tool P 214 or equivalent.
- Remove tensioner mounting nut. Remove tensioner.
- Remove idler sprocket.

◄ If equipped with oil-pressure-fed chain tensioner:

- Compress timing chain tensioner using large pliers.
- Lock tensioner down using 3 mm (1/8 in.) pin through hole in tensioner body (**1**).
- Special tool P 204 (**2**) can be used to brace idler sprocket before removing tensioner.

NOTE —

Check timing chain tensioner piston height before compressing. If piston is fully extended, carefully inspect timing chains and sprockets for wear. See **152 Camshaft Timing Chains**.

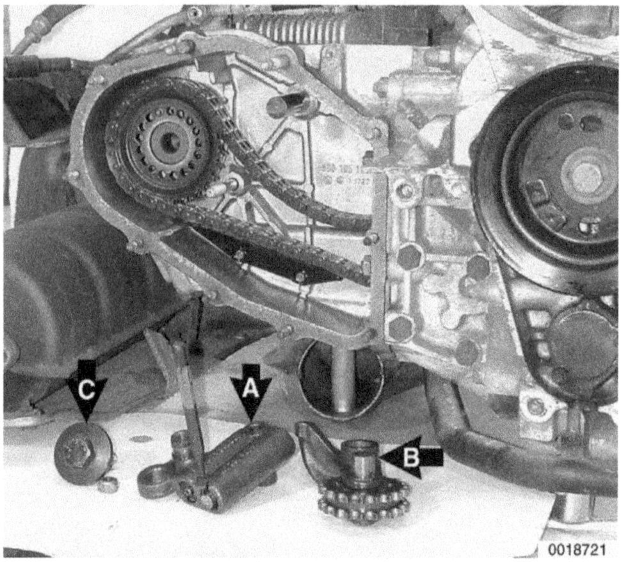

< If equipped with oil-pressure-fed chain tensioner:

- Remove timing chain tensioner (**A**) and idler sprocket assembly (**B**) with special brace P 204 if installed.
- Remove camshaft sprocket fastener from left camshaft (**C**).

< Remove camshaft sprocket lockpin using Porsche special tool P 212 or equivalent.

– Remove camshaft sprocket and disengage from timing chain.

– Remove sprocket flange.

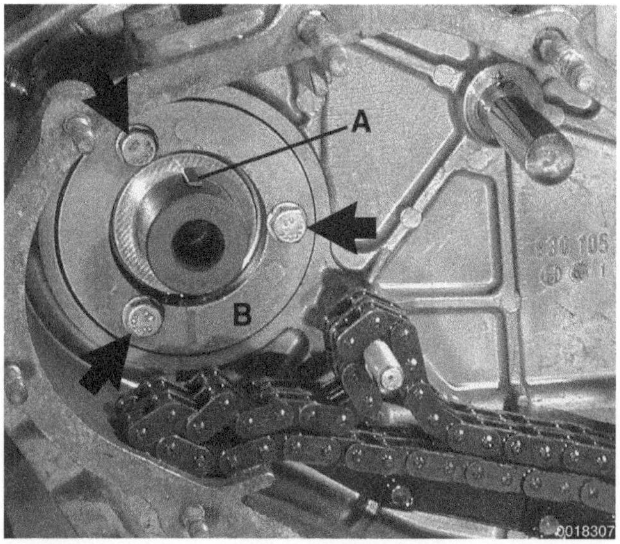

< Remove left camshaft:

- Remove woodruff key (**A**) from camshaft.
- Remove camshaft end cover retaining bolts (**arrows**) and remove camshaft end cover (**B**), shims and thrust spacer.
- Slide camshaft toward pulley end of engine to remove.

NOTE —
Wire tie end cover, shims and thrust spacer together and label.

– Remove left timing chain housing.

– Repeat procedures on right side camshaft.

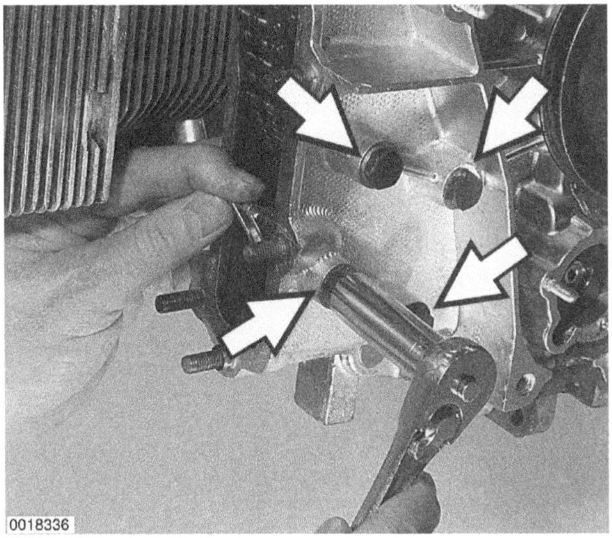

Remove left side upper and lower timing chain guide rails by removing mounting bolts.

– Repeat procedures for right side chain guide rails. Note lower right guide rail is different from all other chain guide rails.

Camshaft cases and cylinder heads, removing

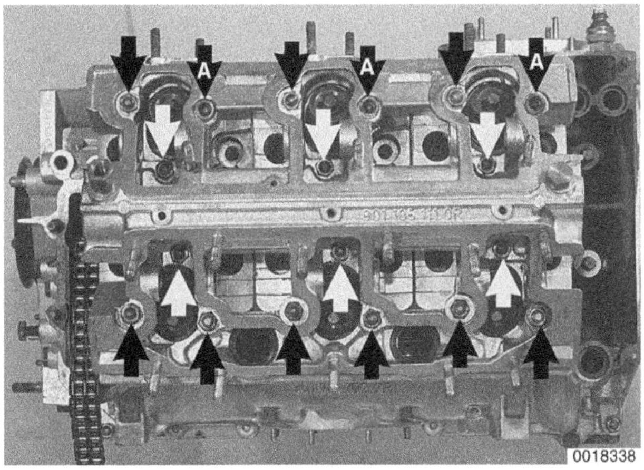

Remove 18 nuts (**arrows**) from left camshaft case.

- Fifteen M8 (13 mm head) nuts
- Three M8 (8 mm allen head) nuts (**A**)

– Remove camshaft case and mark for left side.

– Rotate engine on stand. Working on left side bottom of crankcase, remove oil pressure line and all brackets holding it to crankcase.

– Remove cylinder head nuts from each cylinder head using Porsche special tool P 119 or equivalent (long 10 mm allen wrench).

– Remove cylinder heads, marking installed locations.

– Repeat procedures for right side.

NOTE —

- *Check for broken cylinder head studs when removing the cylinder heads. If any are found, check area around cylinder head and base of stud for signs of blow-by. If blow-by is present, mark that area on both cylinder and head for further inspection once the parts are thoroughly cleaned.*
- *Various head studs have been used on the 911 SC engines. Always check with an authorized Porsche parts department for the latest in parts information.*

Disassemble cylinder heads using Porsche special tool P 200 a (valve spring tool) or equivalent.

- Remove valve keepers, spring retainers, springs, valve stem seals, spacer shims and spring seats.
- Mark location of valves and remove from cylinder head.

Pistons and cylinders, removing

– Remove cylinder air shroud plates.

– Mark cylinder locations on each cylinder barrel.

NOTE —
Label all parts during disassembly. Used parts must be reinstalled in their exact location and position.

Remove cylinders. If necessary, use a soft-faced hammer to tap cylinders off studs.

Set crankshaft to **Z1** (TDC).

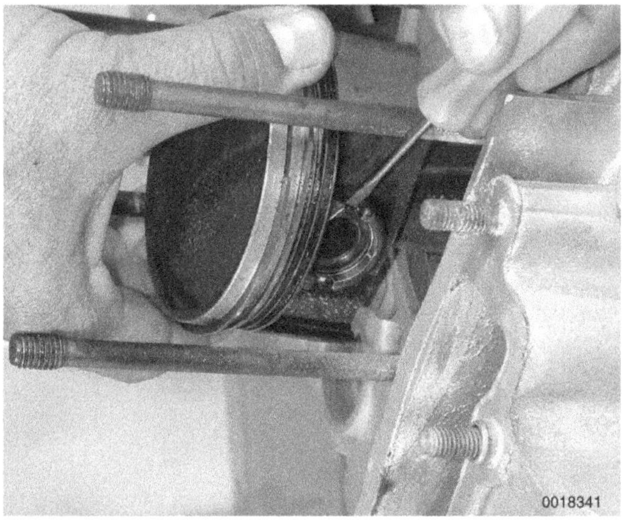

Remove wrist pin circlips from piston 1.

Using a soft drift, tap wrist pin out of piston 1.

- Remove piston, marking its position.

– Rotate crankshaft 240° clockwise. Remove piston 2, marking its position.

> ***CAUTION—***
> *When rotating crankshaft, manually guide pistons in and out of crankcase bores in order to avoid damaging the piston skirts or rings. Use shop towels around the pistons for protection.*

– Rotate crankshaft appropriate amount and remove remaining pistons, marking positions.

NOTE—
The firing order is 1 - 6 - 2 - 4 - 3 - 5.

Crankcase, disassembling

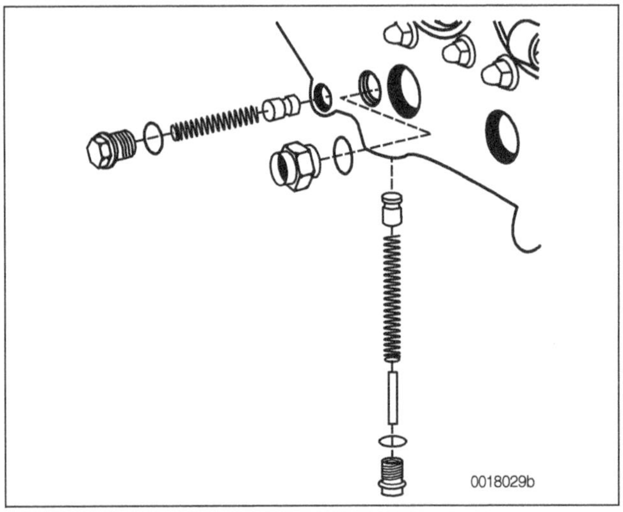

- Remove both oil pressure relief valve assemblies in front bottom of crankcase.

– Remove crankshaft pulley bolt using Porsche special tool 9236 or equivalent to counterhold pulley. Remove pulley.

- Remove M8 nuts (**arrows**) joining case halves together.
  - 3 nuts (**A**) are in flywheel housing.
  - 2 nuts (**B**) face right (bottom of case).

- Remove M10 case nuts (**arrows**) located in oil cooler housing.

◄ Remove M10 case nut (**arrow**) from left timing chain housing.

- Remove remaining M10 case nuts (quantity 11) on left side, counterholding case bolt heads on right side of crankcase.

NOTE—
When using 4 bolt engine stand, remove two mounting bolts prior to separating case.

- Carefully separate crankcase halves.

CAUTION—
Do not pry case apart at sealing surfaces. A damaged surface will result in an oil leak.

- Lift left half of crankcase off and set aside.
- Lift crankshaft assembly out of crankcase and set aside.

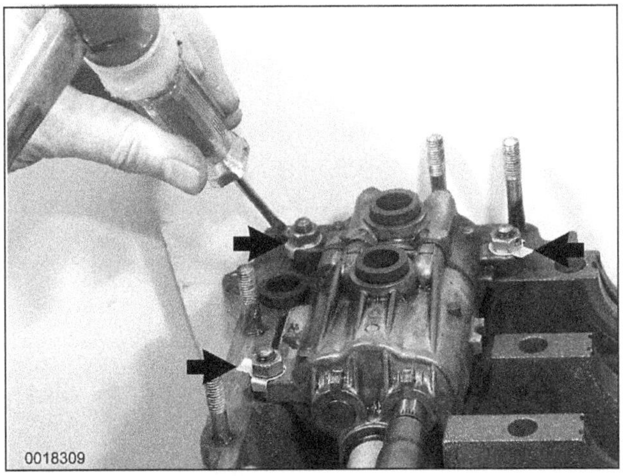

◄ Bend back locking tabs on oil pump mounting nuts (**arrows**). Remove nuts.

- Remove and disassemble intermediate shaft by removing timing chains, oil pump and oil pump drive shaft.

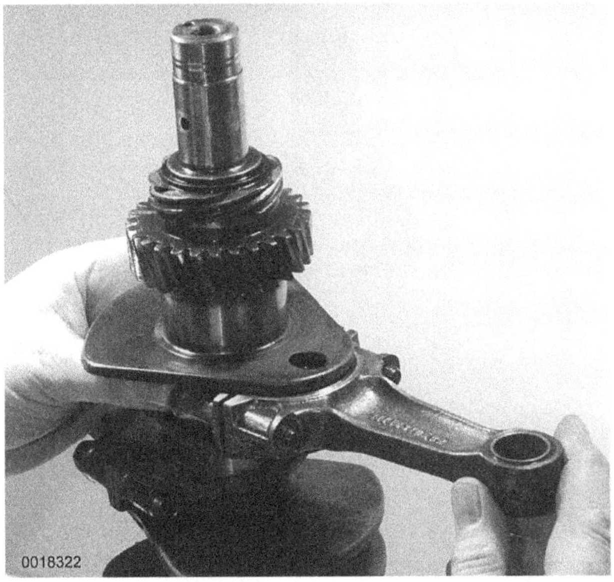

➧Crankshaft, disassembling

- Mount crankshaft to Porsche special tool P 209 a or equivalent and clamp tool with crankshaft in shop vise.

NOTE—
An adequate substitute for special tool P 209 a can be made from an old flywheel.

◄ Mark installed position of connecting rods and remove from crankshaft.

- Slide crankshaft nose bearing off crankshaft.
- Remove circlip holding gear assembly to crankshaft.

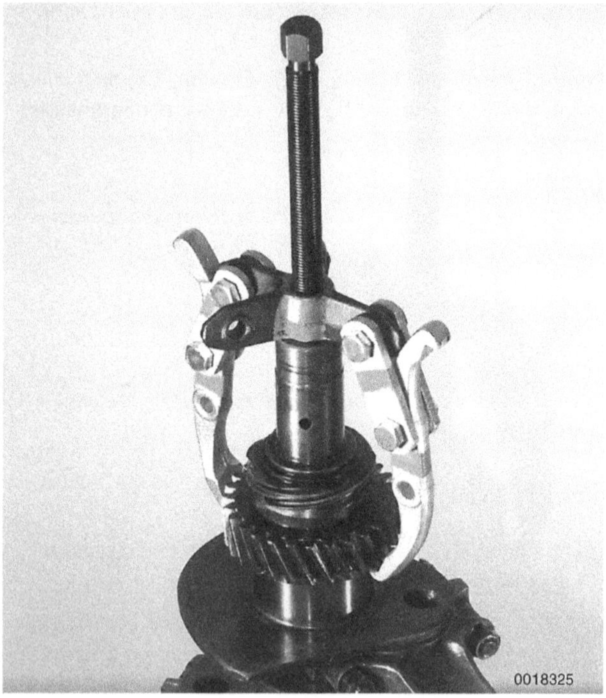

◀ Remove gear assembly on front of crankshaft using suitable gear puller.

ENGINE, ASSEMBLY

All parts must be clean and ready for reassembly at this point. All machine work should be completed and all new replacement parts should be on hand. Use a suitable engine assembling lubricant during assembly.

> ***CAUTION —***
> - *If the engine failed mechanically, it is important to check the intake system for foreign objects.*
> - *If the engine is disassembled because of bearing failure, a thorough flushing of the lubrication system must be carried out prior to reassembly. Refer to* **130 Crankcase and Crankshaft** *and* **170 Engine Lubrication**.

NOTE —
- *Various cylinder head studs have been used on the 911 SC engines. Always check with an authorized Porsche parts department for the latest in parts information.*
- *If either the crankcase, the crankshaft drive gear or the intermediate shaft drive gear are replaced, special installation instructions apply. See* **130 Crankshaft and Crankcase**.

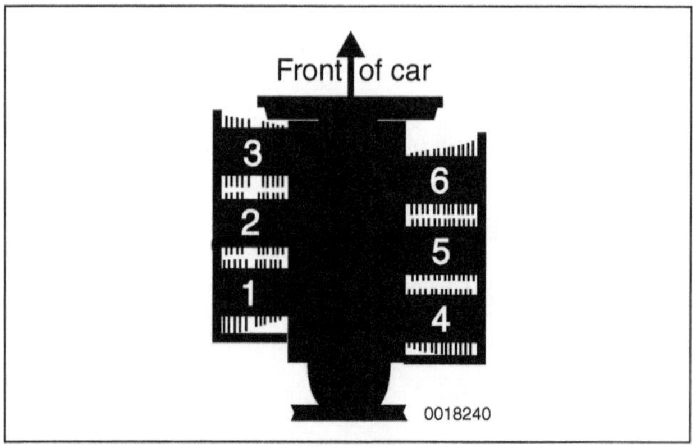

◀ Cylinder numbering is shown in illustration. Cylinder position will be referred to during the assembly process.

The assembly procedure is divided into subsections, as follows:

- **Crankshaft, assembling**
- **Crankcase, assembling**
- **Pistons and cylinders, installing**
- **Cylinder heads, installing**
- **Camshaft cases, camshafts and camshaft timing chains, installing**
- **Camshaft timing, adjusting**
- **Engine peripherals, installing**

Crankshaft, assembling

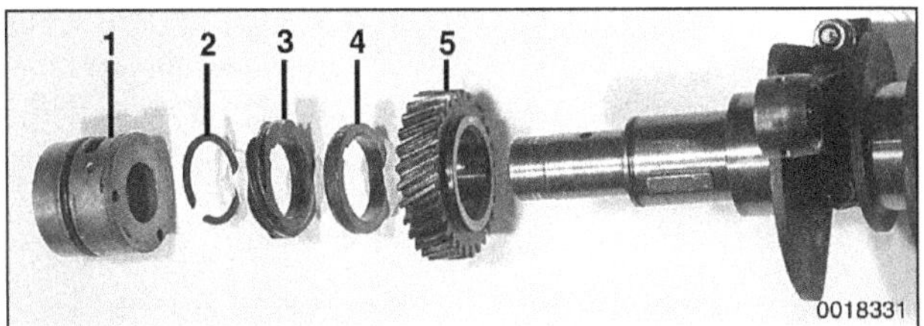

1. Crankshaft nose bearing
2. Circlip
3. Drive gear (for distributor)
4. Spacer
5. Drive gear (with mark **0** or **1**)

Crankshaft reconditioning information is given in **130 Crankcase and Crankshaft**.

NOTE —
Check the crankshaft drive gear for wear prior to installation. If the gear is worn or damaged, the replacement gear must be matched to the crankcase and to the gear on the intermediate shaft. See **130 Crankcase and Crankshaft** *for checking and replacing.*

– Mount crankshaft to Porsche special tool P 209 or equivalent and clamp tool with crankshaft in shop vise.

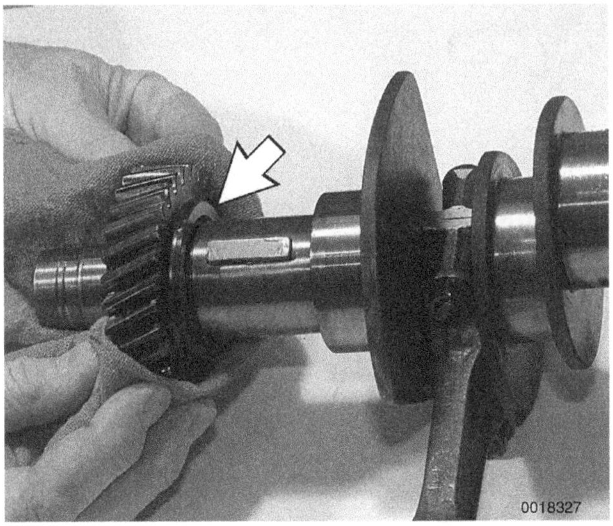

◄ Install drive gear to crankshaft.
- Install woodruff key to crankshaft
- Heat gear to 150°C (300°F) in oil bath.
- With shoulder on gear facing in (**arrow**), slide drive gear fully onto crankshaft.

WARNING —
Use protective gloves or shop rag when handling the hot gear.

– Install spacer to crankshaft.

– Install distributor drive gear to crankshaft.
- Heat gear to 100°C (212°F) in water bath.

WARNING —
Use protective gloves or shop rag when handling the hot gear.

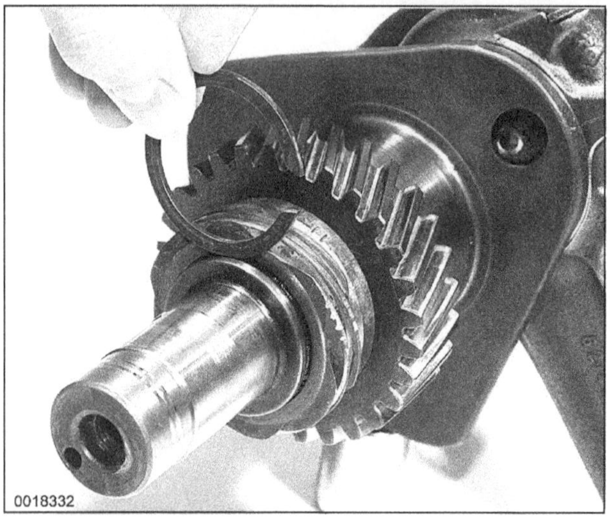
0018332

◀ Install gear retaining circlip to crankshaft.

- Circlip used to hold gears in place on crankshaft comes in different thicknesses. See **Table a**.
- Select circlip to best obtain near zero clearance.

Table a. Crankshaft gear circlip thickness

Porsche P/N	Thickness	Mark
901 102 148 00	2.4 mm	0
901 102 148 01	2.3 mm	1
901 102 148 02	2.2 mm	2
901 102 148 03	2.1 mm	3

– Install new connecting rod bolts in each connecting rod.

– Install connecting rod bearing shells in connecting rods.

0018321

◀ Install connecting rods on crankshaft in their original locations. Matching numbers on side of connecting rod and rod cap should all be facing same side of engine (i.e., all numbers face exhaust side of engine).

– Install new connecting rod nuts on rod bolts.

Tightening torque

- Connecting rod nut . . . . 50 - 55 Nm (37 - 41 ft-lb)

▸Crankcase, assembling

– Mount right side crankcase half on engine stand.

◀ Install intermediate shaft bearing shells in crankcase halves.

0018310

NOTE —

Due to the high loads the timing chains put on the intermediate shaft, it is recommended that the intermediate shaft bearings be replaced during engine overhaul.

◄ Place oil pump sealing O-rings (**arrows**) in right crankcase half.

– Lubricate oil pump gears with engine assembling lubricant.

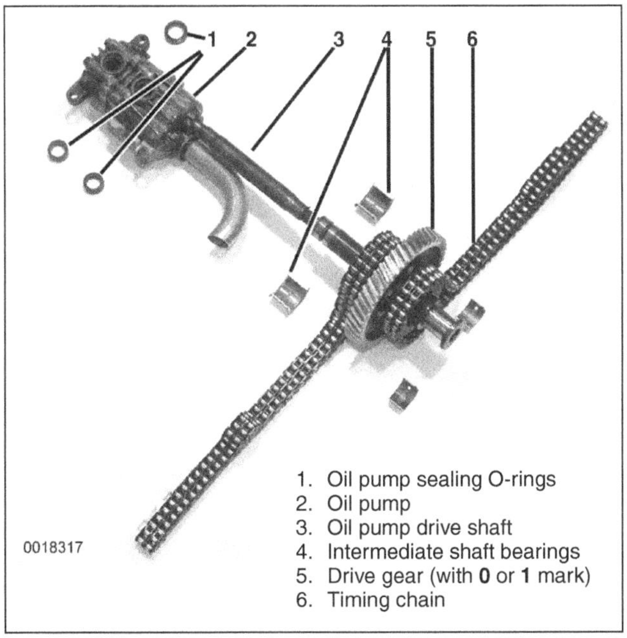

1. Oil pump sealing O-rings
2. Oil pump
3. Oil pump drive shaft
4. Intermediate shaft bearings
5. Drive gear (with **0** or **1** mark)
6. Timing chain

◄ Place both timing chains on intermediate shaft sprockets. Install oil pump drive shaft with oil pump to intermediate shaft.

NOTE —

- *Check intermediate shaft gear for wear prior to installation. If the intermediate shaft gear is worn or damaged, the replacement gear must be matched to the crankcase and crankshaft drive gear. See* **130 Crankcase and Crankshaft** *for checking and replacing.*
- *The intermediate shaft is hollow and should be cleaned out whenever the engine is being reconditioned and in case of engine bearing failure. See* **130 Crankcase and Crankshaft**.

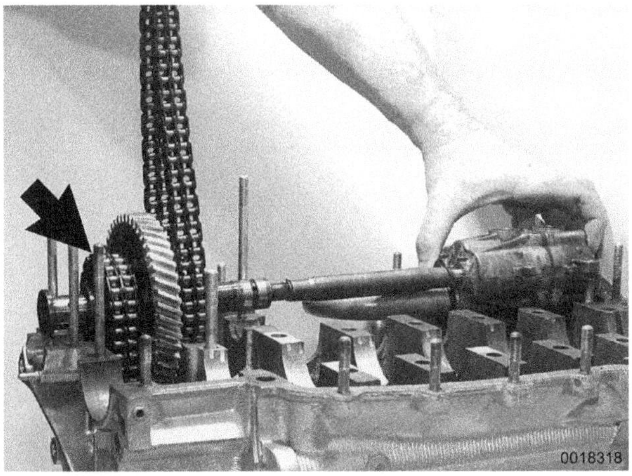

◄ Lower complete intermediate shaft assembly into right crankcase half, guiding front timing chain (**arrow**) through opening in case.

Place new locking tabs (**arrows**) on oil pump mounting studs and install and torque nuts. Bend lock tabs up into place.

Tightening torque

- Oil pump to crankcase .... 25 Nm (18 ft-lb)

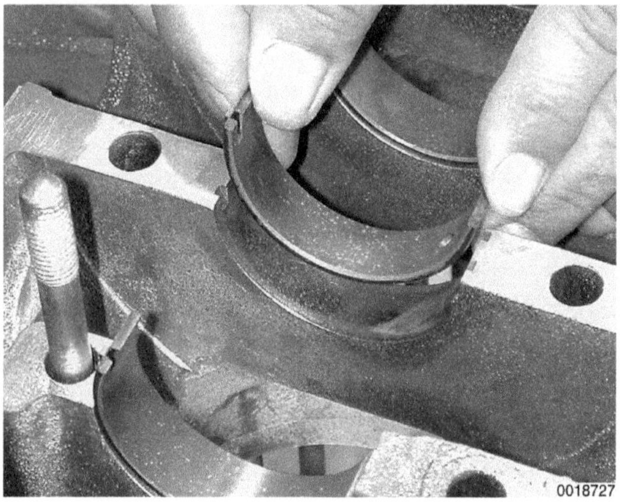

Lubricate crankshaft main bearing shells with engine oil and install in crankcase halves.

- Install crankshaft nose bearing locating pin in crankcase half.

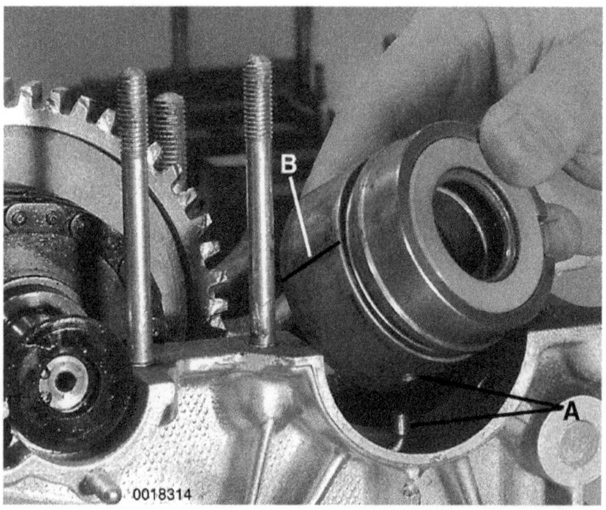

Without nose bearing outer sealing O-ring in place, lay crankshaft nose bearing in crankcase so that it locks into locating pin (**A**). Draw a line (**B**) on each side of bearing to use as a reference when installing crankshaft assembly.

- Remove nose bearing. Install nose bearing outer O-ring.
- Lubricate nose bearing and install on crankshaft.

Place crankshaft assembly into right crankcase half.

- Make sure right side connecting rods for cylinders 4, 5, and 6 are pointing down.
- Check that nose bearing (**A**) is correctly seated in case.
- Slowly turn crankshaft back and forth to check for binding or interference.

NOTE —

- *Porsche special tools P 221 can be used to support the left side connecting rods during crankcase assembly.*
- *Porsche special tool P 222 can be used to support the left timing chain during crankcase assembly.*

Place two sealing O-rings (**A**) in oil pump.

- Check to see that oil passage sealing O-ring previously placed in right crankcase half (**B**) remains undisturbed.
- Lubricate all with engine oil.

NOTE —

Different size seals are used. Be sure to install the correct parts.

– Using Loctite® 574 sealant or equivalent, apply even thin coat on both crankcase mating surfaces.

– Place left case-half in position over right case-half. Tap case-halves together lightly with a rubber mallet.

> ***CAUTION —***
> *Turn the crankshaft back and forth periodically to check for binding.*

– Place sealing O-rings on two case studs in oil cooler housing.

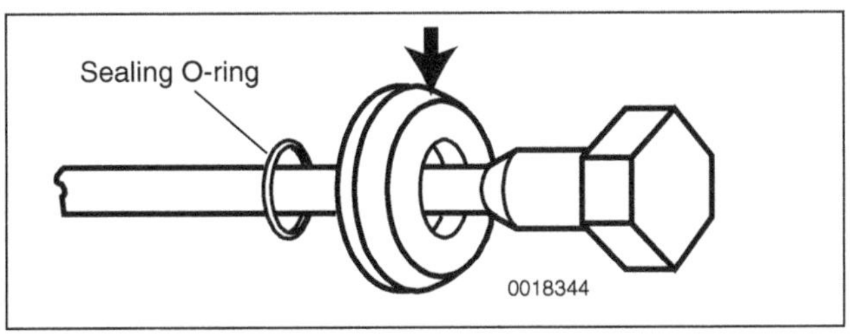

When installing case washers, align bevel on washers (**arrow**) correctly. Bevels should face nut or bolt head.

0018343

Install capped case nuts (**arrows**) in oil cooler housing. Finger tighten nuts.

NOTE —
Photo shows all case nuts installed.

0018326

Install washer and nut in timing chain cut-out on left side of engine. Finger tighten nut.

> ***CAUTION —***
> *Check crankshaft for free movement by turning slowly back and forth.*

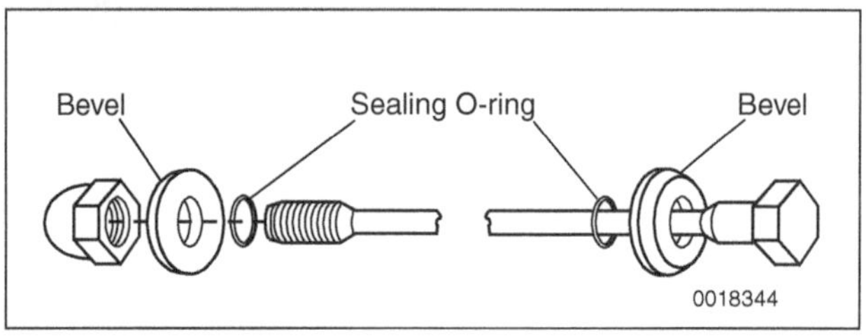

0018344

Insert M10 bolts through case.

- Make sure beveled washers and sealing O-rings are installed correctly. Install mounting bolts and studs from left side of engine stand mount.
- Install remaining case nuts and finger tighten.

977102959

Starting at cross bolt 1 and working sequentially out to bolt 14 (not shown, in timing chain opening next to cyl. no. 1), tighten case halve nuts.

Tightening torque
- Case bolt and nut (M10) . . . . 35 Nm (25 ft-lb)

– Install crankshaft seals on flywheel and pulley ends of engine using Porsche special tools P 216 and 9126 or equivalent.

– Install crankshaft pulley, counterholding with Porsche special tool 9236 or equivalent when torquing bolt.

Tightening torque
- Crankshaft pulley to crankshaft. . . . . 80 Nm (59 ft-lb)

NOTE —
If the crankshaft pulley locating hole or pin is damaged, valve timing may be affected. Replace pulley if hole elongated or damaged.

◄ Install 22 M8 nuts and bolts around outer edge of crankcase.
- Three nuts are in flywheel housing (**A**).
- Two nuts face right side (**B**).

Tightening torque
- Crankcase perimeter fasteners (M8) . . . 25 Nm (18 ft-lb)

– Install oil pressure sending units.

Tightening torques
- Oil warning light sending unit to crankcase . . . . 20 Nm (14 ft-lb)
- Oil gauge sending unit to crankcase. . . . 35 Nm (25 ft-lb)

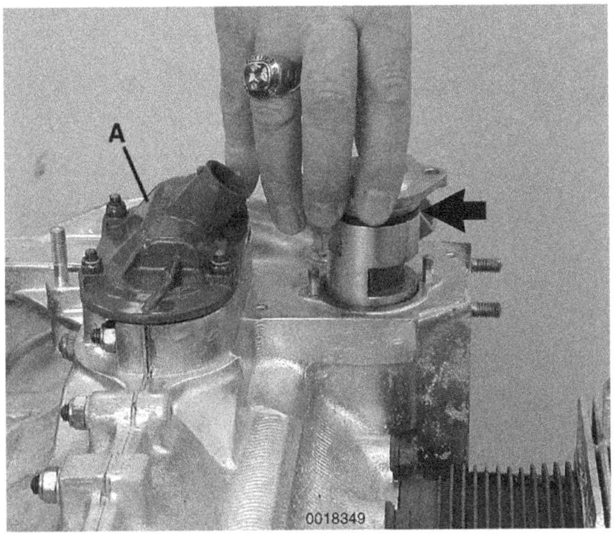

◄ Install oil temperature thermostat with new sealing O-ring (**arrow**).

Tightening torque
- Oil temperature thermostat to crankcase . . . . 8 Nm (6 ft-lb)

– Install breather cap (**A**) with a new gasket.

Tightening torque
- Breather cap to crankcase . . . . 8 Nm (6 ft-lb)

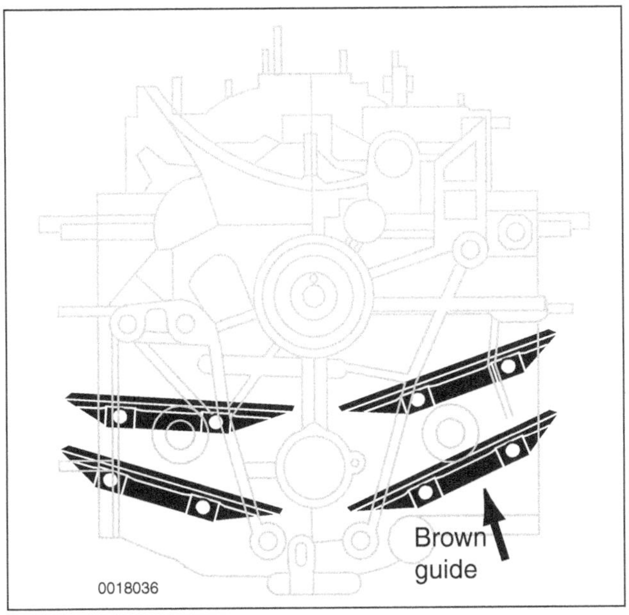

◄ Install chain guide rails inside crankcase chain housings.
- Note location of different guide rail in lower right slot.

Use screwdriver to push chain guide rail toward pulley end of engine (**arrow**) while tightening retaining bolts.

Tightening torque
- Timing chain guide retaining bolt to crankcase . . . . 25 Nm (18 ft-lb)

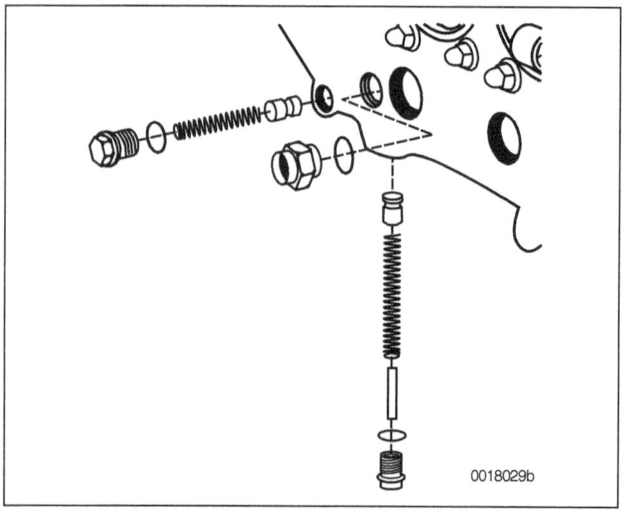

Install oil pressure relief valve assemblies in crankcase. Coat relief valves with assembly lubricant before installing.

Tightening torque
- Oil pressure relief valve plug to crankcase . . . . 60 Nm (44 ft-lb)

Pistons and cylinders, installing

Measurement and reconditioning of pistons and cylinders is covered in **131 Pistons and Cylinders**.

> ***CAUTION —***
> - *Be sure to keep each piston and cylinder set clearly marked and together.*
> - *Be sure used pistons and cylinders are reinstalled in the exact same locations and positions.*

Set crankshaft to **Z1** TDC.

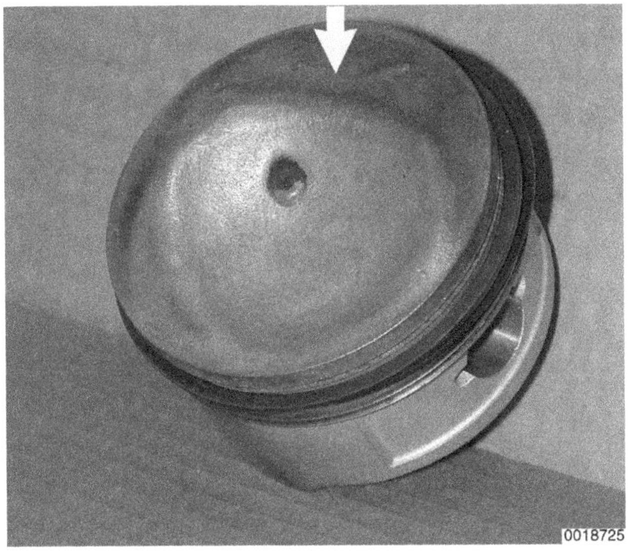

Install piston 1 with flat of intake valve relief (**arrow**) up.

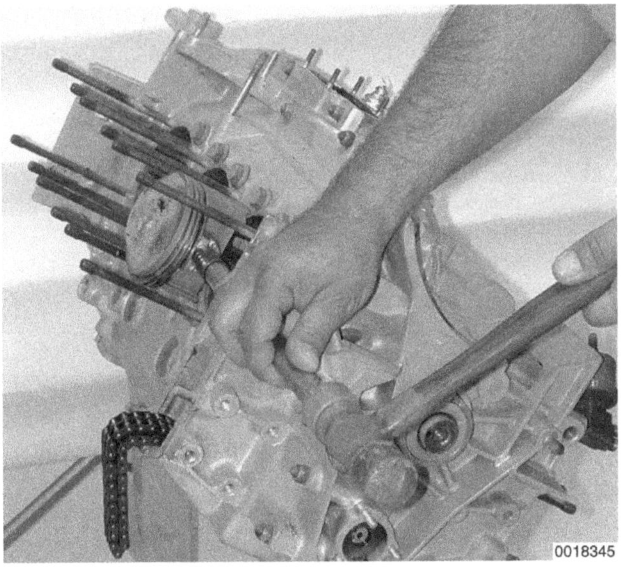

Oil wrist pin and install. If necessary, tap pin in gently using soft punch. Install retaining circlips.

> ***CAUTION—***
>
> - *The wrist pin circlips installed on new pistons are for transportation purposes only. Replace these clips with ones provided separately with the pistons.*
> - *Always be sure that both piston pin clips are in place before installing the cylinder.*

– Rotate crankshaft and install pistons 2 and 3.

> ***CAUTION—***
>
> *When turning crankshaft, manually guide pistons in and out of crankcase bores in order to avoid damaging the piston skirts or rings. Use shop towels around the pistons for protection.*

– Reset crankshaft to **Z1** compression stroke.

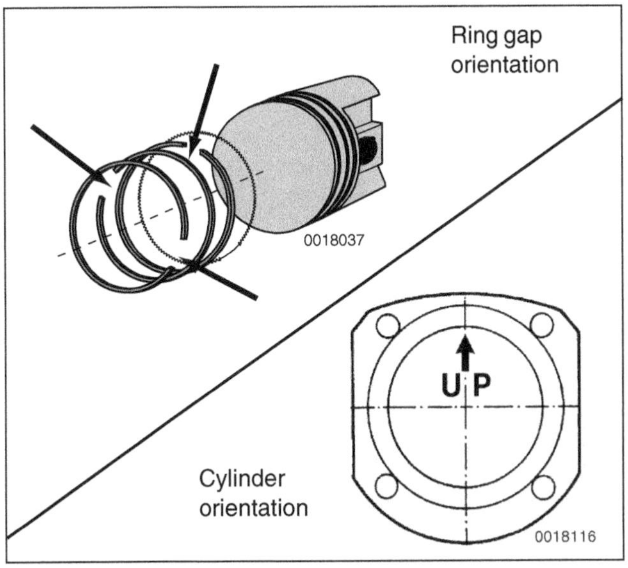

Set piston ring gaps to clock positions 4, 10 and 2 (**arrows**).

– Place a new cylinder base gasket on cylinder 1 and lubricate cylinder wall.

> ***CAUTION—***
>
> *If cylinder heads have been machined, be sure to use 0.50 instead of 0.25 mm thick copper gaskets at cylinder bases. See* ***150 Cylinder Heads****.*

NOTE—

When installing cylinders, make sure the deeper cylinder cooling fins point down.

Using a ring compressor such as Porsche special tool US 1008 a or equivalent, slide cylinder over piston and into side of case.

Temporarily hold cylinder in position using Porsche special tools P 140 (**arrows**).

NOTE —
As a substitute for the special tools, slide two sockets over the cylinder head studs for each cylinder and install two cylinder head nuts finger tight.

- Rotate crankshaft and install cylinders 2 and 3.

> ***CAUTION —***
> *When rotating crankshaft, manually guide pistons in and out of crankcase bores in order to avoid damaging the piston skirts or rings. Use shop towels around the pistons for protection.*

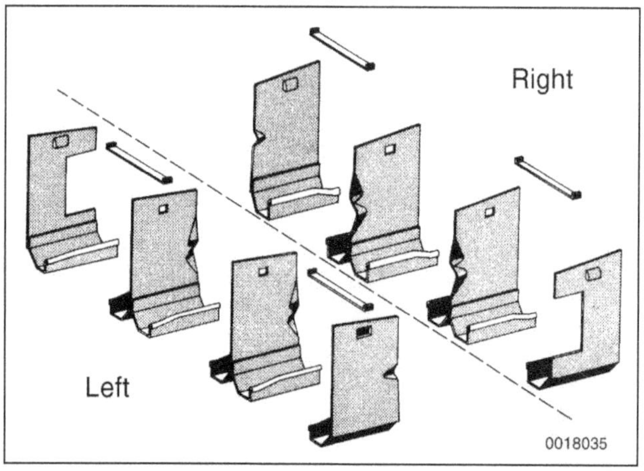

Install left side cylinder air guide plates.

- Install right side (4, 5, 6) pistons and cylinders.
- Install right side air guides.

Cylinder heads, installing

Repairs to cylinder heads and valves are described in **150 Cylinder Heads**.

- With engine on engine stand and pistons, cylinders, gaskets, and air guides installed, rotate engine on stand so that left side cylinders are up.
- Install cylinder head sealing rings in cylinders 1, 2, and 3.

NOTE —
Prior to cylinder head installation be sure to remove temporary cylinder hold down nuts.

- Lubricate cylinder head studs, nuts and washers lightly with Optimoly®HT (Porsche part number 901 104 382 02) or equivalent. Starting with cylinder head 3, install left side cylinder heads.

< Using screwdriver as guide, slide head washers over head studs.

- Install head nuts and hand tighten only. Temporarily fit camshaft case in position on heads. Alignment pins on cylinders heads must line up with alignment holes in camshaft case. Position cylinder heads as required until machined sealing surfaces meet. Temporarily install a few camshaft case mounting nuts (hand tight) to hold case in position.

NOTE —
Sealing surfaces on heads and camshaft cases must be clean, flat, and without burrs.

- Use cross pattern to torque cylinder head nuts.

Tightening Torques
- Cylinder head to crankcase
  - Step 1 .... 15 Nm (11 ft-lb)
  - Step 2 .... +90° (¼ turn)

- Remove camshaft case from cylinder heads 1 - 3. Rotate engine on stand so that right side cylinders 4 - 6 are up.
- Starting with cylinder 6, install right side cylinder heads.
- Slide head washers over head studs.
- Install head nuts and hand tighten only. Temporarily fit camshaft case in position on heads. Alignment pins on cylinders heads must line up with alignment holes in the camshaft case. Position cylinder heads as required until machined sealing surfaces meet. Temporarily install a few camshaft case mounting nuts (hand tight) to hold case in position.
- Use cross pattern to torque cylinder head nuts.

Tightening Torques
- Cylinder head to crankcase
  - Step 1 .... 15 Nm (11 ft-lb)
  - Step 2 .... +90° (¼ turn)

– Remove camshaft case from cylinder heads 4 - 6.

– Rotate engine on stand. Working on left side bottom of crankcase, install oil pressure line.

Camshaft cases, camshafts and camshaft timing chains, installing

Additional camshaft timing chain information can be found in **152 Camshaft Timing Chains**.

> ***CAUTION —***
> *Install camshaft cases and camshafts on left or right side according to marks made during disassembly.*

– Use Loctite® 574 or equivalent to coat machined sealing surfaces on camshaft cases.

– Place new sealing O-rings on oil return tubes.

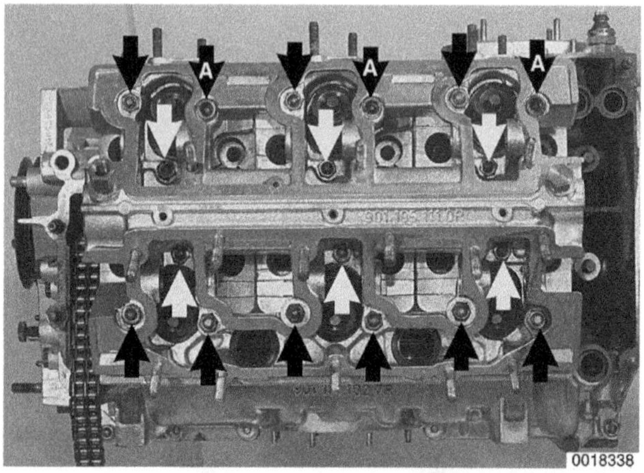

0018338

– Holding oil return tubes in place, guide right camshaft case over its mounting studs.

NOTE —
Two piece oil return tube replacement is covered in **170 Engine Lubrication**.

Install 18 retaining nuts (**arrows**) and tighten after making sure oil return tubes are correctly positioned.

- There are fifteen M8 (13 mm head) nuts.
- There are three M8 (8 mm allen head) nuts (**A**).

Tightening torque
- Camshaft case to cylinder head . . . . 25 Nm (18 ft-lb)

– Install right timing chain housing with new gasket.

Tightening torque
- Timing chain housing to crankcase . . . . 25 Nm (18 ft-lb)

0018350

– Slide right camshaft into camshaft case.

– Install camshaft end cover with new O-ring and gasket. Install end cover retaining bolts and tighten.

– Install right camshaft thrust spacer, sprocket shims, woodruff key and sprocket flange.

– Install right camshaft sprocket to timing chain. Slide sprocket over sprocket flange.

– Install sprocket washer and mounting fastener finger tight.

Install right timing chain guide rail in housing.

– Install right timing chain tensioner and idler sprocket assembly using new O-ring. Tighten tensioner mounting nuts.

– Install left camshaft, camshaft case, timing chain guide rail, camshaft sprocket assembly and timing chain as described in previous steps.

Camshaft sprocket assembly (1982 - 1983)

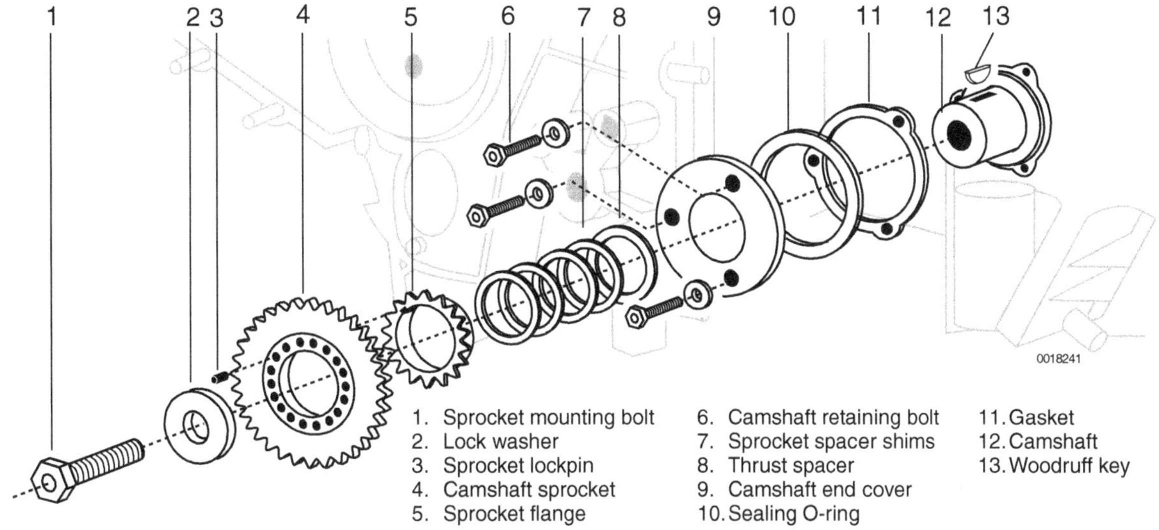

1. Sprocket mounting bolt
2. Lock washer
3. Sprocket lockpin
4. Camshaft sprocket
5. Sprocket flange
6. Camshaft retaining bolt
7. Sprocket spacer shims
8. Thrust spacer
9. Camshaft end cover
10. Sealing O-ring
11. Gasket
12. Camshaft
13. Woodruff key

Camshaft sprocket orientation

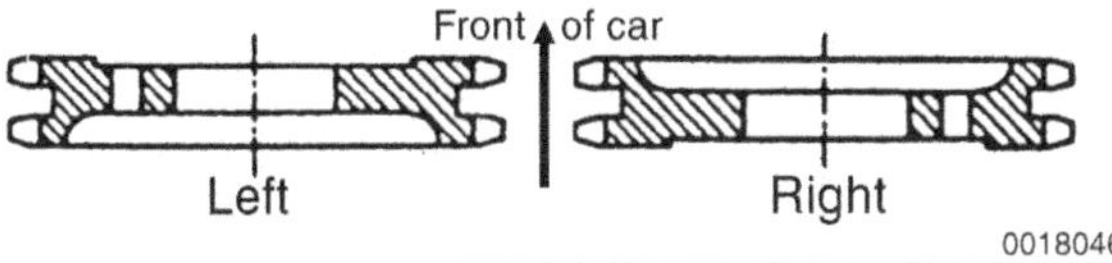

NOTE—

Left and right camshaft sprockets are identical. Dished side of sprocket must face pulley side of engine on left camshaft (cylinders 1 - 3) and flywheel side of engine on right camshaft (cylinders 4 - 6)

– Check parallelism of intermediate shaft and camshaft sprockets by measuring distance from front of intermediate shaft to front edge of gear wheels. Use quality straight edge when making measurement.

NOTE—

- *Normally, if reinstalling camshafts and sprockets, 3 sprocket spacer shims will be necessary under the left sprocket and 4 shims under the right sprocket.*
- *If new camshafts or sprockets are being installed, camshaft sprocket spacing must be adjusted by adding or subtracting spacer shims so that each sprocket is parallel with the corresponding intermediate shaft sprocket. See* **151 Camshafts and Camshaft Cases**.

Timing chain sprocket parallelism

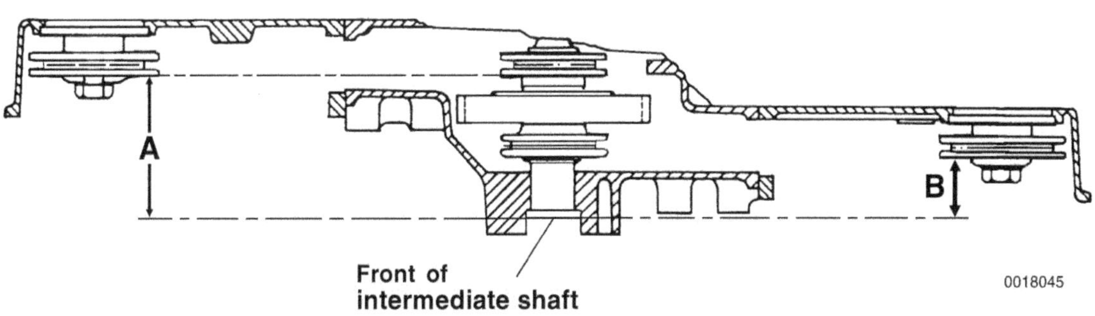

A Factory distance from end of intermediate shaft to face of cylinder 1 - 3 (left) timing chain sprocket = 98.07± 0.25 mm (3.86 ± 0.01 in.)
B Factory distance from end of intermediate shaft to face of cylinder 4 - 6 (right) timing chain sprocket = 43.27 ± 0.25 mm (1.70 ± 0.01 in.)

– Install intermediate shaft cover.

A preliminary setting of camshaft timing must be done as follows:

- Set front crankshaft pulley to **Z1** (TDC).
- Remove camshaft sprocket fastener.
- Rotate left camshaft until punch mark or "930" (**black arrow**) points up.
- Install camshaft sprocket lockpin.
- Reinstall left camshaft sprocket fastener finger-tight.
- Repeat for right camshaft.

Camshaft timing, adjusting

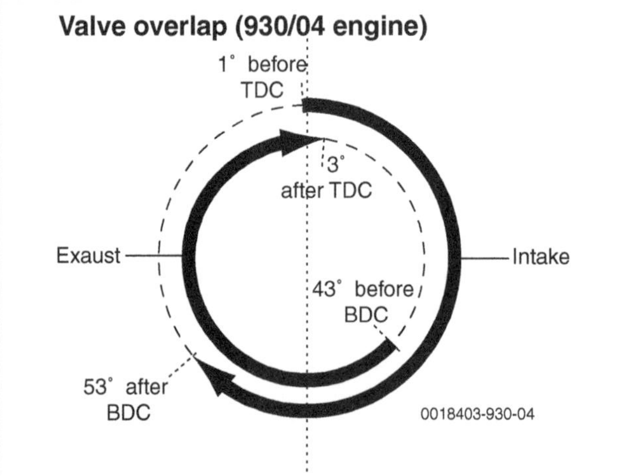

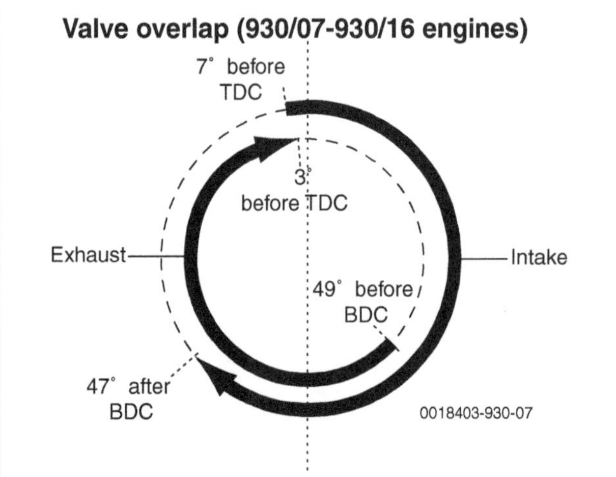

Camshaft timing is adjusted using a dial indicator setup. The dial indicator is used to precisely set the position of each camshaft in relation to partially open ("overlapped") valves.

NOTE —

Camshaft timing specification in degrees of rotation are given below. See **100 Engine–General** *for engine application information.*

930/04 engine valve overlap

- Intake valve opens ....1° before TDC
- Intake valve closes ....53° after BDC
- Exhaust valve opens................. 43° before BDC
- Exhaust valve closes ....3° after TDC

930/07 and 930/16 engines valve overlap

- Intake valve opens ....7° before TDC
- Intake valve closes ....47° after BDC
- Exhaust valve opens................. 49° before BDC
- Exhaust valve closes ....3° before TDC

Rocker shaft and arm, installing

< Set crankshaft to **Z1** cylinder 1 compression stroke.

- Check to see that punch mark or "930" (**black arrows**) on camshaft ends point *up*.

– Assemble rocker shaft.

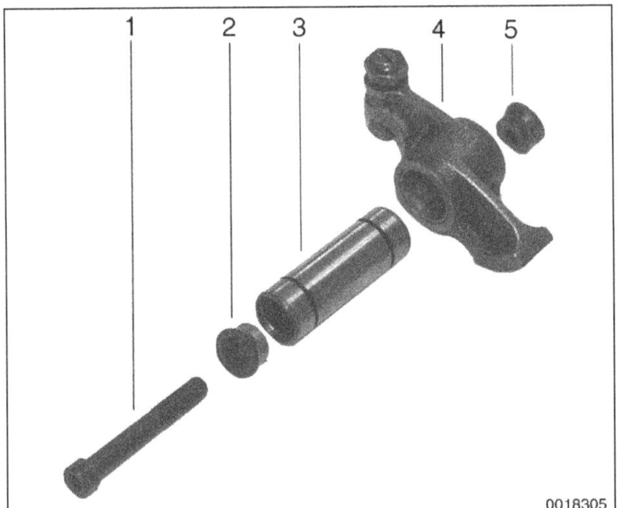

1. Bolt (5 mm allen)
2. Expansion sleeve
3. Rocker shaft
4. Rocker arm
5. Expansion nut (8 mm allen)

< Install cylinder 1 intake and exhaust rocker shafts with rocker arms:

- Install rocker shaft with 5 mm bolt facing cylinder 2.
- Tap rocker shaft into bore until shaft is flush with inner edge of support.
- Place rocker arm in position and tap rocker shaft in until shaft slides into far side support.

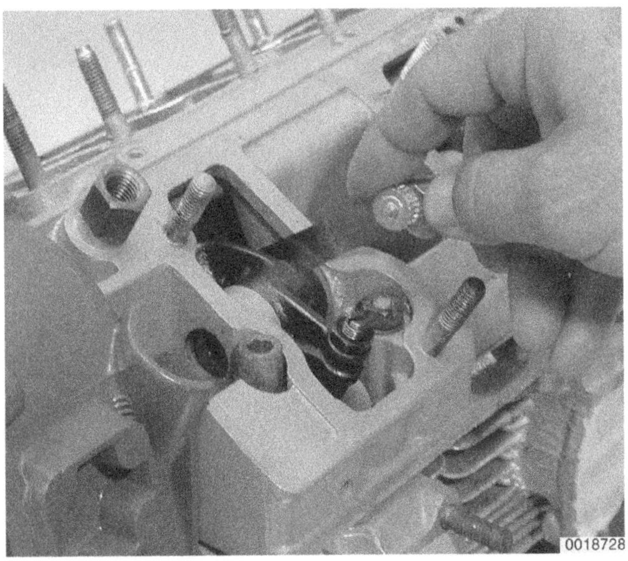

< Place 0.5 mm (0.02 in.) feeler gauge between far side support and rocker arm.

- Continue to tap shaft until feeler gauge engages slot in rocker shaft.

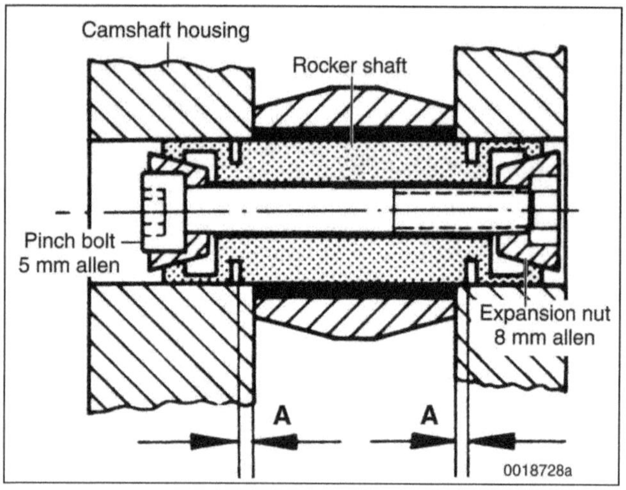

Remove feeler gauge and tap rocker arm shaft into position so that shaft is centered in camshaft housing.

- Grooves in rocker shaft body should be inset approximately 1.5 mm (0.06 in.) into camshaft housing (distance **A**).
- Tighten rocker arm shaft pinch bolt.

Tightening torque

- Rocker shaft pinch bolt . . . . 15 Nm (11 ft-lb)

Use Porsche special tool P 213 b or equivalent feeler gauge to adjust cylinder 1 intake and exhaust valve clearance.

Valve Clearance

- Adjusting value . . . . 0.10 mm (0.004 in.)

Left (cylinders 1 - 3) camshaft timing

Set crankshaft to **Z1** cylinder 1 compression stroke.

- Check to see that punch mark or "930" (**black arrows**) on camshaft ends point *up*.

Using Porsche special tool P 207, mount dial indicator on camshaft case stud next to cylinder 1 intake valve.

- Set dial indicator plunger against valve spring retainer on cylinder 1 intake valve.
- Zero dial indicator with approximately 10 mm (0.4 in.) preload.

– Starting at **Z1** (TDC) mark on pulley, turn crankshaft slowly in clockwise direction approximately 1 full turn while observing dial indicator. Turn crankshaft far enough to reach valve overlap adjusting value.

Valve overlap adjustment

- 930/04 adjusting value . . . . 1.00 mm (0.039 in.)
- 930/07 - 930/16 adjusting value . . . . 1.55 mm (0.061 in.)

– Unscrew and remove left timing chain sprocket mounting fastener.

Pull out camshaft sprocket lockpin (**arrow**) using Porsche special tool P 212 or equivalent.

– Turn crankshaft (either clockwise or counterclockwise, as necessary) just until **Z1** mark on pulley is aligned accurately with joint of crankcase.

– Reinstall lockpin in hole with best alignment. Install camshaft sprocket fastener finger tight.

– Turn crankshaft clockwise two full turns. Recheck timing adjustment. Value must be within tolerance range.

Valve overlap tolerance

- 930/04 range. . . . . 0.9 - 1.1 mm (0.035 - 0.043 in.)
- 930/07 - 930/16 range . . 1.4 - 1.7 mm (0.055 - 0.067 in.)

Torque camshaft sprocket mounting fasteners when settings are correct:

- On 1978 - 1981 models: Use Porsche special tools P 202 and P 203 or equivalent to counterhold camshaft and torque sprocket mounting nut.
- On 1982 - 1983 models: Use Porsche special tool 9191 or equivalent to counterhold sprocket while torquing sprocket bolt.

Tightening torque

- Chain sprocket to camshaft
  1978 - 1981 models (46 mm nut) . . . . 150 Nm (111 ft-lb)
  1982 - 1983 models (M12 bolt). . . . . . . 120 Nm (89 ft-lb)

0018242dn

Right (cylinders 4 - 6) camshaft timing

< Set crankshaft to **Z1** (TDC) cylinder 4 compression stroke.

- Check to see that punch marks or "930" (**black arrows**) on camshaft ends point *down.*

– Install cylinder 4 rocker shafts and arms as described earlier.

- Install 5 mm allen bolt with bolt head facing cylinder 5.

Tightening torque

- Rocker shaft pinch bolt . . . . 18 Nm (16 ft-lb)

– Adjust cylinder 4 intake and exhaust valve clearance.

Valve Clearance

- Adjusting value . . . . 0.10 mm (0.004 in.)

0018598

< With crankshaft at cylinder 4 TDC compression stroke, use Porsche special tool P 207 to mount dial indicator on camshaft case stud next to cylinder 4 intake valve.

- Set dial indicator plunger against valve spring retainer on cylinder 4 intake valve.
- Zero dial indicator with approximately 10 mm (0.4 in.) preload.

– Starting at **Z1** (TDC) mark on pulley, turn crankshaft slowly in clockwise direction approximately 1 full turn while observing dial indicator. Turn crankshaft far enough to reach valve overlap adjusting value.

Valve overlap adjustment

- 930/04 adjusting value . . . . 1.00 mm (0.039 in.)
- 930/07 - 930/16 adjusting value . . . . 1.55 mm (0.061 in.)

– Unscrew and remove right timing chain sprocket mounting fastener.

0018733

< Pull out camshaft sprocket lockpin using special tool P 212 (**arrow**).

– Turn crankshaft just until **Z1** mark on pulley is aligned accurately with joint of crankcase.

– Reinstall lockpin in hole with best alignment. Install camshaft sprocket fastener finger tight.

– Turn crankshaft clockwise two full turns. Recheck timing adjustment. Value must be within tolerance range.

Valve overlap tolerance

- 930/04 range. . . . . 0.9 - 1.1 mm (0.035 - 0.043 in.)
- 930/07 - 930/16 range . . 1.4 - 1.7 mm (0.055 - 0.067 in.)

– Torque camshaft sprocket mounting fastener when settings are correct:

- On 1978 - 1981 models: Use Porsche special tools P 202 and P 203 or equivalent to counterhold camshaft and torque sprocket mounting nut.
- On 1982 - 1983 models: Use Porsche special tool 9191 or equivalent to counterhold sprocket while torquing sprocket bolt.

Tightening torque

- Chain sprocket to camshaft
  1978 - 1981 models (46 mm nut) . . . . 150 Nm (111 ft-lb)
  1982 - 1983 models (M12 bolt). . . . . . . 120 Nm (89 ft-lb)

Set crankshaft to **Z1** cylinder 1 compression stroke.

- Check to see that punch marks or "930" on camshaft ends point up.

NOTE —
The following instructions apply when installing rocker shafts.

- *Cylinder 2 and 5: 5 mm allen bolt head can be installed either way.*
- *Cylinder 3: install 5 mm allen bolt with bolt head facing cylinder 2.*
- *Cylinder 6: install 5 mm allen bolt with bolt head facing cylinder 5.*

– Turn crankshaft 120° in normal direction of rotation. Install cylinder 6 rocker shafts and arms as described earlier.

Tightening torque

- Rocker shaft pinch bolt . . . . 18 Nm (16 ft-lb)

– Adjust cylinder 6 valve clearance.

Valve Clearance

- Adjusting value . . . . 0.10 mm (0.004 in.)

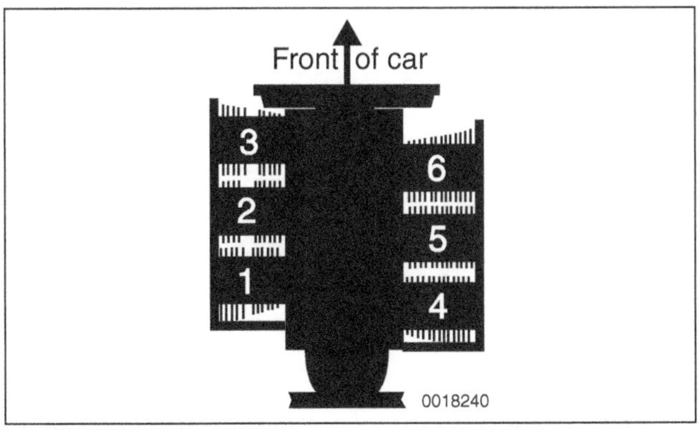

Following firing order, turn crankshaft 120° in normal direction of rotation and install rocker arms for next cylinder and adjust valves.

NOTE —
The firing order is 1 - 6 - 2 - 4 - 3 - 5.

– Repeat until all rocker arms are installed and valves adjusted.

– Install valve covers with new gaskets, sealing washers and lock nuts.

Tightening torque

- Valve cover to camshaft case . . . . 8 Nm (71 in-lb)

– Install timing chain housing covers with new gaskets, sealing washers and lock nuts.

Tightening torque
- Chain housing cover to housing . . . . 8 Nm (71 in-lb)

– Install right and left camshaft and timing chain tensioner oil feed lines with new sealing washers (if applicable).

➧Engine peripherals, installing

◀ With crankshaft at cylinder 1 compression TDC, install distributor so that mark on distributor rotor aligns with mark on distributor housing (**arrows**).

NOTE —
- *To be certain that you are at cylinder 1 compression TDC, place a finger in cylinder 1 spark plug hole. Use a wrench on the crankshaft pulley bolt to rock pulley back and forth at* ***Z1****. You should feel compression escaping past your fingertip*
- *At TDC, the ignition rotor mark lines up to the distributor housing mark approximately, not precisely.*

Tightening torque
- Distributor base clamp (M8) . . . . 18 Nm (14 ft-lb)

– Install new spark plugs.

Tightening torque
- Spark plug to cylinder head . . . . . 25-30 Nm (18-22 ft-lb)

– Install engine mount carrier and cross member.

Tightening torques
- Engine mount carrier to crankcase. . . . . 40 Nm (30 ft-lb)
- Engine mount crossmember to carrier . . . . 40 Nm (30 ft-lb)

– Install A/C compressor mounting brackets.

◀ Install oil cooler from below. Install top and bottom (**arrows**) oil cooler mounting nuts.

CAUTION —
Always use new O-ring seals.

Tightening torques
- Oil cooler to crankcase . . . . 23 Nm (17 ft-lb)

– Install all sheet metal shrouding around engine.

– Set complete fuel system, intake manifold, alternator and cooling shroud in place.

- Reattach alternator ground to top of crankcase under shroud.
- Install intake runner washers and nuts.

Tightening torque
- Intake manifold to cylinder heads . . . . . . 25 Nm (18 ft-lb)

– Connect all spark plug wires and install all wire holders.

– Rotate engine on stand and install heat exchangers, catalytic converter (if applicable) and muffler. See **260 Exhaust System**.

Tightening torques
- Heat exchangers to cylinder heads. . . . . 25 Nm (18 ft-lb)

– Install drain plug (as applicable) with new metal sealing washer and tighten. Install oil sump cover (if applicable) with new gaskets.

Tightening torques
- Oil drain plug to crankcase (from 3/83). . 70 Nm (51 ft-lb)
- Oil drain plug to sump cover (to 3/83). . . 42 Nm (31 ft-lb)
- Sump cover to crankcase (to 3/83)(M6) . . 10 Nm (7 ft-lb)

– Remove engine from engine stand.

– Install flywheel, clutch assembly and transmission. Install engine and transmission in car. See **101 Engine Removal and Installation**.

Tightening torques
- Clutch to flywheel . . . . 25 Nm (18 ft-lb)
- Flywheel to crankshaft . . . . 90 Nm (66 ft-lb)
- Transmission to engine . . . . 45 Nm (33 ft-lb)

NOTE —
Prior to starting engine, fill engine with engine oil. Then disconnect lead from ignition coil and ground. Use starter motor to rotate engine until oil pressure builds up. Reconnect ignition coil lead prior to starting.

130 Crankcase and Crankshaft

TABLES

GENERAL

This repair group provides reconditioning information for the crankcase, crankshaft and intermediate shaft. Engine disassembly is described in **102 Engine Disassembly and Assembly**.

> ***CAUTION—***
> *The information contained here is intended to be used as a reconditioning guide for the professional or experienced automotive technician.*

Special tools

Some of the procedures require special tools.

Special Tools

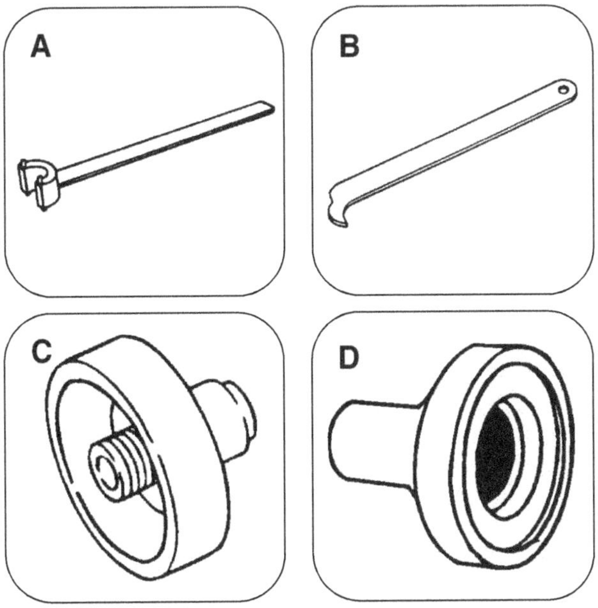

Special tools

- **A** 9236. . . . . Pulley holding spanner wrench (Porsche ordering #000 721 923 60)
- **B** Baum 681 . . . . .Oil seal puller (source: Baum Tools Unlimited)
- **C** P 216/a. . . . . .Pulley end oil seal installer (Porsche ordering #000 721 216 10)
- **D** 9126. . . . . Flywheel end oil seal installer (Porsche ordering #000 721 912 60)

Continued on page 2

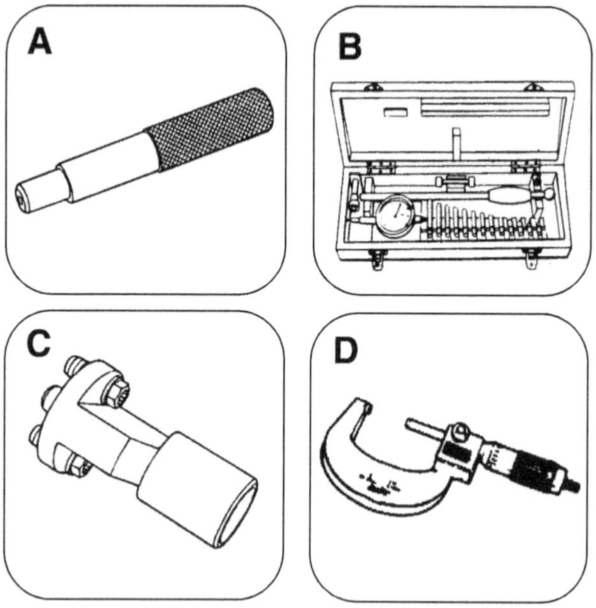

Special tool

A P 370 . . . . Clutch spline aligning tool
B Baum 17. Dial bore gauge 50-150 mm
(source: Baum Tools Unlimited)
C P 209 a. . . . . Bracket for holding crankshaft in vise
(Porsche ordering #000 721 209 10)
D Baum 12. . . . . .Micrometer
(source: Baum Tools Unlimited)

CRANKSHAFT OIL SEALS

NOTE —
*The pulley end of the engine is referred to as the **front**. The flywheel end of the engine is referred to as the **rear**.*

Front crankshaft oil seal (pulley end), replacing

The front crankshaft oil seal can be replaced with the engine installed in the vehicle.

NOTE —
Access to the pulley mounting bolt is very tight, restricted by the presence of the engine mounting bracket. To remove bracket, support engine securely from below, then remove engine mounting crossmember and bracket fasteners.

– Remove alternator and A/C compressor belts. See **270 Battery, Starter, Alternator**.

– Remove pulley mounting bolt and remove pulley.

NOTE —
Use Porsche special tool no. 9236 to counterhold pulley.

– Remove crankshaft oil seal with Porsche special tool 681 or equivalent.

Using Porsche special tool P 216/a or equivalent press new seal into position.

NOTE—
Lubricate inner lip of new seal with engine oil prior to installation.

– Install pulley and bolt. Counterhold pulley when tightening bolt.

Tightening torque
- Pulley to crankshaft. . . . . 80 Nm (52-ft-lb)

– Install accessory drive belts.

Rear crankshaft oil seal (flywheel end), replacing

Replacement of the rear crankshaft oil seal requires that the engine be removed from the car and the transmission be separated from the engine.

– Remove engine and transmission and separate transmission from engine as described in **101 Engine Removal and Installation**.

– Remove clutch mounting bolts and remove clutch pressure plate and disc.

– Remove flywheel bolts and remove flywheel.

– Remove crankshaft oil seal with special tool 681 or equivalent.

Using Porsche special tool 9126 (**arrow**) or equivalent, press seal into crankcase.

NOTE—
Lubricate inner lip of new seal with engine oil prior to installation.

– Reinstall flywheel and clutch using Porsche special tool P 370 or equivalent to align clutch disc.

Tightening torques
- Clutch to flywheel . . . . 25 Nm (18 ft-lb)
- Flywheel to crankshaft . . . . 90 Nm (66 ft-lb)

– Join transmission and engine and reinstall. See **101 Engine Removal and Installation**.

Tightening torques
- Crankcase to transmission . . . . 48 Nm (36 ft-lb)

Crankcase assembly

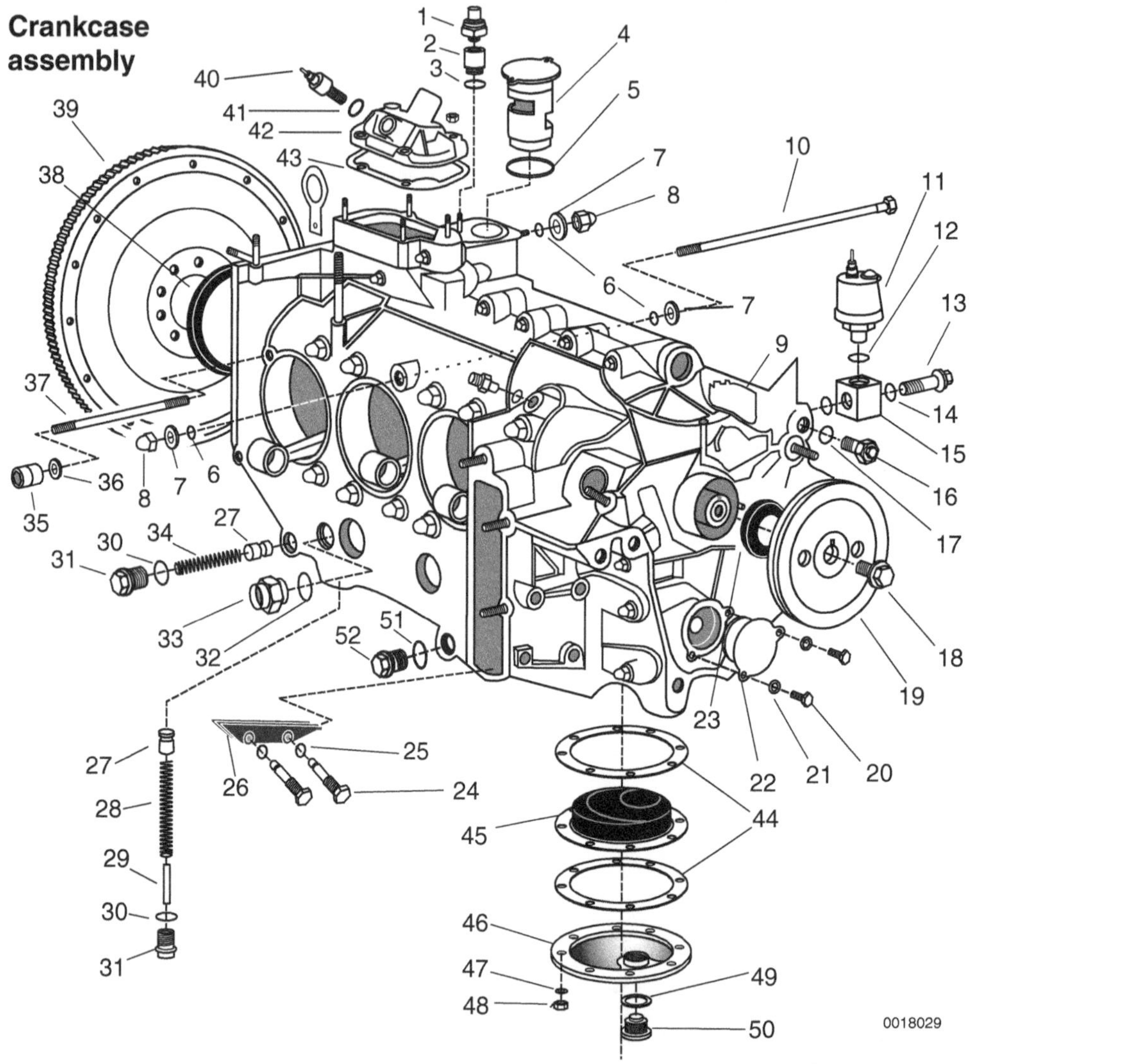

1. Oil pressure switch
   -tighten to 20 Nm (15 ft-lb)
2. Switch adapter
   -tighten to 35 Nm (26 ft-lb)
3. Sealing ring, (18 x 24mm)
4. Oil thermostat
5. Thermostat sealing O-ring (39.2 x 3 mm)
6. Case bolt sealing ring
7. Case bolt beveled washer
8. Case bolt cap nut
   -tighten to 35 Nm (26 ft-lb)
9. Crankcase
10. Case bolt
11. Oil pressure sensor
    -tighten to 35 Nm (26 ft-lb)
12. Oil pressure sensor sealing ring
    (8 x 24 mm)
13. Oil pressure sensor adapter bolt
    -tighten to 35 Nm (26 ft-lb)
14. Adapter bolt sealing ring (12 x 16 mm)
15. Oil pressure sensor adapter housing
16. Oil temperature sensor
    -tighten to 25 Nm (18 ft-lb)
17. Temperature sensor sealing ring
18. Front pulley bolt
    -tighten to 80 Nm (59 ft-lb)
19. Front pulley
20. Intermediate shaft cover bolt
21. Intermediate shaft cover bolt washer
22. Intermediate shaft cover
23. Crankshaft oil seal, front
24. Chain guide rail bolt
    -torque to 25 Nm (18 ft-lb)
25. Guide rail bolt sealing ring (10 x 16 mm)
26. Chain guide rail
27. Oil pressure relief valve piston
28. Oil pressure relief valve spring
29. Oil pressure relief valve sleeve
30. Sealing ring (18 x 24 mm)
31. Plug
    -tighten to 60 Nm (44 ft-lb)
32. Line adapter sealing ring (22 x 27mm)
33. Oil line adapter
    -tighten to 120 Nm (89 ft-lb)
34. Oil pressure safety valve spring
35. Cylinder head allen nut
    -tighten to 32 Nm (24 ft-lb)
36. Cylinder head washer
37. Cylinder head stud
38. Crankshaft oil seal, rear
39. Flywheel
40. Oxygen sensor system
    temperature switch (35°C)
    -tighten to 25 Nm (18 ft-lb)
41. Temperature sensor sealing ring
42. Breather cover
43. Breather cover gasket
44. Oil sump cover gasket (up to 3/83 prod.)
45. Oil sump screen (up to 3/83 prod.)
46. Oil sump cover (up to 3/83 prod.)
47. Washer (M6)
48. Nut (M6)
    -tighten to 10 Nm (7 ft-lb)
49. Oil drain plug sealing ring
50. Oil drain plug (up to 3/83 prod.)
    -tighten to 42 Nm (32 ft-lb)
51. Oil drain plug sealing ring
52. Oil drain plug (from 3/83 prod.)
    -tighten to 70 Nm (51 ft-lb)

CRANKCASE

All disassembled parts should be thoroughly cleaned before inspection. If bearings are to be reused, they should always be reinstalled in their exact positions. Parts should never be interchanged between cylinders.

> ***WARNING —***
> - *When cleaning engine parts always wear eye protection.*
> - *Solvents used for cleaning parts are highly flammable, especially in aerosol form. Use with care. Do not smoke. Do not use these products near any source of sparks or flame.*

Crankcase, inspection

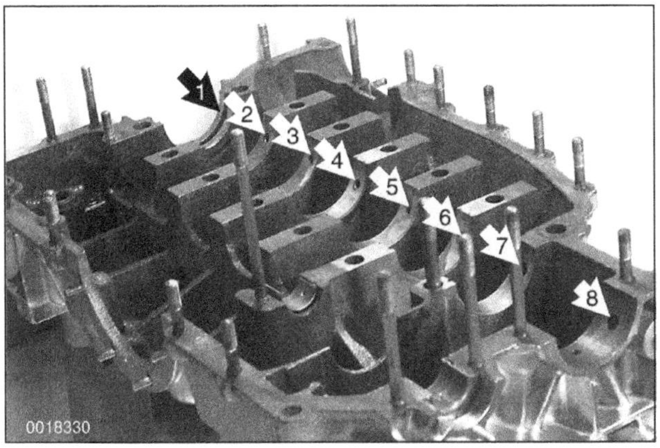

On engines that have had a mechanical failure, the complete lubrication circuit must be cleaned and flushed to remove foreign material before reassembling the engine. This includes cleaning all oil lines, oil pressure valves, crankcase oil passages (**arrows**) and oil tank. In addition to thorough cleaning, the engine oil coolers should be replaced. See **170 Engine Lubrication**.

When replacing cylinder head studs:

- Heat area around base of stud to loosen thread cement.
- Clean threads in crankcase thoroughly. Chase thread holes with tap.
- Install new studs to height of 135 mm (5.315 in.)

NOTE —
Various types of head studs have been used on the 911 SC engines. Always check with an authorized Porsche parts department for the latest in parts information.

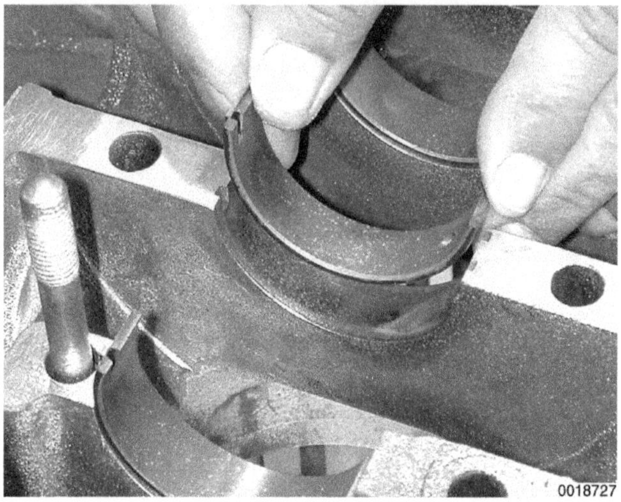

0018727

Remove and label main bearing shells from both case halves and check for wear, scoring or visible damage to case.

0018329

Remove and label intermediate shaft bearing shells from both case halves and check for wear, scoring or visible damage to case.

NOTE —

Always replace intermediate shaft bearing shells. DO NOT re-use intermediate shaft bearing shells.

0018330

Blow out all engine crankcase oil feed bores (**arrows**) with compressed air.

– Install left half of crankcase and tap into place.
  - Install all M10 crankcase bolts, washers and nuts.
  - Install four M8 nuts on bearing 1 (flywheel end) and bearing 8 (pulley end).
  - Finger tighten nuts.
  - Align case sections with each other with light knocks from a plastic hammer. Bearing bore 8 joint must not be offset.
  - Tighten all case bolts and nuts to specified torque.

Tightening torques

- Crankcase bolts and nuts (M10)....... 35 Nm (26 ft-lb)
- No. 1 and 8 main journal nuts (M8)..... 25 Nm (18 ft-lb)

Measure bearing bores 1 to 8 with an internal micrometer. Check for wear, scoring or out of roundness.

Case bore for main bearings (bearings removed)

- Bores 1 - 8
  standard . . . . . 65.000 - 65.019 mm (2.5590 - 2.5598 in.)
  oversize. . . . . . 65.250 - 65.269 mm (2.5689 - 2.5696 in.)

Case bore for intermediate shaft (bearings removed)

- Bearing 1 . . . . 27.500 - 27.521 mm (1.0827 - 1.0835 in.)
- Bearing 2 . . . . 26.500 - 26.521 mm (1.0433 - 1.0441 in.)

NOTE—

If bearing bore sizes are incorrect or out of round, the crankcase must be machined (line-bored) to accept oversized bearing shells. Check with an authorized Porsche dealer or professional machine shop.

Intermediate shaft, inspection

The intermediate shaft gear has an identifying code of either 0 or 1 (**arrow**) stamped on it.

– Inspect intermediate shaft drive gear for wear or damage.

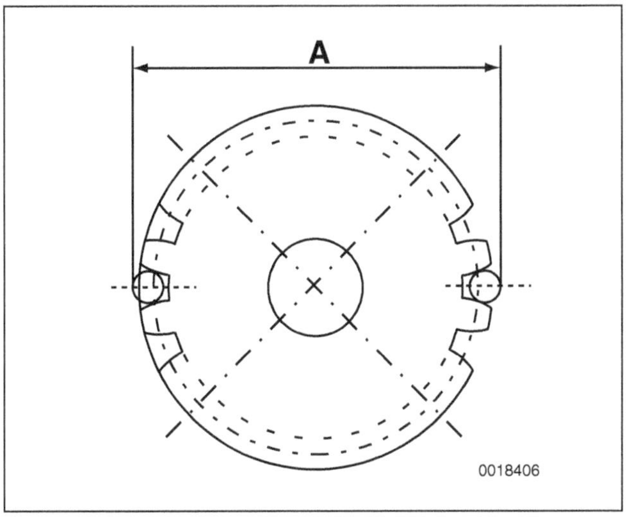

Measure gear by placing two 4.5 mm dowel pins (or drill bits) in gear teeth 180° apart.

- Replace intermediate shaft and drive gear if measurement **A** falls below specifications.

Intermediate shaft drive gear specifications

- Code 0 . . . . 136.50 mm (5.3740 in.)
- Code 1 . . . . 135.55 mm (5.3366 in.)

NOTE—

The intermediate shaft and aluminum drive gear are one replacement part.

– Check shaft for wear or damage. See **Table a**.

Table a. Intermediate shaft clearances

Journal dia. bearing 1 bearing 2	 25.000 - 24.980 mm (0.9842 - 0.9835 in.) 23.980 - 23.967 mm (0.9441 - 0.9436 in.)
Intermediate shaft clearance (Plastigage®)	0.030 - 0.084 mm (0.00012 - 0.0033 in.)
Intermediate shaft axial clearance Wear limit	0.040 - 0.133 mm (0.0016 - 0.0052 in.) 0.16 mm (0.0063 in.)

Crankcase, drive gear on crankshaft, and drive gear on intermediate shaft are assembled as a coded set (**arrows**). This assembly code is based on the case bore centers. When replacing drive gears, note codes and use gears with the same codes. Additionally, the bore centers should be checked before final assembly as described below.

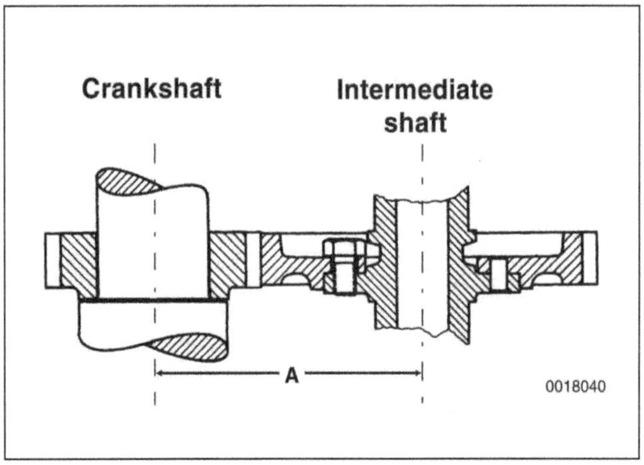

Measure distance between centerline of crankshaft and intermediate shaft bores (**A**).

- Match gears and crankcase as shown in **Table b**.

– The intermediate shaft is hollow and should be cleaned out whenever the engine is being reconditioned and in case of engine bearing failure.

- Drill 6.4 mm (0.252 in.) diameter hole in aluminum plug located in end of intermediate shaft. Tap hole with M8 tap.
- Install M8 X 125 mm bolt into drilled hole and pull aluminum plug out of shaft.
- Flush residue from inside shaft and install new plug.

Table b. Crankcase/gear assembly codes

Distance A	Crankcase code	Crankshaft drive gear code	Intermediate shaft gear code	Backlash
103.975 - 103.990 mm (4.0935 - 4.0941 in.)	0	0	0	0.029 - 0.049 mm (0.0011 - 0.0019 in.)
		installation permitted		
		1	0	0.016 - 0.042 mm (0.0006 - 0.0016 in.)
		0	1	0.017 - 0.043 mm (0.0006 - 0.0017 in.)
103.990 - 104.000 mm (4.0941 - 4.0945 in.)	1	1	1	0.012 - 0.041 mm (0.0004 - 0.0016 in.)
		installation permitted		
		0	1	0.025 - 0.049 mm (0.0009 - 0.0019 in.)
		1	0	0.025 - 0.048 mm (0.0009 - 0.0019 in.)

CRANKSHAFT AND CONNECTING RODS

Crankshaft gears, removing and installing

– Bolt crankshaft to special tool P 209 or equivalent and mount in a shop vise.

– Remove circlip holding gears on crankshaft.

◄ Remove complete gear assembly from crankshaft with a suitable puller. Check gear bore and shaft diameter (interference fit).

Crankshaft drive gear fit

- Timing gear shaft dia. . . . . . 42.002 - 42.013 mm (1.6536 - 1.6540 in.)
- Timing gear bore dia. . . . . . 41.975 - 42.000 mm (1.6526 - 1.6535 in.)
- Fit tolerance . -0.002 to - 0.038 mm (-0.00008 to - 0.0015 in.)

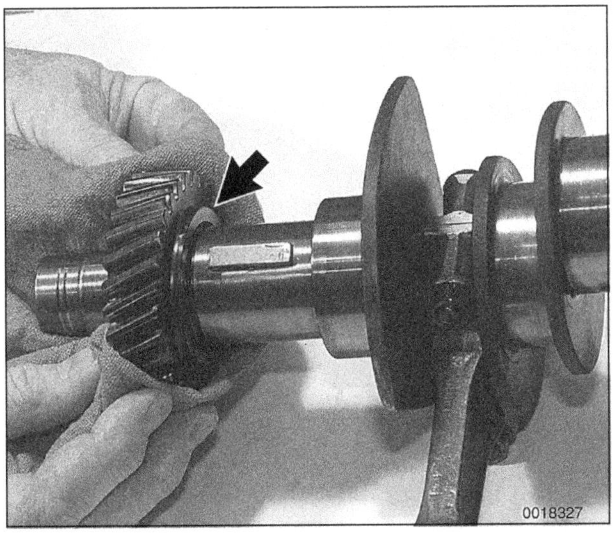

◄ Install drive gear to crankshaft.

- Install woodruff key to crankshaft.
- Heat gear to 150°C (300°F) in oil bath.
- With shoulder on gear facing in (**arrow**), slide drive gear fully onto crankshaft.

> ***WARNING —***
> *Use protective gloves or shop rag when handling the hot gear.*

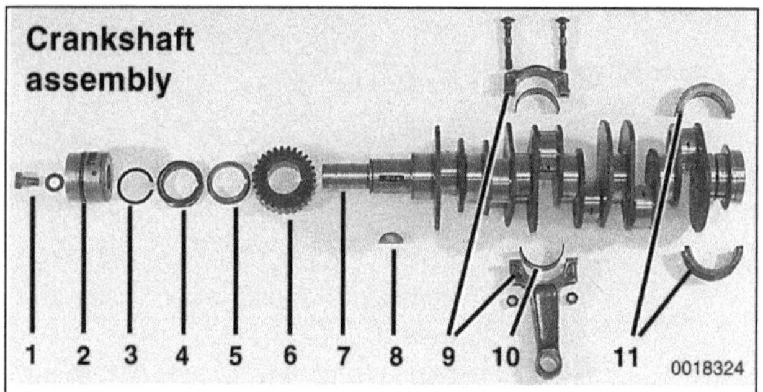

1. Pulley bolt
2. Nose bearing (no. 8)
3. Circlip
4. Distributor drive gear
5. Spacer
6. Intermediate shaft drive gear
7. Crankshaft
8. Woodruff key
9. Connecting rod
10. Connecting rod bearing
11. Main bearing

< Install spacer to crankshaft.

– Install distributor drive gear to crankshaft.

- Heat gear to 100°C (212°F) in water bath. Install gear to crankshaft with gear mark facing out.

> ***WARNING —***
> *Use protective gloves or shop rag when handling the hot gear.*

NOTE —
Be sure marking on distributor drive gear faces V-belt drive pulley end of crankshaft. Incorrect installation will cause initial ignition timing to be off by approximately 13°.

< Install selective circlip to crankshaft so that zero clearance is best obtained.

NOTE —
Circlip used to hold gears in place on crankshaft comes in different thicknesses. See **Table c**. *Select circlip to best obtain near zero clearance.*

Table c. Crankshaft gear circlip thickness

Porsche P/N	Thickness	Marking
901 102 148 00	2.4 mm	0
901 102 148 01	2.3 mm	1
901 102 148 02	2.2 mm	2
901 102 148 03	2.1 mm	3

Crankshaft main and rod bearings

Crankshaft main bearings are numbered from the rear of the engine (flywheel end) forward. Main bearing 8, (nose bearing) is at the pulley end.

Crankshaft main bearing shells, if they are to be reused, should only be installed in their original positions.

To servicing crankshaft, bolt to special mounting tool P 209 and mount in shop vise.

Table d. Crankshaft journal sizes

Size (mm)	Main bearing 8 (nose bearing)	Main bearings 1 - 7	Connecting rod bearing	Crankshaft drive gear	Crankshaft flywheel end
Standard	30.980 - 30.993 mm (1.2197 - 1.2202 in.)	59.971 - 59.990 mm (2.3611- 2.3618 in.)	52.971 - 52.990 mm (2.0855 - 2.0862 in.)	42.002 - 42.013 mm (1.6536 - 1.6541 in.)	89.780 - 90.000 mm (3.5346 - 3.5433 in.)
Undersize 1 (–0.25 mm)	30.730 - 30.743 mm (1.2098 - 1.2103 in.)	59.721 - 59.740 mm (2.3512 - 2.3520 in.)	52.721 - 52.740 mm (2.0756 - 2.0764 in.)	–	–
Undersize 2 (–0.50 mm)	30.480 - 30.493 mm (1.2000 - 1.2005 in.)	59.471 - 59.490 mm (2.3414 - 2.3421 in.)	52.471 - 52.490 mm (2.0658 - 2.0665 in.)	–	–
Undersize 3 (–0.75 mm)	30.230 - 30.243 mm (1.1901 - 1.1907 in.)	59.221 - 59.240 mm (2.3315 - 2.3323 in.)	52.221 - 52.240 mm (2.0559 - 2.0567 in.)	–	–
Undersize 4 (–1.00 mm)	29.980 - 29.993 mm (1.1803 - 1.1808 in.)	58.971 - 58.990 mm 2.3217 - 2.3224 in.)	51.971 - 51.990 mm (2.0461 - 2.0469 in.)	–	–

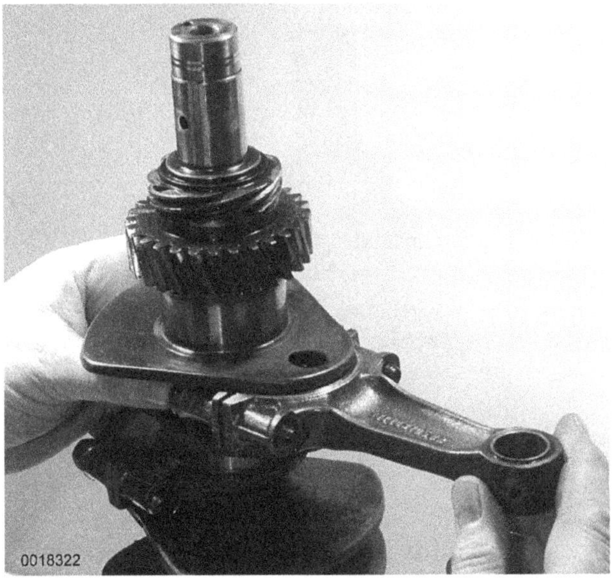

Mark installed position of connecting rods and remove from crankshaft.

Crankshaft journal sizes are listed in **Table d**. Crankshaft clearance specifications are listed in **Table e**. Crankshaft bearings are available in standard size and four undersizes.

Crankshaft axial play (also called end play) is controlled by a two-piece thrust bearing. The thrust bearing is the no. 1 main bearing (flywheel end).

Axial play should be measured at the center main bearing journal with the crankshaft supported on rollers on the outer main bearing journals.

Table e. Crankshaft clearances

Crankshaft main bearing no. 1 - 7 radial clearance (Plastigage®)	0.010 - 0.072 mm (0.0003 - 0.0028 in.)
Crankshaft main bearing no. 8 radial clearance (Plastigage®)	0.048 - 0.104 mm (0.0019 - 0.0041 in.)
Connecting rod bearing radial clearance (Plastigage®)	0.020 - 0.088 mm (0.0008 - 0.0035 in.)
Crankshaft axial clearance Wear limit	0.110 - 0.195 mm (0.0043 - 0.0078 in.) 0.20 mm (0.008 in.)
Maximum permissible crankshaft runout	0.15 mm (0.0060 in.)

Connecting rods

– Check weight of each connecting rod. Weigh a complete connecting rod less the bearing shells. See **Table f**.

NOTE —

Replacement connecting rods are available in different weight classes. Each class can be identified by the last digit of the part number. The weight class is stamped on the shaft of the connecting rod.

CAUTION —

In a complete set of connecting rods installed in one engine no two must differ in weight by more than 9 grams.

Table f. Connecting rod weights

Weight in grams	Weight class	Porsche P/N
633 to 642	3	930 103 015 51
642 to 651	4	930 103 015 52
651 to 660	5	930 103 015 53
660 to 669	6	930 103 015 54
669 to 678	7	930 103 015 55
678 to 687	8	930 103 015 56
687 to 696	9	930 103 015 57
696 to 705	10	930 103 015 58
705 to 714	11	930 103 015 59

– Hold connecting rod in soft-jawed vise and torque old rod bolts/nuts for measurement.

NOTE —
Do not use new fasteners when making this check. Instead, use the old rod bolts.

Torque check
- Connecting rod bolt/nut. . . . . . . 50 - 55 Nm (37 - 41 ft-lb)

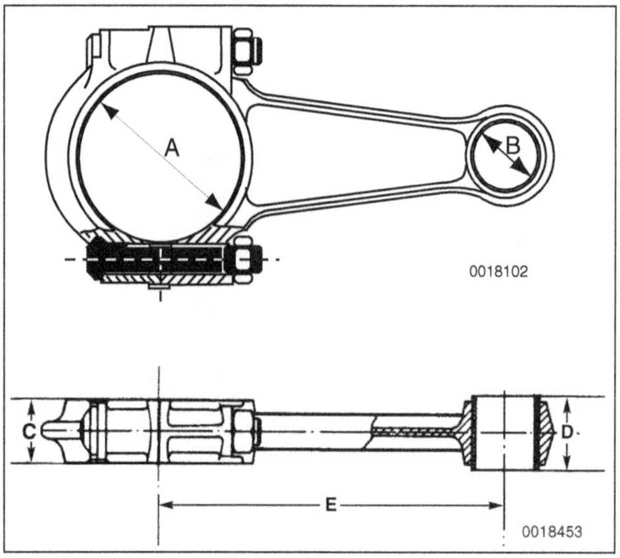

◄ Use micrometer and bore gauge to check connecting rod dimensions. See **Table g**.

Table g. Connecting rod dimensions

Big end bore (**A**)	56.000 - 56.019 (2.2047 - 2.2055 in.)
Small end bore (**B**)	22.020 - 22.033 mm (0.8669 - 0.8674 in.)
Width at big end (**C**)	21.70 - 21.80 mm (0.8543 - 0.8583 in.)
Width at small end (**D**)	24.5 - 25.0 mm (0.9646 - 0.9842 in.)
Distance between centers (**E**)	127.75 - 127.80 mm (5.0295 - 5.0315 in.)
Connecting rod bushing/piston pin clearance Wear limit	0.020 - 0.037 mm (0.0008 - 0.0014 in.) 0.055 mm (0.0022 in.)

– Check wrist pin fit to small end bushing. Pin should slide through bushing with light pressure.

– Check that numbers stamped on side of cap and rod are the same (cap and rod are a matched set).

– Insert bearing shells in connecting rod and corresponding rod cap.

– Lubricate bearing surfaces with suitable engine assembling lubricant.

0018321

During reinstallation, matching numbers on side of connecting rod and rod cap should all be facing same side of engine (for example, all marked numbers should face exhaust side of engine).

> ***CAUTION —***
> *Always use new connecting rod nuts and bolts during final assembly.*

0018073d

With all connecting rods installed, torque each connecting rod nut.

Tightening torques

- Connecting rod bolt/nut
  Stage 1 .... 20 Nm (14 ft-lb)
  Stage 2 .... 90°± 2°

NOTE —
Connecting rod must rotate easily under its own weight when the bearing is lubricated with engine oil.

131 Pistons and Cylinders

TABLES

GENERAL

This repair group provides reconditioning information for the pistons and cylinders. Engine disassembly is described in **102 Engine Disassembly and Assembly**.

> ***CAUTION—***
> *The information contained here is intended to be used as a reconditioning guide for the professional or experienced automotive technician.*

Special tools

Some of the procedures require special tools.

Special Tools

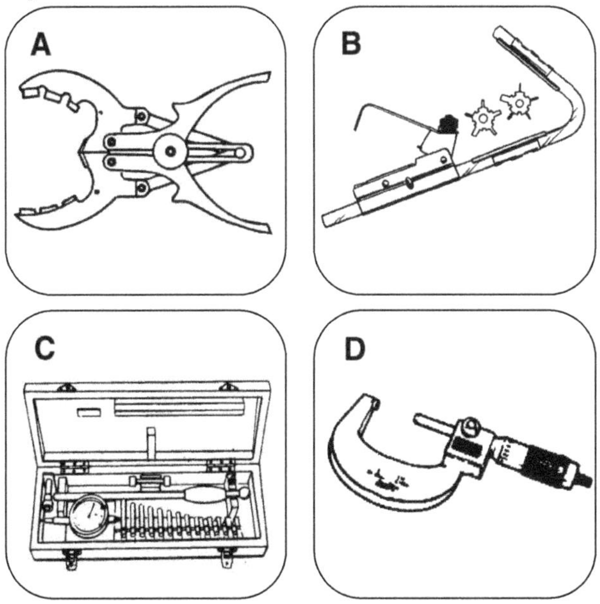

Special tool

A VW 121 b . . . . Piston ring expander
(source: Baum Tools Unlimited)
B Baum 65. . . . . Piston ring groove cleaner
(source: Baum Tools Unlimited)
C Baum 17. . . . . Inside bore gauge
(source: Baum Tools Unlimited)
D Baum 12. . . . . Micrometer, 75 - 100 mm (3 - 4 in) range
(source: Baum Tools Unlimited)

PISTONS AND CYLINDERS

Pistons must be matched by weight. When replacing pistons and cylinders, always check with an authorized dealer for the latest in replacement parts information.

Mark all parts during engine disassembly. If a part is to be reused, it must be reinstalled in its previously installed position. Parts must not be interchanged between cylinders.

If the engine is equipped with Nikasil® pistons and cylinders which meet all critical dimensions, hone the cylinders lightly. Replace piston rings with Göetz® rings. If equipped with Alusil® pistons and cylinders, replace with Nikasil® pistons and cylinders.

Piston weight, checking

Piston weight specifications

- Weight tolerance
  New piston: match to within 4 grams (0.14 oz.)
  Replacement piston: match to within 8 grams (0.28 oz.)

– Piston is weighed with wrist pin, wrist pin clips and rings installed.

Table a. Weight classes for Mahle® pistons

Weight class	Engine types: 930/07, 930/16	Code
Standard	636 - 640 g (22.43 - 22.58 oz.) 640 - 644 g (22.58 - 22.72 oz.)	– – –
New piston max. weight difference: 4 grams (0.14 oz)	644 -648 g (22.72 - 22.86 oz.) 648 - 652 g (22.86 - 23.00 oz)	+ ++
Replacement piston max. weight difference: 8 grams (0.28 oz.)	636 - 644 g (22.43 - 22.72 oz.) 644 - 652 g (22.72 - 23.00 oz)	– – or – + or ++

Table b. Weight classes for KS® pistons

Weight class	Engine type 930/16	Code
Standard	673 - 677 g (23.74 - 23.88 oz.) 677 - 681 g (23.88 - 24.02 oz.)	– – –
New piston max. weight difference: 4 grams (0.14 oz)	681 - 685 g (24.02 - 24.16 oz.) 685 - 689 g (24.16 - 24.30 oz.)	+ ++
Replacement piston max. weight difference: 8 grams (0.28 oz.)	673 - 681 g (23.74 - 24.02 oz.) 681 - 689 g (24.02 - 24.30 oz.)	– – or – + or ++

Cylinder height, checking

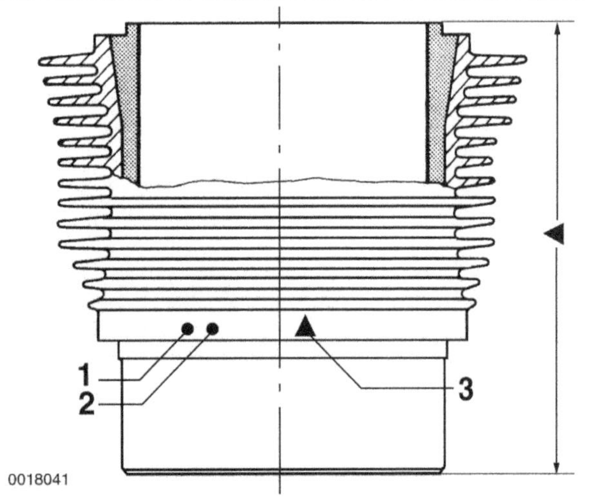

1. Manufacturer identification
2. Cylinder diameter tolerance group (see **Table e** or **Table f**).
3. Cylinder height tolerance group (5 or 6)

◄ Cylinder manufacturer identification and dimensions are indicated by marks on cylinder base.

Table c. Cylinder height

Tolerance Group	Cylinder height
5	85.400 – 85.425 mm (3.3621 – 3.3631 in.)
6	85.425 – 85.450 mm (3.3631 – 3.3641 in.)

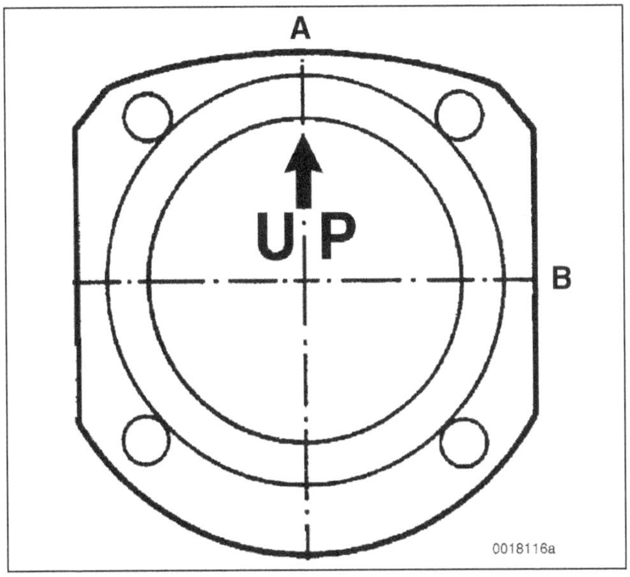

Cylinder roundness, checking

Check cylinder roundness by measuring bore in two directions perpendicular to stud holes (**A** and **B**).

- Measure cylinder at a point 30 mm (1.181 in.) below upper edge.

Cylinder roundness wear limit

- Difference between **A** and **B**. . . . . . 0.04 mm (0.0016 in.)

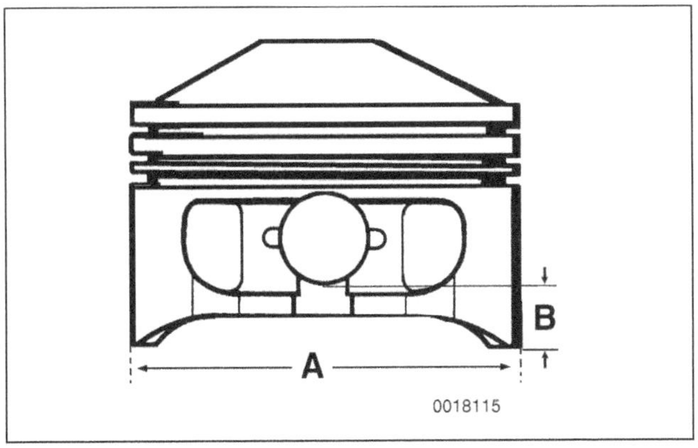

Piston and cylinder wear, checking

Check piston for wear by measuring across piston skirt (**A**).

- Measure Mahle® pistons 18 mm (0.71 in.) from bottom (**B**).
- Measure KS® pistons 10 mm (0.39 in.) from bottom (**B**).

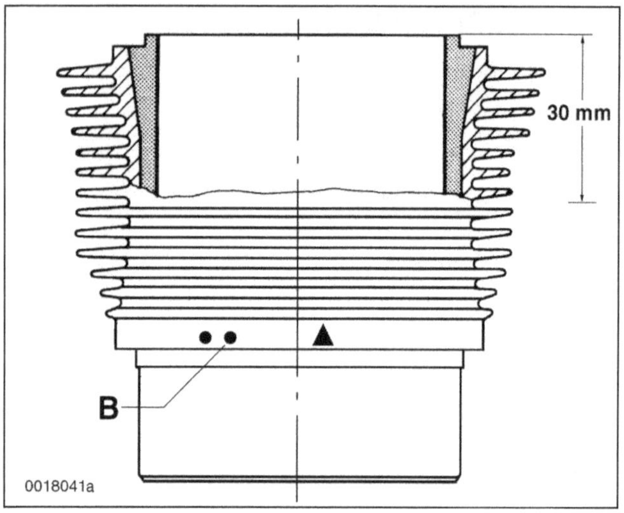

◀ Check cylinder for wear by measuring bore 30 mm (1.18 in.) below upper edge.

– Replace cylinders if measured diameter exceeds 0.08 mm (0.0031 in.) from installed diameter.
  - Determine installed diameter by checking code **B** vs. specifications in **Table e** or **Table f**.

– On 1978 - 1981 engine: Replace pistons and cylinders if running clearance exceeds 0.15 mm (0.059 in.)

– On 1982 - 1983 engine: Replace pistons and cylinders if running clearance exceeds 0.12 mm (0.0047 in.)

Table d. Mahle® piston and cylinder sizes

Code (B)	Cylinder diameter	Piston diameter
0	95.000 – 95.007 mm (3.7401 – 3.7404 in.)	94.965 – 94.975 mm (3.7387 – 3.7391 in.)
1	95.007 – 95.014 mm (3.7404 – 3.7407 in.)	94.972 – 94.982 mm (3.7390 – 3.7394 in.)
2	95.014 – 95.021 mm (3.7407 – 3.7409 in.)	94.979 – 94.989 mm (3.7393 – 3.7397 in.)
3	95.021 – 95.028 mm (3.7409 – 3.7413 in.)	94.986 – 94.996 mm (3.7395 – 3.7399 in.)

Table e. KS® piston and cylinder sizes

Code (B)	Cylinder diameter	Piston diameter
0	95.000 – 95.007 mm (3.7401 – 3.7403 in.)	94.963 – 94.977 mm (3.7391 – 3.7393 in.)
1	95.007 – 95.014 mm (3.7403 – 3.7405 in.)	94.970 – 94.984 mm (3.7393 – 3.7395 in.)
2	95.014 – 95.021mm (3.7405 – 3.7407 in.)	94.977 – 94.991 mm (3.7395 – 3.7397 in.)
3	95.021– 95.028 mm (3.7407 – 3.7409 in.)	94.984 – 94.998mm (3.7397 – 3.7399 in.)

New piston and cylinder clearance
- Mahle® . . . . . . 0.025 - 0.042 mm (0.00098 - 0.00165 in.)
- KS®. . . . . 0.023 - 0.044 mm (0.00091 - 0.00173 in.)

Piston rings

– Check and adjust piston ring end gaps if reusing old pistons and cylinders.

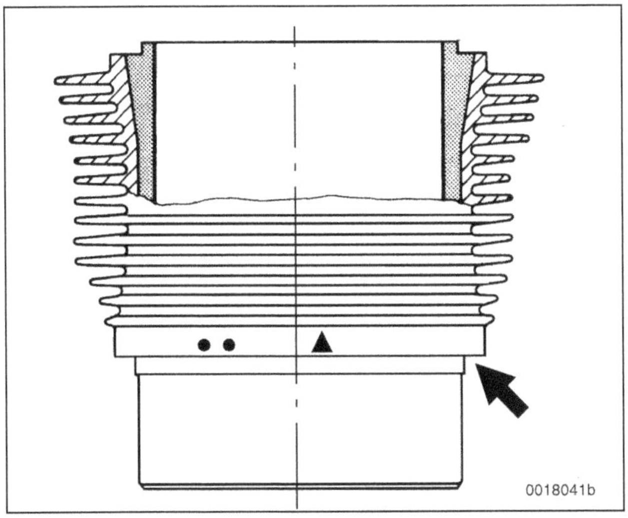

Measure ring end clearance by pushing ring into bottom of cylinder to approximate height of cylinder base gasket surface (**arrow**).

Piston ring side and end clearance tolerances are given in **Table g** and **Table h**.

Table f. Piston ring end gap clearance

Year (ring)	Gap at installation	Wear limit
1978 - 1981:		
Compression 1 & 2	0.15 - 0.30 mm (0.006 - 0.012 in.)	0.8 mm (0.031 in.)
Oil control (scraper)	0.15 - 0.30 mm (0.006 - 0.012 in.)	1.0 mm (0.039 in.)
Oil control (3-piece)	0.40 - 1.40 mm 0.016 - 0.055 in.)	2.0 mm (0.078 in.)
1982 - 1983:		
Compression 1	0.20 - 0.40 mm (0.008 - 0.016 in.)	0.8 mm (0.031 in.)
Compression 2	0.20 - 0.40 mm (0.008 - 0.016 in.)	1.0 mm (0.039 in.)
Oil control	0.30 - 0.60 mm (0.012 - 0.024 in.)	2.0 mm (0.078 in.)

Table g. Piston ring side clearance

Ring (groove)	Clearance	Wear limit
Compression 1	0.070 – 0.102 mm (0.0027 – 0.0040 in.)	0.2 mm (0.0078 in.)
Compression 2	0.040 – 0.072 mm (0.0016 – 0.0028 in.)	0.2 mm (0.0078 in.)
Oil control	0.020 – 0.052 mm (0.0008 – 0.0020 in.)	0.1 mm (0.0039 in.)

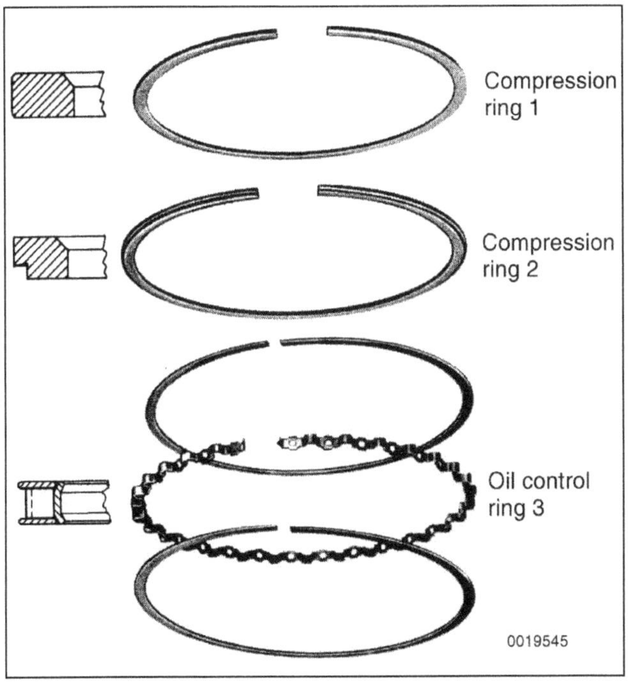

Install rings on piston according to illustration. (3-piece oil-control ring shown.)

150 Cylinder Heads

TABLES

GENERAL

This repair group provides reconditioning information for the cylinder heads, including valvetrain reconditioning specifications.

Cylinder head repair and reconditioning require that the engine be removed from the car and disassembled on an appropriate engine stand. Engine disassembly is described in **102 Engine Disassembly and Assembly**.

> ***CAUTION —***
> *The information contained in the repair group is intended to be used as a reconditioning guide for the professional, experienced automotive technician or machine shop operator.*

Special tools

Some of the procedures require special tools.

Special Tools

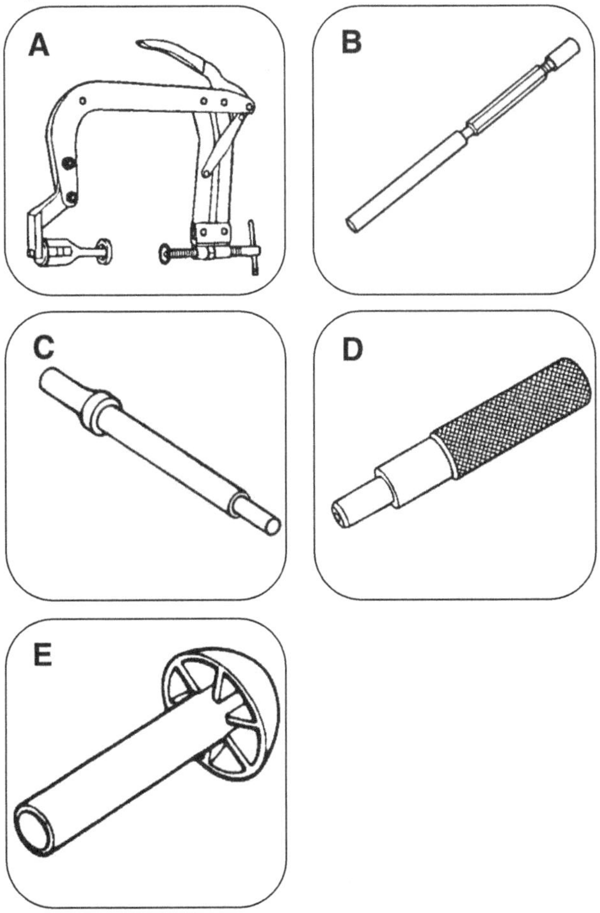

Special tool

- **A** 1020 (source: Baum Tools Unlimited) Valve spring compressor
- **B** P 206 . . . . Valve guide go/nogo gauge (Porsche ordering #000 721 206 00)
- **C** 87a (source: Baum Tools Unlimited) . Valve guide removal drift
- **D** 83 (source: Baum Tools Unlimited) Valve guide installation drift
- **E** 10-204 (source: Baum Tools Unlimited)Valve stem seal installer

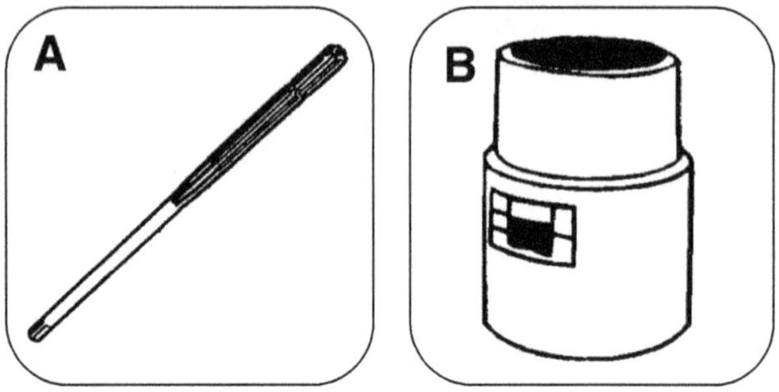

Special tool

- **A** 78HSS . . . . 9 mm valve guide reamer (source: Baum Tools Unlimited)
- **B** P 10 c. . . . . Valve spring height gauge (Porsche ordering #000 721 010 30)

CYLINDER HEADS

All six cylinder heads are identical.

Cylinder head assembly

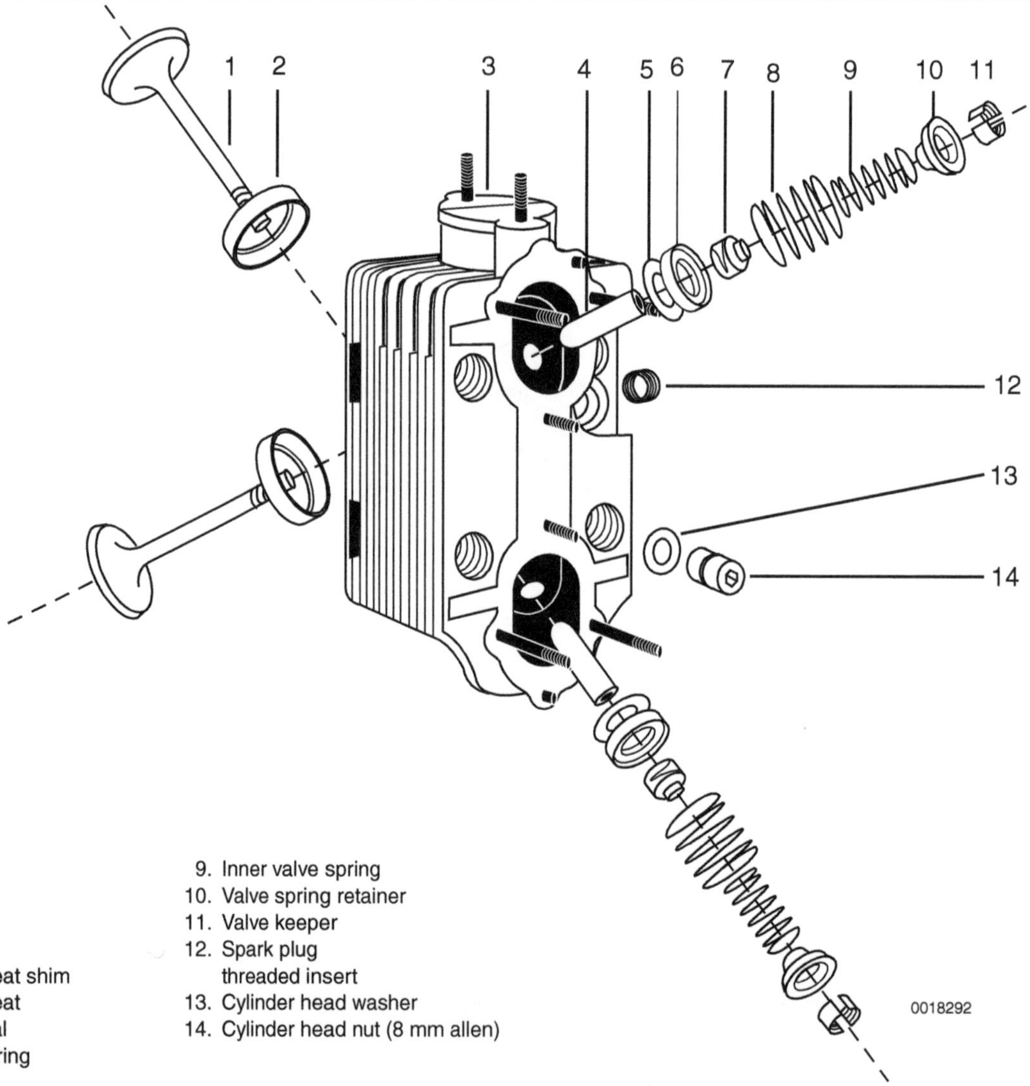

1. Valve
2. Valve seat
3. Cylinder head
4. Valve guide
5. Valve spring seat shim
6. Valve spring seat
7. Valve stem seal
8. Outer valve spring
9. Inner valve spring
10. Valve spring retainer
11. Valve keeper
12. Spark plug threaded insert
13. Cylinder head washer
14. Cylinder head nut (8 mm allen)

Cylinder head service

0018711

◄ Remove and install valve springs using Porsche special tool P 200 a or equivalent.

NOTE —

Keep parts for each valve assembly (keepers, springs, spring retainer, etc.) together.

– Check sealing surfaces on cylinder head prior to cleaning. If any head studs were found to be broken during disassembly, check carefully for damage to castings caused by blow-by.

– Wash cylinder head and clean thoroughly by bead blasting.

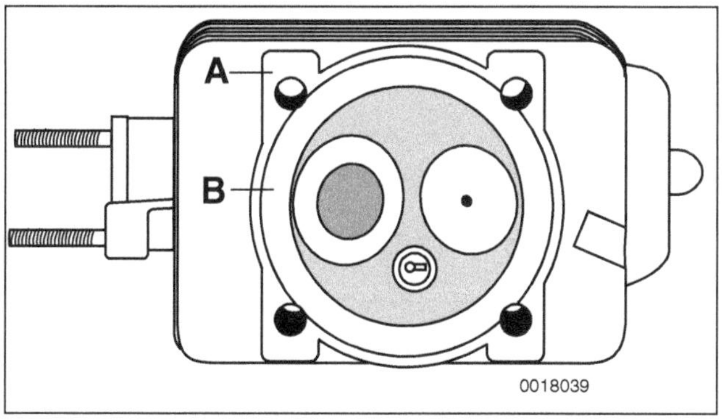

0018039

◄ Cylinder heads can be machined, providing the following rules are adhered to:

- Mill surfaces **A** and **B** equal amounts.
- Mill all three heads on each side an equal amount.
- Bevel edges of machined surfaces slightly.
- Mark cylinder head with "– 25", indicating that head was machined.

Cylinder head machining limit

- Mill surfaces **A** and **B** . 0.25 ± 0.02 mm (0.01 ± 0.001 in.)

CAUTION —

- *Cylinder head surface may only be machined once.*
- *When assembling an engine with machined cylinder heads, be sure to use thicker 0.50 mm (0.02 in.) cylinder base gaskets, available from an authorized Porsche parts department.*

VALVES

Valve guide, checking

– Check valve guides for wear.

NOTE —
Special tool P 206 can be used for a quick and easy check of valve guide wear.

Valve guide to valve stem clearances
- Intake valve .... 0.15 mm (0.006 in.)
- Exhaust valve .... 0.20 mm (0.008 in.

– If valve stem to guide clearance exceed tolerance, valve guide must be replaced.
- Valve guide and cylinder head bore diameters are listed in **Table a**.
- Refer to illustration for guide to dimensions listed.

Valve guide, replacing

– Working from camshaft side, machine away top of valve guide flush with top of cylinder head.

– Using special valve guide drift 87a or equivalent, dislodge each valve guide with sharp blow from a hammer.

– Press guide out through combustion chamber.

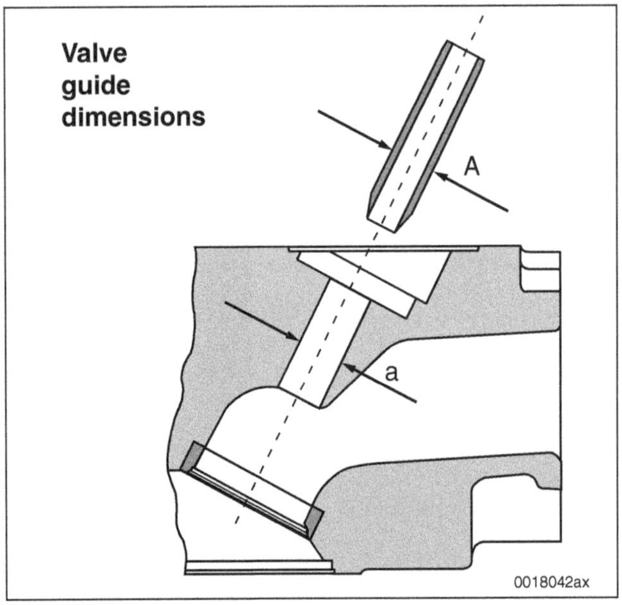

◄ Measure guide bores in cylinder head. Machine oversized replacement valve guides so that valve guide diameter is 0.06–0.09 mm (0.0023–0.0035 in.) larger than valve guide hole.

NOTE —
When the valve guides are removed, the bores in the cylinder head will most likely be widened. Replacement valve guides from Porsche are oversized and must be machined (turned on a lathe) so that an interference fit is obtained.

Table a. Valve guide and cylinder head bore diameters

Valve guide	Guide outside diameter (A)	Cylinder head bore inside diameter (a)
Standard size	13.060 mm (0.5142 in.)	13.000 - 13.018 mm (0.5118 - 0.5125 in.)
Oversize 1	13.260 mm (0.5220 in.)	13.000 - 13.200 mm (0.5118 - 0.5197 in.)
Oversize 2 (1978 - 1981 only)	13.460 mm (0.5299 in.)	13.201 - 13.400 mm (0.5197 - 0.5276 in.)

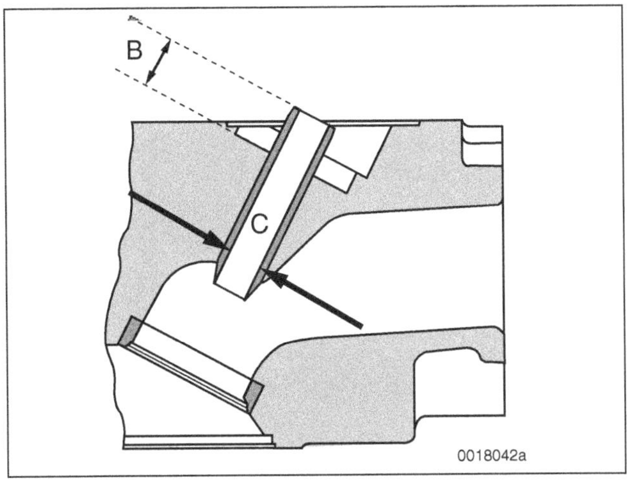

Apply lubricant (tallow-based) to valve guide and press new guides in to specified installed height using special tool 83 or equivalent (9 mm stepped drift).

Valve guide installed height

- Height above cylinder head (**B**) . . . . 13.2 mm (0.520 in.)

– Ream new valve guides with special tool 78HSS (9 mm).

Valve guide inner dimension

- Reamed diameter (**C**) 9.00 - 9.015 mm (0.354 - 0.355 in.)

Valve seat, checking

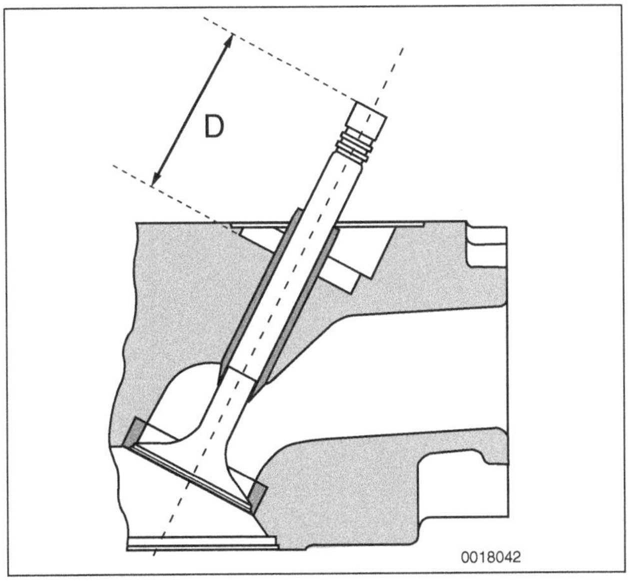

Place valve in cylinder head and measure distance between top of valve stem and valve spring seat (shims removed).

Valve seat check

- Valve seat checking height **D** . . . . 46.0 ± 0.3 mm (1.811 ± 0.011 in.)

– If dimension **D** exceeds specifications, recheck with new valve. If measurement is still out of specifications, replace valve seat or cylinder head.

Valve seat, replacing

- Use rough cut stone to carefully grind away valve seat until seat becomes loose in cylinder head. Remove seat.

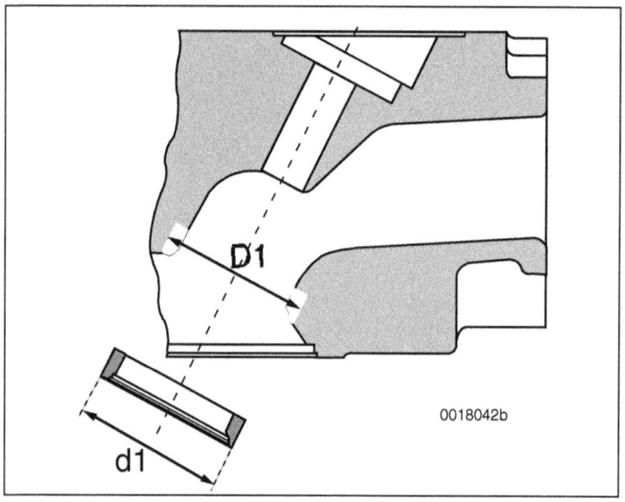

- Measure valve seat outer diameter **d1**. Measure cylinder head bore diameter **D1**.
- Machine replacement valve seats so that seat diameter is larger than valve seat hole according to fit specifications. See **Table b**.
- Heat cylinder head slowly to 200°C (400°F).
- Install valve seat using appropriate drift.
- Allow cylinder head to slowly cool to room temperature.
- Reheat cylinder head to 200°C (400°F) and maintain temperature for 2 hours.
- Allow cylinder head to slowly cool to room temperature.

Table b. Valve seat dimensions

Valve Seat	Valve seat diameter (d1)	Cylinder head bore diameter (D1)
Standard size		
Intake seat	51.680 - 51.661mm (2.0346 - 2.0339 in.)	51.500 - 51.530 mm (2.0276 - 2.0287 in.)
Exhaust seat	44.200 - 44.184 mm (1.7402 - 1.7395 in.)	44.000 - 44.025 mm (1.7323 - 1.7333 in.)
Oversize 1		
Intake seat	52.000 - 51.981mm (2.0472 - 2.0465 in.)	51.820 - 51.850 mm (2.0401 - 2.0413 in.)
Exhaust seat	44.760 - 44.744 mm (1.7622 - 1.7616 in.)	44.560 - 44.585 mm (1.7543 - 1.7553 in.)
Valve seat to cylinder head clearance (interference fit):		
Intake valve	–0.14 to –0.18 mm (–0.0055 to –0.0071 in.)	
Exhaust valve	–0.16 to –0.20 mm (–0.0063 to –0.0078 in.)	

Valve seat, cutting

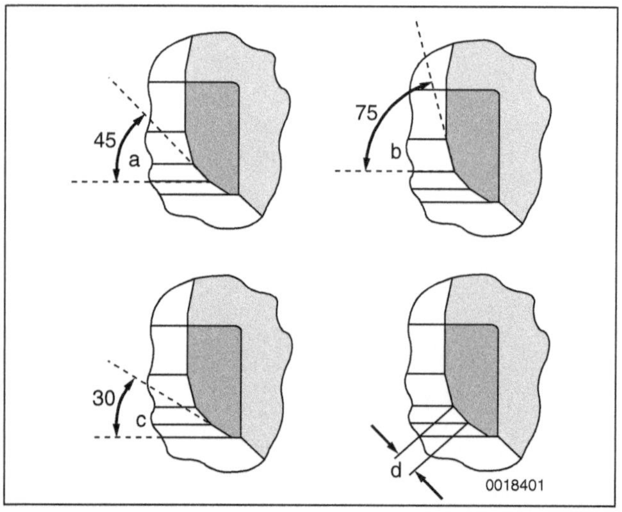

Use illustration and **Table c** as guide in cutting valve seat.

Table c. Valve seat cutting guide

	Intake valve/exhaust valve
Valve seat angle (**a**)	45°
Valve seat lower edge (**b**)	75°
Valve seat upper edge (**c**)	30°
Seat width (**d**): Intake Exhaust	1.3 ± 0.1 mm (0.051 ± 0.004 in.) 1.5 ± 0.1 mm (0.059 ± 0.004 in.)

Valves

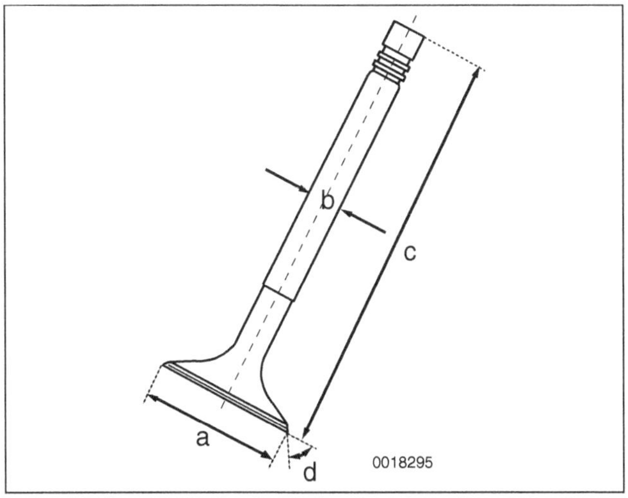

Use illustration and **Table d** for valve specifications and measurements.

Table d. Valve specifications

Dimension	Intake	Exhaust
Valve head diameter (**a**)	49.00 ± 0.10 mm (1.929 ± 0.004 in.)	41.5 ± 0.10 mm (1.634 ± 0.004 in.)
Valve stem diameter (**b**)	8.97 - 0.012 mm (0.3531 - 0.0004 in.)	8.95 - 0.012 mm (0.3523 - 0.0004 in.)
Total valve length (**c**)	110.1 ± 0.25 mm (4.335 ± 0.010 in.)	108.4 ± 0.25 mm (4.268 ± 0.010 in.)
Valve seat angle (**d**)	45°	

Valve spring height, setting

Keep each set of valves, springs, keepers and spring retainers with the cylinder head that they came out of. Check both inner and outer springs for damage. The outer valve spring is progressively wound and should be installed with the tighter coils resting on the cylinder head. The inner spring can be installed either way.

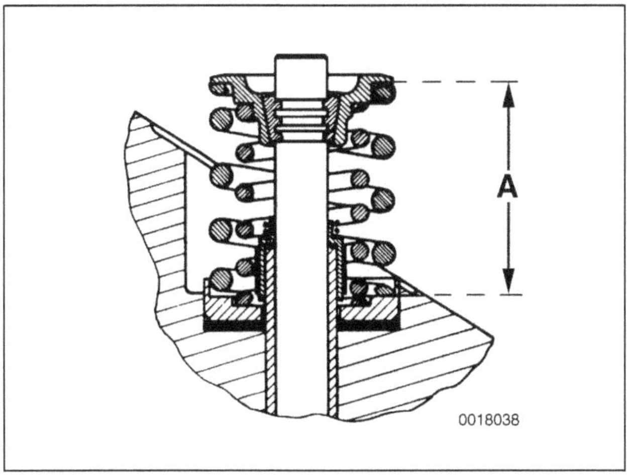

Use Porsche special tool P 10 c or equivalent to measure distance **A**.

- Adjust valve spring height by adding or subtracting shims under valve spring.
- Correct valve spring height to specifications listed.

Table e.

Year/model	911 T 2.4	911 E 2.4
Intake valve	35 .0 ± 0.3 mm (1.3779 i± 0.001in.)	34.0 ± 0.3 mm (1.3385 ± 0.001in.)
Exhaust valve	35.0 ± 0.3 mm (1.3779 i± 0.001in.)	34.0± 0.3 mm (1.3385 ± 0.001in.)
Year/model	911 S 2..4	911 2.7
Intake valve	35.5 ± 0.3 mm (1.3976 ± 0.001in.)	35 .0 ± 0.3 mm 1.3779 i± 0.001in.)
Exhaust valve	34.5 ± 0.3 mm (1.3582 ± 0.001in.)	35.5 ± 0.3mm (1.3976 ± 0.001in.)

151 Camshafts and Camshaft Cases

TABLES

Special Tools

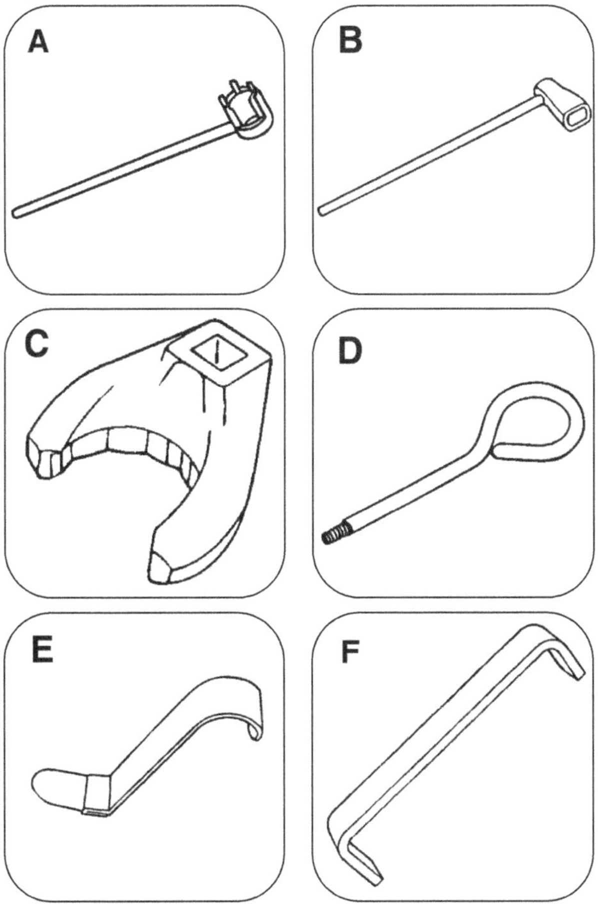

Special tool

A 9191 . . . . . Cam sprocket counterholding wrench (1982 - 1983) (Porsche ordering #000 721 919 10)
B P 202 . . . . . Camshaft counterholding tool (1978 - 1981) (Porsche ordering #000 721 202 00)
C P 203 . . . . . 46 mm camshaft nut tool (1978 - 1981 (Porsche ordering #000 721 203 00)
D P 212 . . . . . Camshaft sprocket lockpin tool (Porsche ordering #000 721 203 00)
E P 213 b. . . . . . Valve adjusting feeler gauge (Porsche ordering #000 721 213 20)
F P 214 . . . . . Spring-loaded tensioner lock (1978 - 1981) (Porsche ordering #000 721 214 00)

GENERAL

This section covers information and repairs to the camshaft cases, camshafts, rocker shafts and rocker arms. Engine disassembly, including camshaft and camshaft case removal, is covered in **102 Engine Disassembly and Assembly**.

Timing chain replacement or adjustment is covered in **152 Camshaft Timing Chains**.

Special tools

Some of the procedures require special tools.

CAMSHAFTS

Camshafts and rocker shafts can be removed from the camshaft cases with the engine installed.

Camshafts, removing and installing

– Raise car and support in a safe manner.

> ***WARNING —***
> *Make sure the car is stable and well supported at all times. Use a professional automotive lift or jack stands designed for the purpose. A floor jack is not adequate support.*

– Disconnect negative (–) battery cable and cover battery terminal to keep cable from accidentally contacting terminal.

> ***WARNING —***
> *Battery electrolyte is a corrosive acid. Always wear eye protection when working on or near the battery.*

– Remove muffler as described in **260 Exhaust System**. Remove rear engine metal shroud.

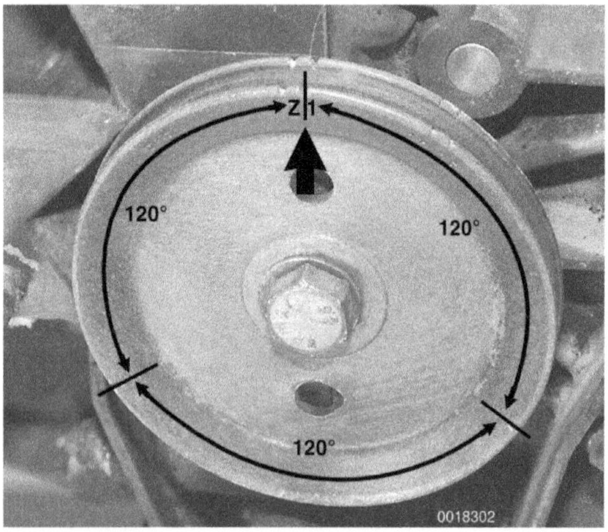

◄ Set crankshaft to **Z1** cylinder 1 compression TDC (**arrow**).

NOTE—
At cylinder 1 compression TDC, the intake and exhaust valves for cylinder 1 will be closed.

◄ Confirm cylinder 1 compression TDC by checking that distributor rotor points to scribe mark on distributor housing (**arrow**).

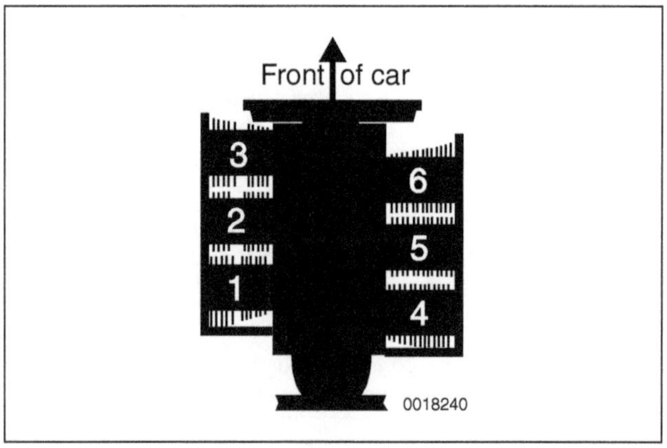

◄ Note cylinder numbering.

NOTE—
The firing order is 1 - 6 - 2 - 4 - 3 - 5.

– Remove intake and exhaust valve covers on both sides.

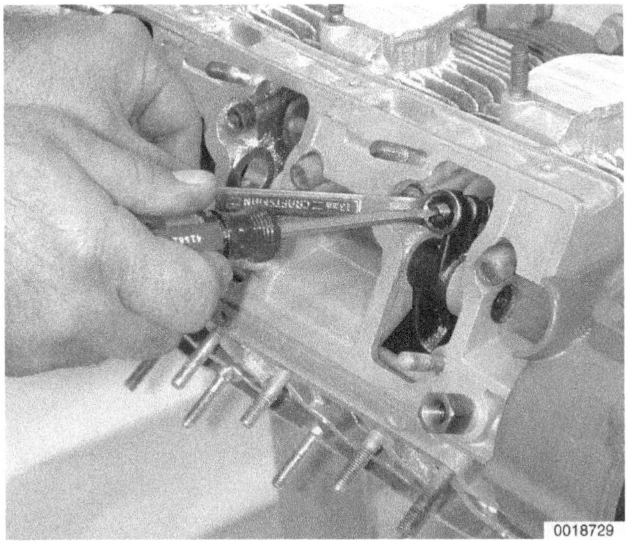

◀ Loosen cylinder 1 valve adjuster lock nuts and fully back out valve adjusting screws.

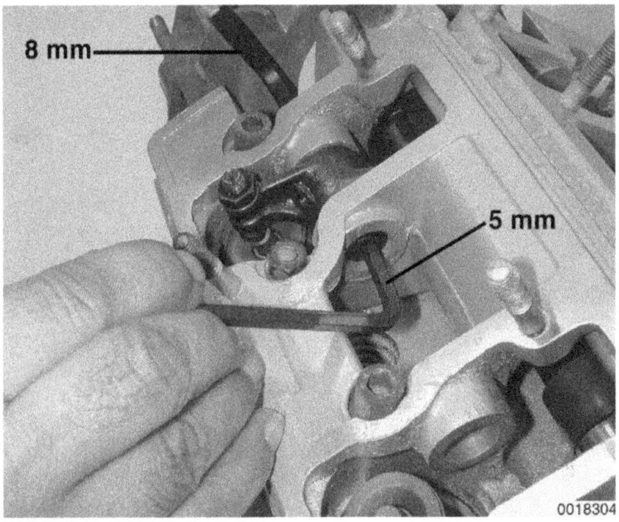

◀ Counterhold cylinder 1 intake rocker shaft expansion nut with 8 mm allen wrench and loosen other end with 5 mm allen wrench.

NOTE —
Mark location of each rocker arm. Do not interchange used parts.

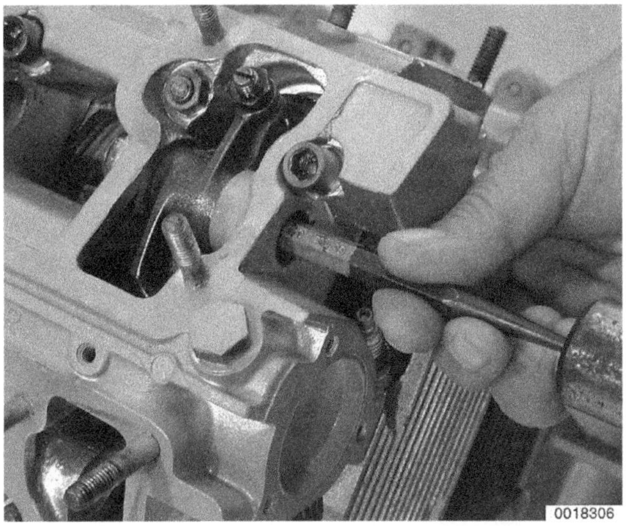

◀ Tap rocker shaft out, removing rocker arm.

– Turn crankshaft 120° in normal direction of rotation.

- Loosen and back off valve adjusting screws fully.
- Remove cylinder 6 rocker shafts and arms.

– Following firing order, turn crankshaft 120°, back off appropriate valve adjusters, and remove rocker shafts and arms for next cylinder. Repeat until all rocker shafts and arms are removed.

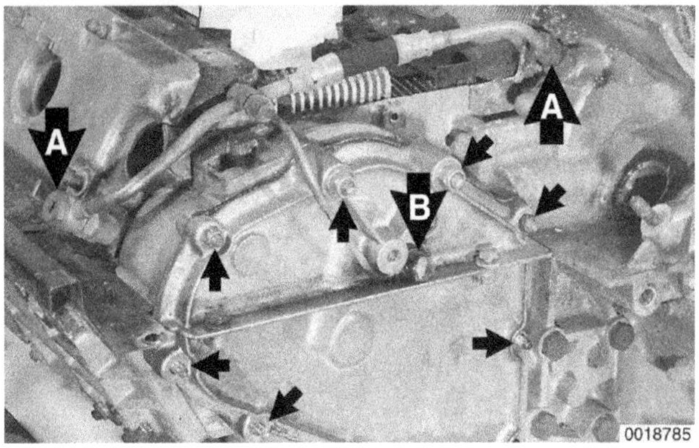

If equipped with Carrera type (oil-pressure-fed) timing chain tensioners: Remove left camshaft and timing chain tensioner oil feed line pressure fittings (**A** and **B**).

- Remove left chain housing cover fasteners (**small arrows**), and remove cover.

Loosen left camshaft sprocket:

- On 1978 - 1980 models: Use Porsche special tools P 202 and P 203 or equivalent to counterhold camshaft and loosen sprocket mounting nut.
- On 1981 - 1983 models: Use Porsche special tool 9191 or equivalent to counterhold sprocket while loosening sprocket bolt.

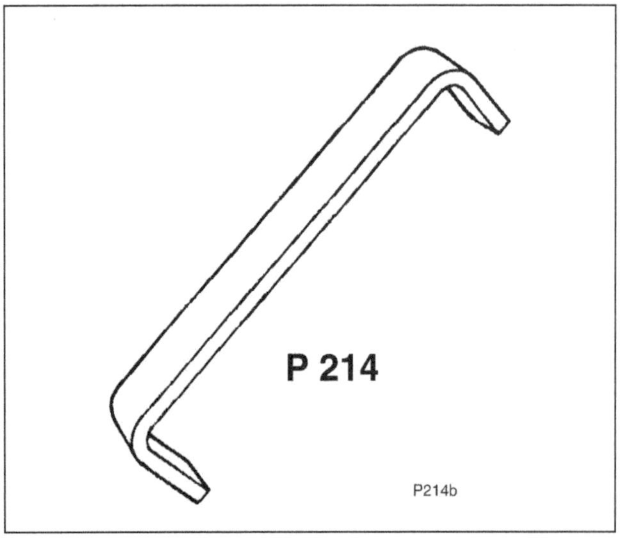

If equipped with spring-loaded timing chain tensioner:

- Compress spring-loaded chain tensioner and lock down using Porsche special tool P 214 or equivalent,
- Remove tensioner mounting nut. Remove tensioner.

◀ If equipped with oil-pressure-fed tensioner:

- Compress timing chain tensioner using large pliers.
- Lock tensioner down using 3 mm (1/8 in.) pin through hole in tensioner body (**1**).
- Special tool P 204 (**2**) can be used to brace idler sprocket before removing tensioner.

NOTE —

Check timing chain tensioner piston height before compressing. If piston is fully extended, carefully inspect timing chains and sprockets for wear. See **152 Camshaft Timing Chains**.

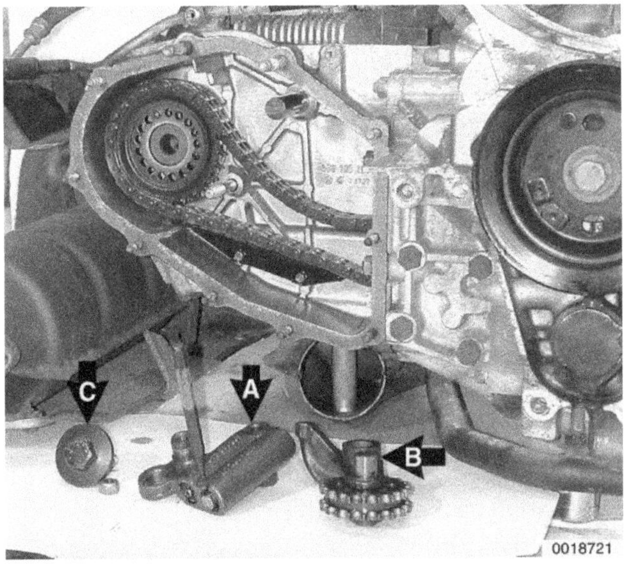

◀ Remove timing chain tensioner (**A**) and idler sprocket assembly (**B**).

– Remove camshaft sprocket fastener from left camshaft (**C**).

◀ Remove camshaft sprocket lockpin using Porsche special tool P 212 or equivalent.

– Remove camshaft sprocket and disengage from timing chain.

– Remove camshaft sprocket flange.

Camshaft sprocket assembly (1982 - 1983)

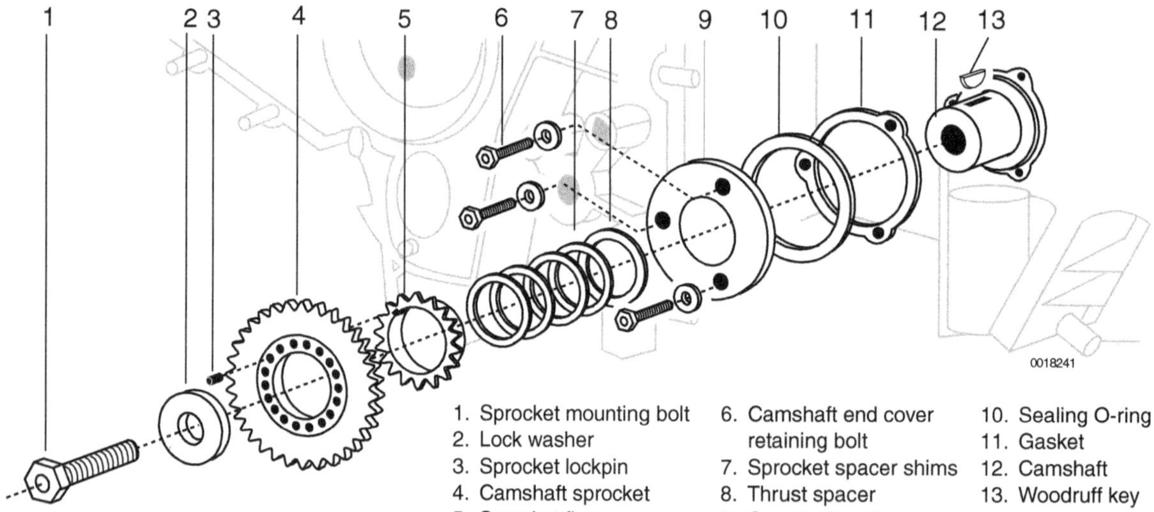

1. Sprocket mounting bolt
2. Lock washer
3. Sprocket lockpin
4. Camshaft sprocket
5. Sprocket flange
6. Camshaft end cover retaining bolt
7. Sprocket spacer shims
8. Thrust spacer
9. Camshaft end cover
10. Sealing O-ring
11. Gasket
12. Camshaft
13. Woodruff key

Remove left camshaft:

- Remove woodruff key (**A**) from camshaft.
- Remove camshaft end cover retaining bolts (**arrows**).
- Remove camshaft end cover (**B**), shims and thrust spacer.
- Slide camshaft toward pulley end of engine to remove.

NOTE —
Wire tie end cover, shims and thrust spacer together and label "left".

— Repeat procedures on right side camshaft.

— To reinstall, slide left camshaft into camshaft case.

- Install camshaft end cover with new O-ring and gasket.
- Install camshaft end cover retaining bolts and tighten.

— Reinstall left side cam sprocket spacer shims, woodruff key and camshaft sprocket flange.

— Put left timing chain over camshaft sprocket.

- Slide sprocket and lock washer over sprocket flange.
- Install sprocket mounting fastener finger tight.

— Install right camshaft, camshaft sprocket assembly and timing chain as described in previous steps.

Camshaft sprocket orientation

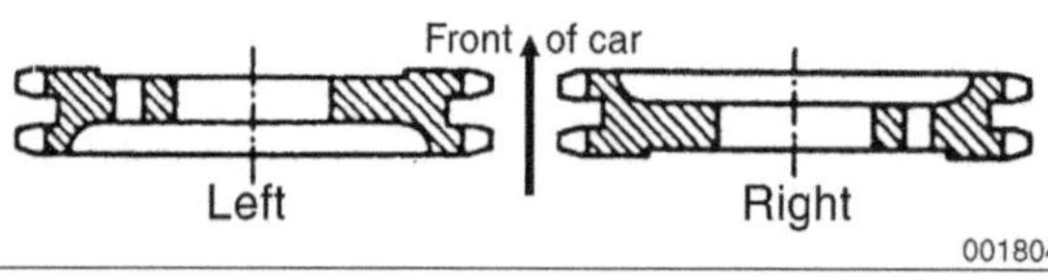

NOTE —
Left and right camshaft sprockets are identical. Dished side of sprocket must face pulley side of engine on left camshaft (cylinders 1 - 3) and flywheel side of engine on right camshaft (cylinders 4 - 6).

— Check that cam sprockets run parallel with intermediate shaft sprocket as described later in this section.

NOTE—
Under most normal circumstances, if reinstalling previously used camshafts and sprockets, 3 sprocket spacer shims will be necessary under the left sprocket and 4 shims under the right sprocket.

– Install left timing chain tensioner and idler sprocket assembly.
  - Remove locking 3 mm (1/8 in.) pin from chain tensioner, if applicable.

◄ Set crankshaft to **Z1**. Make sure it is at cylinder 1 compression TDC by aligning distributor rotor with scribe mark on distributor housing (**arrows**).

◄ Turn camshafts so that punch marks or "930" (**black arrows**) point up.

– For initial timing adjustment:
  - Remove left camshaft sprocket mounting fastener and lock washer.
  - In this position one hole in camshaft sprocket will be aligned with one hole in sprocket flange. Insert camshaft sprocket lockpin.
  - Reinstall sprocket retaining fastener finger tight.

– Repeat initial timing procedures for right camshaft.

– Turn crankshaft clockwise two full turns and recheck initial timing.

> ***CAUTION—***
> *Be sure to set precise camshaft timing as described in* **152 Camshaft Timing Chains**.

Torque camshaft sprocket mounting fasteners when camshaft timing settings are correct:

- On 1978 - 1981 models: Use Porsche special tools P 202 and P 203 or equivalent to counterhold camshaft and torque sprocket mounting nut.
- On 1982 - 1983 models: Use Porsche special tool 9191 or equivalent to counterhold sprocket while torquing sprocket bolt.

Tightening torque

- Chain sprocket to camshaft
  1978 - 1980 models (46 mm nut) .... 150 Nm (111 ft-lb)
  1981 - 1983 models (M12 bolt)....... 120 Nm (89 ft-lb)

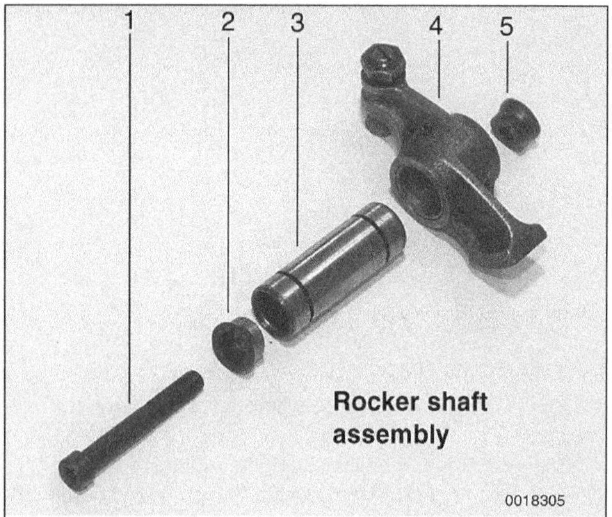

1. Pinch bolt (5 mm allen)
2. Expansion sleeve
3. Rocker shaft
4. Rocker arm
5. Expansion nut (8mm allen)

Install rocker shafts and arms for each cylinder with camshafts set for TDC compression for that cylinder.

- Install cylinder 1 and 3 rocker shafts with 5 mm allen bolt facing cylinder 2.
- Install cylinder 4 and 6 rocker arms with 5 mm bolt facing cylinder 5.
- Cylinder 2 and 5 rocker shaft can be installed either way.

– Tap rocker shaft into bore until shaft is flush with inner edge of support.

– Place rocker arm in position and tap rocker shaft in until shaft slides into far side support.

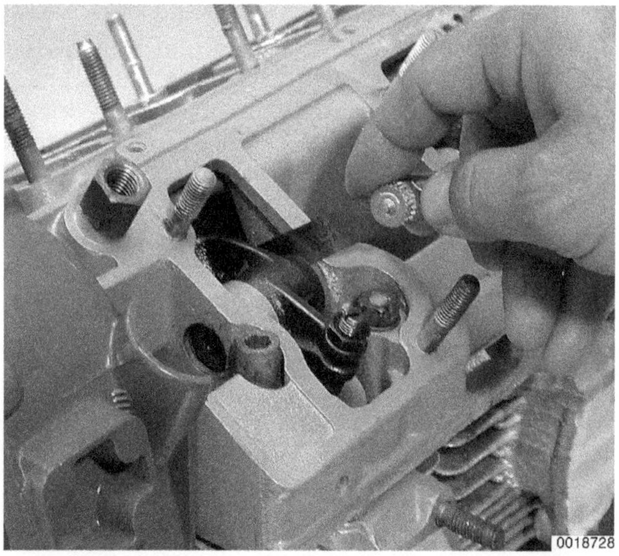

Place 0.5 mm (0.02 in.) feeler gauge between far side support and rocker arm.

- Continue to tap shaft until feeler gauge engages slot in rocker shaft.

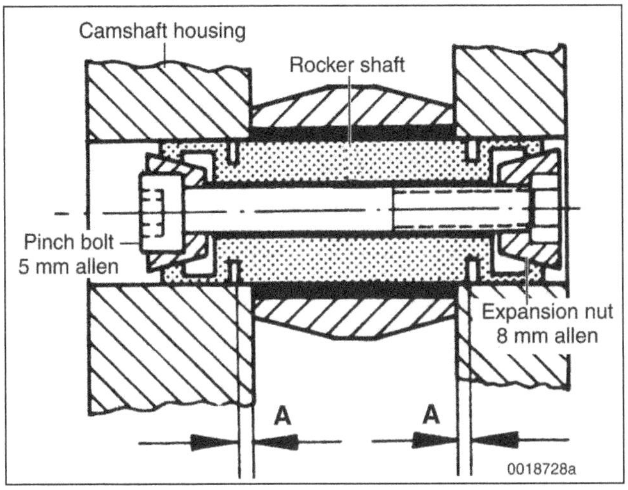

◀ Remove feeler gauge and tap rocker arm shaft into position so that shaft is centered in camshaft housing.

- Grooves in rocker shaft body should be inset approximately 1.5 mm (0.06 in.) into camshaft housing (distance **A**).
- Tighten rocker arm shaft pinch bolt.

Tightening torque
- Rocker shaft pinch bolt . . . . 15 Nm (11 ft-lb)

◀ Adjust all intake and exhaust valves to 0.1 mm (0.004 in.) as described in **030 Maintenance**.

– Install valve covers with new gaskets, sealing washers and lock nuts.

Tightening torque
- Valve cover to camshaft case . . . . 10 Nm (89 in-lb)

– Install timing chain housing covers with new gaskets, sealing washers and lock nuts.

Tightening torque
- Chain housing cover to housing . . . . 10 Nm (89 in-lb)

– If applicable, install right and left camshaft oil feed lines with new sealing washers.

– Install rear engine metal shroud.

– Install muffler, using a new gasket and hardware.

Timing chain sprocket parallelism

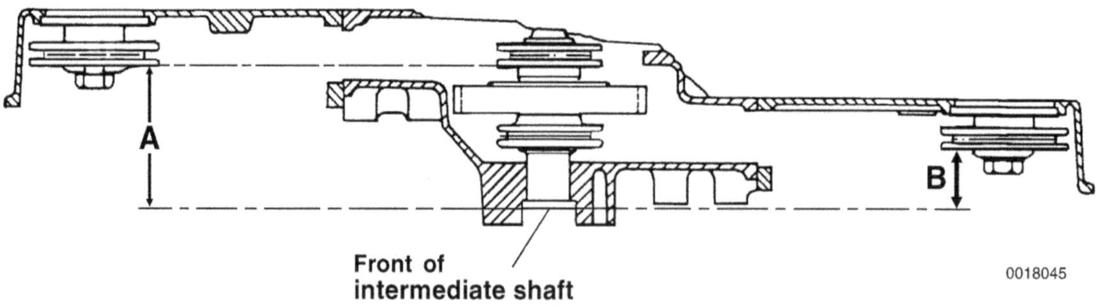

A Factory distance from end of intermediate shaft to face of cylinder 1 - 3 (left) timing chain sprocket = 98.07 ± 0.25 mm (3.86 ± 0.01 in.)
B Factory distance from end of intermediate shaft to face of cylinder 4 - 6 (right) timing chain sprocket = 43.27 ± 0.25 mm (1.70 ± 0.01 in.)

Timing chain sprockets, checking parallelism

The timing chain sprockets must be aligned parallel to each other to prevent premature wear of the timing chain and chain components.

If new camshafts or sprockets are installed, camshaft sprocket parallelism must be checked. Use vernier caliper and a quality straight edge to check sprocket depth in relation to the intermediate shaft. If adjustment is needed, shims can be added or subtracted between cam sprockets and camshafts.

NOTE —
Normally, when removing camshaft sprockets, 3 spacer shims will be found under the left sprocket hub and 4 shims under the right sprocket hub.

Valve overlap (930/04 engine)

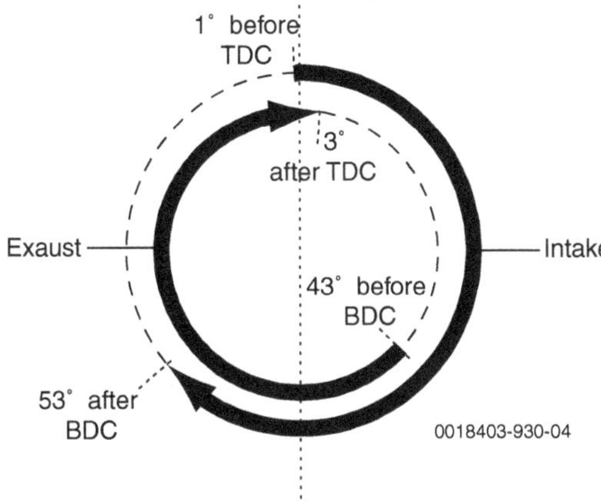

Valve overlap (930/07-930/16 engines)

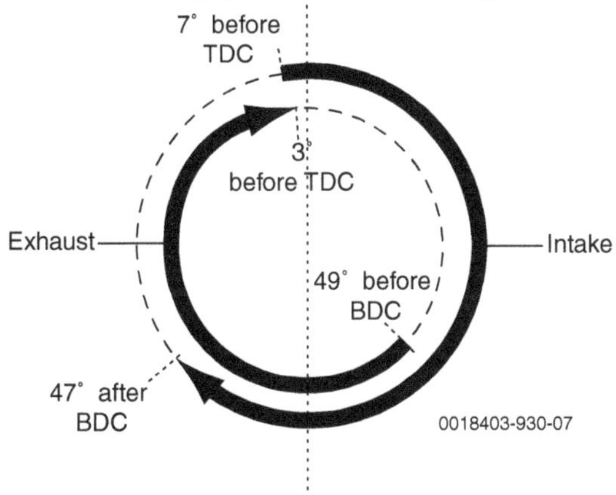

Camshaft specifications

Camshaft timing is adjusted using a dial indicator setup. The dial indicator is used to precisely set the position of each camshaft in relation to partially open ("overlapped") valves.

NOTE —

- *Camshaft timing specification in degrees of rotation are given below.*
- *Camshaft timing adjustment is covered in* **152 Camshaft Timing Chains**.

930/04 engine valve overlap

- Intake valve opens .1° before TDC
- Intake valve closes . . . . 53° after BDC
- Exhaust valve opens. . . . . 43° before BDC
- Exhaust valve closes . . . . 3° after TDC

930/07 and 930/16 engines valve overlap

- Intake valve opens .7° before TDC
- Intake valve closes . . . . 47° after BDC
- Exhaust valve opens. . . . . 49° before BDC
- Exhaust valve closes . . . . .3° before TDC

Camshaft and valve train specifications are given in **Table a**.

Table a. Camshaft and valvetrain clearances

Measuring point	Tolerance	Wear limit
Camshaft bearing bore	48.967 - 48.992 mm (1.9278 - 1.9288 in.)	0.10 mm (0.004 in.)
Camshaft journal	48.926 - 48.942 mm (1.9262 - 1.9268 in.)	0.10 mm (0.004 in.)
Camshaft run-out	0.02 mm (0.0008 in.)	
Camshaft axial play	0.150 - 0.200 mm (0.059 - 0.079 in.)	0.40 mm (0.016 in.)
Rocker shaft	17.992 - 18.000 mm (0.7083 - 0.7086 in.)	
Camshaft housing bore (for rocker shaft)	18.000 - 18.018 mm (0.7086 - 0.7093 in.)	
Rocker arm bore	18.016 - 18.027 mm (0.7093 - 0.7097 in.)	0.080 mm (0.0031 in.)
Rocker arm axial play	0.100 - 0.350 mm (0.004 - 0.014 in/)	0.50 mm (0.020 in.)

CAMSHAFT CASES

The camshaft cases can only be removed with the engine out of the vehicle. See **102 Engine Disassembly and Assembly**.

Camshaft oil spray tube, replacing

The left and right camshaft cases each contain an oil spray tube for valvetrain lubrication. On high mileage engines and engines that have had catastrophic bearing failure, the tubes should be removed and cleaned prior to engine assembly.

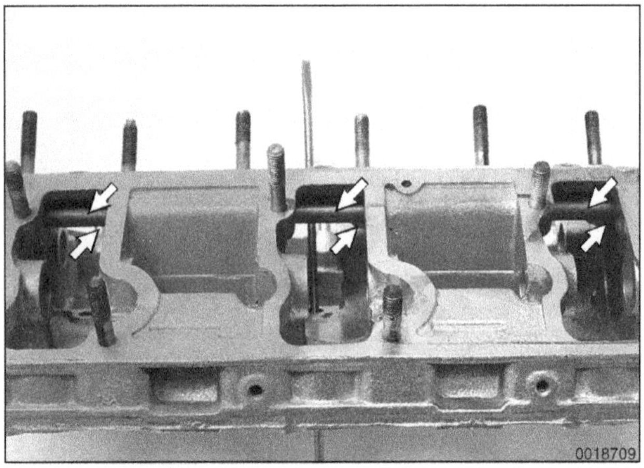

With camshaft case removed and cleaned, blow compressed air in each camshaft tower oil feed line hole. Check oil spray tube orifices (**arrows**) for air flow.

If a blocked hole is found, drill 5 mm (0.2 in.) hole approximately 9 mm (0.35 in.) deep in oil passage plug (**arrow**).

- Using an M6 tap, thread drilled hole.
- Pull out plug with 6 mm screw.
- Remove oil spray tube and clean thoroughly.
- Reinstall tube and replace end plug.

> ***CAUTION —***
> *Oil spray tubes must always be removed and cleaned thoroughly when engine bearings have failed.*

152 Camshaft Timing Chains

Special Tools

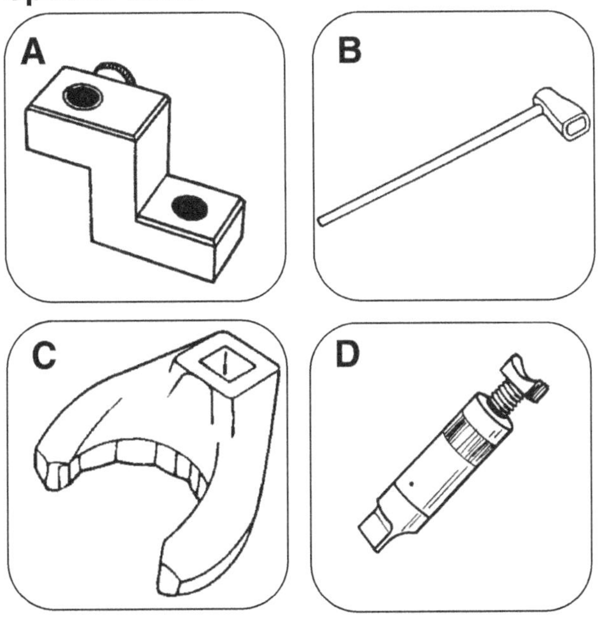

Special tool
A P 207 . . . . Dial indicator bracket
(Porsche ordering #000 721 207 00) . . . .
B P 202 . . . . Camshaft counterholding tool
(Porsche ordering #000 721 202 00)
C P 203 . . . . 46 mm camshaft nut tool
(Porsche ordering #000 721 203 00)
D P 204 . . . . Chain idler sprocket brace
(source: Baum Tools Unlimited)

Continued on page 2

GENERAL

This section covers repairs to the camshaft timing chain assembly. Timing chain replacement or adjustment can be carried out with the engine installed in the engine bay.

Special tools

Most of the procedures in this section require special tools.

Camshaft timing chains

The original Porsche 911 timing chains are continuous one-piece chains, without master links. Replacing the one-piece chain with another of the same design requires that the engine be removed and completely disassembled and the crankcase split. As an alternative, replacement chains with master links are available from Porsche. The master link makes it possible to replace a timing chain with the engine installed.

Replacing the one-piece timing chain with the engine removed is covered in **102 Engine Disassembly and Assembly**. When performing an engine overhaul, the continuous chain is the recommended replacement.

Camshaft timing

Setting camshaft timing correctly is essential after the timing chain assembly has been disassembled for any reason. Precise setting of camshaft timing results in smooth, vibration-free engine operation.

> ***CAUTION—***
> *There is a real danger of valve heads contacting piston tops during normal engine operation unless camshaft timing is set precisely.*

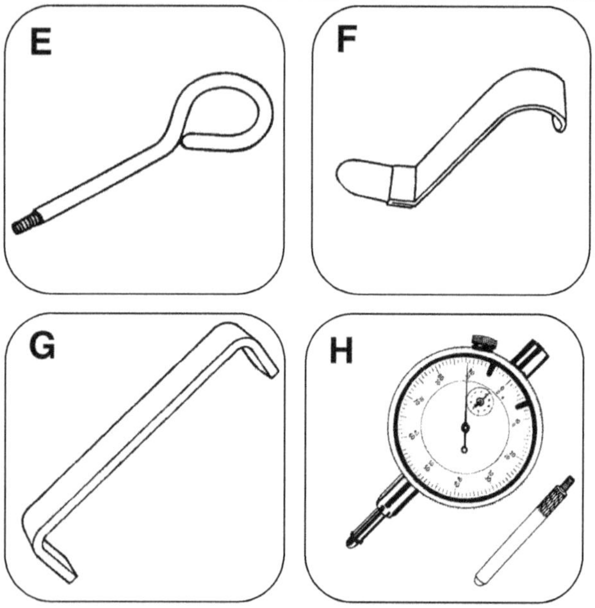

Special tool

- **E** P 212 .... Camshaft sprocket lockpin tool (Porsche ordering #000 721 203 00)
- **B** P 213 b.................... Valve adjusting feeler gauge (Porsche ordering #000 721 213 20)
- **G** P 214Lock-down tool for spring-loaded tensioner (1978 - 1981) (Porsche ordering #000 721 214 00)
- **H** Baum 20A and 21 ....Dial indicator and extension (source: Baum Tools Unlimited)

Timing chain sprocket parallelism

The sprockets on the left and right camshafts must be set parallel to the corresponding intermediate shaft sprockets. Timing chain replacement will not disturb this setting, but camshaft replacement should be followed by a resetting of timing chain sprocket parallelism.

Uneven wear of the plastic timing chain guides may indicate faulty parallelism in the sprocket assembly.

Information for setting sprocket parallelism is covered in **151 Camshafts and Camshaft Cases**.

0019637

Timing chain tensioners

SC engines were originally equipped with mechanical timing chain tensioners. This type of tensioner is spring-loaded and oil dampened. Engine oil splash fills a reservoir in the bottom of the tensioner. This dampens tensioner piston movement and prevents vibration and shaking of the timing chain.

In 1984 Porsche introduced an oil-pressure-fed (hydraulic) chain tensioner in the Carrera models. It is possible to update the early tensioners to the later. The timing chain housing covers and oil feed lines must also be updated.

CAMSHAFT TIMING CHAINS

Timing chains, replacing (engine installed)

– Disconnect negative (–) battery cable and cover battery terminal to keep cable from accidentally contacting terminal.

> ***WARNING —***
> *Battery electrolyte is a corrosive acid. Always wear eye protection when working on or near the battery.*

– Raise car and support in a safe manner.

– Remove spark plugs so engine will turn easily.

> ***WARNING —***
> *Make sure the car is stable and well supported at all times. Use a professional automotive lift or jack stands designed for the purpose. A floor jack is not adequate support.*

– Remove muffler as described in **260 Exhaust System**. Remove rear engine shroud.

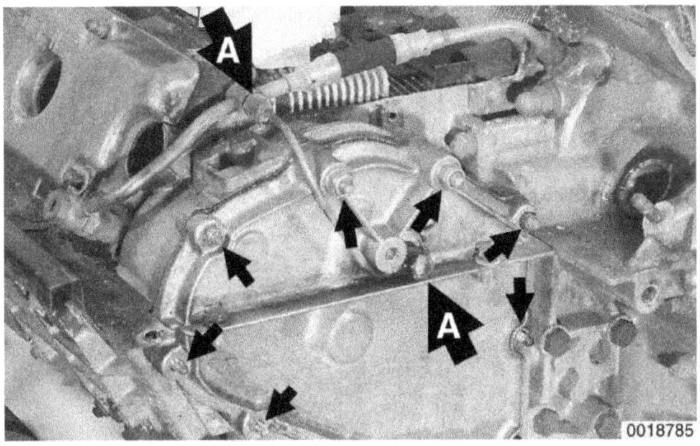

◄ Remove left chain housing cover fasteners (**small arrows**), and remove cover.

- If equipped with Carrera type (oil-pressure-fed) timing chain tensioners: Remove left timing chain tensioner oil feed line pressure fittings (**A**).

– Repeat steps for right chain housing cover.

◄ Loosen left camshaft sprocket:

- Use Porsche special tools P 202 and P 203 or equivalent to counterhold camshaft and loosen sprocket mounting nut.

– Repeat steps for right camshaft sprocket.

< Set crankshaft pulley to cylinder 1 compression TDC by lining up **Z1** mark with mark on fan housing.

- Check to see that punch marks or "930" (**black arrows**) on camshaft ends point *up*.

NOTE —
On 1982 - 1983 models: Remove camshaft sprocket mounting bolts in order to see the "930".

< Rotate crankshaft counterclockwise until punch mark or "930" (**arrow**) on end of left camshaft points approximately 45° outward.

NOTE —
When the camshaft mark points 45° outwards, no valves are completely open and the crankshaft can be rotated safely without the pistons contacting the valves.

– Temporarily reinstall camshaft sprocket bolt finger tight, if applicable

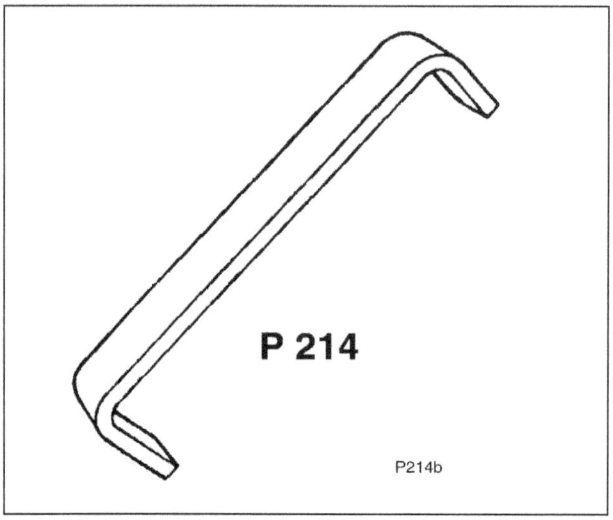

< If equipped with original equipment (spring-loaded) chain tensioner:

- Compress tensioner and lock down using Porsche special tool P 214 or equivalent,
- Remove tensioner mounting nut. Remove tensioner.
- Remove idler sprocket.

◀ If equipped with oil-pressure-fed chain tensioner:

- Use special tool P 204 (**1**) to brace idler sprocket before removing tensioner.
- Compress tensioner using large pliers.
- Lock tensioner down using 3 mm (1/8 in.) pin through hole in tensioner body (**2**).

NOTE —

- *Check timing chain tensioner piston height before compressing. If piston is fully extended, carefully inspect timing chains sprockets for wear.*
- *Compressing the tensioner may take a little time. Oil flow out of the tensioner is restricted to a small oil relief hole.*
- *If the tensioner collapses easily it is most likely faulty.*

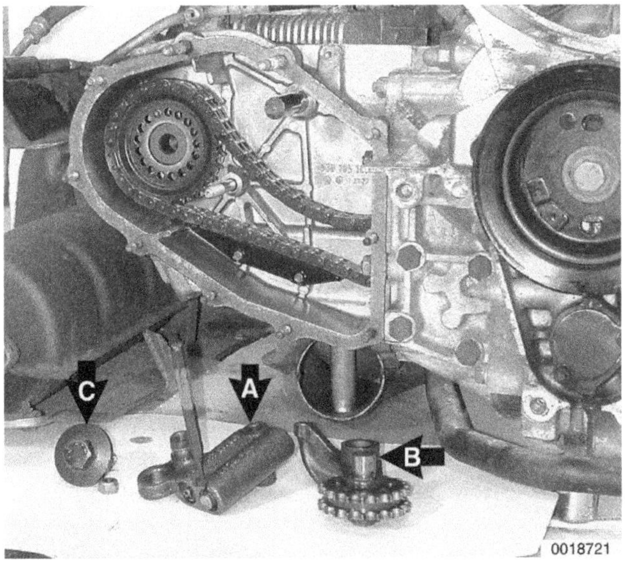

◀ If equipped with oil-pressure-fed chain tensioner:

- Remove timing chain tensioner (**A**) and idler sprocket assembly (**B**).
- Remove camshaft sprocket nut from left camshaft (**C**).

◀ Remove camshaft sprocket lockpin using Porsche special tool P 212 or equivalent.

– Remove camshaft sprocket and disengage sprocket from chain.

NOTE —

If necessary, the plastic chain rail can be removed by prying with a large screwdriver.

– Grind rivet heads off one chain link and remove link.

> ***CAUTION —***
>
> *Place a shop towel in the crankcase opening to prevent metal filings from falling into crankcase.*

– Attach new chain with connecting link to end of old chain.

– Rotate crankshaft slowly in direction of rotation (clockwise as viewed from pulley end), keeping tension on chain at all times.

NOTE—

When rotating the crankshaft, make sure the valves do not contact the pistons. If any resistance is felt, reverse rotation direction of the crankshaft slightly. Then rotate the camshaft enough to avoid contact and continue the operation.

– Rotate engine until master link can be installed on new chain. Remove old chain.

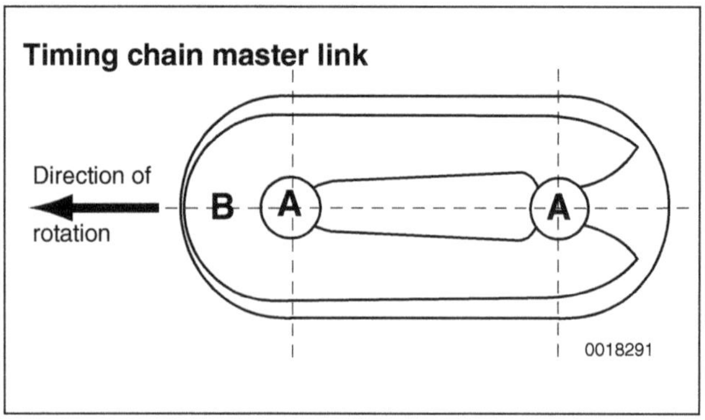

◄ Install master link. Connecting pins (**A**) should come in from back side of chain (flywheel side).

- Install spacers.
- Closed end of lock clip (**B**) must point in direction of rotation.

– Reinstall camshaft sprocket to timing chain. Slide sprocket over sprocket flange.

– Install sprocket mounting fastener finger tight.

– Install chain guide rail, chain tensioner and idler sprocket carrier.

◄ Set crankshaft pulley to cylinder 1 compression TDC by lining up **Z1** mark with mark on fan housing.

- Manually turn left camshaft so that punch mark on camshaft end points *up*.
- Check to see that punch mark on right camshaft end points *up*.

◀ Lock camshaft sprocket to camshaft hub by inserting sprocket lock-pin in hole with best alignment. Reinstall sprocket mounting fastener finger tight.

– Rotate crankshaft forward (clockwise) until punch mark on end of right camshaft points approximately 45° outward.

NOTE—

When the camshaft mark points 45° outwards, no valves are completely open and the crankshaft can be rotated safely without the pistons contacting the valves.

– Repeat replacement steps for right timing chain.

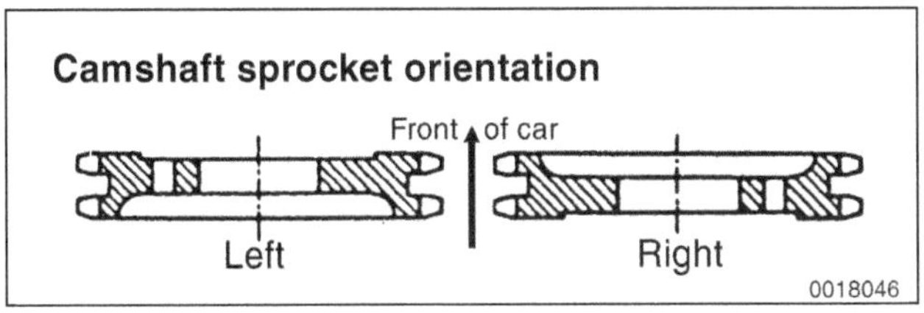

NOTE—

- *Left and right camshaft sprockets are identical. Dished side of sprocket must face pulley side of engine on left camshaft (cylinders 1 - 3) and flywheel side of engine on right camshaft (cylinders 4 - 6).*
- *If timing chains, sprockets or guide rails show uneven or lopsided wear, be sure to check timing chain sprocket parallelism as described in* **151 Camshaft and Camshaft Cases**.

– Adjust camshaft timing as described later.

– Engine reassembly is reverse of disassembly. Use new gaskets, seals and nuts.

Tightening torques
- Chain sprocket to camshaft 46 mm nut. . . . . 150 Nm (111 ft-lb)
- Chain housing cover to chain housing . . . 8 Nm (71 in-lb)

Camshaft timing, checking and adjusting

> ***CAUTION —***
> *Setting camshaft timing by using crankshaft and camshaft marks gives, at best, approximate results. There is a real danger of valve heads contacting piston tops during normal engine operation unless camshaft timing is set precisely.*

Camshaft timing is adjusted using a dial indicator setup. The dial indicator is used to precisely set the position of each camshaft in relation to partially open ("overlapped") valves. Valve overlap comes two strokes (one full crankshaft revolution) after the TDC compression stoke.

– Raise car and support in a safe manner.

> ***WARNING —***
> *Make sure the car is stable and well supported at all times. Use a professional automotive lift or jack stands designed for the purpose. A floor jack is not adequate support.*

– For access to timing chains:
- Remove spark plugs so engine turns easily.
- Remove muffler as described in **260 Exhaust System**.
- Remove rear engine shroud.
- Remove timing chain tensioner oil feed lines, if applicable.
- Remove timing chain housing covers.
- Remove valve covers.

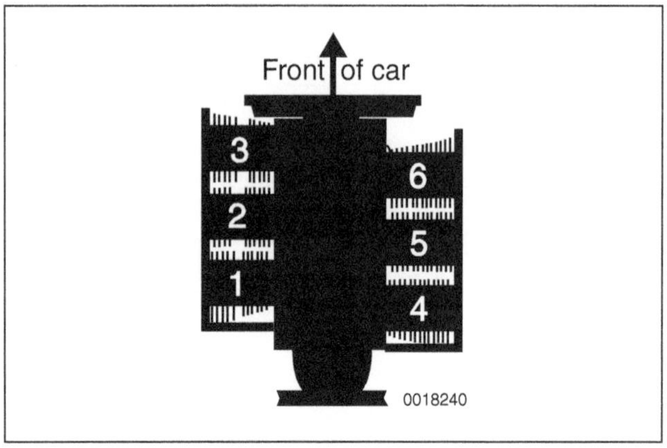

◄ Note cylinder numbering.

◄ Loosen left camshaft sprocket:
- On 1978 - 1980 models: Use Porsche special tools P 202 and P 203 or equivalent to counterhold camshaft and loosen sprocket mounting nut.
- On 1981 - 1983 models: Use Porsche special tool 9191 or equivalent to counterhold sprocket while loosening sprocket bolt.

– Repeat steps for right camshaft sprocket.

Set crankshaft pulley to cylinder 1 compression TDC by lining up **Z1** mark with mark on fan housing.

- Check to see that punch marks or "930" (**black arrows**) on camshaft ends point *up*.

NOTE—

On 1981 - 1983 models: Remove camshaft sprocket mounting bolts in order to see the "930".

Adjust cylinder 1 intake and exhaust valve clearance.

Valve clearance

- Adjusting value . . . . 0.10 mm (0.004 in.)

Using Porsche special tool P 207, mount dial gauge on camshaft case stud next to cylinder 1 intake valve.

- Set dial gauge plunger against valve spring retainer on cylinder 1 intake valve.
- Zero dial gauge with approximately 10 mm (0.4 in.) preload.

– Starting at **Z1** (TDC) mark on pulley, turn crankshaft slowly in clockwise direction approximately 1 full turn while observing dial gauge. Turn crankshaft far enough to reach valve overlap adjusting value.

Valve overlap adjustment

- 930/04 adjusting value . . . . 1.00mm (0.040 in.)
- 930/07 or 930/16 adjusting value . . . 1.55 mm (0.061 in.)

– Remove left timing chain sprocket mounting fastener.

◄ Pull out left camshaft sprocket lockpin with Porsche special tool P 212 or equivalent.

– Turn crankshaft either clockwise or counterclockwise, as necessary, just until **Z1** mark on pulley is aligned accurately with mark on fan housing.

– Reinstall lockpin and install camshaft sprocket fastener finger tight.

– Turn crankshaft clockwise two full turns. Recheck timing adjustment. Value must be within tolerance range.

Valve overlap tolerance

- 930/04 range. . . . . 0.9 - 1.1 mm (0.035 - 0.043 in.)
- 930/07 or 930/16 range . 1.4 - 1.7 mm (0.046 - 0.067 in.)

◄ Torque left camshaft sprocket mounting fastener when settings are correct:

- On 1978 - 1980 models: Use Porsche special tools P 202 and P 203 or equivalent to counterhold camshaft and torque sprocket mounting nut.
- On 1981 - 1983 models: Use Porsche special tool 9191 or equivalent to counterhold sprocket while torquing sprocket bolt.

Tightening torques

- Chain sprocket to camshaft
  1978 - 1980 models (46 mm nut) . . . . 150 Nm (111 ft-lb)
  1981 - 1983 models (M12 bolt). . . . . . . 120 Nm (89 ft-lb)

◄ Set crankshaft pulley to cylinder 4 compression TDC by lining up **Z1** mark with mark on fan housing.

- Check to see that punch marks or "930" (**black arrows**) on camshaft ends point *down*.

– Adjust cylinder 4 intake and exhaust valve clearance.

Valve clearance

- Adjusting value . . . . 0.10 mm (0.004 in.)

– Repeat camshaft timing adjusting procedures on cylinder 4 intake valve as described in steps for left camshaft timing adjustment.

– Reassemble engine using new gaskets, seals and nuts.

Tightening torques

- Chain housing cover to chain housing . . . 8 Nm (71 in-lb)
- Valve cover to camshaft case . . . . 8 Nm (71 in-lb)

CAMSHAFT TIMING CHAIN TENSIONERS

When a timing chain tensioner fails (collapses), usually there will be a rattling noise coming from the timing housing area due to the loose chain. In addition, the engine may show a drop in power output.

Timing chain tensioner, checking and replacing

– Disconnect negative (–) battery cable and cover battery terminal to keep cable from accidentally contacting battery terminal.

> ***WARNING —***
> *Avoid getting tools or clothing near the battery. Battery electrolyte is a corrosive acid. Always wear eye protection when working on or near the battery.*

– Raise car and support in a safe manner.

> ***WARNING —***
> *Make sure the car is stable and well supported at all times. Use a professional automotive lift or jack stands designed for the purpose. A floor jack is not adequate support.*

– Remove muffler. See **260 Exhaust System**. Remove rear engine shroud.

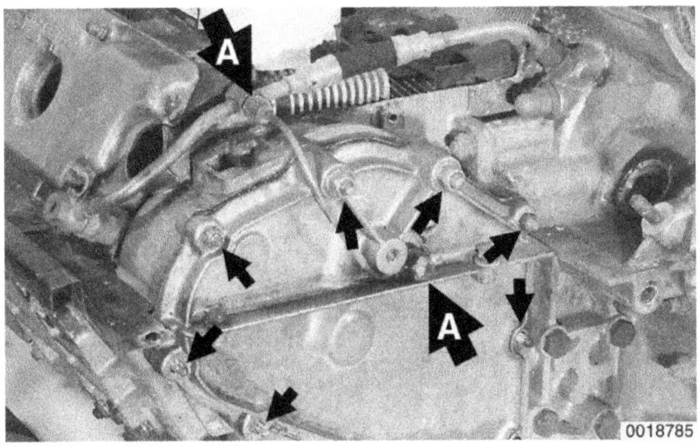

◄ Remove left chain housing cover fasteners (**small arrows**), and remove cover.

- If equipped with Carrera type (oil-pressure-fed) timing chain tensioners: Remove left timing chain tensioner oil feed line pressure fittings (**A**).

– Repeat for right side.

With a large screwdriver, check tensioner by prying down on idler sprocket bracket.

- If tensioner piston moves easily, it is faulty and should be replaced.

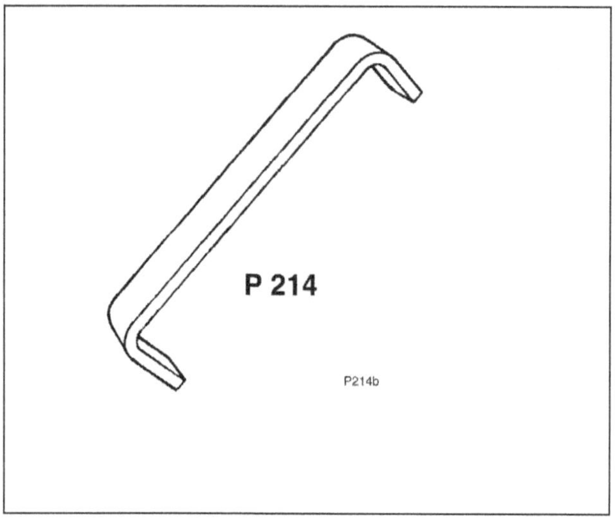

If equipped with original equipment (spring-loaded) chain tensioner:

- Compress tensioner and lock down using Porsche special tool P 214 or equivalent,
- Remove tensioner mounting nut. Remove tensioner.
- Remove idler sprocket.

If equipped with oil-pressure-fed chain tensioner:

- Use special tool P 204 (**1**) to brace idler sprocket before removing tensioner.
- Compress tensioner using large pliers.
- Lock tensioner down using 3 mm (1/8 in.) pin through hole in tensioner body (**2**).

NOTE —

- *Check timing chain tensioner piston height before compressing. If piston is fully extended, carefully inspect timing chains sprockets for wear.*
- *Compressing the tensioner may take a little time. Oil flow out of the tensioner is restricted to a small oil relief hole.*
- *If the tensioner collapses easily it is most likely faulty.*

– Remove timing chain tensioner mounting nut and slip off tensioner.

– Install new tensioner.

NOTE —

- *If the original style (spring-loaded) tensioners are being replaced with the oil-pressure-fed type, it is necessary to also replace the timing chain housing covers and oil feed lines.*
- *Reset camshaft timing as described earlier.*

– Reassembly is reverse of disassembly. Use new gaskets, seals and hardware.

Tightening torque

- Chain housing cover to chain housing . . . 8 Nm (71 in-lb)

170 Engine Lubrication

GENERAL

This section covers the engine lubrication system, including troubleshooting and component replacement.

If the engine is disassembled because of bearing failure, a thorough flushing of the lubrication system must be carried out prior to reassembly. That procedure is also covered here.

NOTE—
- *Oil change and oil filter replacement is covered in* **030 Maintenance**.
- ***Front*** *of engine is the pulley side.* ***Rear*** *of engine is the flywheel side.*

Special tools

Some of the procedures require special tools.

Special Tools

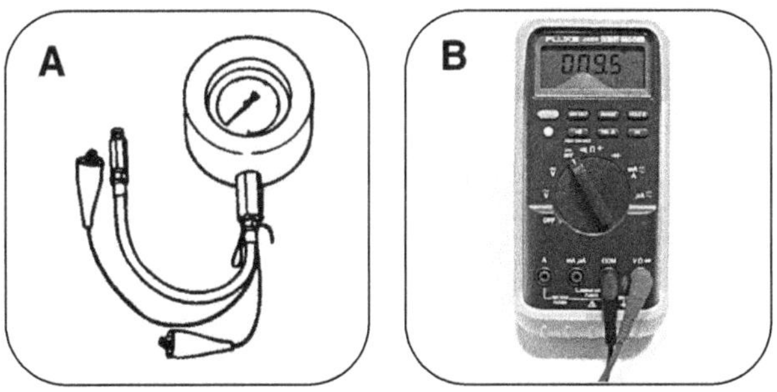

Special tool

A VW 1342 . . . . Oil pressure gauge
B Fluke 87 . . . . Automotive digital multimeter

Lubrication system

The dry sump lubrication system contains the tandem oil pump, oil supply tank, oil cooler, pressure safety system (including oil pressure gauge), oil filter, and connecting oil lines.

The tandem oil pump is in the base of the crankcase. As the name implies, the tandem oil pump contains two pumps. One pump draws oil from the oil tank and pumps it to the main oil circuit and main bearings. The second pump draws circulated oil from the crankcase and delivers it back to the oil tank via the oil filter. Oil pump removal requires that the crankcase halves be separated. This operation is described in **102 Engine Disassembly and Assembly**.

Engine lubrication system schematic

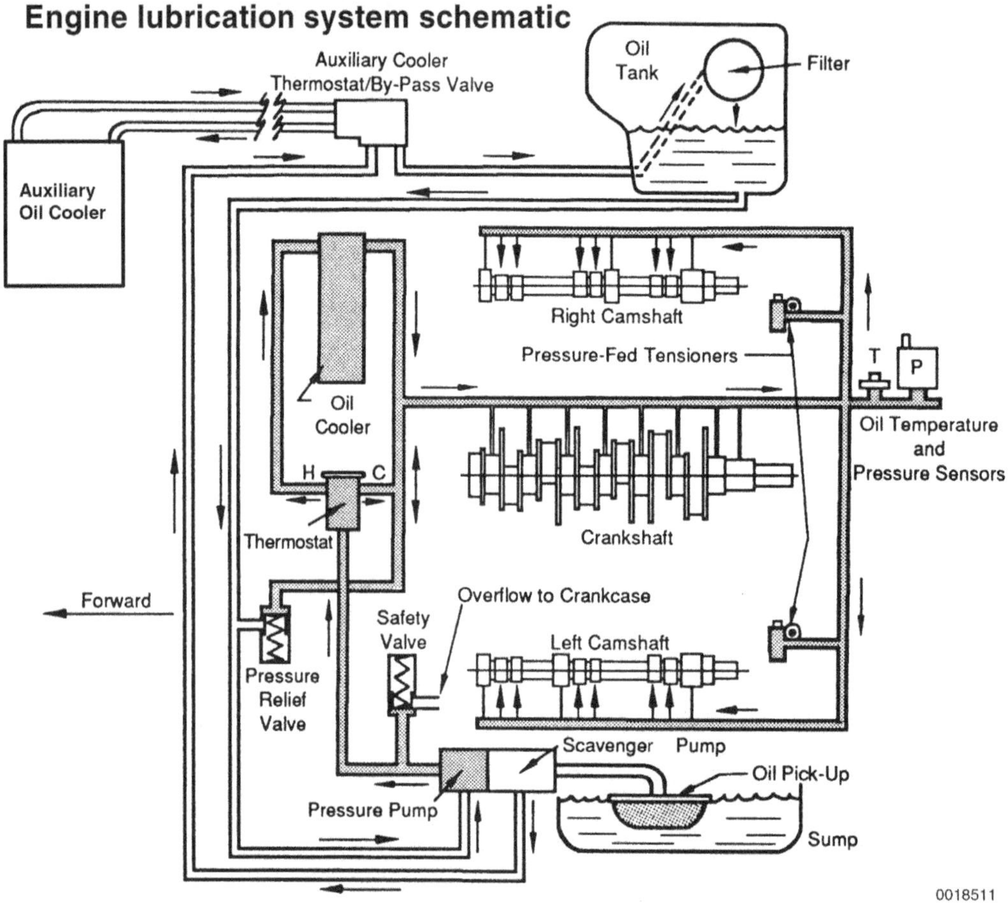

The main bearings, rod bearings and camshaft bearings are supplied with pressurized oil. In addition, the undersides of the pistons are cooled using oil spray jets. Oilways in the camshaft cases provide splash lubrication for the rocker shafts and valve stems.

A thermostat in the main oil circuit regulates the flow of oil to the engine oil cooler. If the oil temperature is below 80°C (176°F), a thermostatically-operated flap is held closed to prevent oil from circulating through the cooler. When the flap is open, oil is circulated through the cooler before reaching the bearings.

An auxiliary oil cooler mounted behind the front right fender provides additional cooling.

Porsche 911 SC engines are equipped with an oil pressure warning system and an oil level warning system to help prevent engine damage. Other safety features include a filter bypass to guard against bursting the filter due to over pressure and three different oil pressure relief valves.

Engine lubrication system components

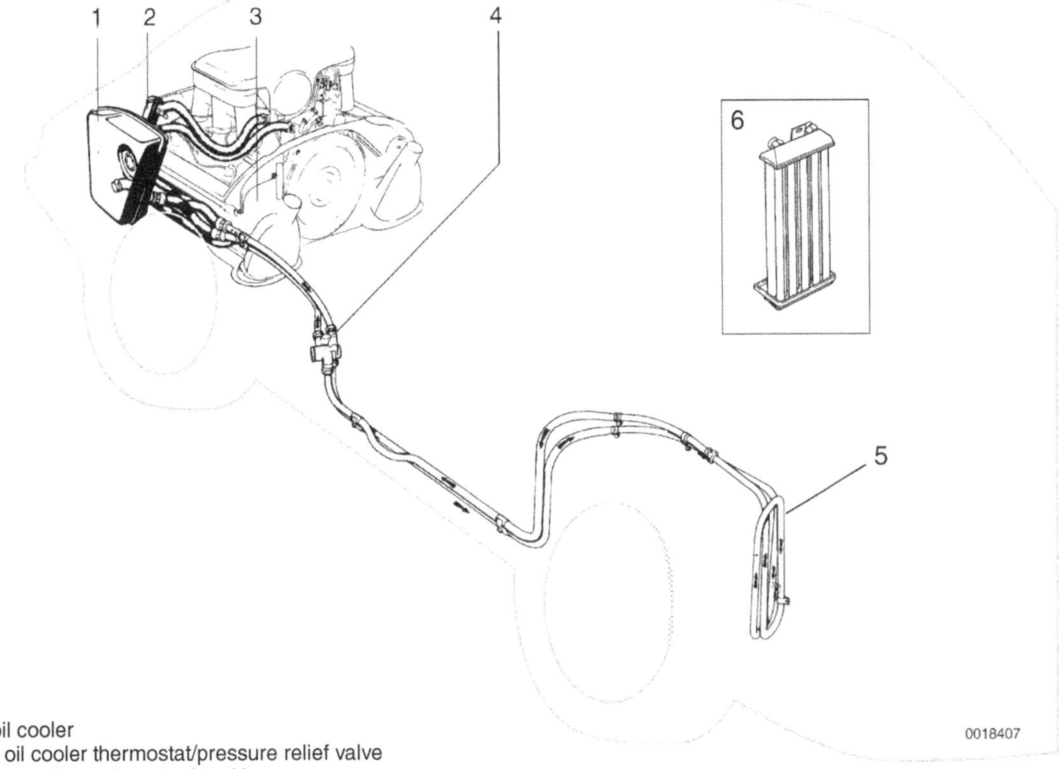

1. Oil tank
2. Oil filler
3. Engine oil cooler
4. Auxiliary oil cooler thermostat/pressure relief valve
5. Tube-type auxiliary oil cooler (stock)
6. Finned auxiliary oil cooler (optional)

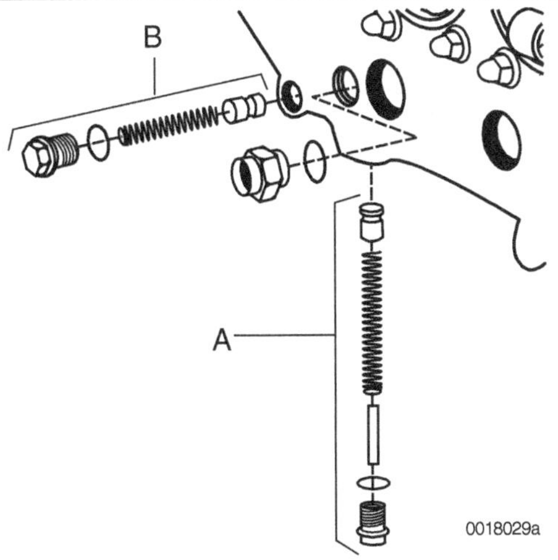

Oil pressure relief valves

The crankcase incorporates two different oil pressure relief valves. The main oil pressure relief valve (**A**) is mounted at the bottom of the crankcase. The safety relief valve (**B**) is mounted in the left side of the crankcase.

NOTE—
The original oil pressure relief valve has been updated with superceded parts. When replacing the valve, be sure to use the newest parts. Consult an authorized Porsche parts department for the latest parts information.

The oil pressure relief valve opens and allows oil to pass directly into the crankcase when oil pressure exceeds 6.2 ± 0.8 bar. (88.2 ± 11.3 psi). The oil pressure relief valve spring is 86 mm (3.39 in.) long.

In the event that the oil pressure relief valve doesn't open when oil pressure is excessive, the safety relief valve will open to prevent damage to the oil cooler, oil filter and oil lines. The oil pressure safety valve spring is 70 mm (2.75 in.) long. The safety valve is calibrated to open at approximately 8.0 bar (113.7 psi).

The auxiliary oil cooler thermostat/pressure relief valve (**arrow**) controls oil flow to the auxiliary oil cooler at the front of the car. The thermostat/pressure relief valve is located in the right rear wheel well in front of the tire.

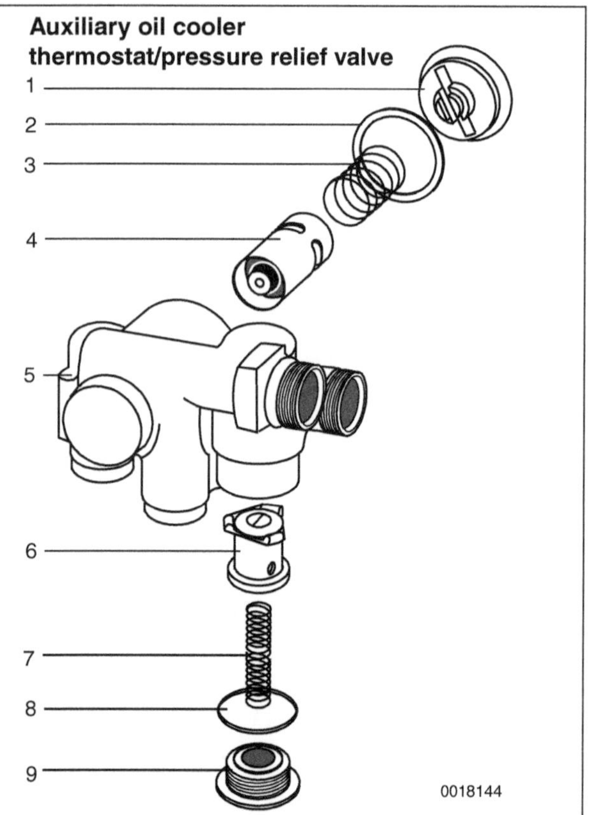

1. End plug
2. Gasket
3. Spring
4. Thermostatic valve
5. Valve housing
6. Pressure relief valve
7. Spring
8. Gasket
9. End plug

The thermostat/pressure relief valve contains two spring-loaded valves (plungers). One valve is thermostatically controlled and opens to allow pressurized oil to enter the front cooler when the temperature of the returning oil exceeds approximately 83°C (113°F). The other valve is pressure controlled and will open to bypass the cooler when the oil pressure is excessive.

TROUBLESHOOTING

Oil pressure, checking

> ***CAUTION —***
> *If the red oil pressure warning light comes on or flashes on while driving, always assume that the oil pressure is low.*

The oil pressure testing procedure given here works well using standard automotive oil pressure testing equipment.

Oil pressure sensor location
- Gauge sensor . . . . on top right of engine

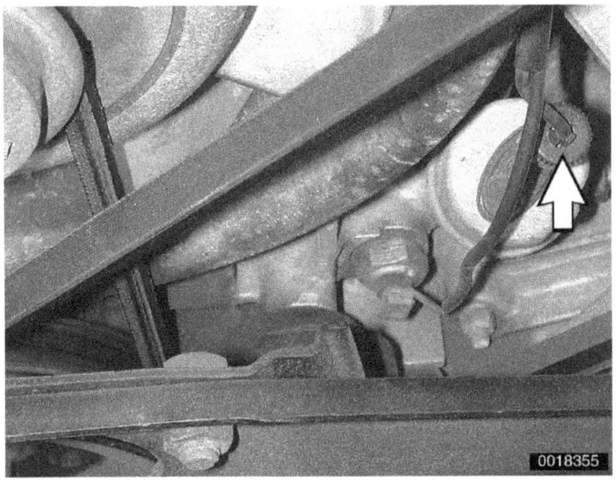

◄ Disconnect harness connector (**arrow**) from oil pressure gauge sensor and remove sensor.

NOTE —
Thoroughly clean around the oil pressure sensor before removing it.

– Install pressure gauge in place of sensor.

NOTE —
The pressure gauge will need a 22 mm male thread adapter.

– With gauge installed, start engine and allow to reach operating temperature.

NOTE —
- *For the most accurate test results, the engine oil (and filter) should be new and of the correct viscosity grade.*
- *As a general guide, oil pressure should rise approximately 10 psi for every 1,000 rpm.*

Oil pressure specifications
- 1978 - 1980. . . . . 4.5 bar (66 psi)
- 1980 - 1983. . . . . 4.0 bar (58 psi)

– Remove pressure gauge and reinstall sensor.
- Lubricate sensor threads prior to installation.

Tightening torque
- Oil gauge sensor to crankcase . . . . 35 Nm (26 ft-lb)

If testing shows low oil pressure, one or more of the following conditions may be indicated:

- Worn or faulty oil pump assembly
- Worn or faulty engine bearings
- Severe engine wear

All of the conditions indicate the need for major repairs.

Lubrication warning system

The lubrication warning system consists of the following:

- Oil pressure warning light
- Oil pressure gauge
- Oil level gauge
- Oil temperature gauge

Oil pressure warning light, testing

When the ignition is turned on, the oil pressure warning light comes on. When the engine is started and the oil pressure rises slightly, the oil pressure warning switch contacts open and the warning light goes out. Make sure the oil level is correct before making tests.

> ***CAUTION—***
> *If the oil warning light remains on while the engine is running, check the oil pressure as described earlier. Do not drive the car until the problem is corrected. The engine may be severely damaged.*

– Turn ignition switch on.

- Warning light on instrument panel must light up.

If warning light does not light when ignition is on, remove connector (**arrow**) from oil pressure warning switch and ground harness connector to a clean metal surface.

- If warning light comes on, check switch as described later.
- If warning light does not come on, wiring to instrument cluster or to light itself is faulty.

– Remove connector from oil pressure switch.

- Warning light on instrument panel must go out.
- If warning light does not go out, wiring to switch is grounded somewhere between switch terminal and warning light. See **970 Electrical Wiring Diagrams**.

– To test switch, connect ohmmeter between switch terminal and ground.

- With engine off, there should be continuity.
- With engine running, oil pressure should open switch and there should be no continuity.

– Replace a faulty switch. Lubricate switch threads during installation.

Tightening torque
- Oil warning switch to crankcase . . . . 20 Nm (15 ft-lb)

Oil level sensor, testing

The oil level sensor is similar in operation to the fuel level sensor in the fuel tank. Check that the oil level is correct before testing the sensor.

The oil level sensor is mounted in the oil tank in the right rear of the engine bay. It is accessed from below.

- Raise car and support in a safe manner.

> ***WARNING —***
> *Make sure the car is stable and well supported at all times. Use a professional automotive lift or jack stands designed for the purpose. A floor jack is not adequate support.*

- Remove right rear wheel.

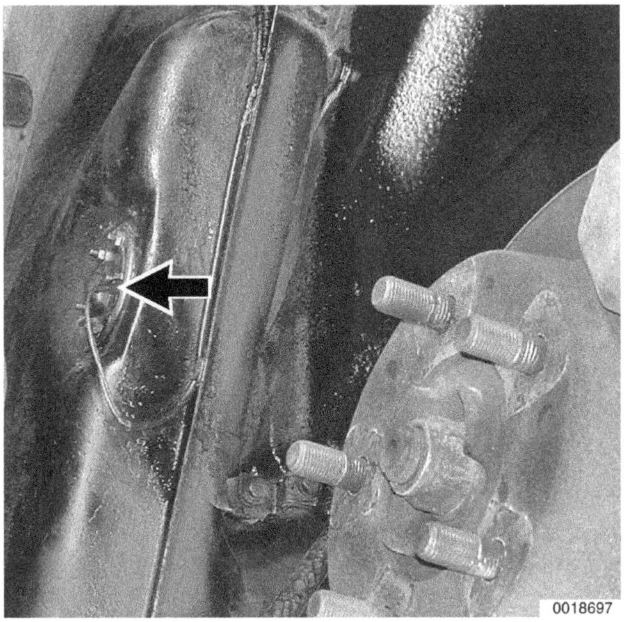

- Working behind right rear wheel housing, disconnect electrical connectors at oil level sensor (**arrow**).
- Check electrical resistance between sensor terminal and ground. If sensor is electrically open, it is faulty and should be replaced.

NOTE —
The oil level sensor can be removed and installed with the oil tank in its installed position. It is not necessary to drain the engine oil to replace the oil level sensor.

- If no faults are found, ground disconnected wire. Turn ignition key on and check that gauge moves to full position.
  - If gauge reacts correctly, sensor is faulty and should be replaced. If gauge does not respond, check electrical wiring to gauge. See **970 Electrical Wiring Diagrams**.

Lubrication system, flushing

After an engine bearing failure, the oil passages in the crankcase and crankshaft, cylinder heads and camshaft cases must be thoroughly flushed. See **130 Crankcase and Crankshaft**. In addition, the entire lubrication system must be flushed clean.

NOTE —
It is impossible to remove all metal partices from the engine oil cooler. It must be replaced after an engine bearing failure.

- Remove oil filter and oil tank as described earlier. Rinse oil tank with engine cleaner.

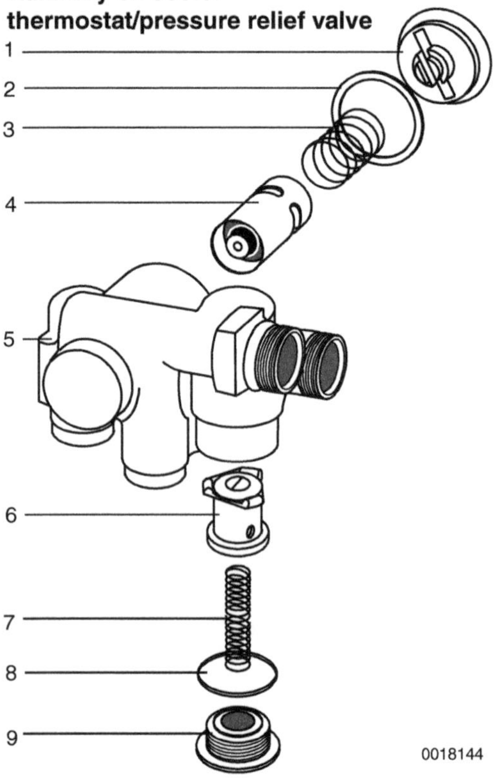

1. End plug
2. Gasket
3. Spring
4. Thermostatic valve
5. Valve housing
6. Pressure relief valve
7. Spring
8. Gasket
9. End plug

◄ Remove oil pressure relief valve for auxiliary oil cooler as described later.

- Disassemble and clean valve.

NOTE —
Apply heat to the pressure fitting nuts to avoid damaging the threads on the relief valve.

– Remove engine mounted oil cooler and replace it with a new cooler as described later.

– Remove auxiliary oil cooler as described later.

- Back-flush with engine parts cleaner.

– Flush oil lines that run front to rear with engine cleaner and carefully blow clean and dry with compressed air.

– Flush oil lines from oil tank to engine and from engine to auxiliary oil cooler pressure regulator.

COMPONENT REPLACEMENT

NOTE —
Engine oil pump replacement is covered in **102 Engine Disassembly and Assembly**. *Access to the oil pump requires that the engine/transmission be removed and separated. The engine can then be disassembled.*

Engine oil tank, removing and installing

The engine oil supply tank is located in the right rear fender well and can be accessed from below.

– Raise car and support in a safe manner.

WARNING —
Make sure the car is stable and well supported at all times. Use a professional automotive lift or jack stands designed for the purpose. A floor jack is not adequate support.

– Remove right rear wheel.

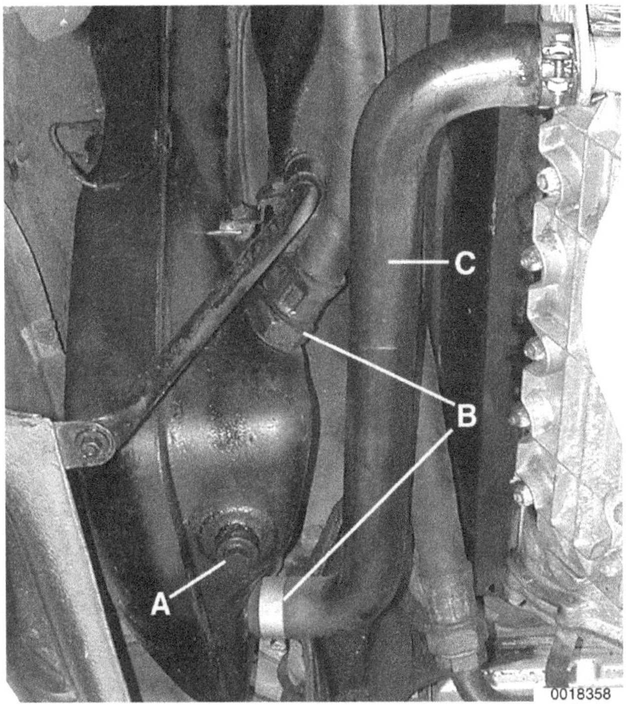

Remove oil tank drain plug (**A**) and drain oil.

- Disconnect both oil hoses (**B**) from tank.

NOTE—

- *Have a drain pan ready when disconnecting the hoses. The hoses will contain some residual oil. Do not let oil drip onto brake disc.*
- *Inspect hose (**C**) for cracks and replace if necessary.*

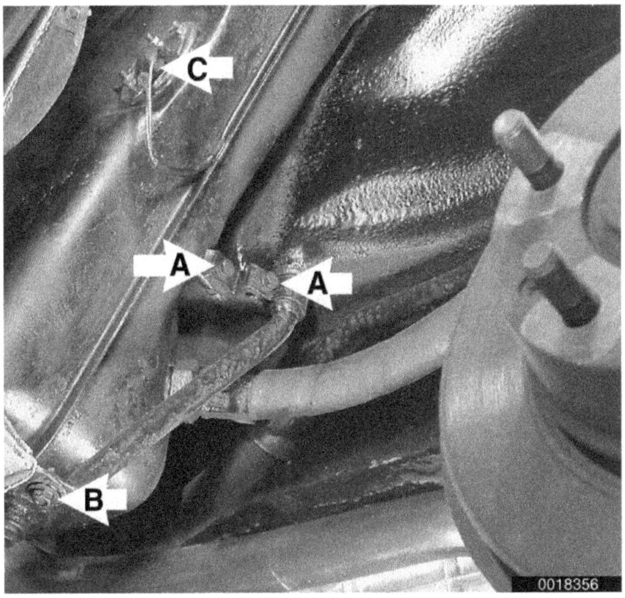

Remove bolts (**A**) from body panel support in right rear wheel housing and position bracket downward.

- Remove screw (**B**) holding support to quarter panel.
- Disconnect wires (**C**) from oil level sensor on oil tank.

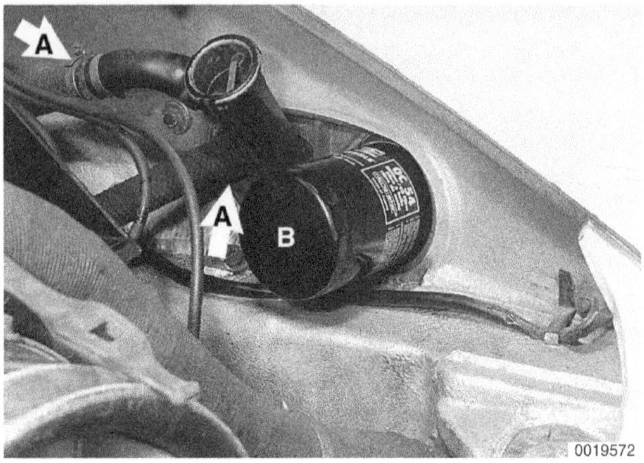

Remove oil tank attachments.

- Remove oil filler cap.
- Detach two breather hoses in oil filter area (**A**).
- Remove oil filter (**B**).

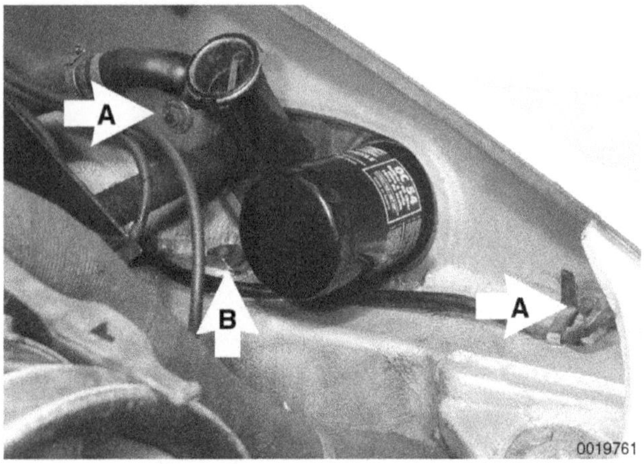

Support oil tank while removing mounting fasteners:

- Remove oil tank mounting nuts (**A**).
- Remove upper mounting nut (**B**) from support tube and remove tube from below.

NOTE —

Support tube upper nut is locked in place. Bend back lock plate before loosening nut. Always replace lock plate.

– Remove oil tank by pulling it forward and down.

– Installation is reverse of removal.

- Be sure wheel housing liner gasket is correctly positioned on oil tank before installing tank.
- Top up engine and fill tank with oil.
- Run engine and check for leaks.
- Check oil level. See **030 Maintenance**.

Engine oil cooler, removing and installing

The engine oil cooler is mounted to the right side of the engine crankcase, next to the flywheel housing. Be sure to have new O-rings on hand before removing cooler.

– Raise car and support in a safe manner.

WARNING —

Make sure the car is stable and well supported at all times. Use a professional automotive lift or jack stands designed for the purpose. A floor jack is not adequate support.

– Drain engine oil from crankcase and oil tank. See **030 Maintenance**.

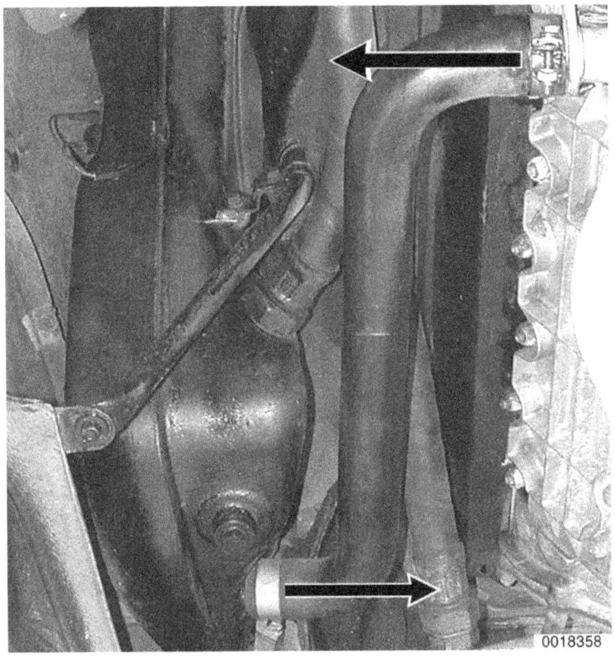

0018358

◀ Working underneath right side of engine, detach S-shaped oil return hose from oil cooler and oil tank (**arrows**).

NOTE —

Have a catch pan ready when disconnecting hose.

– Without disconnecting A/C pressure lines, remove A/C belt, A/C compressor and compressor bracket as described in **870 Air Conditioner**. Rest compressor on pad to right of engine compartment.

0019585

◀ Supporting engine securely from below, remove rear engine support crossmember bolts. Right side shown (**arrow**).

- Lower engine about 3 in. and support in that position.

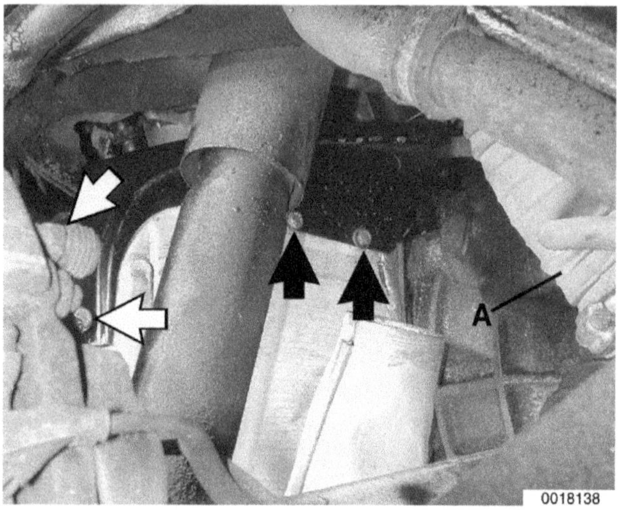

– Remove connector from spark plug 6.

◄ Working underneath right side of engine, remove right rear (flywheel-side) sheet metal shrouding after removing screws (**arrows**).

NOTE —

*To access shroud screws, detach right heat exchanger hose (**A**) and push forward.*

◄ Remove oil cooler plastic air shroud mounting screws (**arrow**) and remove shroud.

◄ Remove upper oil cooler mounting nuts (**arrows**).

Remove lower oil cooler mounting nuts (**arrows**).

NOTE —
For better access to oil cooler fasteners, remove right heat exchanger. See **260 Exhaust System**.

– Slide oil cooler out and down to remove.

– Installation is reverse of removal.

- Use new O-rings at oil cooler.
- Add engine oil and check oil level.

NOTE —
Always use new gaskets and seals.

Tightening torques

- Engine oil drain plug to oil tank. . . . . 42 Nm (31 ft-lb)
- Heat exchanger to crossover pipe . . . . . 25 Nm (18 ft-lb)
- Heat exchanger to cylinder head . . . . . . 25 Nm (18 ft-lb)
- Oil cooler to crankcase . . . . 25 Nm (18 ft-lb)
- Oil drain plug to crankcase (from 3/83). . 70 Nm (51 ft-lb)
- Oil drain plug to sump cover (to 3/83). . . 42 Nm (31 ft-lb)
- Sump cover to crankcase (to 3/83)(M6) . . 10 Nm (7 ft-lb)
- Rear engine mount to crossover support . . . . 80 Nm (58 ft-lb)

Auxiliary oil cooler thermostat/pressure relief valve, removing and installing

The auxiliary oil cooler thermostat/pressure relief valve is located under the car, ahead of the right rear wheel.

– Raise car and support in a safe manner. Remove right rear wheel.

> ***WARNING —***
> *Make sure the car is stable and well supported at all times. Use a professional automotive lift or jack stands designed for the purpose. A floor jack is not adequate support.*

– Drain engine oil tank.

Loosen oil line mounting bracket (**A**). Detach oil lines (**arrows**) from relief valve.

NOTE —
Apply heat to the pressure fitting nuts to avoid damaging the threads on the relief valve.

– Installation is reverse of removal.

Tightening Torques

- Engine oil drain plug to oil tank. . . . . 42 Nm (31 ft-lb)

Auxiliary oil cooler, removing and installing

The auxiliary oil cooler is mounted in front of the right front wheel housing, just behind the front bumper. Oil flow to the cooler is controlled via the auxiliary oil cooler pressure relief valve.

– Raise front of car and support in a safe manner. Remove right front wheel.

> ***WARNING —***
> *Make sure the car is stable and well supported at all times. Use a professional automotive lift or jack stands designed for the purpose. A floor jack is not adequate support.*

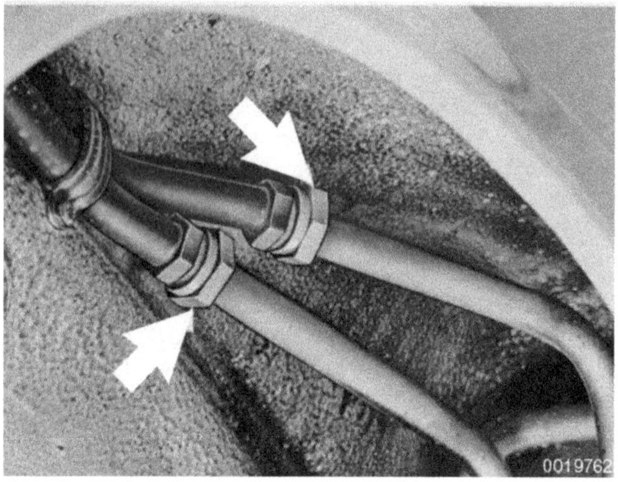

◄ Separate oil cooler lines from cooler fittings (**arrows**). Use a wrench to counterhold cooler and line fittings (if applicable).

NOTE —
- *Apply heat to the pressure fitting nuts to avoid damaging threads.*
- *Plug all disconnected oil lines to keep dirt out of system.*
- *Have a drain pan ready to catch oil that will spill out.*

– Remove oil cooler mounting bracket(s) and lower cooler.

– Installation is reverse of removal. Clean threads on oil cooler fittings before reinstalling oil lines.

• Coat cooler line fittings with Optimoly® TA.

> ***CAUTION —***
> *Always check the auxiliary oil cooler and its connections for leaks after allowing the engine to reach operating temperature. The auxiliary oil cooler pressure relief valve must be open for oil to flow to the cooler.*

Auxiliary oil cooler lines, removing and installing

– Raise car and support in a safe manner. Remove front and rear right side wheels.

> ***WARNING —***
> *Make sure the car is stable and well supported at all times. Use a professional automotive lift or jack stands designed for the purpose. A floor jack is not adequate support.*

Detach front oil lines (**arrows**) from auxiliary oil cooler relief valve.

NOTE—

Apply heat to the pressure fitting nuts to avoid damaging the threads on the relief valve.

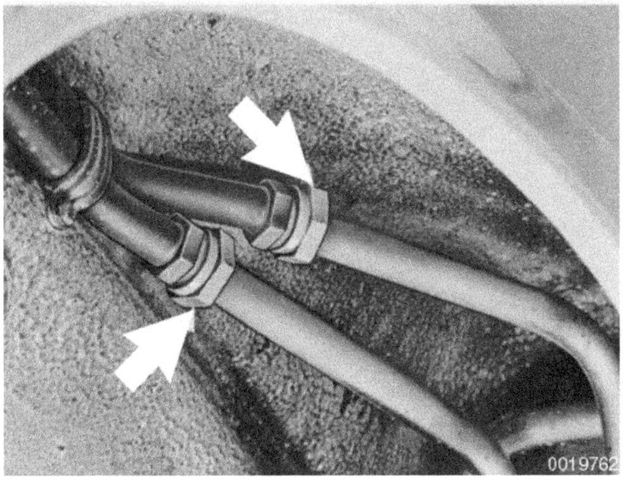

Separate oil cooler lines from cooler fittings (**arrows**). Use a wrench to counterhold cooler and line fittings (if applicable).

NOTE—

- *Apply heat to the pressure fitting nuts to avoid damaging threads.*
- *Plug all disconnected oil lines to keep dirt out of system.*
- *Have a drain pan ready to catch oil that will spill out.*

– Installation is reverse of removal.

Oil return tube, replacing

The procedure below describes replacing the one-piece oil return tube with an expandable two-piece tube. This procedure can be done with the engine installed.

Oil return tubes can be replaced one at a time, although replacing in pairs is recommended.

If removing a left side tube, remove left heat exchanger. See **260 Exhaust System**.

- Disconnect heater hoses from front and rear of heat exchanger (**A**).
- Remove nuts holding heat exchanger to cylinder head (**B**).
- Remove catalytic converter mounting hardware (**C**).

NOTE—

If removing a right side tube, the right side heat exchanger can remain in place.

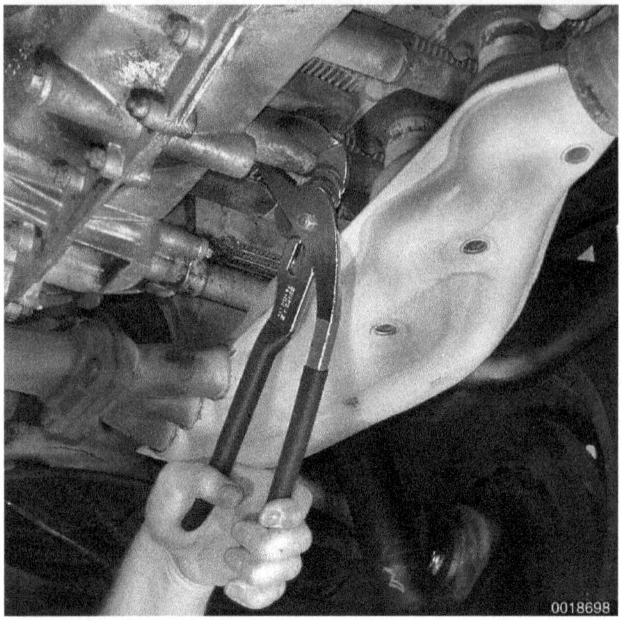

0018698

◄ Crush center of oil return tube with large pliers and bend.

NOTE —
Have a drain pan ready to catch any engine oil that may spill.

– Remove destroyed return tube and clean sealing surfaces on crankcase and camshaft case.

Two-piece oil return tube

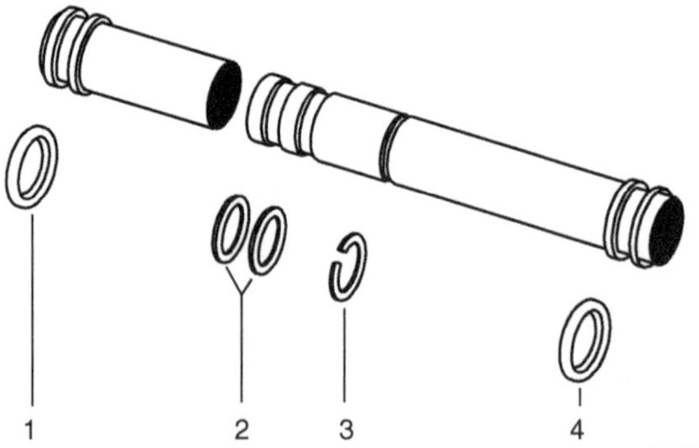

1. O-ring
2. O-rings
3. Locking ring
4. O-ring

◄ Mount O-ring seals to oil return tube sections and assemble.

– Push tube locking ring out of locking groove and slide approximately 40 mm (1.6 in.) away from locking groove.

– Apply a light film of tire mounting paste to seals and collapse sectioned oil return tube.

– Align telescoping end of tube with crankcase and camshaft housing holes. Expand tube so that it fits tightly into both holes.

– Slide locking ring into locking groove.

– Reinstall heat exchanger, if necessary.

Tightening Torques
- Heat exchanger to catalytic converter. . . 25 Nm (18 ft-lb)
- Heat exchanger to cylinder head . . . . . . 25 Nm (18 ft-lb)

– Run engine and check for oil leaks.

200 Engine Management–General

TABLE

GENERAL

The Porsche 911 SC is equipped with the Bosch Continuous Injection System (CIS) and Capacitor Discharge Ignition (CDI) system. This section describes the operation of both systems.

Refer to the following sections for additional repair information:

- **201 Fuel Supply**
- **240 Fuel Injection**
- **280 Ignition System**

Engine management systems

In the Bosch Continuous Injection System (CIS), fuel is injected continuously whenever the engine running. The quantity of fuel metered to the engine is based on a regulated fuel pressure and controlled mechanically by the air flow sensor.

There are three variants of CIS used in SC models. This information is summarized in **Table a**.

Table a. CIS variants

Model Year	CIS variant
1978 - 1979	Basic
1980	Oxygen sensor system
1981 -1983	Oxygen sensor system with Cold acceleration enrichment

In addition to oxygen sensor control, there are several systems designed to reduce harmful emissions:

- Evaporative emission controls
- Exhaust gas recirculation (EGR) (California only)
- Crankcase ventilation
- Secondary air injection (where applicable)

The ignition system is a breakerless Bosch Capacitor Discharge Ignition (CDI). The main components of CDI are described later in this section.

CONTINUOUS INJECTION SYSTEM (CIS)

The main components of the fuel injection system are shown in the accompanying illustration:

- The **air flow sensor** is moved by the flow of intake air, in effect measuring the volume of air entering the engine.
- The **fuel distributor** meters a proportional amount of fuel to the fuel injectors based directly on the movement of the air flow sensor. All fuel metering is done by the fuel distributor.
- The **fuel injectors** are precise nozzles through which fuel flows continuously while the engine is running. They have no other fuel metering function.

Several auxiliary components are used to further adapt the fuel mixture to various operating conditions:

- The **control pressure regulator** enriches fuel mixture during start-up and cold running.
- The **cold start valve** injects additional fuel during a cold start.
- The **auxiliary air regulator** allows additional air to bypass the throttle valve during cold engine operation to provide an elevated cold idle speed.
- The **auxiliary air valve** allows additional air to bypass the throttle valve during starting, independent of engine temperature.
- The **frequency valve (1980 and later models),** controlled by the oxygen sensor control unit based on input from the oxygen sensor, interacts with the fuel distributor to bleed off fuel pressure for warm running mixture control.

Continuous Injection System (CIS) components (1978 - 1979)

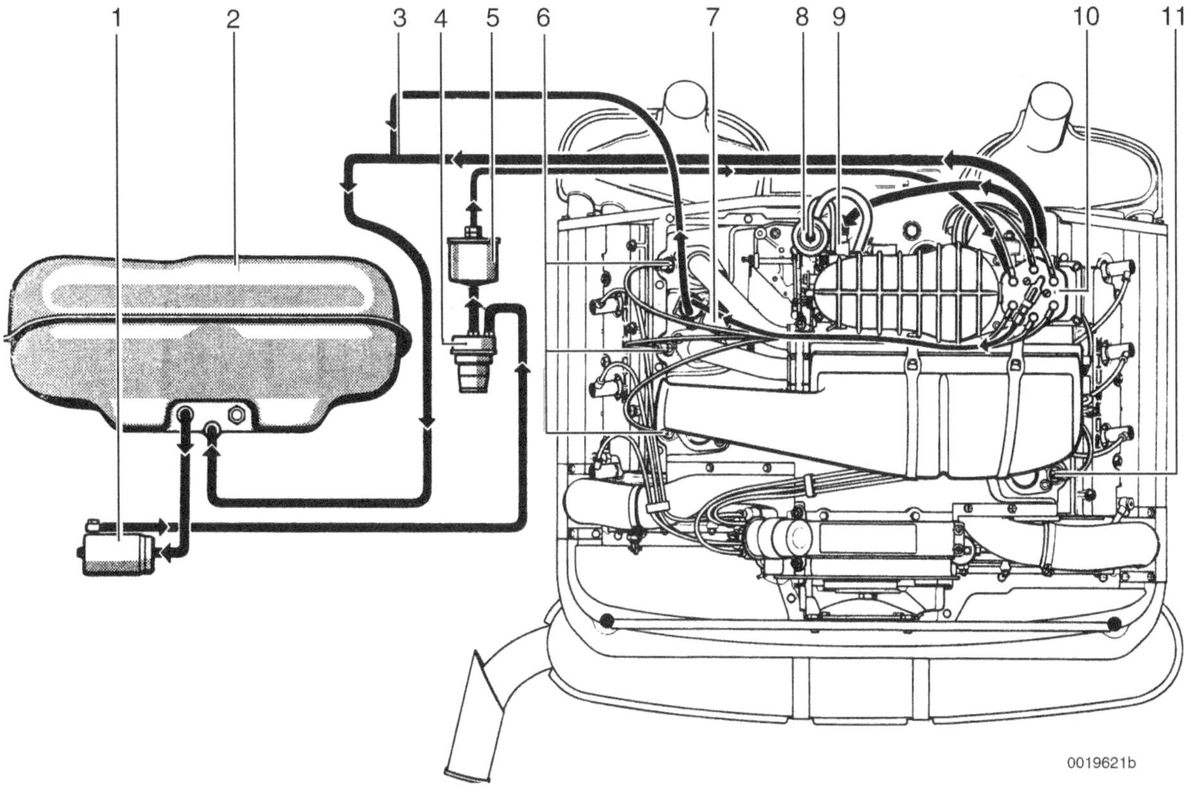

1. Fuel pump
2. Fuel tank
3. Fuel return plumbing
4. Fuel pressure accumulator
5. Fuel filter
6. Fuel injectors
7. Control pressure regulator
8. Decel air valve
9. Throttle switch (1980 and later)
10. Fuel distributor
11. Auxiliary air regulator

CIS fuel circuit

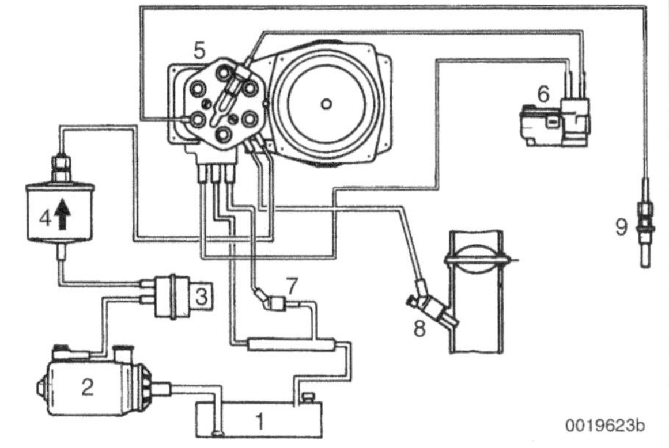

1. Fuel tank
2. Fuel pump
3. Fuel accumulator
4. Fuel filter
5. Fuel distributor
6. Control pressure regulator
7. Frequency valve (1980-83)
8. Cold start valve
9. Fuel injector

Fuel supply and fuel delivery

The fuel tank is located in the floor of the luggage compartment, behind the front bumper. The fuel pump is mounted to the front suspension cross-member behind the fuel tank. The fuel pump is controlled by a relay located in the main fuse/relay box in the luggage compartment. The pump is cooled by fuel and therefore may be damaged if allowed to run dry. The pump runs whenever the engine is being cranked or is running.

The fuel accumulator dampens fuel pulsations and reduces pump noise. The accumulator also helps to maintain fuel pressure when the engine is switched off to prevent vapor lock.

The fuel pump delivers pressurized fuel to the fuel distributor via the fuel accumulator and fuel filter.

CIS fuel injector

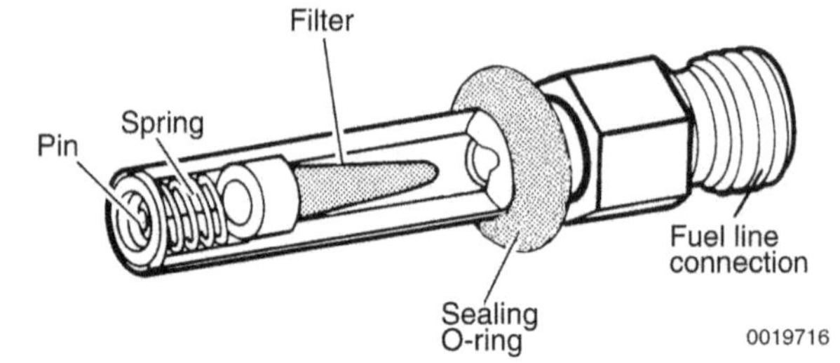

Fuel injectors

The mechanical fuel injectors are precision spray nozzles. Above a minimum pressure, the injector nozzles open and continuously deliver atomized fuel into the engine intake ports. The injectors are held in place in the intake manifold runners by thick rubber O-rings.

CIS mixture control unit

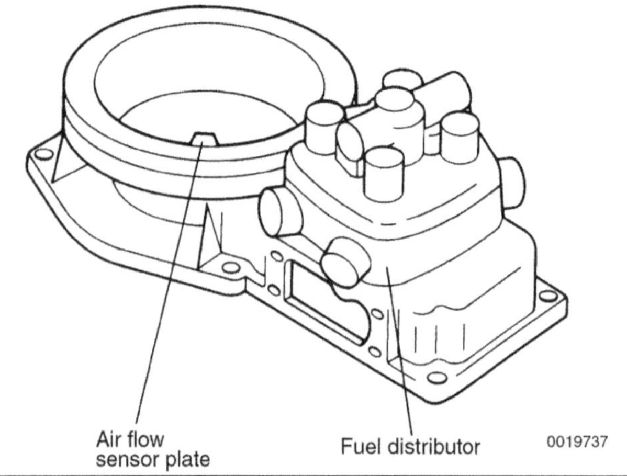

Mixture control unit

The mixture control unit is the heart of the injection system. The fuel distributor meters fuel to the injectors based on the position of the air flow sensor. The main components of the mixture control unit are the **air flow sensor assembly** and the **fuel distributor**.

Air flow sensor

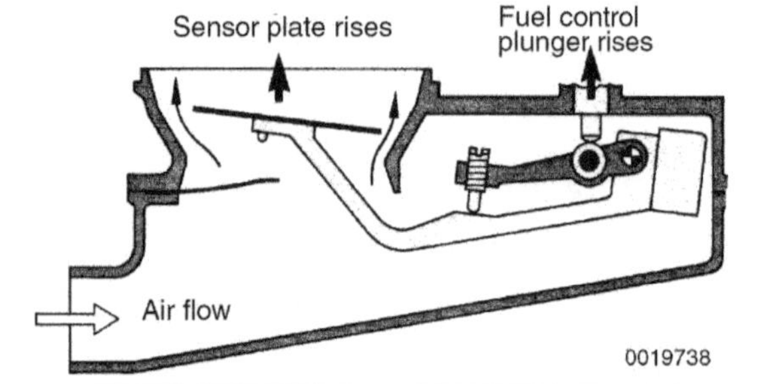

The circular air-flow sensor plate is positioned in the intake tract so that all air entering the engine flows past it. The sensor plate and its cone-shaped venturi are carefully designed to achieve the proper balance of air flow and fuel metering for all driving conditions, from idle to full throttle.

The plate is attached to a lever which pivots, allowing the plate to move up and down. Intake air flowing through the housing raises the sensor plate. The movement of the sensor plate is in direct proportion to the volume of incoming air.

Fuel control plunger action

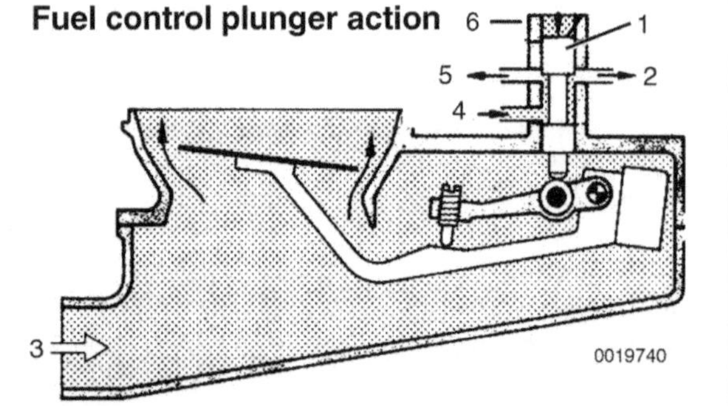

1. Control plunger
2. Fuel out
3. Air flow
4. Fuel in
5. Fuel out
6. Fuel distributor

Air measurement is turned into an injection quantity by the control plunger in the fuel distributor. The plunger rests on the air-flow sensor lever and rises and falls with the sensor plate. When air flow into the engine increases and raises the air-flow sensor, the rise of the plunger increases fuel flow to the injectors proportionally.

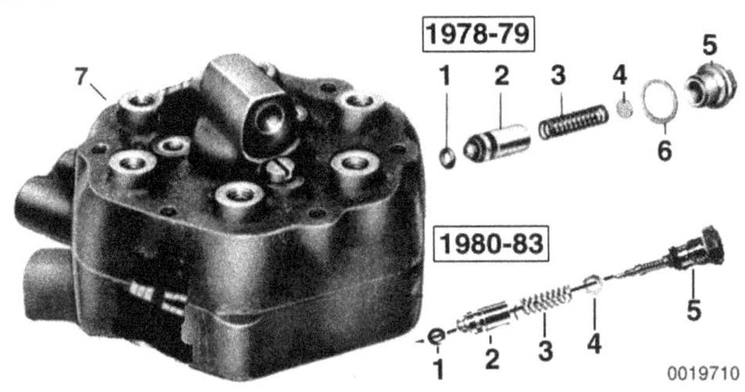

1. O-ring
2. Plunger
3. Spring
4. Shim(s)
5. End plug
6. Sealing ring
7. Fuel distributor

For precise fuel metering, fuel pressure must remain constant. A change in fuel pressure results in a change in the amount of fuel delivered to the engine.

Fuel pressure is maintained by a pressure relief valve built into the fuel distributor. When the fuel pressure exceeds a predetermined level, the valve opens and surplus fuel is returned to the fuel tank. System pressure is adjusted by adding or subtracting shims on the relief valve.

Control pressure regulator fuel circuit

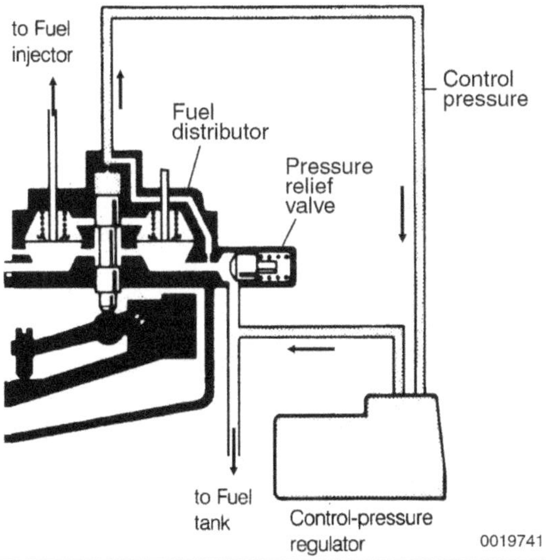

Control pressure regulator

The control pressure regulator, also called the warmup regulator, adapts the fuel mixture during the warmup phase by adjusting the control pressure on top of the control plunger. It is located between air intake runners for cylinder 1 and 2. The regulator adjusts control pressure by bleeding off system pressure from the top of the control plunger and returning the excess fuel back to the fuel tank.

A lower control pressure allows the sensor plate to lift higher for a given air flow, resulting in a richer mixture.

NOTE —

- *Different control pressure regulators have been used on the 911 SC models. The regulators are differentiated by the last two digits of the part number on the regulator body. Be sure to use the correct part number when replacing the regulator.*
- *There is a vacuum connection on the early style control pressure regulators. This is used for supplemental enrichment during acceleration.*

Control pressure regulator construction

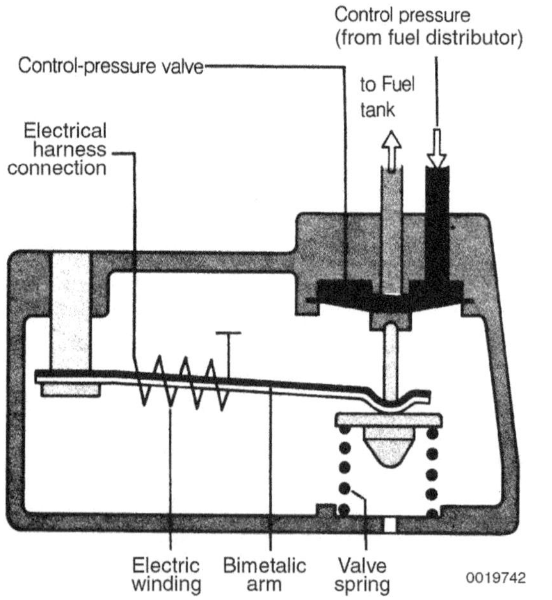

The control pressure regulator consists of a valve mounted on an electrically heated bimetallic arm. The two different metals have different expansion rates. The arm bends one way when cold and the other way when warm, changing control pressure in response to temperature.

When cold, the natural shape of the bimetallic arm pulls the diaphragm away from the valve. As a result, the valve increases fuel return to the tank to reduce the control pressure and richen the mixture.

As the regulator warms up from the current supplied to the heater, the bimetallic arm bends the other way. The spring partially closes the valve, which decreases fuel return to tank to increases control pressure for warm running operation.

Cold start injector

Cold start injector electrical circuit

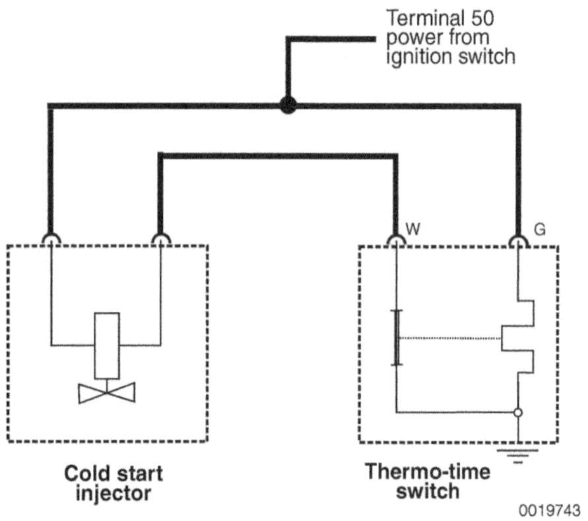

The electrically operated cold start injector (also called the cold start valve) sprays extra fuel into the intake manifold for several seconds when the starter is actuated and the engine is cold.

When the starter operates, current is supplied to the cold start injector. If the thermo-time switch is cold (closed), it completes the ground to the cold start injector. The same current is also supplied to a heating element in the thermo-time switch. If the engine is cranked for a prolonged period, the heating element cuts off (opens) the ground to the cold start injector to prevent engine flooding.

NOTE —

The thermo-time switch determines the operating time of the cold start injector based on temperature. At temperatures below -20°C (-4°F), the switch remains closed for about 8 seconds. When engine temperature is above 35°C (95°F), the switch remains open. Between these extremes, the switch remains closed for a varying amount of time.

Auxiliary air regulator

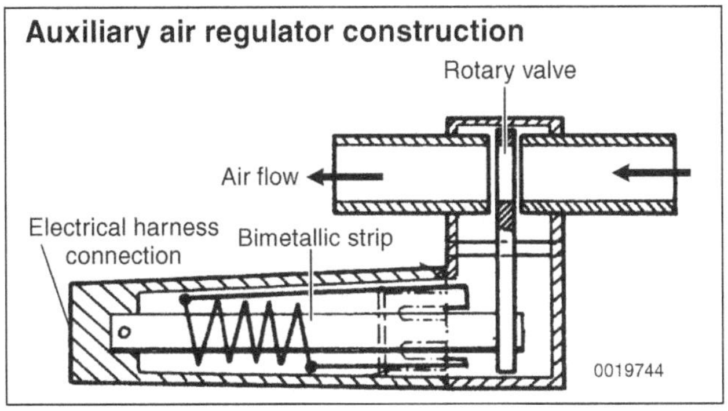

During cold engine operation, the auxiliary air regulator lets additional air bypass the throttle valve to raise idle speed. Electrical current warms the heating element and causes the bimetallic strip to deflect gradually, closing the rotary valve and cutting off the additional air. The auxiliary air regulator is mounted between intake runners 5 and 6.

Auxiliary air valve

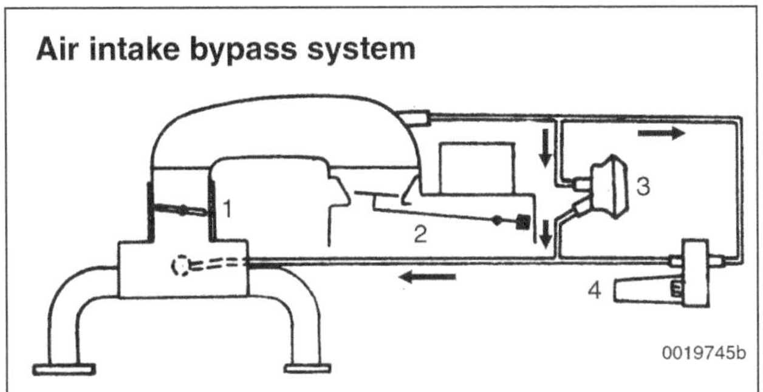

1. Throttle housing
2. Air flow sensor
3. Auxiliary air valve
4. Auxiliary air regulator

The auxiliary air valve is mounted behind the mixture control unit and provides additional bypass air to the engine during starting, regardless of the operational state of the auxiliary air regulator. The additional air flow allows more reliable air flow sensor plate operation during starting.

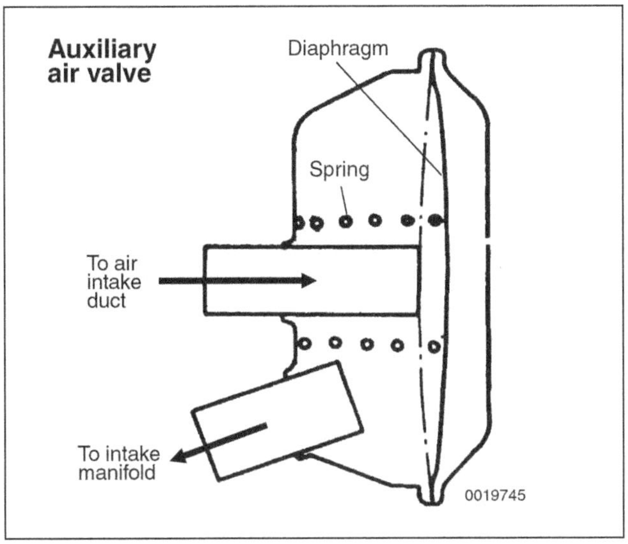

The auxiliary air valve contains a diaphragm with two hose connections. When the engine is stopped and during starting, the diaphragm is held open by a spring, allowing bypass air to enter the engine. Once the engine has started, vacuum acts on the lower connection to pull the diaphragm closed and cutoff off the bypass.

1. Auxiliary air valve
2. Deceleration air valve
3. Auxiliary air regulator vacuum line
4. to intake manifold

Deceleration air valve

The deceleration air valve allows additional air to bypass the throttle and temporarily maintain an elevated idle speed during sudden deceleration. This helps control exhaust emissions during deceleration.

OXYGEN SENSOR CONTROL SYSTEM (1980 – 1983 MODELS)

1980–1983 911 SC models are equipped with an oxygen sensor system that continuously corrects the air-fuel mixture when the engine is warm. Mixture is held near the optimal (stoichiometric) ratio necessary for catalytic converter operation.

NOTE —
The stoichiometric ratio is 14.7 parts air to 1 part fuel (by weight). The greek letter lambda (λ) is generally used to denote air:fuel mixture. $\lambda = 1$ is stoichiometric.

The oxygen sensor system can correct the fuel mixture by approximately ±20%.

The oxygen sensor control system includes:

- Oxygen sensor control unit
- Oxygen sensor
- Electric frequency control valve
- Temperature switch(es).

Closed loop operation (oxygen sensor feedback)

The oxygen sensor is mounted in the exhaust stream ahead of the catalytic converter. Once the sensor warms up, it begins to generate a small voltage (0 - 1 VDC) based on the oxygen content in the exhaust. The sensor produces a larger voltage when the mixture is rich and smaller voltage when the mixture is lean. The sensor output for the optimal mixture ($\lambda = 1$) is approximately 0.45 VDC.

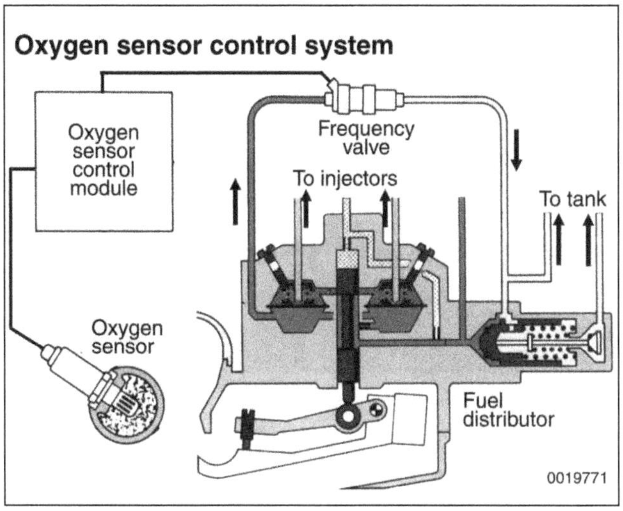

Based on the input from the oxygen sensor, the oxygen sensor control module modulates the on-off signal to the frequency valve to bleed off fuel pressure in the lower chambers of the fuel distributor. The pulse width of the on-off signal is known as the "duty cycle."

NOTE—

Duty cycle is measured on a scale of 0 to 100%. When the duty cycle is 65%, that means the frequency valve is open 65% of the time and closed 35% of the time.

When the oxygen sensor system is leaning out the mixture, the duty cycle is less than 50%. When the system is enriching the mixture, the duty cycle is more than 50%.

By controlling the pressure in the lower chambers, the amount of fuel delivery can be precisely controlled. The control module is designed to continually cycle between lean and rich conditions within close tolerances. This is known as closed loop operation.

Open loop operation

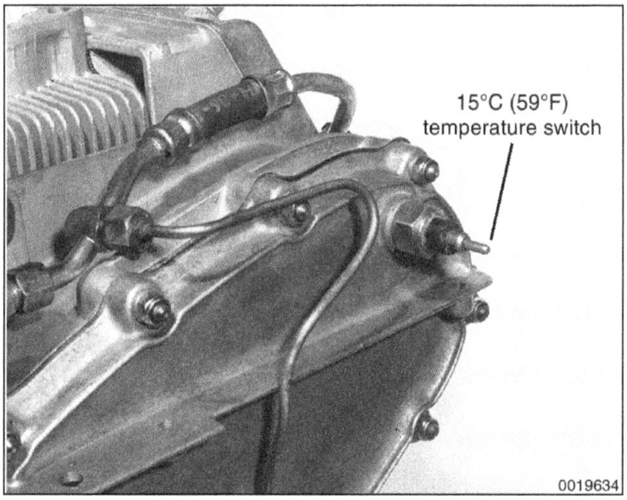

Because oxygen sensor signal is not dependable until the sensor gets hot, and because a cold engine needs a slightly richer mixture for good driveability, a fixed frequency valve duty cycle (65 - 85%, depending on model year) is programmed into the oxygen sensor control module for good cold engine driveability. Cold engine operation is in effect whenever the engine oil temperature is below (15°C/59°F temperature switch closed). The 15°C temperature switch is in the right timing case cover.

NOTE—

A constant duty cycle means that the oxygen sensor system is in open loop mode.

Cold acceleration enrichment (1981 – 1983 models)

On 1981 and later cars, a cold acceleration enrichment system was added to the basic oxygen sensor control system. This system operates when the engine and/or the oxygen sensor is below 35°C (95°F).

During cold engine operation, additional enrichment is provided by briefly increasing the frequency valve duty cycle during acceleration. The system uses an acceleration enrichment control module, an additional throttle microswitch (15°throttle angle) and a 35°C (95°F) temperature switch. See **240 Fuel Injection** for additional repair information.

NOTE—

The acceleration enrichment control module is beneath the passenger's seat, next to the oxygen sensor control module. The 35°C temperature switch is located in the crankcase vent cover at the rear of the engine.

EMISSION CONTROLS

Secondary air injecton

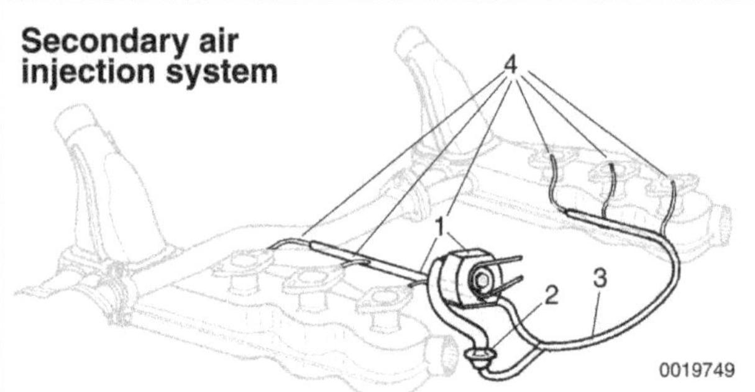

1. Secondary air pump
2. Check valve
3. Distribution tube
4. Exhaust ports

The air injection system uses a belt driven mechanical pump to reduce exhaust emissions. Air is injected into the exhaust ports right at the exhaust valves to reduce HC and CO during the engine warm-up phase, before the catalytic converter reaches operating temperature.

The secondary air pump (**arrow**) is mounted on the left side of the engine. The pump is maintenance free and requires no lubrication.

Evaporative emission controls

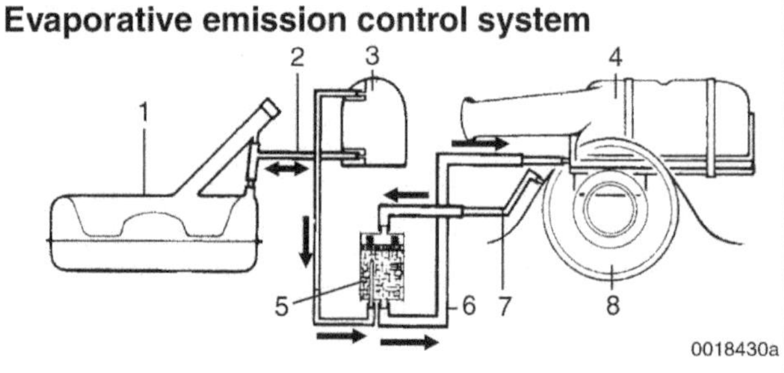

1. Fuel tank
2. Return hose (fuel tank to expansion tank)
3. Expansion tank
4. Air filter housing
5. Charcoal canister
6. Hose (charcoal canister to air filter housing)
7. Hose (cooling fan to charcoal canister)
8. Cooling fan with hose connector

This system provides venting for the fuel tank while at the same time trapping the fuel vapors that would otherwise be vented into the atmosphere. The evaporative emission system collects fuel vapors from the tank in a charcoal canister. During certain engine operation conditions, the canister is purged, drawing the fuel vapors into the engine to be burned.

Exhaust gas recirculation (EGR) (1978 - 1979 California models)

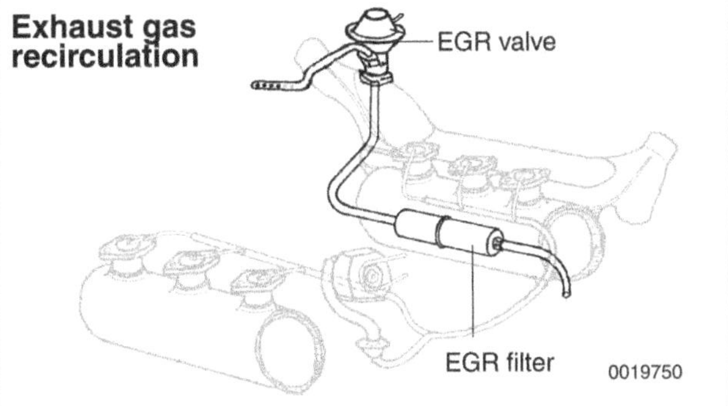

◀ The exhaust gas recirculation (EGR) system is used to lower the combustion chamber temperatures to help reduce NO_x exhaust emissions

The EGR valve is opened and closed based on throttle position. When the EGR valve is open, exhaust gases are routed through a connector pipe between the heat exchanger and catalyst to the throttle housing.

Crankcase and oil tank ventilation

Crankcase ventilation

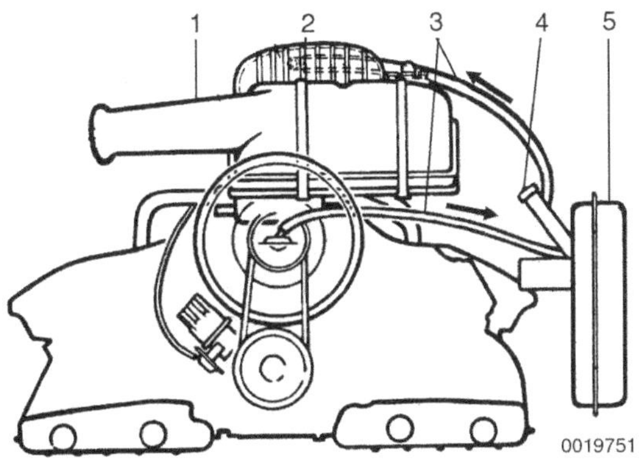

1. Air filter housing
2. Intake boot
3. Breather lines
4. Oil filler cap
5. Oil tank

◀ This system provides venting for the crankcase and oil tank while at the same time trapping fuel vapors that would otherwise be vented to the atmosphere.

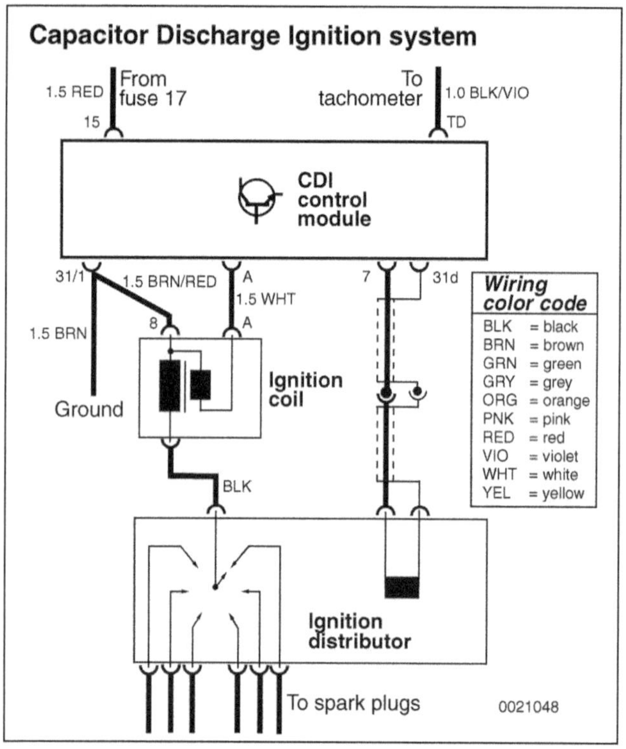

CAPACITOR DISCHARGE IGNITION (CDI)

The main components of the CDI system include:

- Ignition control module
- Ignition coil
- Breakerless ignition distributor
- Spark plugs

NOTE —

An electronic speed control relay shuts off the fuel pump at approximately 7,000 rpm. The relay is located under the instrument panel to the left of the turn signal flasher relay.

Ignition control module

The ignition control module (**arrow**) has six terminals and is mounted in the left side of the engine compartment. Control module testing is given in **280 Ignition System**.

Ignition distributor

The ignition distributor (which turns counterclockwise) uses an inductive pulse generator (**A**).

A toothed rotor (**B**) causes a magnetic flux change for each tooth while rotating in the field created by permanent magnets. One complete turn of the rotor in the distributor will produce six control pulses.

The pulse generator produces an AC current that is supplied to the ignition control module. The control module converts and amplifies these control pulses and releases the ignition voltage in synchronization with the control pulses.

The centrifugal and vacuum advance mechanisms, as well as the vacuum retard mechanism (1980 and later models) operate like a conventional distributor.

TROUBLESHOOTING

Before troubleshooting a poorly running engine or an engine that will not start, determine the general condition of the engine, especially if it has high mileage. If the engine is severely worn or has mechanical problems, the only remedy is overhaul or repair.

General engine condition can be assessed by checking the compression at each cylinder. Make sure that the valves are correctly adjusted before making the test. The compression check is described in **100 Engine–General**.

NOTE —
A compression test requires special test equipment. If the equipment is not available, most automotive repair shops can do these tests quickly and at a reasonable cost.

Carbon deposits on the fuel injectors and the intake valves will affect the way the engine idles and runs. Even a 10% decrease in the amount of fuel that injectors deliver can cause driveability problems. Short trips around town or in city traffic increase the likelihood of deposits forming.

Light carbon can usually be removed using a fuel additive. Specialized cleaning equipment is required to remove heavier carbon deposits. Check with an authorized Porsche dealer for information on carbon deposits and the best methods used to remove them.

CAUTION —
Always follow the manufacture's directions when using fuel additives designed to remove carbon deposits and clean injectors. The excess amount of cleaner in the fuel can dilute engine oil and accelerate engine wear. Check with an authorized Porsche dealer before using fuel additives together with high detergent fuel.

Preventive maintenance

The condition of the tune-up and emission control components can affect engine performance and driveability. Extending maintenance schedules beyond those recommended by the manufacturer can adversely affect the way the engine runs. When experiencing driveability problems, a good starting point is performing the manufacturer-specified maintenance. Many driveability problems are eliminated simply by replacing worn out or heavily fouled spark plugs.

Replacement schedules for the spark plugs, spark plug wires, distributor cap, ignition rotor, fuel filter, air filter, oxygen sensor (where applicable), and oil and filter are given in **030 Maintenance.**

Basic settings

In addition to tune-up component replacement, it is important that all of the basic settings and adjustments are within allowable limits. Check throttle plate adjustment, idle speed, idle mixture (CO%), and ignition timing to be sure they are all within the specified limits. See **240 Fuel Injection**.

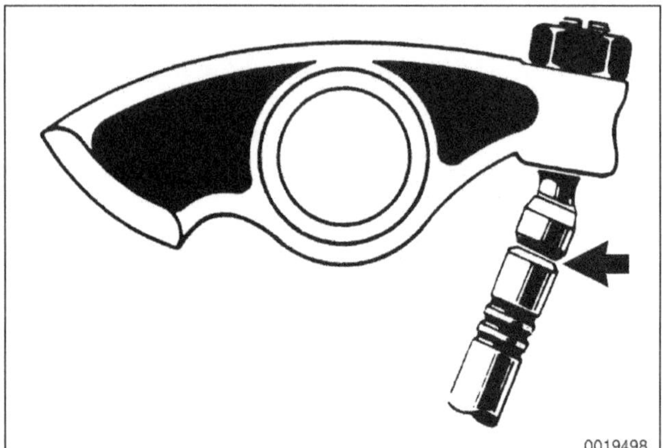
0019498

Correctly adjusted valves are necessary for efficient engine operation. Procedures for checking and adjusting valve clearances are described in **030 Maintenance**.

Electrical system

Although CIS is largely a mechanical system, some electrical components are used to adapt the basic fuel mixture, especially for cold and warmup operation. The fuel pump, the cold start valve, the control pressure regulator, and the auxiliary air regulator are the main components that rely on electrical power to operate. In addition, 1980 and later models use a sensitive low-voltage oxygen sensor system.

Make sure all ground connections and harness connections are tight, free of corrosion and damage-free. Make sure the battery is fully charged and in good condition.

Fuel pressure

Correct fuel pressure is probably the most important factor for troublefree CIS operation. If either system pressure or control pressure is incorrect, driveability problems will be encountered. Another important item is residual fuel pressure. Residual pressure is necessary for quick restarts and the prevention of vapor lock. Complete fuel pressure tests are given in **240 Fuel Injection**. Fuel pump and fuel pump relay troubleshooting is given in **201 Fuel Supply**.

Table b lists additional probable causes and corrective actions for CIS system driveability problems.

Table b. CIS troubleshooting

Cold start-Engine starts hard or fails to start when cold	Hot start-Engine starts hard or fails to start when warm	Engine misses and hesitates under load	Engine starts but runs poorly when cold	Engine idles too fast	Hesitation on acceleration	Poor fuel mileage	Engine starts but stalls at idle	Low power	Probable Cause of Fault	Corrective action
X									Cold start valve or thermo-time switch faulty	Test cold-start valve and thermo-time switch. Replace faulty parts.
X									Fuel pump not running	Check fuel pump fuse and relay.
X	X		X		X		X		Air-flow sensor plate rest position incorrect	Inspect air-flow sensor plate and adjust if necessary.
X	X	X	X		X	X	X	X	Fuel pressure incorrect	Test fuel pressure.
	X					X			Cold start valve leaking or operating continuously	Test cold-start valve and thermo-time switch.
									Insufficient residual fuel pressure	Test residual fuel pressure. Replace fuel pump check valve or fuel accumulator.
			X			X			Fuel leaks	Inspect fuel lines and connections.
						X		X	Oxygen sensor control faulty (1980 - 1983)	Check oxygen sensor control system
		X	X		X		X	X	Fuel injector clogged.	Test fuel injectors. Check for clogged injector lines. Replace faulty injectors.
X			X	X			X		Auxiliary-air regulator faulty	Test and, if necessary, replace auxiliary air regulator.
	X		X		X		X		Vacuum leaks (intake manifold)	Inspect intake air component for leaking hoses, hose connections.
				X					Air leaking past throttle valve	Inspect throttle valve. Adjust or replace as required.
	X	X	X		X		X	X	Control plunger in fuel distributor binding or fuel distributor faulty.	Check air-flow sensor plate movement. If necessary, replace fuel distributor.
			X	X	X	X	X	X	Idle speed, ignition timing, and idle mixture (CO%) out of adjustment	Check and adjust to specification.
		X	X	X	X	X	X	X	Incorrect ignition timing or faulty ignition system	Check ignition system.
						X		X	Full throttle switch faulty or incorrectly adjusted (1980 - 1983)	Check throttle switch and adjust if necessary. Replace faulty switch.
X	X		X				X		Control pressure regulator faulty	Check control-pressure regulator full-load function.

201 Fuel Supply

TABLES

Special Tools

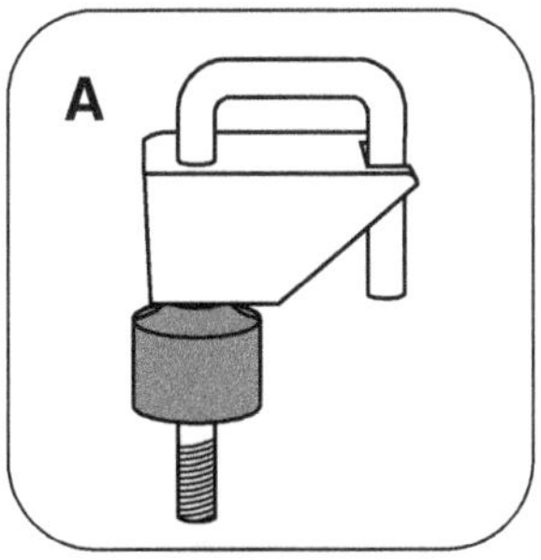

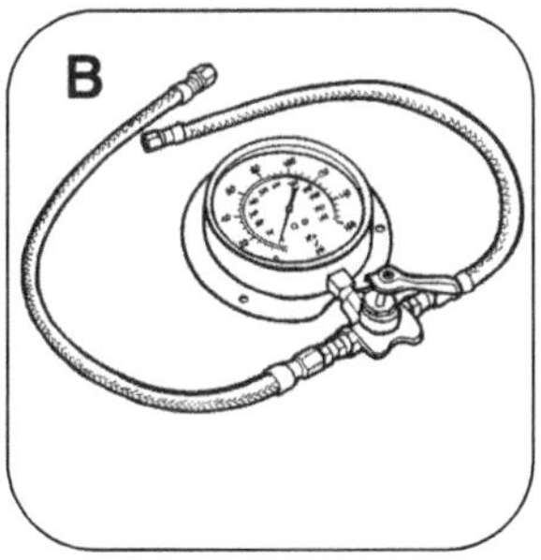

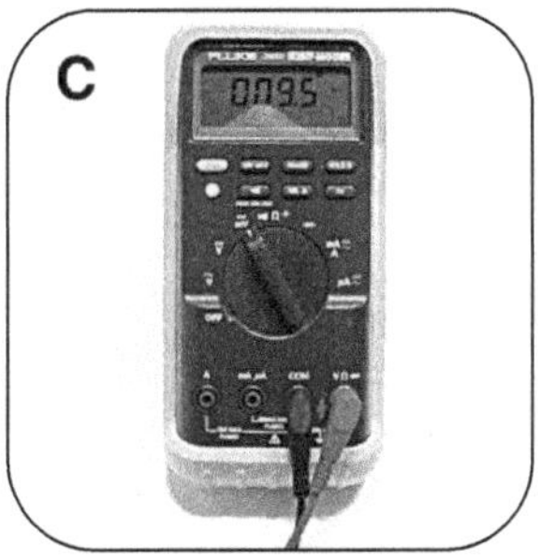

Special tool

A 647b . . . . Fuel hose pinch clamp
(source: Baum Tools Unlimited)
B 1318 . . . . Fuel pressure gauge
(source: Baum Tools Unlimited)
C Fluke 87 . . . . Automotive digital multimeter

GENERAL

This repair group covers repairs to the fuel tank, fuel pump, fuel level sender and connecting lines. The Continuous Injection System (CIS) and emission control systems are covered in **240 Fuel Injection**.

NOTE—
Fuel filter replacement is covered in **030 Maintenance**.

Test equipment

< Some of the procedures require special tools.

> *CAUTION—*
> *A high impedance digital multimeter should be used for all voltage and resistance tests.*

Warnings and cautions

Please observe the following warnings and cautions when performing any service or repair on the fuel system.

WARNING —

- *When working on an open fuel system, wear suitable hand protection, as prolonged contact with fuel can cause illnesses and skin disorders.*
- *Gasoline is highly flammable and its vapors are explosive. Do not smoke or work on a car near heaters or other fire hazards when diagnosing and repairing fuel system problems. Have a fire extinguisher available in case of an emergency.*

CAUTION —

- *Before making any electrical tests with the ignition turned on, disable the ignition system as described in* **280 Ignition System**.
- *Always unscrew the fuel tank cap to release pressure in the tank before working on the tank or lines.*
- *To prevent damage to the ignition system or the electronic fuel system components, including the ECU, always connect and disconnect wires and test equipment with the ignition off.*
- *Fuse and relay locations are subject to change. Use care when troubleshooting the electrical system at the fuse/relay panels. To resolve problems in identifying a relay, consult an authorized Porsche dealer for the latest information*
- *Absolute cleanliness is essential when working with the fuel system. Even a minute particle of dirt can cause trouble if it reaches an injector. Thoroughly clean the fuel line unions before disconnecting any of the lines.*
- *Use only clean tools. Keep removed components clean and sealed or covered with a clean, lint-free shop towel, especially if completion of the repair is delayed.*
- *When replacing parts, install only new, clean components.*
- *Avoid nearby use of compressed air, and do not move the car while the fuel system is open.*
- *Do not use compressed air above 40 psi to blow out lines or components. Internal damage may result to components.*
- *Always replace seals and O-rings.*

FUEL SUPPLY TROUBLESHOOTING

The fuel pump delivers fuel at high pressure to the fuel injection system. Checking fuel pressure delivery volume is a fundamental part of troubleshooting and diagnosing the fuel supply system. An accurate fuel pressure gauge with an integral 3-way valve will be needed to make the tests.

– Two significant fuel supply values can be measured:
  - System pressure is the **pressure** created by the fuel pump and maintained by the pressure regulator.
  - Fuel delivery volume is the **quantity** of fuel that the pump is capable of delivering to the injection system.

NOTE —
Two additional pressures are also influential in how the engine runs: control pressure and residual pressure. See **240 Fuel Injection** *for additional information on these fuel pressures.*

Relieving fuel pressure and connecting fuel pressure gauge

WARNING —
- *When working on an open fuel system, wear suitable hand protection, as prolonged contact with fuel can cause illnesses and skin disorders.*
- *Gasoline is highly flammable and its vapors are explosive. Do not smoke or work on a car near heaters or other fire hazards when diagnosing and repairing fuel system problems. Have a fire extinguisher available in case of an emergency.*

Remove control pressure regulator line (**arrow**) from center port on fuel distributor.

WARNING —
To prevent fuel from spraying on a hot engine, system pressure should be relieved before disconnecting fuel lines. One method is to tightly wrap a shop towel around a the control pressure regulator line where it connects to the fuel distributor and than loosen the fuel line fitting.

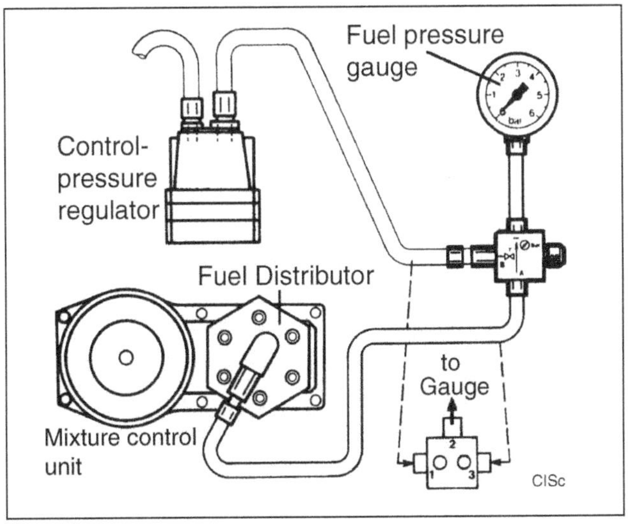

Connect fuel gauge between open port on fuel distributor and disconnected line.

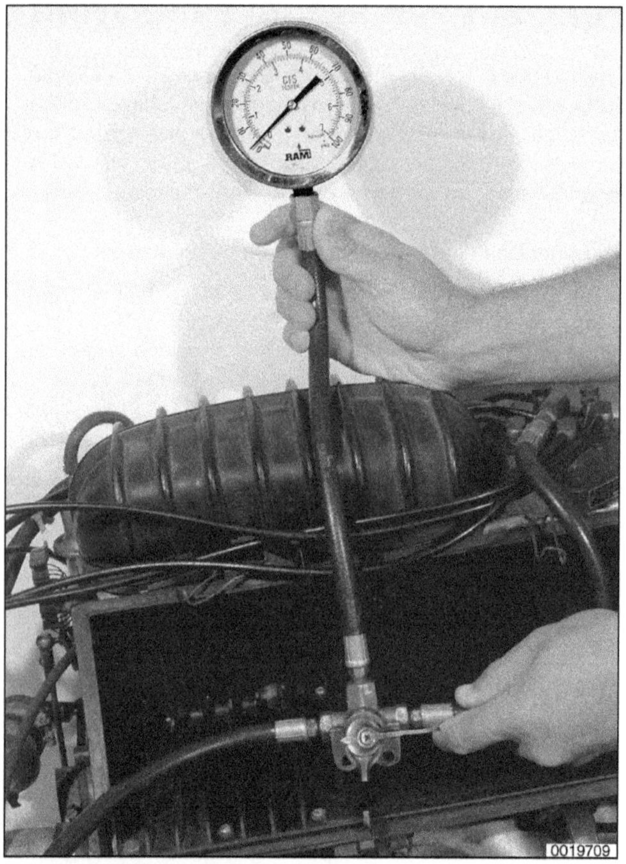

With fuel pump operating (see **Operating fuel pump for tests**), turn pressure gauge upside down. Open and close valve on gauge several times to bleed trapped air. When finished, turn gauge up-right.

NOTE —
The pressure gauge should have a range of 0 to 7 bar (0 to 100 psi).

Operating fuel pump for tests

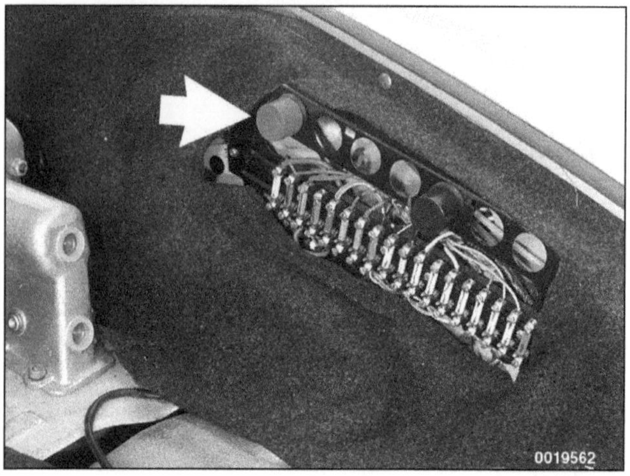

To operate fuel pump for testing purposes without having to run engine, the fuel pump relay, located in luggage compartment fuse box, can be bypassed to power pump directly.

– With ignition switched off, remove fuel pump relay (**arrow**).

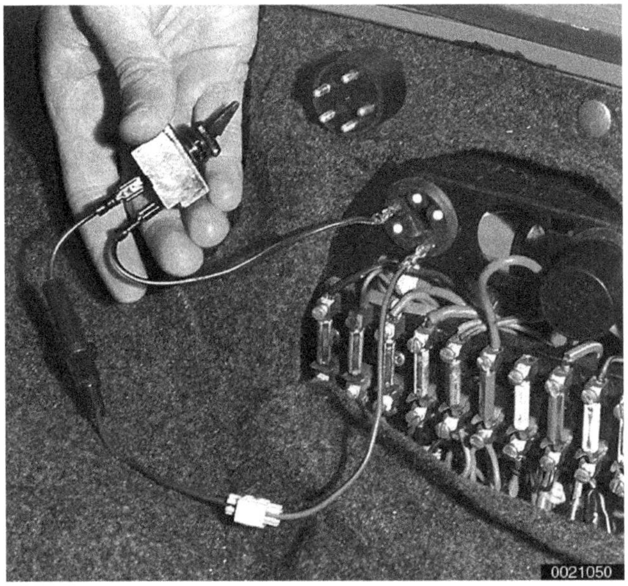

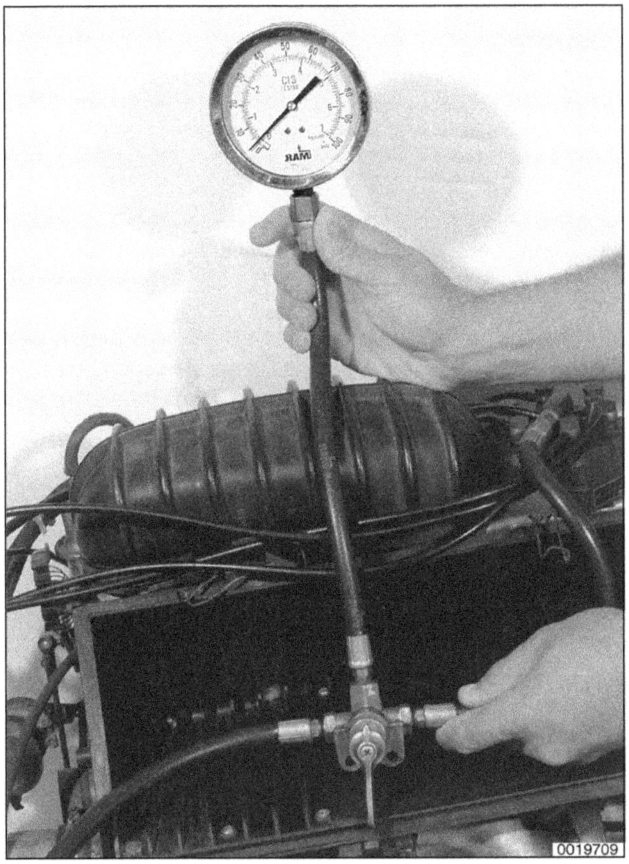

◄ Connect socket for relay terminal **30** to the socket for relay terminal **87a** with fused jumper wire.

NOTE —
The jumper wire should be 1.5 mm² (14 gauge.) and include an in-line fuse holder with a 25 amp fuse. To avoid harness connector damage from repeated connecting and disconnecting, also include a toggle switch.

– Turn ignition ON to run fuel pump.

– After completing tests, turn ignition off, remove jumper wire and reinstall relay.

> ***CAUTION —***
> *See* **970 Electrical Wiring Diagrams** *for additional wiring information.*

System pressure, testing

System pressure is created by the fuel pump and maintained by the line pressure regulator valve. The proper function of CIS fuel injection is dependent on system fuel pressure.

– Relieve fuel pressure and install and bleed pressure gauge as described earlier under **Relieving fuel pressure and connecting fuel pressure gauge**.

> ***WARNING —***
> - *Gasoline is highly flammable and its vapors are explosive. Do not smoke or work on a car near heaters or other fire hazards, when diagnosing and repairing fuel system problems. Have a fire extinguisher available in the event of an emergency.*
> - *To prevent fuel from spraying on a hot engine, system pressure should be relieved before disconnecting fuel lines. One method is to tightly wrap a shop towel around a the control pressure regulator line where it connects to the fuel distributor and then loosen the fuel line fitting.*

◄ Close valve on gauge as shown in photo.

– Operate fuel pump without running engine. See **Operating fuel pump for tests**. Check that fuel pressure corresponds to specifications listed in **Table a**.

Table a. System pressure specifications

Delivery condition	Pressure
Engine off, pump running	4.5 - 5.2 bar (65 - 75 psi)

– If system pressure is too high, check the following:

- Restricted fuel return line
- Faulty or incorrectly adjusted pressure regulator.

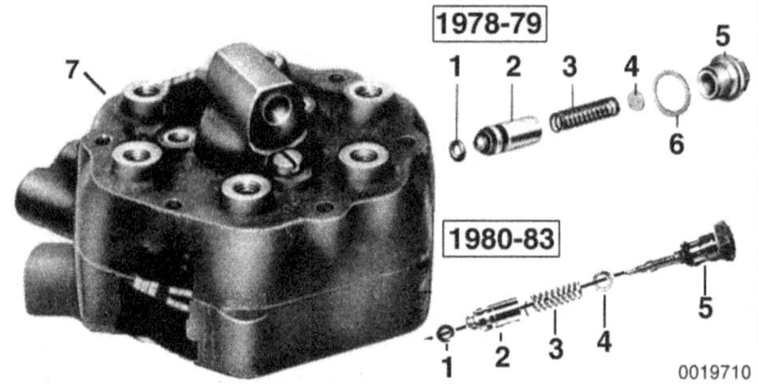

1. O-ring
2. Plunger
3. Spring
4. Shim(s)
5. End plug
6. Sealing ring
7. Fuel distributor

– If system pressure is too low, check the following:
  - External fuel leaks
  - Restricted/kinked fuel supply line
  - Clogged fuel filter or fuel pickup in tank
  - Faulty or incorrectly adjusted pressure regulator.

NOTE —
The fuel pressure relief valve can be adjusted by adding or subtracting shims at the pressure relief valve. Adding a 0.50 mm (.020 in.) shim will increase system pressure by about 0.30 bar (4 psi). Subtracting a 0.50 mm (.020 in.) shim will decrease system pressure by about 0.30 bar (4 psi).

Fuel delivery volume, testing

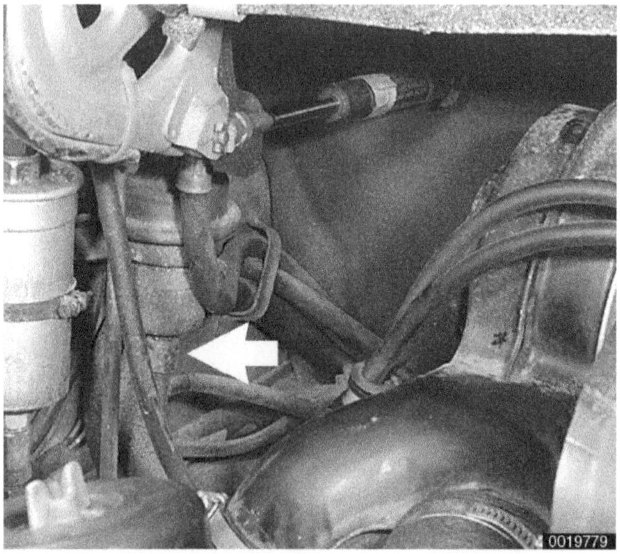

Disconnect fuel line from accumulator on left side of engine (**arrow**).

WARNING —
To prevent fuel from spraying on a hot engine, system pressure should be relieved before disconnecting fuel lines. One method is to tightly wrap a shop towel around a fuel line fitting and loosen or disconnect the fitting.

– Connect a length of hose to disconnected fuel line. Place open end of hose in a suitable container for catching fuel (2-quart capacity).

– Run fuel pump for exactly 30 seconds as described earlier under **Operating fuel pump for tests** and measure fuel collected. Compare to specification in **Table b**.

Table b. Fuel pump delivery volume

Delivery condition	Delivery rate
30 seconds @ 12V	1.0 liter (1.06 qt.), minimum

– When finished testing, reconnect fuel line. Tighten all hose clamps.

Fuel pump relay circuit

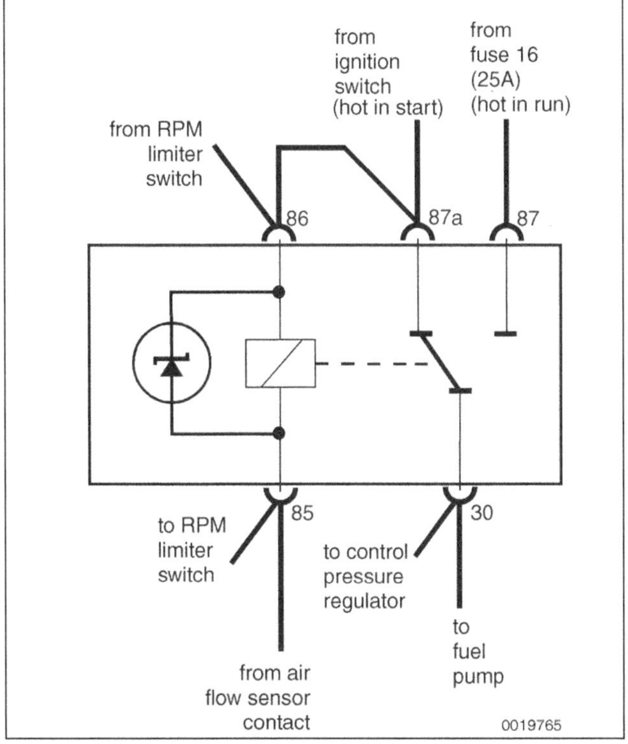

FUEL PUMP ELECTRICAL TESTS

During engine cranking, the fuel pump is supplied power from terminal 50 of the ignition switch (hot in start position) through the fuel pump relay. Incoming air then causes the air flow sensor plate to lift and open the sensor plate contact. This de-energizes the fuel pump relay coil and routes terminal 15 power of the ignition switch (hot in run position) directly to the pump.

NOTE —
If the engine starts only during cranking, check the air flow sensor contact (contact stuck closed).

CAUTION —
On cars with a factory alarm (alarm key on door edge), the alarm system interrupts power to terminal 85 of the fuel pump relay when the system is armed. When diagnosing a no-start condition be sure to check for key-on power at terminal 86.

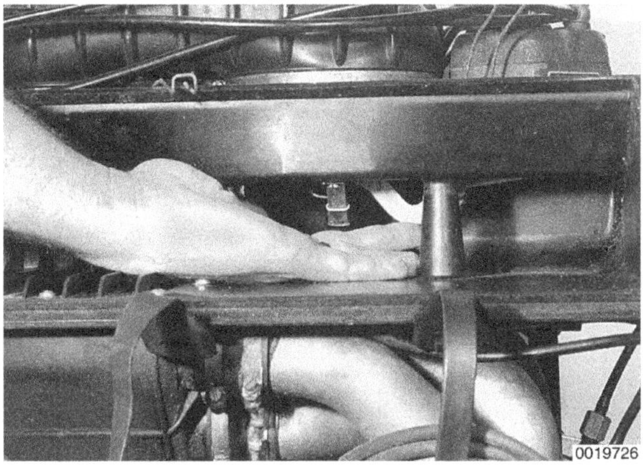

Fuel pump circuit, quick check

– Turn ignition switch in ON.

◄ Remove air filter housing. Lift up on air flow sensor plate lever. Pump should audibly run.

– If pump does not run:
  - Check fuel pump fuses. See **971 Electrical Component Locations**.
  - Test fuel pump relay and fuel pump circuit as described later.

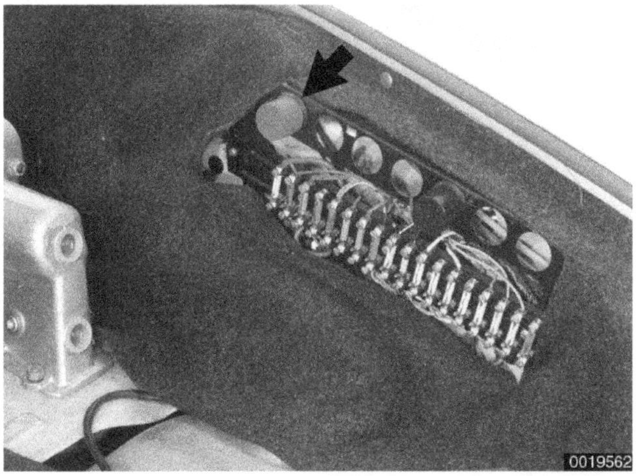

Fuel pump relay circuit, checking

◄ With ignition switch in OFF position, remove fuel pump relay (**arrow**) from main fuse panel in luggage compartment.

– With key ON, check for battery voltage at terminal 87a and terminal 86 in relay socket.

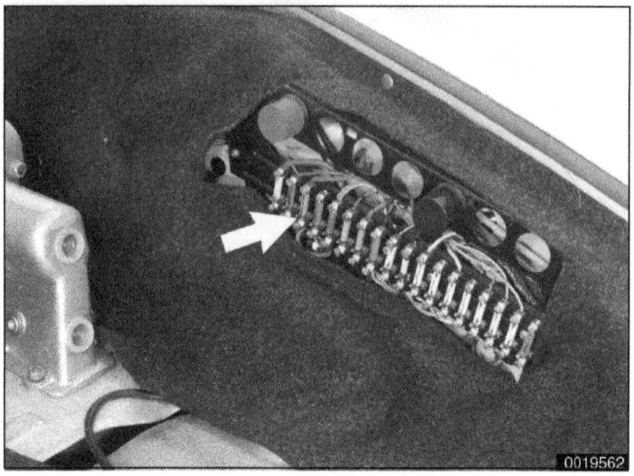

If battery voltage is not present, fuse (no. 16). Check for breaks in red wire between relay socket and fuse panel. See **970 Electrical Wiring Diagrams**.

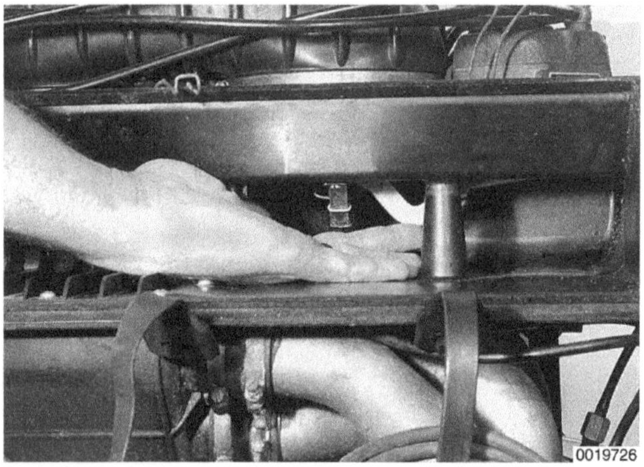

With key OFF, check for ground at terminal 85 in relay socket. Have a helper raise air flow sensor plate lever and check the ground is no longer present at terminal 85.

- If any faults are found, check air flow sensor plate contact (contact should be closed (continuity) with plate in rest position and open (no continuity) with plate raised).
- Check wiring between air flow sensor contact and fuel pump relay. See **970 Electrical Wiring Diagrams.**

– Turn ignition OFF and reinstall relay.

NOTE —
If no faults are found with wiring at the relay socket, but the pump operates when the relay is bypassed (see **Operating fuel pump for tests**)*, the relay is faulty and should be replaced.*

Fuel pump wiring, checking

If fuel pump fuse and fuel pump relay are OK, but fuel pump will not run when the relay is bypassed, check for voltage at pump.

– Raise car and support in a safe manner.

CAUTION —
Make sure the car is stable and well supported at all times. Use a professional automotive lift or jack stands designed for the purpose. A floor jack is not adequate support.

Working underneath front of car, remove stone guard mounting bolts (**arrow**). Remove stone guard.

– Working at fuel pump, pull back two rubber boots that protect fuel pump electrical connectors.

With fuel pump relay bypassed as described earlier under **Operating fuel pump for tests,** check for voltage between fuel pump connectors.

- If battery voltage and ground are present and connectors are not loose or corroded, then fuel pump is most likely faulty.
- If there is no voltage or ground, look for a faulty wire or connector.

Fuel pump current draw, checking

– Raise car and support in a safe manner.

> ***CAUTION —***
> *Make sure the car is stable and well supported at all times. Use a professional automotive lift or jack stands designed for the purpose. A floor jack is not adequate support.*

Working underneath front of car, remove stone guard mounting bolts (**arrow**). Remove stone guard.

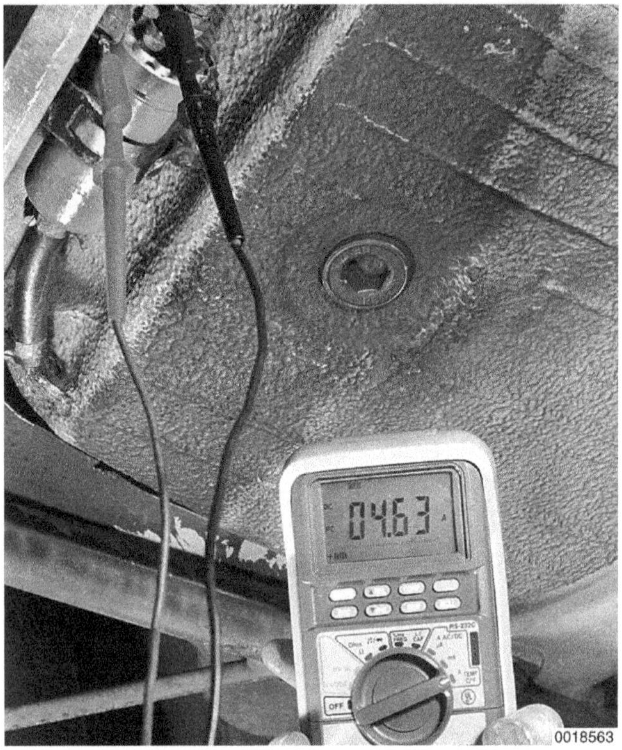

- Pull back two protective rubber boots from pump connectors. Disconnect black wire from pump.

◄ Connect ammeter between disconnected wire and fuel pump terminal as shown. The other wire remains connected to pump.

> ***CAUTION —***
> *Do not allow the test leads to short to ground.*

- Bypass fuel pump relay as described earlier under **Operating fuel pump for tests**. Fuel pump current should be in the range of 4 to 6 amps.

NOTE —
A higher than normal power consumption usually indicates a worn fuel pump, which may cause intermittent fuel starvation due to pump overheating and seizure. The only remedy is to replace the pump.

FUEL PUMP

◄ The fuel pump is mounted beneath the car, behind the fuel tank (**arrow**).

Fuel pump, replacing

- Disconnect negative (–) battery cable and cover battery terminal to keep cable from accidentally contacting terminal.

> ***CAUTION —***
> *Avoid getting tools or clothing near the battery. Battery electrolyte is a corrosive acid. Always wear eye protection when working on or near the battery.*

- Raise car and support in a safe manner.

> ***CAUTION —***
> *Make sure the car is stable and well supported at all times. Use a professional automotive lift or jack stands designed for the purpose. A floor jack is not adequate support.*

Working underneath front of car, remove stone guard mounting bolts (**arrow**). Remove stone guard.

Pinch off fuel feed hose (**arrow**) from tank with a fuel line clamp.

- Loosen hose clamp at pump end of hose.
- Detach hose from pump.

> ***WARNING —***
> - *Fuel will be expelled. Wear protective gloves and eye protection.*
> - *Gasoline is explosive. Have a fire extinguisher available in case of an emergency.*

– Disconnect two electrical wires from pump.

– Remove pressure side fuel line cap nut while counter holding pump fitting with wrench.

– Loosen pump mounting clamp and remove pump.

– Installation is reverse of removal.

- When replacing fuel pump inspect rubber insulators on fuel pump clamping bracket. Replace any that are cracked or damaged.
- Always replace sealing washers.

Tightening torques
- Electrical connector nuts . . . . 1.5 Nm (13 in-lbs)
- Fuel pump clamp bolt . . . . 7 Nm (62 in-lbs)
- Hose clamps . . . . 2 Nm (18 in-lbs)
- Stone guard to body (M10 bolt) . . . . 47 Nm (35 ft-lb)
- Stone guard to subframe (M10 nut) . . . . 28 Nm (21 ft-lb)

Fuel pump check valve, replacing

– Access fuel pump and pinch off fuel feed line as described earlier.

– Remove pressure side fuel line cap while counterholding pump fitting with wrench.

< Remove check valve (**arrow**) from end of pump.

- Replace copper sealing rings during reassembly.
- Start engine and check for leaks after tightening fittings, before reinstalling stone guard.

FUEL TANK AND FUEL LEVEL SENDER

Fuel Tank Capacity

- 1978 - 1983 Porsche SC. . . . . . . 21.1 US gals. (80 liters)

The fuel tank is located at the front of the car beneath the luggage compartment. The fuel level sending unit is mounted through the top of the tank. Connecting lines for the evaporative emission control system and fuel supply and return lines are also attached to the tank.

Fuel tank, draining

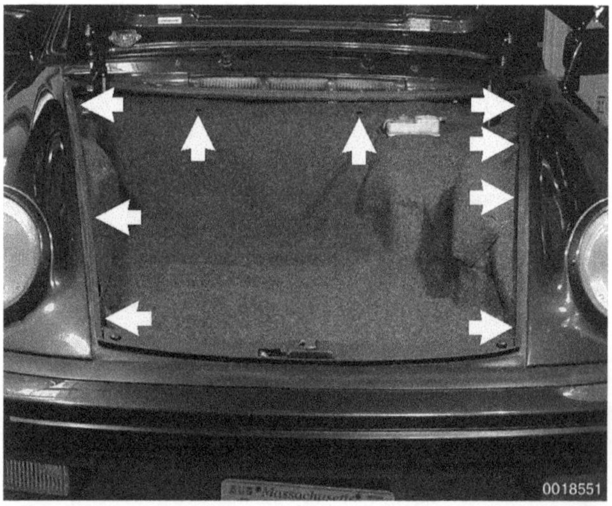

< Unsnap luggage compartment floor carpet (**arrows**) and lift off. Remove spare tire.

– Disconnect negative (–) battery cable and cover battery terminal to keep cable from accidentally contacting terminal.

> ***CAUTION —***
> *Avoid getting tools or clothing near the battery. Battery electrolyte is a corrosive acid. Always wear eye protection when working on or near the battery.*

– Remove fuel level sending unit as described later.

– Insert hose into tank opening and use an approved transfer pump and container to pump out fuel.

> ***WARNING —***
> - *Fuel will be spilled. Wear protective gloves.*
> - *Gasoline is explosive. Have a fire extinguisher available in case of an emergency.*

> ***CAUTION —***
> *Be sure the container is large enough to hold the volume of fuel that is to be pumped out of the tank.*

NOTE —
As an alternative, the drain plug (allen head) at the bottom of the fuel tank can be removed to drain tank.
– Use a large funnel and drum to catch draining fuel.
– Clean out filter element before reinstalling.

Tightening torque
- Fuel tank drain plug to tank . . . . 25 Nm (18 ft-lb)

Fuel tank assembly

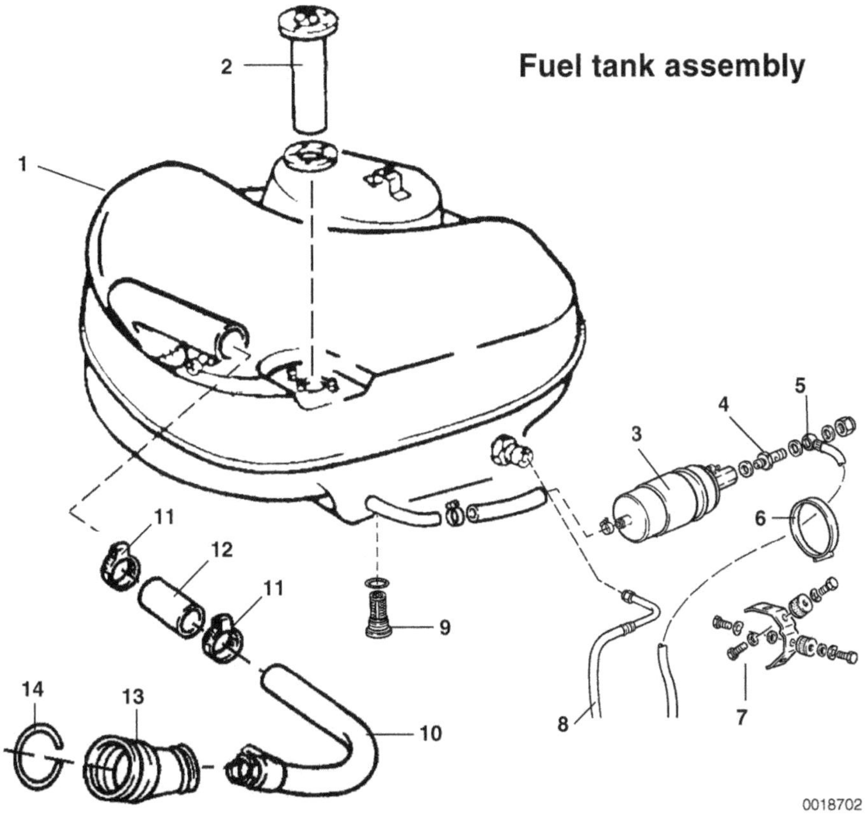

1. Fuel tank
2. Fuel level sending unit
3. Fuel pump
4. Fuel pump check valve
5. Fuel supply line (to fuel injection system)
6. Clamp
7. Fuel pump mounting assembly
8. Return line
9. Drain plug/screen filter
10. Connecting hose
11. Clamp
12. Connecting hose
13. Filler neck bellows
14. Circlip

Fuel tank, removing and installing

The fuel tank must be completely drained as described earlier before it is removed from the car.

- Disconnect battery negative (–) cable.

- Raise car and support in a safe manner.

> ***CAUTION —***
> *Make sure the car is stable and well supported at all times. Use a professional automotive lift or jack stands designed for the purpose. A floor jack is not adequate support.*

< Working underneath front of car, remove stone guard mounting bolts (**arrow**). Remove stone guard.

< Pinch off fuel feed hose (**arrow**) from tank with a fuel line clamp.

- Loosen hose clamp at tank end of hose.
- Detach hose from tank.

> ***WARNING —***
> - *Fuel will be expelled. Wear protective gloves and eye protection.*
> - *Gasoline is explosive. Have a fire extinguisher available in case of an emergency.*

- Lower car to ground.

< Working inside luggage compartment, detach hose from tank vent (**arrow**).

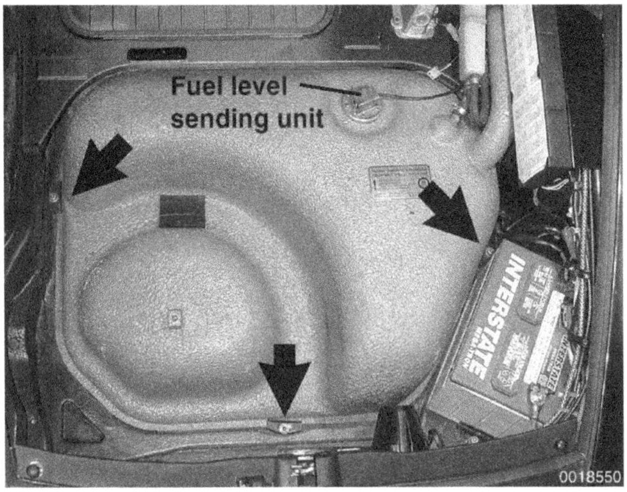

< Remove tank mounting bolts (**arrows**).

– Loosen filler neck clamps.

– Remove fuel tank by lifting it up and out through luggage compartment.

– Installation is reverse of removal.

- Always use new tank mounting gasket and hose clamps.
- Inspect all rubber parts including hoses and replace any that are worn, chafed, or cracked.

Tightening torques

- Fuel tank to body . . . . 20 Nm (15 ft-lb)
- Fuel drain plug . . . . 25 Nm (18 ft-lb)
- Stone guard to body
  M10. . . . . 45 Nm (33 ft-lb)
  M8. . . . . 25 Nm (18 ft-lb)

Fuel level sending unit, removing and installing

The fuel level sending unit is mounted in the top of the fuel tank. The unit has a float connected to a variable resistance for fuel level, and a switch that closes to turn on the low fuel light. When replacing the sending unit always replace hose clamps, gaskets and O-rings.

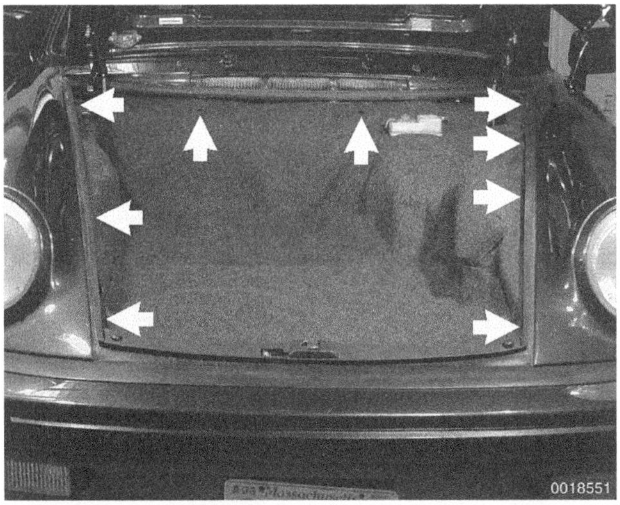

< Unsnap luggage compartment floor carpet (**arrows**) and remove carpet.

– Disconnect negative (–) battery cable and cover battery terminal to keep cable from accidentally contacting terminal.

> ***CAUTION —***
> *Avoid getting tools or clothing near the battery. Battery electrolyte is a corrosive acid. Always wear eye protection when working on or near the battery.*

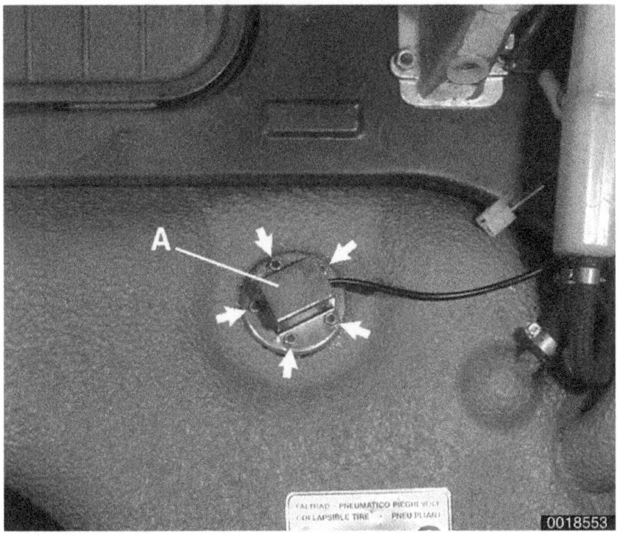

< Remove nuts and washers (**arrows**) mounting sending unit to tank. Remove harness connector cover (**A**).

– Disconnect sending unit harness connector.

◄ Slowly withdraw sending unit from fuel tank, allowing it to drain as it is being removed.

> ***WARNING —***
> - *Fuel will be spilled. Wear protective gloves.*
> - *Gasoline is explosive. Have a fire extinguisher available in case of an emergency.*

– Test unit as described below.

– When reinstalling use a new gasket.

NOTE —
New replacement fuel level sending units from Porsche may be fitted with a guard for protection during shipping. This guard must be removed prior to installing the unit.

Tightening torque
- Fuel level sending unit
  to fuel tank . . . . 2 Nm (18 in-lb)

– Reconnect harness connector. Fill fuel tank and check for leaks.

Fuel level sending unit, testing

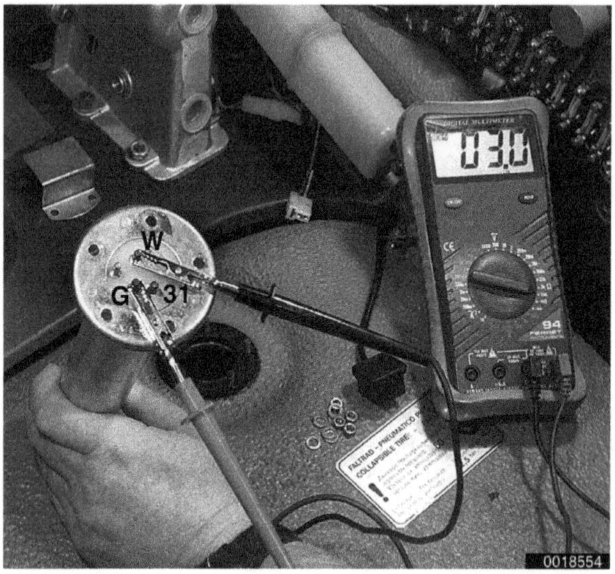

◄ Test sending unit using an ohmmeter connected across terminals. Compare results to specification in **Table c**.

Table c. Fuel gauge sending unit specifications

Connector terminals	Float position	Test results in Ω
G and **31**	Empty	71.7 ± 2.3
G and **31**	Full	3.2 ± 0.7
G and **31**	Slowly tilting removed sending unit to move internal float from empty to full	Resistance continuously changing
W and **31**	Empty low fuel light on	Continuity

NOTE —
To check the fuel gauge in the instrument cluster, see **901 Instruments**.

240 Fuel Injection

TABLES

240

GENERAL

This repair group covers specifically the fuel injection components of the CIS (Continuous Injection System), including the Oxygen Sensor Control System on 1980 and later models.

NOTE —
- *For fuel pump and related fuel supply component testing and repair see* **201 Fuel Supply.**
- *Wiring diagrams for the CIS system can be found in* **970 Electrical Wiring Diagrams.**

Test equipment

> ***CAUTION —***
> *A high impedance digital multimeter should be used for all voltage and resistance tests. An LED test light should be used in place of an incandescent-type test lamp.*

◄ Special equipment is necessary for some of the procedures given in this repair group. If you do not have the equipment required to do the job, it is recommended that these repairs be left to an authorized Porsche dealer.

Special Tools

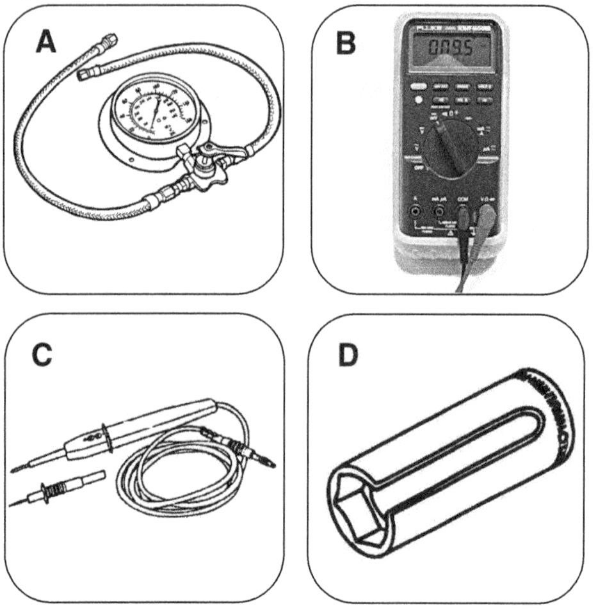

Special tool

A 1318 .... Fuel pressure gauge (source: Baum Tools Unlimited)
B Fluke 87.................... Automotive digital multimeter
C 1115................... LED tester with thin spade probes (source: Baum Tools Unlimited)
D 9222...................... Oxygen sensor slotted socket (Porsche ordering #000 721 922 21)

Warnings and cautions

Please observe the following cautions and warnings when performing any service or repair on the engine.

> ***WARNING —***
> - *The fuel system is designed to retain pressure even when the ignition is off. When working with the fuel system, loosen the fuel lines very slowly to allow residual fuel pressure to dissipate gradually. Avoid spraying fuel.*
> - *Fuel is highly flammable. When working around fuel, do not disconnect any wires that could cause electrical sparks. Do not smoke or work near heaters or other fire hazards. Keep an approved fire extinguisher handy.*
> - *The ignition system produces high voltages that can be fatal. Avoid contact with exposed terminals. Use extreme caution when working on a car with the ignition switched on or the engine running.*
> - *Do not touch or disconnect any high voltage cables from the coil, distributor, or spark plugs while the engine is running or being cranked by the starter.*

> ***CAUTION —***
> - *Before making any electrical tests that require the engine to be cranked using the starter, disable the ignition system as described in* **280 Ignition System**.
> - *Wire colors may vary from those given in text and schematics. Always check component terminals to verify wire colors.*

CIS fuel injection, introduction

The 911 SC models covered in this manual are equipped with a continuous fuel injection system called CIS. Fuel is injected continuously whenever the engine is running or being turned by the starter.

NOTE—
An operational overview of the CIS fuel injection system and system components, including emission controls, is given in **200 Engine Management–General**.

Continuous Injection System (CIS) components (1978 - 1979)

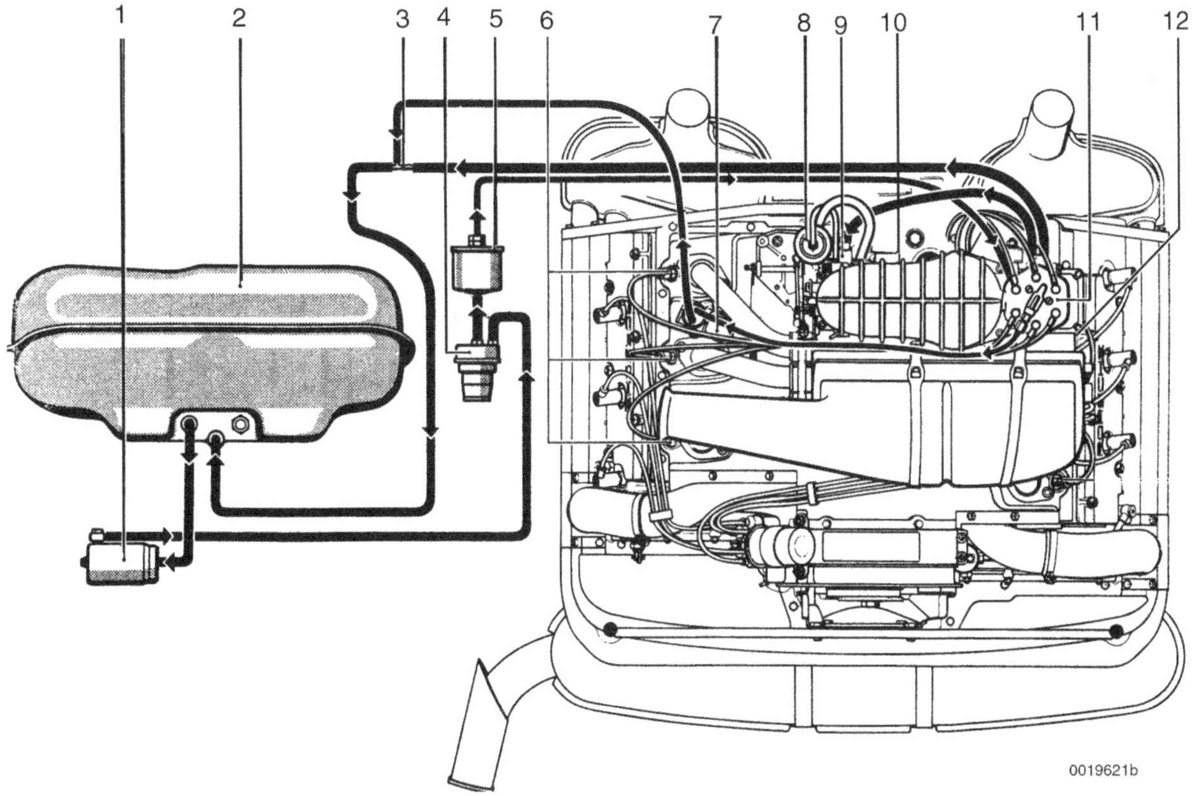

1. Fuel pump
2. Fuel tank
3. Fuel return plumbing
4. Fuel pressure accumulator
5. Fuel filter
6. Fuel injectors
7. Control pressure regulator
8. Decel air valve
9. Throttle switch (1980 and later)
10. Air intake boot
11. Fuel distributor
12. Auxiliary air regulator

BASIC SETTINGS

Before beginning diagnosis and troubleshooting of CIS problems, check to make sure all basic settings are correct.

Throttle plate basic setting, checking and adjusting

The factory-set throttle plate stop screw is not used to adjust idle speed. Its function is to provide a mechanical stop to prevent the throttle plate from contacting the throttle housing and causing wear. It should be adjusted only if the factory setting has been altered.

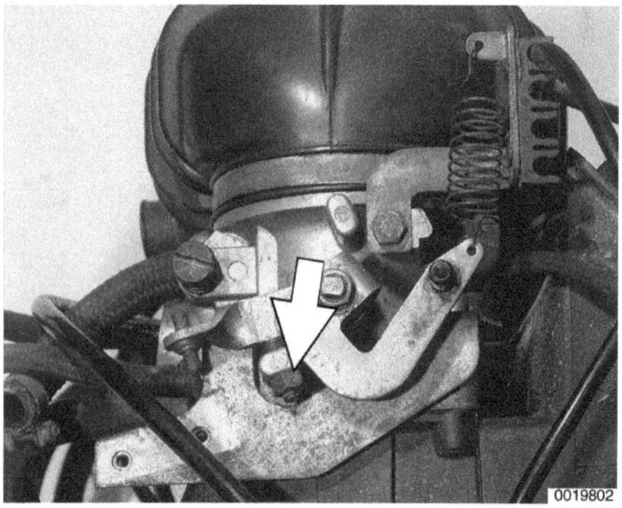

– Remove intake rubber boot from throttle housing.

< Loosen throttle plate adjusting screw locking nut (**arrow**) at left side of throttle housing.

– Turn screw counterclockwise just until throttle plate contacts throttle housing.

– Slowly turn screw clockwise just until throttle plate no longer contacts the throttle housing.

– Tighten throttle plate adjusting screw locking nut.

Air flow sensor basic setting, checking and adjusting

The sensor plate and its cone-shaped venturi are carefully designed to achieve the proper balance of air flow and fuel metering for all flow rates form idle to full throttle. If the sensor plate is off-centered or binding, or the lever has too much resistance, the fuel distributor cannot respond properly to the incoming air.

– Remove rubber intake air boot from top of air flow sensor.

– Disconnected electrical connector from CDI ignition control unit to disable ignition function. See **280 Ignition System**.

– Unhook air filter housing retaining straps and remove air filter housing horn and filter element.

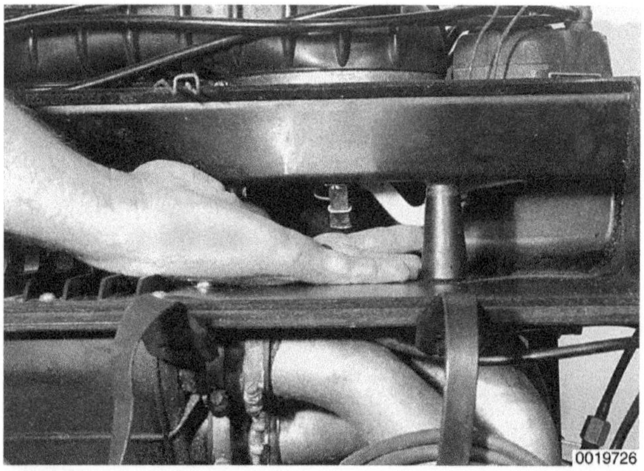

< Working from below, raise air flow sensor operating lever by hand.

- Resistance should be uniform throughout entire range of travel.

– Allow sensor plate to return to rest position.

- There should be no resistance. Plate should fall quickly and smoothly.

– If lever binds or is sluggish, remove mixture control unit assembly. Clean and lubricate pivot pin and operating lever. See **Mixture Control Unit** given later.

– Gently push down on top of sensor plate to check for binding at base of cone-shaped venturi.

◄ If plate binds, loosen sensor plate center bolt. Center sensor plate and tighten sensor plate bolt.

◄ Relieve fuel pressure by loosening control pressure fitting at fuel distributor (**arrow**). Wrap shop towel around fitting before loosening.

> ***WARNING —***
> *Fuel is highly flammable. When working around fuel, do not disconnect any wires that could cause electrical sparks. Do not smoke or work near heaters or other fire hazards. Keep an approved fire extinguisher handy.*

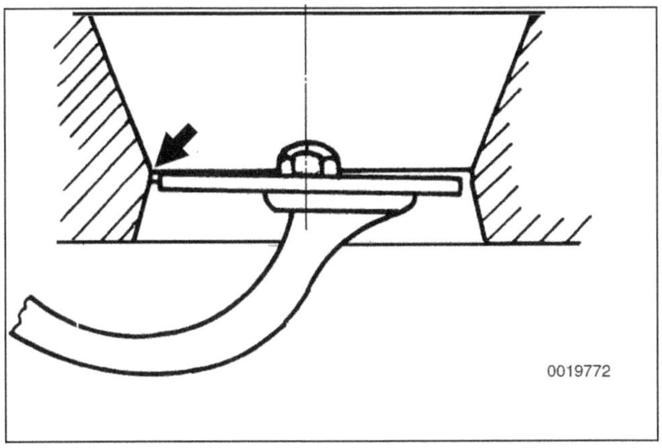

◄ Check sensor plate height with plate in rest position.

- At rest, front edge of plate should be aligned with narrowest point in venturi (**arrow**).

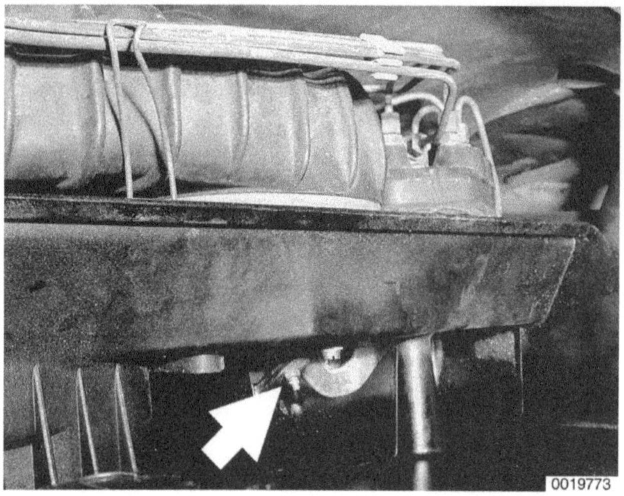

< Adjust basic height by adjusting spring-loaded stop at base of plate. Loosen adjusting locknut and screw (**arrow**) and reposition stop.

– Tighten control pressure fitting, reinstall rubber intake boot, air filter element and air filter housing.

– Reconnect CDI control unit connector.

– Start engine and adjust idle speed and mixture (%CO) as described below.

NOTE—
Be sure that control plunger moves freely in the fuel distributor bore. To check this, the fuel distributor with control plunger should be removed from the mixture control unit. See **Mixture Control Unit**, *given later.*

Idle speed, checking and adjusting

The procedure given below should be done on a fully warm engine. This will ensure that the control pressure regulator is not influencing idle speed.

This procedure assumes that all idle adapting components (control pressure regulator, auxiliary air regulator, auxiliary air valve, oxygen sensor system) are functioning correctly. If in doubt, be sure to check these components/systems prior to adjusting the idle speed.

NOTE—
- *Checking the control pressure regulator requires measuring the control pressure using a fuel pressure gauge with an integral 3-way valve. See* **Fuel Delivery Tests**.
- *If warm idle problems are encountered, check for vacuum leaks and places that would allow unmetered air to enter the engine. In particular, check that large intake air boots and intake hoses for cracks or leaks. Also make sure the crankcase is capable of holding a vacuum. Something as simple as a faulty oil tank cap can cause a low or erratic idle speed. Crankcase vacuum leaks will result in a poor idle quality.*

– Run engine until it reaches operating temperature.

– Connect external tachometer according to manufacturer's instructions.

< Adjust idle bypass screw until idle specification is obtained.

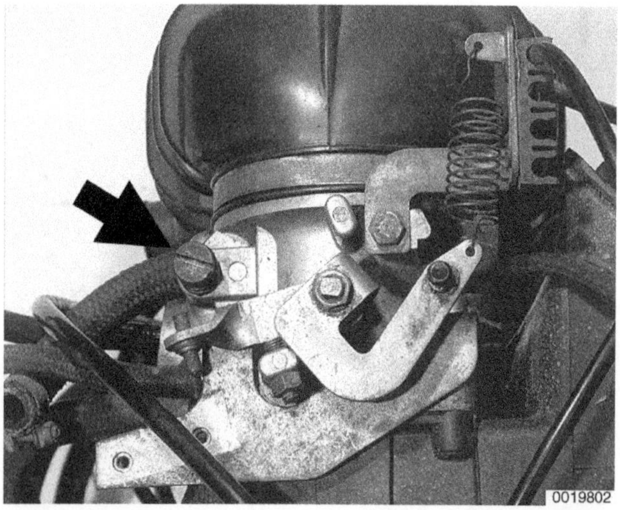

Table a. Idle speed

Specification	Adjusting Value	Checking conditions
Idle speed	900±50 rpm	Engine at operating temperature

Idle mixture, checking and adjusting

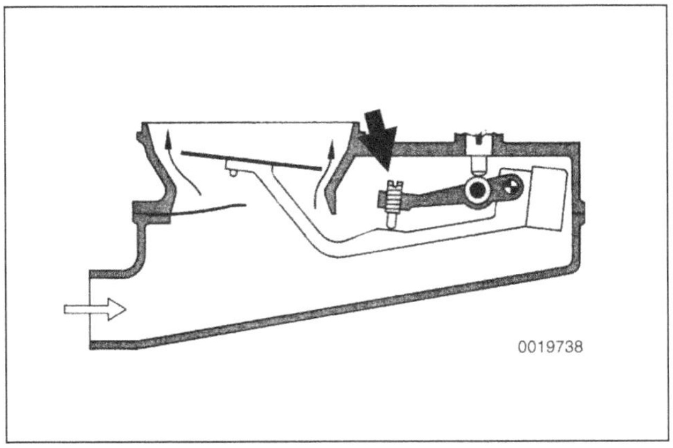

0019738

The idle mixture is adjusted via a 3 mm allen screw in the air flow sensor assembly (**arrow**). Turning the screw changes the relationship between the sensor plate and the control plunger. This adjustment has the greatest effect at idle.

The CO adjustment should be made by an authorized Porsche dealer or other qualified shop with the appropriate exhaust gas analyzer test equipment.

- Remove test connection plug from exhaust pipe ahead of catalytic converter. Connect exhaust pickup line from gas analyzer to exhaust pipe.
- Set parking brake. Start engine and run until engine is fully warmed up (oil temperature gauge on upper end of white range, oil temperature about 90°C).

0019804

On cars fitted with secondary air pump: disconnect and plug pressure hose from the rear of pump (**arrow**).

0019805

On 1980 and later models, disconnect oxygen sensor connector at left side of engine compartment (**arrow**).

NOTE —

- *On 1980 and later models, the idle mixture (%CO) is adaptive via the oxygen sensor feedback system. By disconnecting the harness connector, the oxygen sensor control unit substitutes a fixed default value.*

– With engine fully warmed up and idling at the specified idle speed, check that CO content is within allowable limits.

Table b. Basic settings

Parameter	Adjusting Value	Checking conditions
Idle speed	900 ± 50 rpm	Engine at operating temperature
Idle mixture (%CO) 1978 - 1979 models 1980 - 1983 models	 1.5 to 3.5% 0.4 to 0.8%	-Measured at test port on front of catalytic converter -Oxygen sensor disconnected (1980 - '83) -Air pump to check valve hose disconnected and plugged.
HC	<300 ppm	

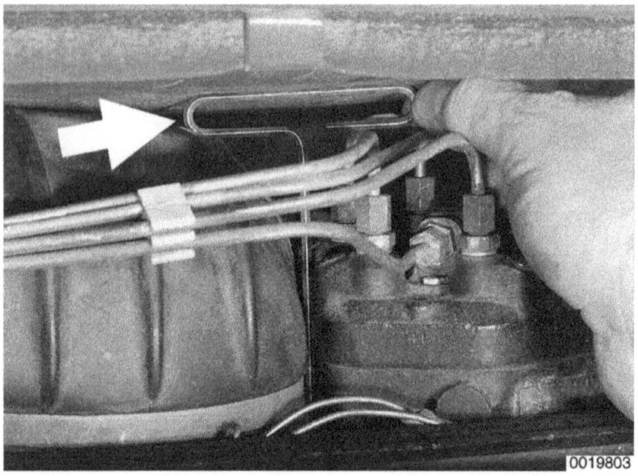

If adjustment is necessary, remove plug from mixture control unit. Insert 3 mm Allen wrench (**arrow**) into hole in mixture control unit to engage Allen screw. Adjust screw until mixture is correct, making final adjustments always in rich (clockwise) direction.

NOTE —

- *Do not push down on allen wrench during adjustment or the engine will stall. Make small adjustments. Remove the tool and raise the engine speed a few times before rechecking the adjustment. Do not raise the idle speed with the tool in place.*
- *On 1981 and later cars, the CO adjusting hole is capped off with an anti-tamper plug, To remove the plug, drill a 2 mm hole 3.5 mm in depth in the aluminum core of the steel plug. Screw a sheet medal screw into the drilled hole and pull the plug out. Be sure to replace the plug when finished.*

Ignition timing, checking and adjusting

Before proceeding with further CIS testing, be sure the ignition timing is correctly adjusted. Ignition timing is covered in **280 Ignition System**.

FUEL DELIVERY TESTS

Checking fuel delivery is a fundamental part of troubleshooting and diagnosing the fuel injection system. Fuel pressure directly influences fuel delivery. An accurate fuel pressure gauge with an integral 3-way valve will be needed to make the tests.

Three significant fuel delivery values can be measured:

- **System pressure** is created by the fuel pump and maintained by the pressure regulator.
- **Control pressure** is created by the fuel pump and affected by restrictions, such as clogged fuel filter.
- **Residual pressure** is maintained in the closed system after the engine and fuel pump are shut off.

> ***WARNING —***
> - *When working on an open fuel system, wear suitable hand protection, as prolonged contact with fuel can cause illnesses and skin disorders.*
> - *Gasoline is highly flammable and its vapors are explosive. Do not smoke or work on a car near heaters or other fire hazards when diagnosing and repairing fuel system problems. Have a fire extinguisher available in case of an emergency.*

> ***CAUTION —***
> - *Cleanliness is essential when working with fuel circuit components. Thoroughly clean the unions before disconnecting fuel lines.*
> - *Renew fuel system hoses, clamps and O-rings any time they are removed.*

Relieving fuel pressure and connecting fuel pressure gauge

0019801

◄ Remove controlled pressure regulator line from center port on fuel distributor.

> ***WARNING —***
> *To prevent fuel from spraying on a hot engine, system pressure should be relieved before disconnecting fuel lines. One method is to tightly wrap a shop towel around a the control pressure regulator line where it connects to the fuel distributor and than loosen the fuel line fitting.*

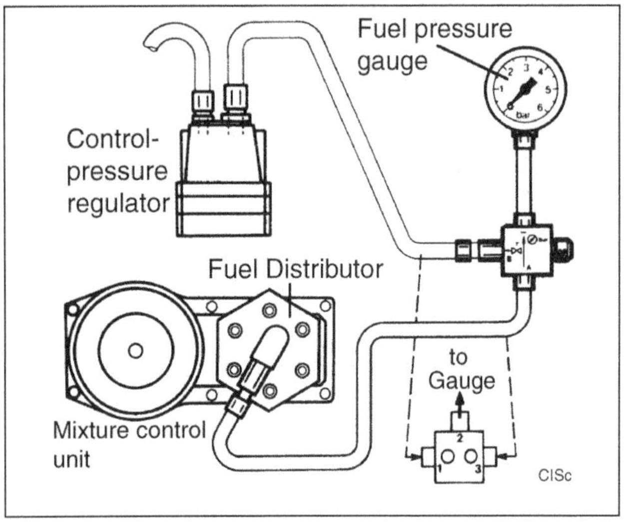

Connect fuel gauge (special tool 1318, or equivalent) between open port on fuel distributor and disconnected line.

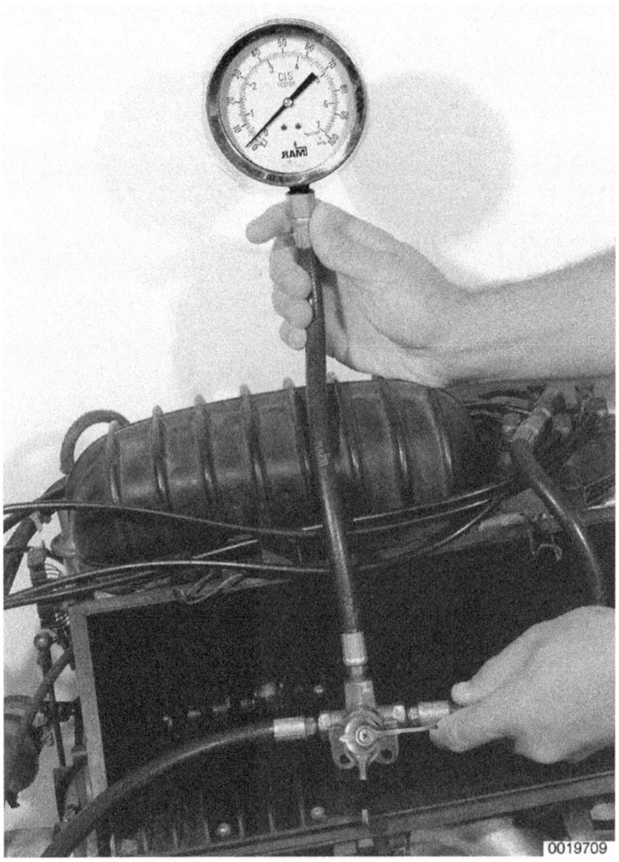

With fuel pump operating (see **Operating fuel pump for tests**), turn pressure gauge upside down open and close valve on gauge several times to bleed trapped air. When finished, turn gauge up-right.

NOTE —
The pressure gauge should have a range of 0 to 7 bar (0 to 100 psi).

Operating fuel pump for tests

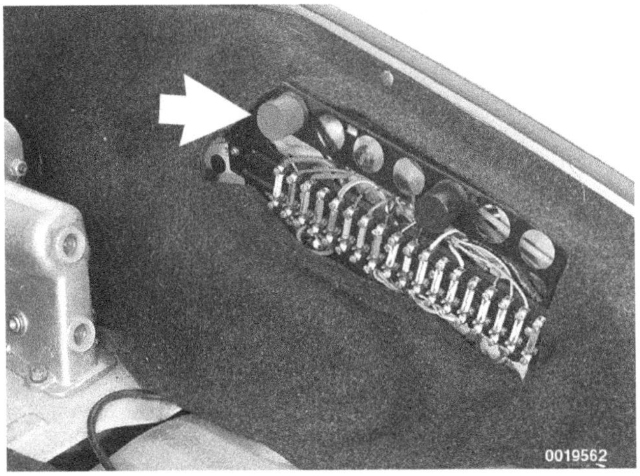

To operate the fuel pump for testing purposes without having to run the engine, the fuel pump relay can be bypassed to power pump directly.

– With ignition switched off, remove fuel pump relay (**arrow**) in luggage compartment.

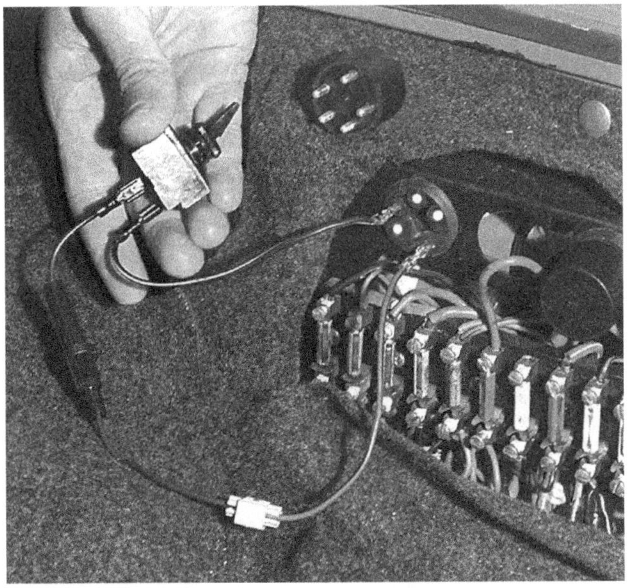

Using a fused jumper wire, connect socket for relay terminal **30** to the socket for relay terminal **87a** with fused jumper wire.

NOTE —

The jumper wire should be 1.5 mm² (14 gauge.) and include an in-line fuse holder with a 25 amp fuse. To avoid harness connector damage from repeated connecting and disconnecting, also include a toggle switch.

– Turn ignition on to run fuel pump.

– After completing tests, turn ignition off, remove jumper wire and reinstall relay.

> ***CAUTION —***
> *See* **970 Electrical Wiring Diagrams** *for additional wiring information.*

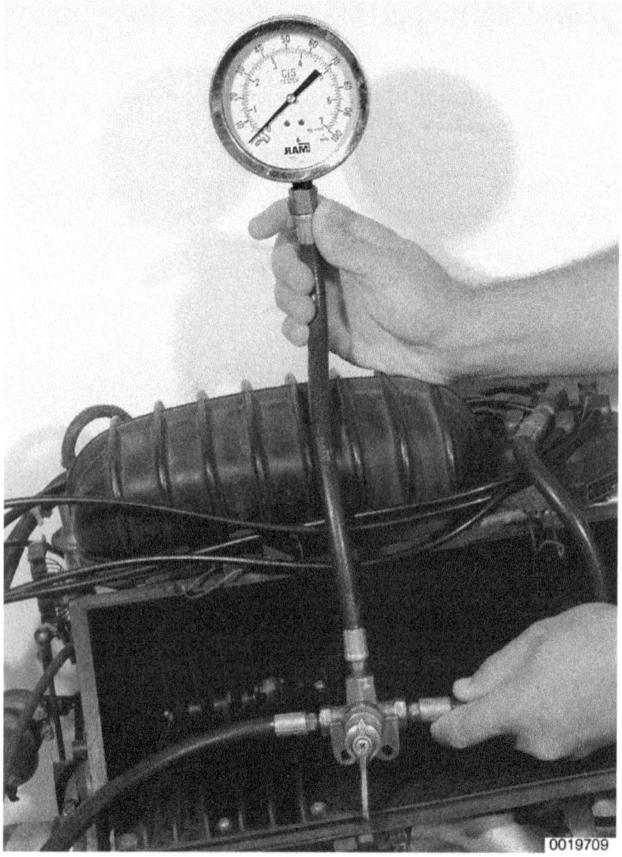

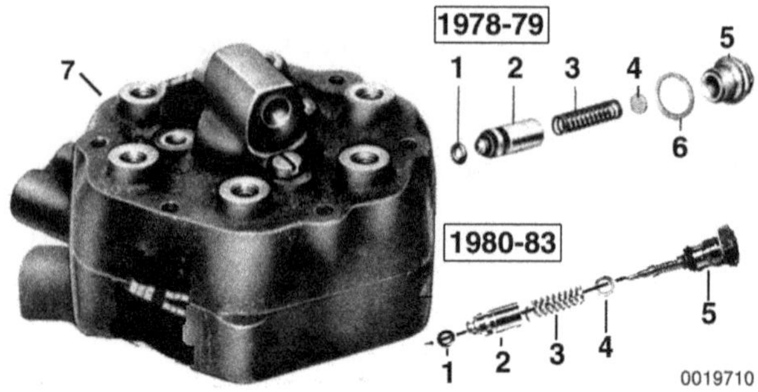

1. O-ring
2. Plunger
3. Spring
4. Shim(s)
5. End plug
6. Sealing ring
7. Fuel distributor

System pressure, checking

System pressure is created by the fuel pump and maintained by the line pressure regulator valve. The proper function of CIS fuel injection is dependent on system fuel pressure. This is the fundamental pressure that all other system functions rely on. Be sure system pressure is correct before any other diagnostic work is carried out.

– Relieve fuel pressure and install and bleed pressure gauge as described earlier under **Relieving fuel pressure and connecting fuel pressure gauge**.

> ***WARNING —***
> - *Gasoline is highly flammable and its vapors are explosive. Do not smoke or work on a car near heaters or other fire hazards, when diagnosing and repairing fuel system problems. Have a fire extinguisher available in the event of an emergency.*
> - *To prevent fuel from spraying on a hot engine, system pressure should be relieved before disconnecting fuel lines. One method is to tightly wrap a shop towel around a the control pressure regulator line where it connects to the fuel distributor and than loosen the fuel line fitting.*

< Close valve on gauge as shown in photo.

– Operate fuel pump without running engine. See **Operating fuel pump for tests**. Check that fuel pressure corresponds to specifications listed in **Table c**.

Table c. System pressure specifications

Delivery condition	Pressure
Engine off, pump running	4.5 - 5.2 bar (65 - 75 psi)

– If system pressure is too high, check the following:
- restricted fuel return line
- incorrectly adjusted pressure regulator.

– If system pressure is too low, check the following:
- external fuel leaks
- restricted/kinked fuel supply line
- clogged fuel filter or fuel pickup in tank
- leaking or incorrectly adjusted pressure regulator.

NOTE —
The fuel pressure can be adjusted by adding or subtracting shims at the pressure regulator. Adding a 0.50 mm (.020 in.) shim will increase system pressure by about 0.30 bar (4 psi). Subtracting a 0.50 mm (.020 in.) shim will decrease system pressure by about 0.30 bar (4 psi).

Control pressure, checking

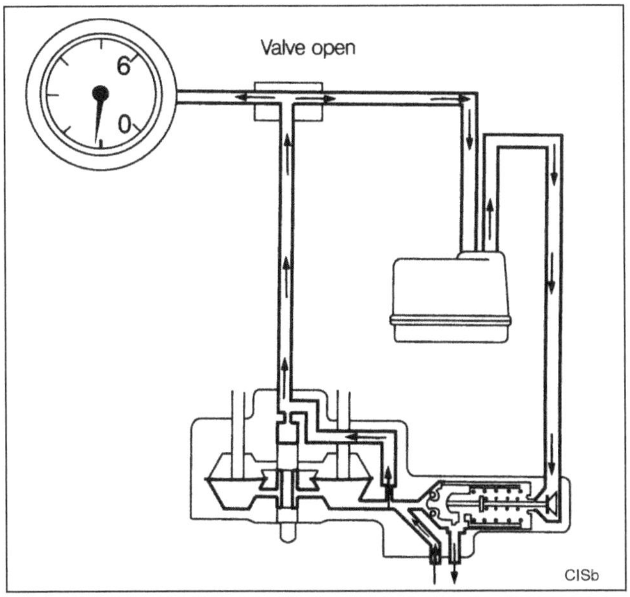

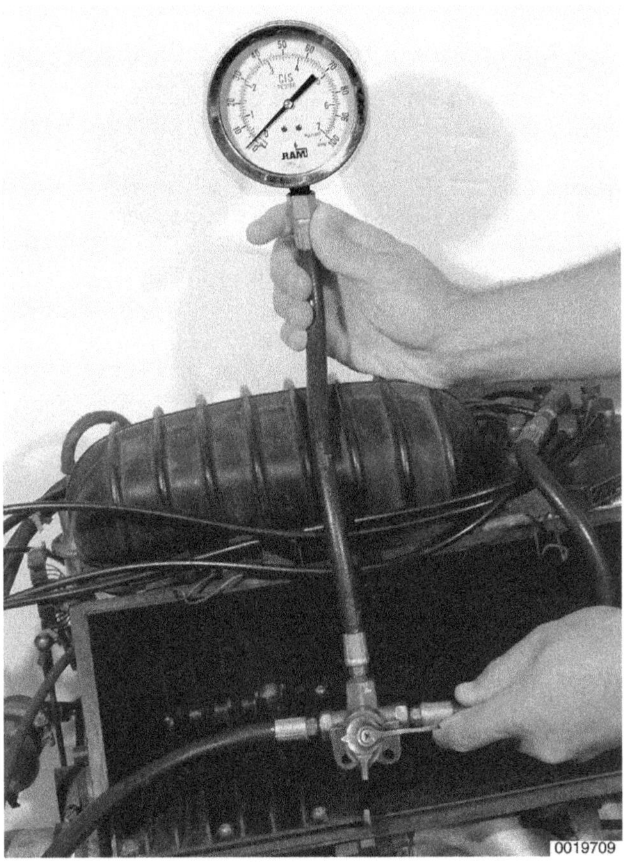

Control pressure is the pressure supplied to the top of the control plunger. This pressure is the basis for cold and warm-up enrichment. When the engine is cold, this control pressure is reduced to allow the air flow sensor/control plunger to lift higher for a given air flow and richer mixture.

When experiencing cold engine driveability and drive-off hesitation, check the cold control pressure. Always make this test on a cold engine. For the most accurate test, the vehicle should be left to sit overnight.

NOTE —

System pressure must be correct before checking control pressure and the engine temperature should be below 20°C (68°F) for the most accurate check of the control pressure regulator.

< Relieve fuel pressure and install and bleed pressure gauge as described earlier under **Relieving fuel pressure and connecting fuel pressure gauge**.

WARNING —

- *Gasoline is highly flammable and its vapors are explosive. Do not smoke or work on a car near heaters or other fire hazards, when diagnosing and repairing fuel system problems. Have a fire extinguisher available in the event of an emergency.*
- *To prevent fuel from spraying on a hot engine, system pressure should be relieved before disconnecting fuel lines. One method is to tightly wrap a shop towel around a the control pressure regulator line where it connects to the fuel distributor and than loosen the fuel line fitting.*

< Open valve on gauge as shown in photo.

– Operate fuel pump without running engine. See **Operating fuel pump for tests**. Check that control pressure corresponds to specifications listed in graphs that follow. See **Table d** through **Table g**.

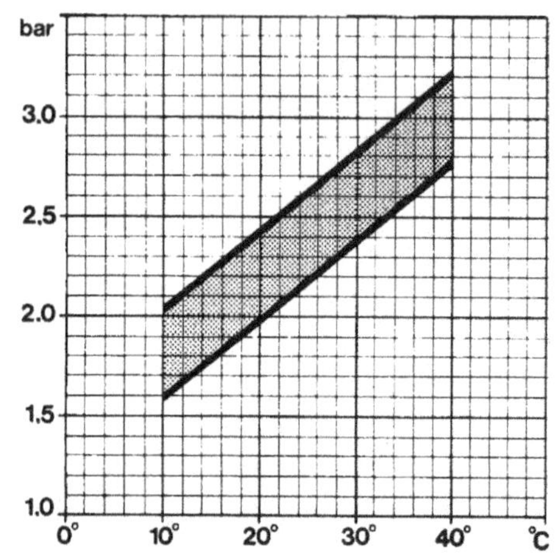

Control pressure graph for SC models **manufactured from start of production 1978 to December 1979** equipped with 911 606 105 05 regulator (Bosch # 0 438 140 045).

Table d. Control pressure specifications

Start of Prod. 1978 to Dec. 1978	Fuel pressure
Engine cold and running	See graph
Engine warm and running with vacuum line attached with vacuum line removed	 3.2 - 3.6 bar (46 - 53 psi) 2.7 - 3.1 bar (39 - 45 psi)

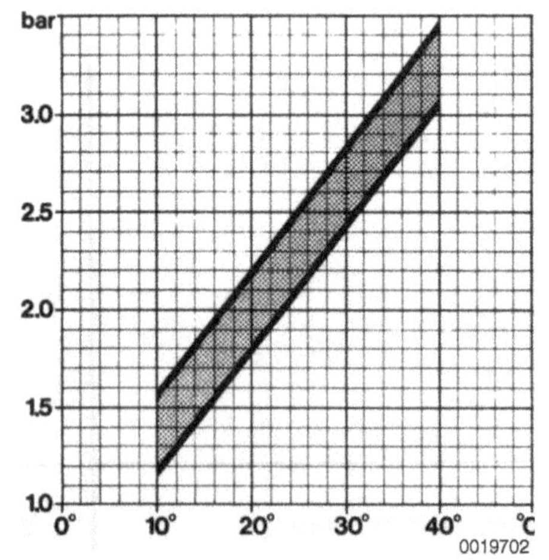

Control pressure graph for SC models **manufactured from January 1979 to August 1979** equipped with 911 606 105 06 control pressure regulator (Bosch # 0 438 140 069).

Table e. Control pressure specifications

Jan. 1979 to Aug. 1979	Fuel pressure
Engine cold and running	See graph
Engine warm and running with vacuum line attached with vacuum line removed	 3.2 - 3.6 bar (46 - 53 psi) 2.7 - 3.1 bar (39 - 45 psi)

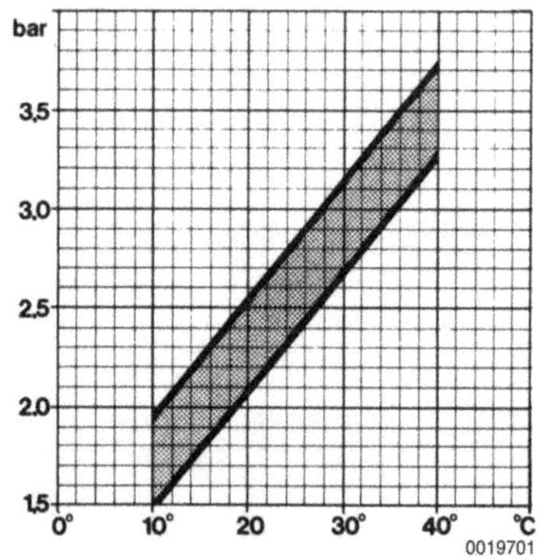

Control pressure graph for SC models for **1980 model year with oxygen sensor** equipped with 911 606 105 07 or 911 606 105 010 control pressure regulator (Bosch # 0 438 140 072).

Table f. Control pressure specifications

1980 with oxygen sensor	Fuel pressure
Engine cold and running	See graph
Engine warm and running	3.4 - 3.8 bar (50 - 55 psi)

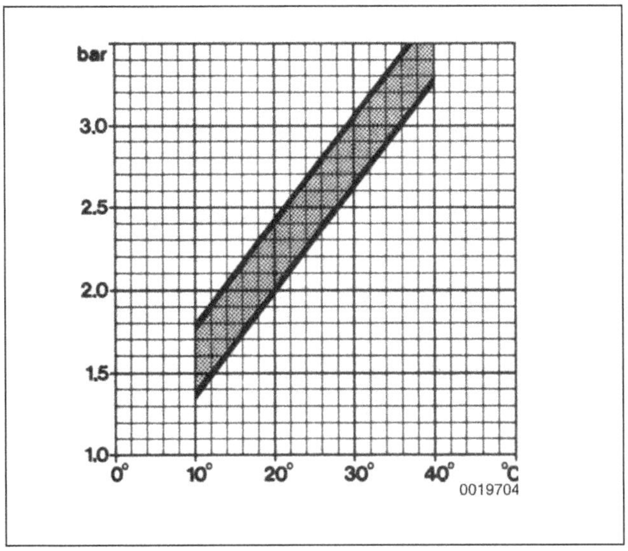

Control pressure graph for SC models for **1981-1983 model years oxygen sensor** equipped with 911 606 105 08 control pressure regulator (Bosch # 0 438 140 090).

Table g. Control pressure specifications

1981 to 1983	Fuel pressure
Engine cold and running	See graph
Engine warm and running	3.4 - 3.8 bar (50 - 55 psi)

Residual pressure, checking

Residual pressure is the pressure maintained in the fuel system after the engine is turned off. This pressure helps to prevent vapor lock in the fuel lines and provides for faster restarts. Residual fuel pressure should be checked if the engine is difficult to start when hot.

– Relieve fuel pressure and install and bleed pressure gauge as described earlier under **Relieving fuel pressure and connecting fuel pressure gauge**.

> ***WARNING —***
> *To prevent fuel from spraying on a hot engine, system pressure should be relieved before disconnecting fuel lines. One method is to tightly wrap a shop towel around a the control pressure regulator line where it connects to the fuel distributor and than loosen the fuel line fitting.*

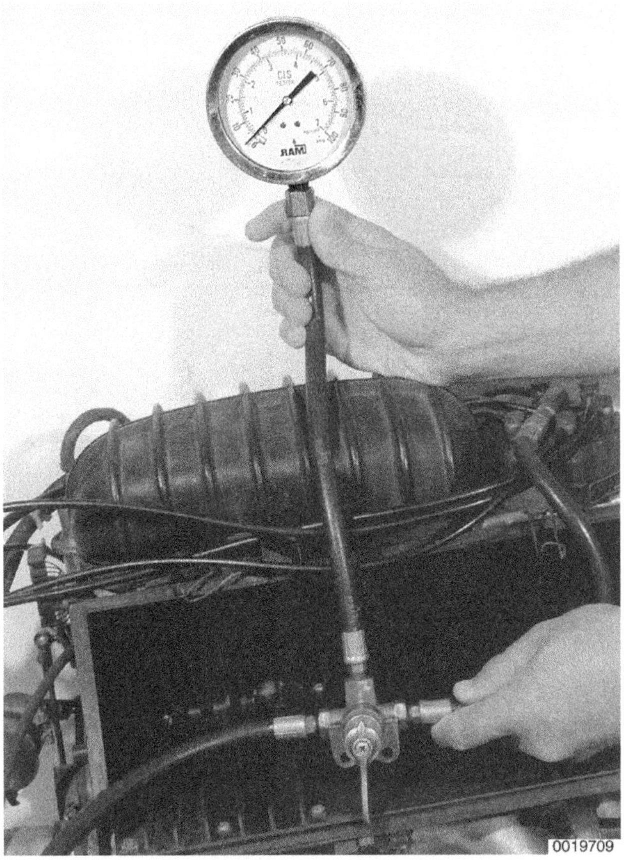

Close valve on gauge as shown in photo.

– Operate fuel pump without running engine. See **Operating fuel pump for tests**. Check that system pressure corresponds to specifications listed in **Table h**.

Table h. System pressure specifications

Engine off, pump running	4.5 - 5.2 bar (65 - 75 psi)

> ***WARNING —***
> - *Gasoline is highly flammable and its vapors are explosive. Do not smoke or work on a car near heaters or other fire hazards, when diagnosing and repairing fuel system problems. Have a fire extinguisher available in the event of an emergency.*
> - *To prevent fuel from spraying on a hot engine, system pressure should be relieved before disconnecting fuel lines. One method is to tightly wrap a shop towel around a the control pressure regulator line where it connects to the fuel distributor and than loosen the fuel line fitting.*

– Run engine for at least three minutes and switch off ignition. See **Table i** for residual pressure specifications.

Table i. Residual pressure

Leak test	Fuel pressure
after 10 minutes	1.3 bar (19 psi) minimum
after 30 minutes	1.1 bar (16 psi) minimum

NOTE —
If residual value is OK with the gauge valve closed, open valve to check control pressure regulator circuit. If residual pressure drops, when the valve is open, check the line pressure regulator for faults first. If no faults are found there, check the control pressure regulator for internal leaks.

FUEL INJECTORS

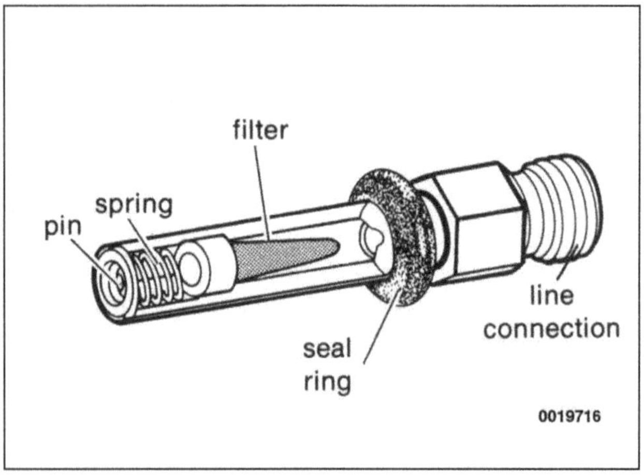

The mechanical fuel injectors used are precision nozzles. Above a certain pressure, the injector nozzles open and continuously deliver atomized fuel into the intake ports.

The injectors a mounted in inserts that are screwed into the intake runners. The injector are held in place in the inserts by thick rubber O-ring.

The most common problems associated with fuel injectors are low fuel flow rate or uneven spray pattern, both caused by carbon deposits built up around the discharge nozzle or by contaminated fuel. Some signs of faulty injectors are hard starting, rough idle, and detonation under load.

Fuel injector, quick-check

Each individual injector can be quick-checked to evaluate its spray pattern and to check for leakage.

– Remove fuel injector as described under **Fuel injectors, removing and installing**, Leave fuel line connected.

WARNING —
- *Gasoline is highly flammable and its vapors are explosive. Do not smoke or work on a car near heaters or other fire hazards, when diagnosing and repairing fuel system problems. Have a fire extinguisher available in the event of an emergency.*
- *To prevent fuel from spraying on a hot engine, system pressure should be relieved before disconnecting fuel lines. One method is to tightly wrap a shop towel around a the control pressure regulator line where it connects to the fuel distributor and than loosen the fuel line fitting.*

– Place injector in a clear fuel-resistant container.

– Disconnect rubber boot from air flow sensor. Unclip and separate air filter housing to access underside of air flow sensor.

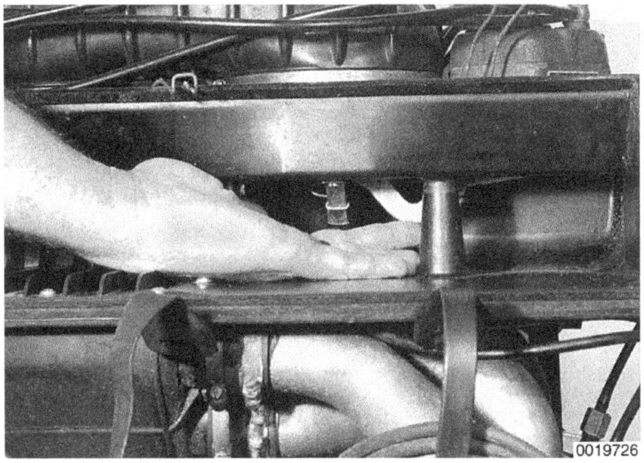

0019726

◄ Turn ignition on and slowly lift air sensor plate from below. Observe injector spray pattern.

- The injector should spray a fine mist in a uniform cone-shaped pattern.

NOTE —
An injector with an irregular spray pattern, or one that drips fuel when the sensor plate is at rest, should be replaced.

– Release sensor plate to stop flow of fuel. Hold injector in horizontal position to make sure that it does not drip fuel.

> ***WARNING —***
> *Fuel will be discharged from all injectors during this check. Keep the test time short to discharge as little fuel as possible. Do not smoke or work near fire hazards. Keep a fire extinguisher handy.*

– Repeat test on remaining injectors.

– Reinstall fuel injectors with new O-ring seal.

– Turn engine over by hand to clear fuel from cylinders before starting.

Fuel injectors, bench testing

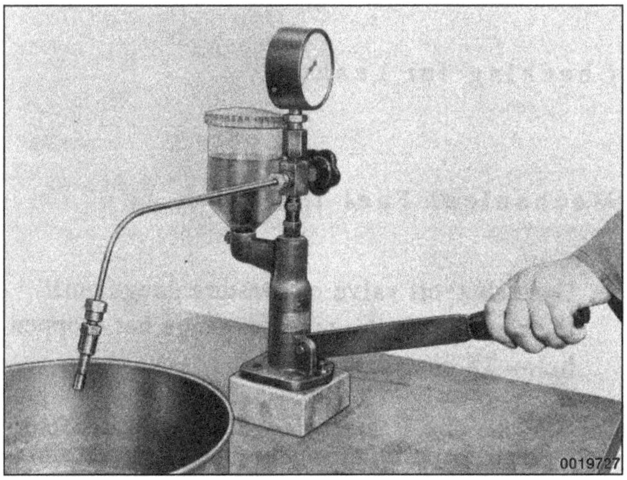

0019727

◄ A special injector tester can be used to measure opening pressure or check and flush new injectors prior to installation.

– Remove injector as described under **Fuel injectors, removing and installing**.

– Install injector on tester line.

– Pump tester lever several times to purge air from line.

– Pump tester lever slowly while reading pressure gauge as injector starts to open.

- Any injector which opens or leaks below the specified opening pressure should be replaced.
- If injector continues to have an uneven spray pattern, it should be replaced.

– Check that opening pressure corresponds to specifications listed in **Table j**.

Table j. Injector opening pressure specifications

Bench test	2.5 - 3.6 bar (34 - 52 psi)

> ***WARNING —***
> *Use Shell mineral spirit 135 or equivalent when testing or cleaning injectors. Shell mineral spirit 135 is flammable. Do not smoke or work near fire hazards. Keep a fire extinguisher handy.*

Fuel injectors, removing and installing

The fuel injectors must be removed for inspection or testing. The injectors are held in place O-rings.

- To remove an injector, disconnect fuel lines and pull straight out, forcing it out of retaining O-ring. Pull straight out on injector body and not fuel line.

- Installation is reverse of removal. Always replace injector seals. Soak new seals in fuel for several minutes before installing. Press injector into insert until O-ring is fully seated in groove.

Injector insert, removing and installing

It is not normally necessary to remove the injector inserts, except to replace the O-rings and cure air leaks which can cause rough idling. The inserts have an internal hex-shaped socket and can be removed using the proper size hex wrench. The O-rings seals should be replaced whenever the inserts are removed.

Remove the insert and clean the threads, replace the insert and coat the insert threads with Loctite 620 or equivalent. Moisten O-ring seal with fuel before installing into intake runner.

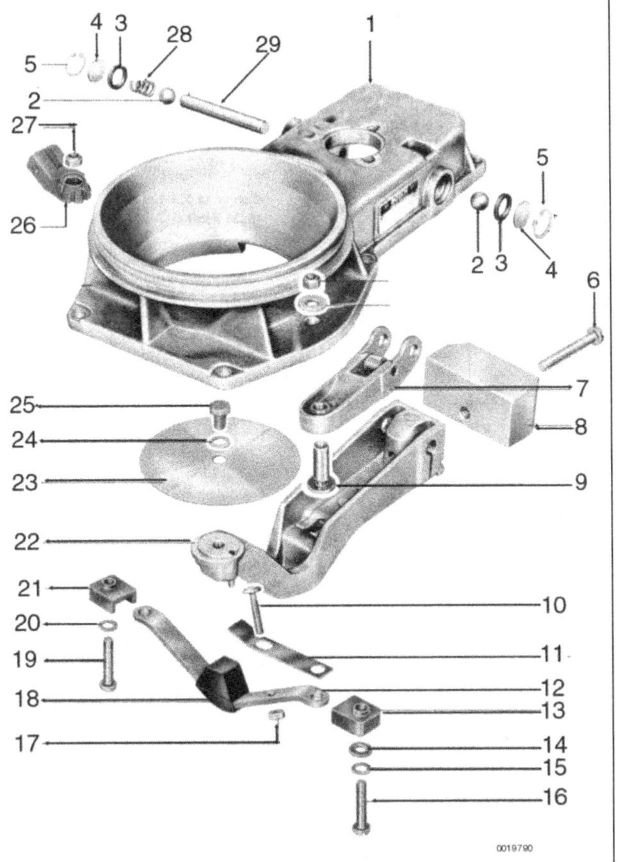

1. Mixture control unit housing
2. Ball
3. Seal
4. End cover
5. Circlip
6. Cap screw
7. Mixture adjusting lever
8. Counterweight
9. Mixture adjusting screw
10. Sensor plate stop adjusting screw
11. Leaf spring
12. Stop bracket
13. Bracket
14. Washer
15. Washer
16. Cap screw
17. Nut
18. Buffer
19. Cap screw
20. Washer
21. Bracket
22. Air flow sensor plate lever
23. Air flow sensor plate
24. Lock washer
25. Bolt
26. Plug connector for sensor plate contact
27. Nut
28. Spring
29. Pivot pin
30. Pivot pin

MIXTURE CONTROL UNIT

◀ The mixture control unit assembly contains the air flow sensor assembly and the fuel distributor.

– The air flow sensor measures all air being drawn in by the engine. As air flows past the air flow sensor plate, the plate is lifted together with the control plunger in the fuel distributor to meter the fuel.

Mixture control unit, removing and installing

- Disconnect battery negative (–) cable.
- Relieve fuel pressure from fuel distributor.f

> ***WARNING —***
> *To prevent fuel from spraying on a hot engine, system pressure should be relieved before disconnecting fuel lines. One method is to tightly wrap a shop towel around a the control pressure regulator line where it connects to the fuel distributor and than loosen the fuel line fitting.*

- Mark location of fuel lines on fuel distributor and remove all lines.
- Loosen intake boot clamps, disconnect hose from rear of boot and remove boot.

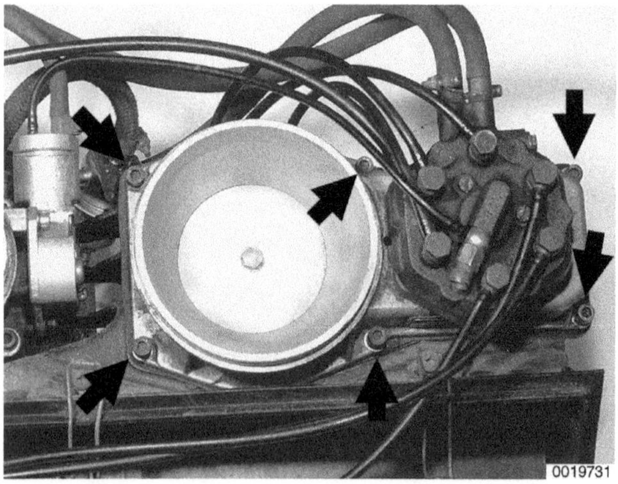

- Remove allen bolts holding mixture control unit to air box and remove complete mixture control unit with fuel distributor.
- Remove three screws holding fuel distributor to mixture control unit housing and remove distributor and base O-ring.

> ***CAUTION —***
> *When removing the fuel distributor from the mixture control unit, be careful not to allow the control plunger to fall out.*

- Remove fuel distributor plunger from fuel distributor and clean plunger. Check that plunger moves smoothly and freely in fuel distributor bore.

> ***NOTE —***
> *If the control plunger hangs up in fuel distributor or does not move freely in bore, the complete fuel distributor must be replaced. Fuel distributors cannot be serviced. Due to the precision fit required, faulty control plungers cannot be replaced separately.*

- Installation is reverse of removal. Use a new O-ring when installing fuel distributor to mixture control unit. Replace fuel line sealing rings and air flow housing gasket to mixture control unit. Tighten mixture control unit mounting bolts fully, then loosen each bolt one full turn.

AUXILIARY AIR REGULATOR

Auxiliary air regulator construction

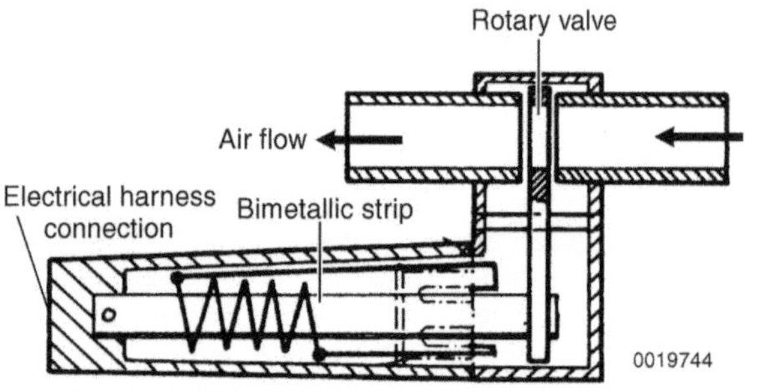

During cold engine operation, the auxiliary air regulator lets additional air bypass the throttle valve to raise idle speed. Electrical current warms the heating element and causes the bimetallic strip to deflect gradually, closing the rotary valve and cutting off the additional air. The auxiliary air regulator is mounted between cylinder intake runners 5 and 6.

AUXILIARY AIR REGULATOR

Auxiliary air regulator, checking

Check auxiliary air regulator (**arrow**) function by its effect on engine idle speed.

With the engine cold, disconnect the electrical harness from the auxiliary air regulator. Start the engine and pinch off the air hose from the auxiliary air regulator to the intake manifold. The regulator valve should be open so the engine should lose rpms with the hose pinched off.

With the engine idling at operating temperature and the auxiliary air regulator electrical harness connected. Pinch off the regulator hose, The engine rpm should not change

NOTE—
- *If the idle speed does not drop when cold, remove auxiliary air regulator. Cool off regulator and visually check that regulator valve is open. If the valve is not open when cold, the regulator is faulty and should be replaced.*
- *If the idle speed drops when the engine is warm, remove harness connector from auxiliary air regulator and check for battery voltage at harness connector. If battery voltage is not present, check for ground at brown wire in connector. Check red/white wire for positive (+) voltage coming from terminal 30 of the fuel pump relay. If the valve does not close when battery voltage is supplied, the regulator is faulty and should be replaced.*

CONTROL PRESSURE REGULATOR

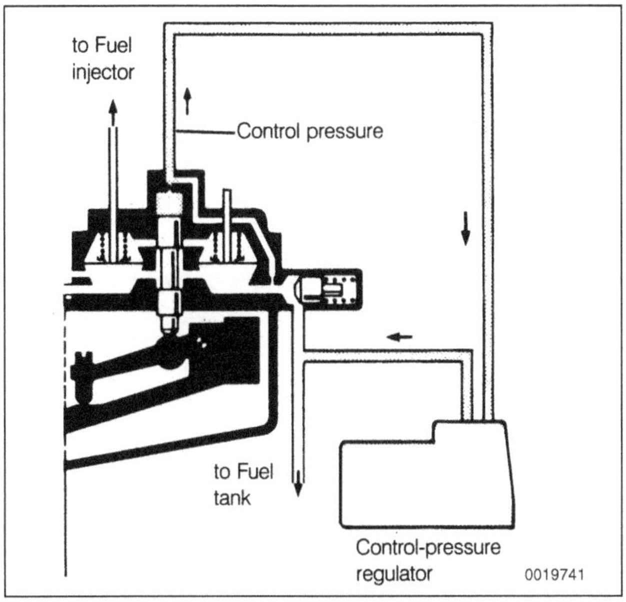

The control pressure regulator, also called the warmup regulator, adapts the fuel mixture during the warmup phase by adjusting the 'control pressure' on top of the control plunger.

Control pressure regulator, checking

– The function of the control pressure regulator is checked by measuring the control pressure with the engine cold and as it warms up. See **Control pressure, checking**.

NOTE—
If the control pressure does not increase as the engine is warmed up, the heating element within the regulator may not be functioning. Check that the heating element in the regulator is receiving voltage. If battery voltage is not present at the connector, check for ground at the brown wire in the connector. Check red/white wire for positive (+) voltage coming from terminal 30 of the fuel pump relay. If no faults are found the, heating element in the regulator may be faulty.

Control pressure regulator, replacing

- Disconnect battery negative (–) cable.
- Relief fuel pressure from control pressure regulator by carefully loosening fuel lines.

> **WARNING —**
> *To prevent fuel from spraying on a hot engine, system pressure should be relieved before disconnecting fuel lines. One method is to tightly wrap a shop towel around a the control pressure regulator line where it connects to the fuel distributor and than loosen the fuel line fitting.*

- Remove both fuel lines.
- Remove two 5 mm allen bolts and remove regulator.
- Installation is reverse of removal.

NOTE —
Always use new sealing washers on fuel line fittings.

COLD START VALVE

The electrically operated cold-start valve sprays extra fuel into the intake manifold for several seconds when the starter is actuated and the engine is cold. The valve is controlled by a thermo-time switch, located in the left timing chain housing cover.

- When the engine is cold, the switch is closed allowing power to reach the valve and open it. When the engine is warm, the switch is open and the cold start valve does not operate. To limit valve operation and to prevent flooding, electrical current also warms the switch and opens the circuit after a few seconds.

If the cold start valve fails to inject fuel during cold starting it will be difficult or impossible to start the engine when cold. If the cold start valve leaks, the engine may receive extra fuel at the wrong time and become flooded, especially if the engine is hot.

Cold-start valve, checking

- Make sure engine temperature is below 45°C (113°F).
- Remove two screws holding cold start valve to manifold.
- Remove cold start valve without disconnecting fuel line or electrical harness.
- Disconnect electrical connector from CDI control unit. See **280 Ignition System**.

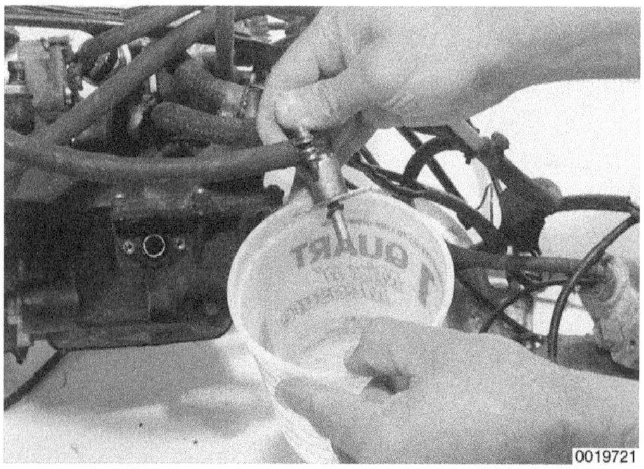

Point valve into a transparent container.

– Have a helper actuate starter while observing cold start injector spray pattern. The valve should spray in an even, cone shaped pattern until thermo-time switch interrupts circuit.

• An irregular spray pattern indicates a dirty or faulty cold start valve.

– If valve does not spray, disconnect harness connector from cold start valve. Connect test light across terminals in connector. With engine cranking, test light should light-up.

• If test light does not illuminate, check thermo time switch operation as described below.
• If test light does illuminate, but valve does not spray, valve is faulty and should be replaced.

– Wipe nozzle dry. Valve should not drip for at least a minute.

• If valve drips, it is faulty and needs replacement.

– Reinstall cold start valve and reconnect CDI control unit connector.

Thermo-time switch, checking

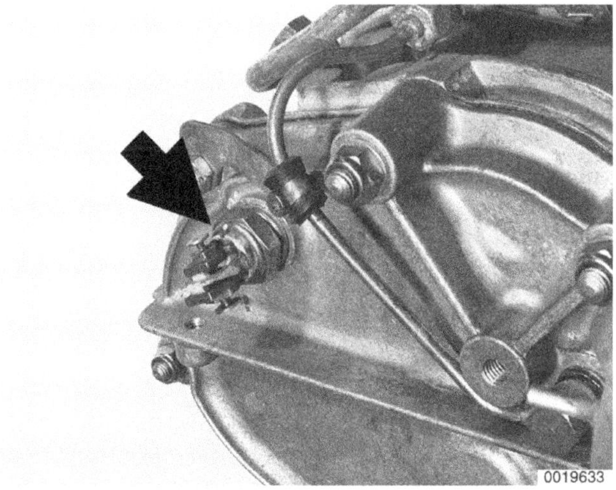
0019633

< The thermo-time switch senses cold engine temperature and controls the cold start valve opening time. The switch is mounted in the left timing chain cover (**arrow**).

– Make sure engine temperature is below 45°C (113°F). Preferably engine should sit for several hours.

– Disconnect electrical connector from CDI control unit. See **280 Ignition System**.

– Disconnect cold-start valve harness connector and connect a test light across connector terminals.

– Have a helper actuate starter while observing test light.

• Test light should light for several seconds, then go out. (The amount of time the light is on depends on engine temperature.)
• If test light does not illuminate, check for voltage at yellow wire at thermo-time switch with starter actuated. If voltage is not present, check yellow wire between thermo-time switch and starter See **970 Electrical Wiring Diagrams**.
• If test light does light but cold start valve does not receive voltage, thermo-time switch is faulty and should be replaced.

– Reconnect cold start wiring connector and CDI control unit connector when finished testing.

OXYGEN SENSOR SYSTEM

The oxygen sensor system provides the Oxygen Sensor Control Unit with feedback information about combustion efficiency. The signal is used to continuously control the air-fuel mixture via the frequency valve. Two versions of the oxygen systems were used on the SC models. The electrical schematics for each system are shown below.

Oxygen sensor system components

- Oxygen sensor control unit (beneath passenger seat)
- Oxygen sensor
- Oxygen sensor relay (beneath passenger seat)
- Temperature switch (15°C)
- Frequency valve
- Throttle switch

Additional components on 1981 and later models only

- Temperature switch (35°C)
- Additional throttle switch (15°)
- Cold acceleration control unit (beneath passenger seat)

Oxygen sensor wiring diagram 1980 models

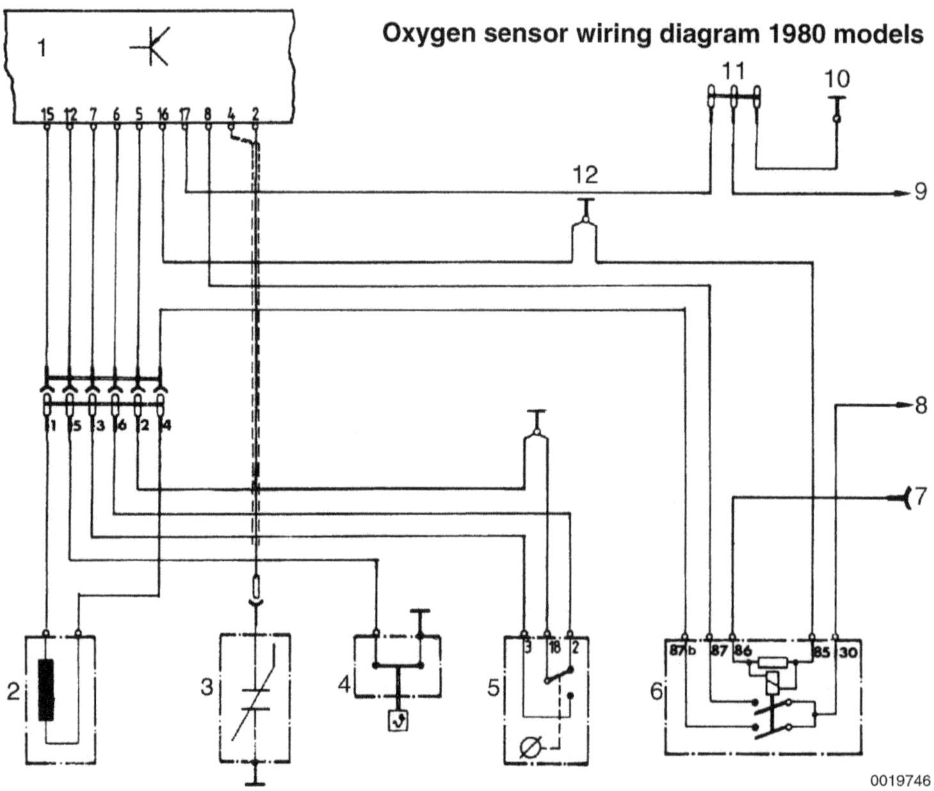

1. Oxygen sensor control unit
2. Frequency valve
3. Oxygen sensor
4. Temperature switch (15°C)
5. Throttle switch
6. Oxygen sensor relay
7. to Control pressure regulator
8. from Fuse box (fuse no. 1)
9. from Fuse box (fuse no. 1)
10. Test connector ground
11. Test connector
12. Control unit ground

Oxygen sensor wiring diagram
1981 - 1983 models

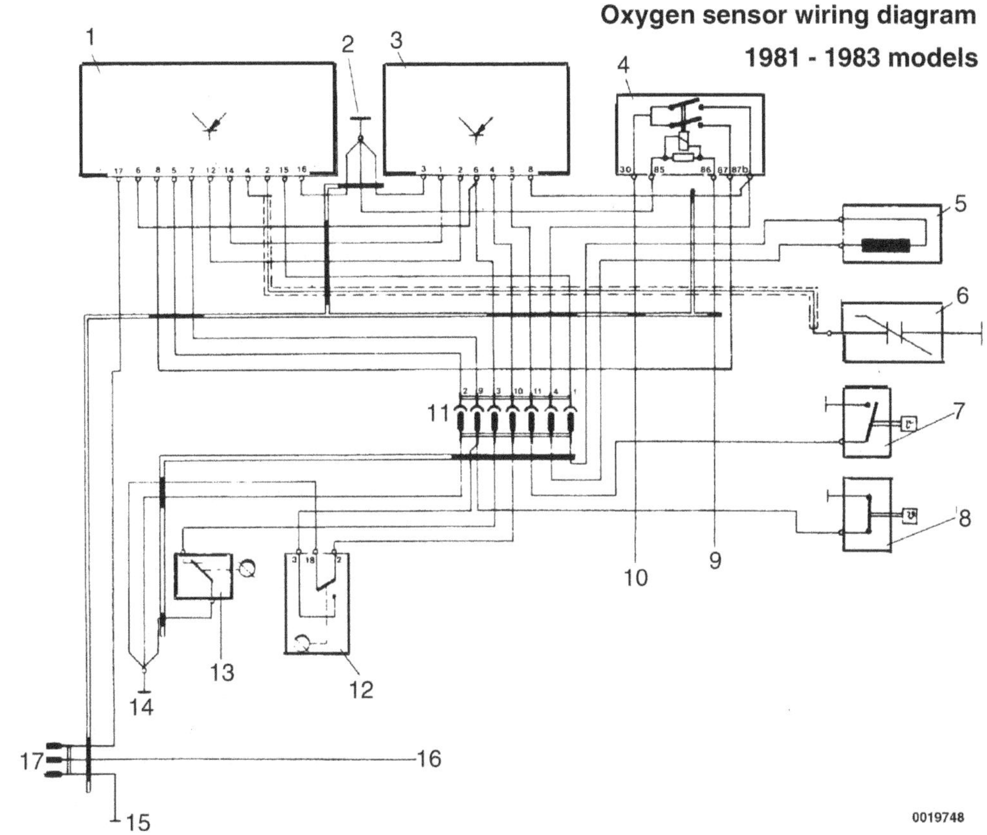

1. Oxygen sensor control unit
2. Control unit ground
3. Cold acceleration control unit
4. Oxygen sensor relay
5. Frequency valve
6. Oxygen sensor
7. Temperature switch (35°C)
8. Temperature switch (15°C)
9. to Control pressure regulator
10. from Fuse box (fuse no. 1)
11. Multi-point connector
12. Throttle switch
13. Microswitch (15° throttle angle switch)
14. Engine ground
15. Sensor ground
16. from Fuse box (fuse no. 1)
17. Test connector

Oxygen sensor system, checking

> ***WARNING —***
> *Exhaust system parts can be hot enough to cause serious burns. When working near hot pipes or mufflers, use heavy gloves and other appropriate protection.*

NOTE —
For the most accurate test results, the engine should be fully cold, the exhaust system must be free of leaks, and all electrical consumers must be off. Use a high quality digital multimeter.

◄ Locate test connector in left side of engine compartment (**arrow**). Connect duty cycle or dwell meter to test connector.

NOTE —
When using a dwell meter, the scale range is 90% and must be converted to correspond to the 100% duty cycle scale. For example, a 50% duty cycle is equivalent to 45% dwell.

– Start engine and let it idle. Check that frequency valve duty cycle is as specified.

Frequency valve duty cycle (engine temp. below 15°C)
- 1980 models 85%
- 1981-1983 models 65%

NOTE —
If duty cycle is not as specified, check for faulty wiring to test connector. Check power and ground inputs to control unit. Check all wiring to and from control unit. See **970 electrical Wiring Diagrams**. *Check that the frequency valve is functioning (buzzing) with the engine running. If no faults can be found, the control unit may be faulty.*

– Disconnect test equipment.

– Drive vehicle for a few miles until engine is warmed up.

– Reconnect test equipment. Check that frequency valve duty cycle is as specified.

Frequency valve duty cycle (engine temp. above 15°C)
- All models . . . . .45-65%, fluctuating

NOTE —
If duty cycle is not as specified, check that the oxygen sensor is producing a voltage as described later. If sensor is OK, check that 15°C temperature switch is closing.

– If oxygen sensor system does not function as specified, check system components as described below.

Oxygen sensor, checking

WARNING —
Exhaust system parts can be hot enough to cause serious burns. When working near hot pipes or mufflers, use heavy gloves and other appropriate protection.

NOTE —
For the most accurate test results, the engine must be fully warmed up, the exhaust system must be free of leaks, and all electrical consumers must be off. Use a high quality digital multimeter.

◄ Locate oxygen sensor harness connector (**arrow**) in left side of engine compartment. Without disconnecting harness connector, connect positive lead (+) of voltmeter to oxygen sensor wire. Connect negative (–) test lead to ground.

– Start engine and let it idle. After a maximum of two minutes oxygen sensor output voltage should begin fluctuating.

Oxygen sensor voltage output
- Engine warm. . . . . 0.1–0.9 VDC fluctuating

– With engine running, loosen oil filler cap to admit unmeasured intake air and simulate a lean running condition.

- Oxygen sensor output voltage should drop momentarily and then return to normal

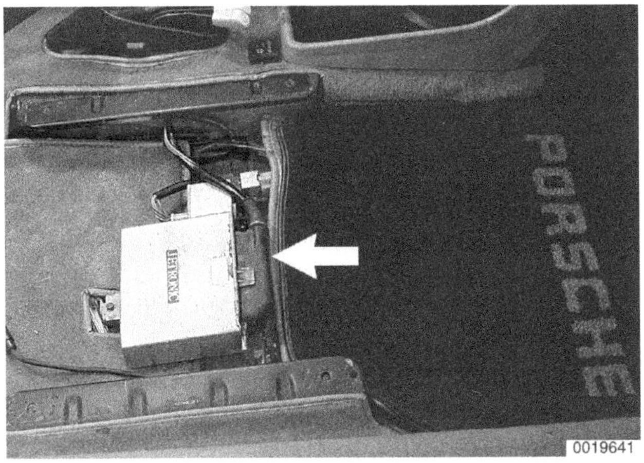

0019641

– If voltage signal does not fluctuate or if there is no change when oil filler cap is removed, disconnect oxygen sensor connector (**arrow**) and check for voltage at sensor wire with the engine running. There should be a non-fluctuating voltage between 0-1 VDC. If sensor does not output any voltage it is probably faulty.

NOTE—

- *If the sensor produces a voltage when disconnected (open-loop), but does not fluctuate when connected, the oxygen sensor control unit may be faulty. Check for a faulty sensor wire in harness or other wiring faults before condemning control unit.*
- *The oxygen sensor control unit (**arrow**) is mounted beneath the driver's seat. For wiring schematics for the oxygen sensor control system, see* **970 Electrical Wiring Diagrams**.

Oxygen sensor, replacing

0018764

< The sensor (**arrow**) is mounted in the front of the catalytic converter.

– Disconnect oxygen sensor plug in engine compartment.

– Push plug and wiring grommet through engine side cover.

– Using special socket remove oxygen sensor.

– Installation is reverse of removal.

NOTE—

- *Three wire oxygen sensor shown, 911 SC models use a single wire sensor*
- *Special sockets for replacing the oxygen sensor are available from most automotive parts stores. The socket has a groove cut down one side to allow the sensor to be installed without damaging the wire harness.*

Temperature switch (15°C), checking

The 15°C temperature is the main temperature input to the oxygen sensor control unit. When the engine is cold, this switch is normal open. When the engine reaches the switching temperature, the switch closes and provides a ground input to the control unit and the system then goes into closed loop (feedback) operation.

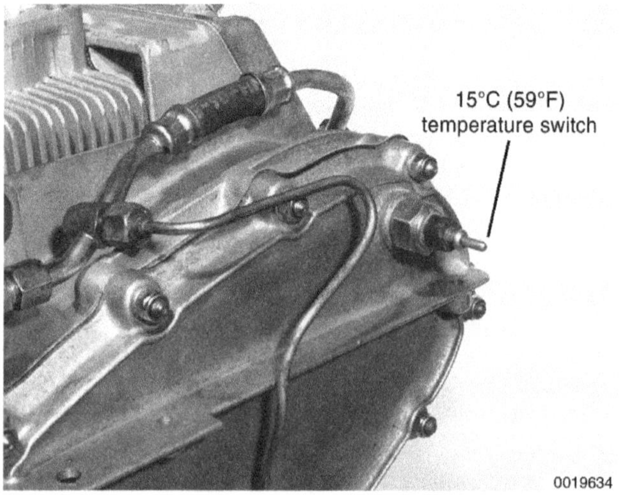

The 15°C (59°F) temperature switch is mounted in the right front timing chain cover.

– Test the switch using an ohmmeter between switch terminal and ground (wire disconnected). Check the switching point with the engine temperature below the switching temperature and again with engine temperature above the switching temperature. For a more accurate check, the switch can be removed and tested in a bath of water.

Temperature switch (35°C), checking (1981 - 1983 models)

The 35°C (95°F) temperature switch is an additional switch used on 1981 and later cars as part of the cold acceleration enrichment system.

When the engine oil temperature is below 35°C, additional enrichment is provided briefly (about 2 seconds) when accelerating off idle (approximately 2-3° throttle angle) and again at part throttle (approximately 15° throttle angle). This switch is normally open. When the engine oil temperature reaches the switching temperature, the switch opens and acceleration enrichment is no longer provided.

The 35°C temperature switch (**arrow**) is the mounted in the crankcase vent cover on the rear of the engine.

– Test the switch using an ohmmeter between switch terminal and ground (wire disconnected). Check the switching point with the engine temperature below the switching temperature and again with engine temperature above the switching temperature. For a more accurate check, the switch can be removed and tested in a bath of water.

Throttle position switch, checking

The throttle switch contains two integral microswitches. One switch provides a closed throttle input to the oxygen sensor control unit and one switch provides full throttle position (closed at 35° throttle angle).

When the idle contact switch is closed, the control unit delays the oxygen feedback cycle to prevent an erratic idle. When the full throttle contact is closed, the oxygen sensor control unit switches to an open-loop, fixed (richer) control mode and the A/C compressor is turned off.

◄ Disconnect throttle switch electrical harness connector (**arrow**).

– Connect ohmmeter between terminals 2 and 18 of throttle switch and check that switch is closed (continuity) with throttle is rest position.

– Opens throttle switch slightly (between 2° and 3°) and check that switch opens (no continuity).

– Connect ohmmeter between terminals 3 and 18 of throttle switch.

– Open throttle all the way.

- Meter should read infinity from 35° throttle opening to wide open throttle.

NOTE —

If any faults are found, loosen the throttle switch mounting screws and try adjusting the switch by rotating left or right. If the switch can not be made to function correctly, the switch is faulty and should be replaced.

Additional 15° throttle position switch, checking (1981 - 1983)

The 15° throttle switch provides a 15° throttle angle input to the oxygen sensor control unit. When the engine is cold, this switch signals for momentary enrichment (75% duty cycle for 2 seconds) when the throttle moves past 15°.

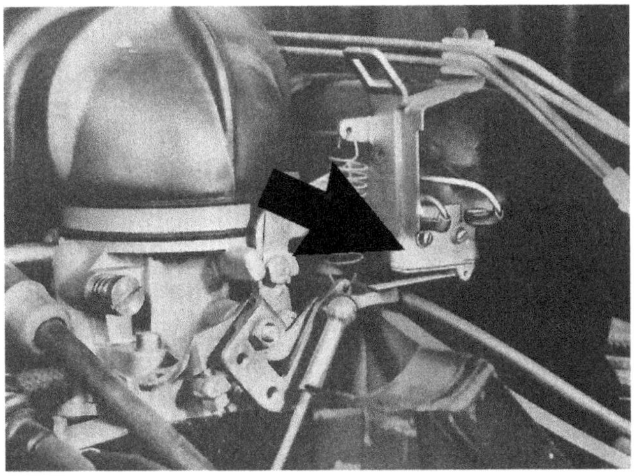

◄ Disconnect two single wires from switch on throttle housing.

– Connect ohmmeter between contacts on switch (**arrow**). Check that switch is closed with throttle in closed position (continuity). Open throttle approximately 15° and check that switch opens (no continuity). If any faults are found, replace microswitch.

Frequency valve, testing

The frequency valve is an electrically operated valve that functions to bleed off fuel pressure in the lower chamber in the fuel distributor. The longer the valve is held open (higher duty cycle), the lower the pressure in the lower chamber. When the lower chamber pressure is reduced, fuel flow to the injectors is reduced.

Based on input from the oxygen sensor, the oxygen sensor control unit corrects the fuel/air mixture by varying the frequency valve duty cycle.

To quickly check frequency valve, start engine and place a finger on valve. It should be buzzing.

– If valve is not buzzing, remove electrical harness plug from valve. Connect ohmmeter between frequency valve terminals and check for an open winding. See **table k.**

Table k. Frequency valve specifications

	Resistance
Frequency valve winding	2 to 3 ohms

– If no faults are found and valve does not operate, check wiring to valve and power and ground supplies to the oxygen sensor control unit. See **970 Electrical Wiring Diagrams**.

260 Exhaust System

GENERAL

The exhaust system is designed to be maintenance free. Regular inspection is warranted due to harsh operating conditions. The only scheduled maintenance is replacement of the oxygen sensor (1980 and later cars) at specified time and mileage intervals.

This section covers maintenance, troubleshooting, and repair of the exhaust system. For information on testing the oxygen sensor system, see **240 Fuel Injection.**

> ***WARNING —***
> - *To avoid injury, wear eye protection and heavy gloves when working on rusty parts of an old exhaust system.*
> - *Do not use flammable workshop chemicals near a hot catalytic converter. They may ignite.*

Exhaust system components

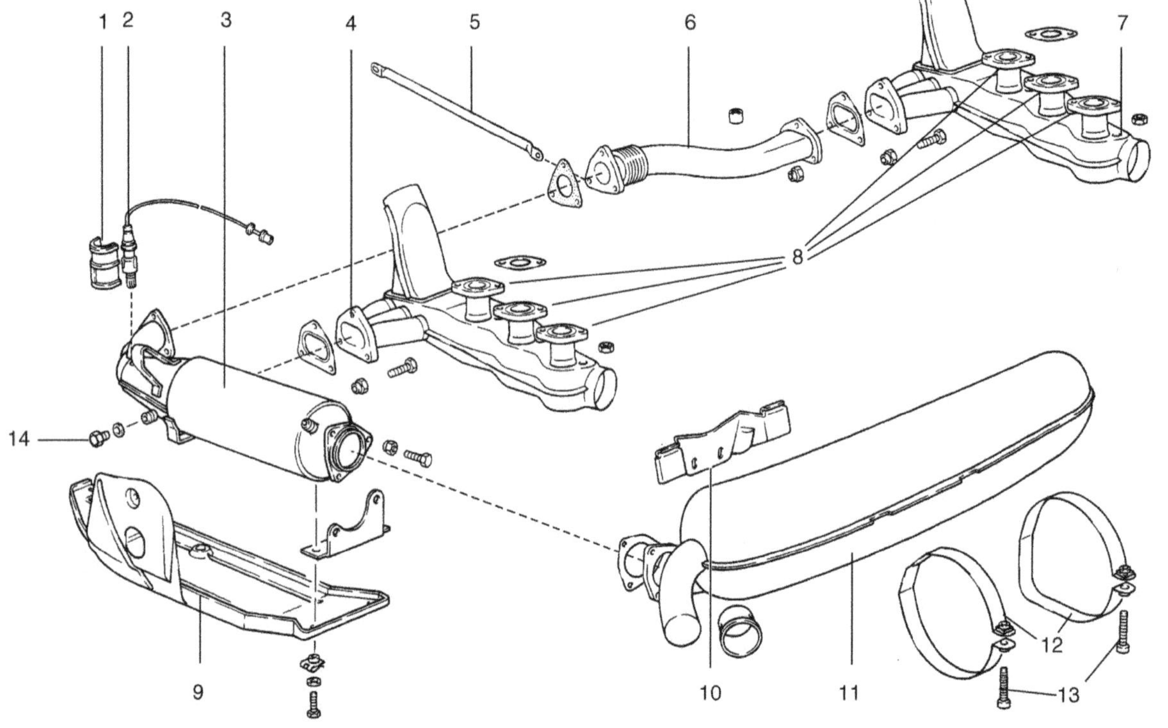

1. Oxygen sensor heat shield (1983)
2. Oxygen sensor (1980 - 1983)
3. Catalytic converter
   Monolith (1978-1979)
   3-way (1980-1983)
4. Left heat exchanger
5. Support bracket (optional)
6. Crossover pipe
7. Right heat exchanger
8. Exhaust headers
9. Catalytic converter heat shield
10. Muffler support bracket
11. Muffler
12. Muffler support straps
13. Muffler strap bolts
14. Exhaust gas analyzer test port

Exhaust system

The basic exhaust system components are the exhaust headers that are encased within the heat exchangers, a cross-over pipe, a catalytic converter and a muffler.

The exhaust headers from each bank of cylinders are housed in a heat exchanger that supplies heated air for car interior heating and windshield defrosting.

A cross-over (connector) pipe brings the right side exhaust gases across to the left side. Both the cross-over pipe and left exhaust header connect to the catalytic converter. The muffler connects directly to the rear of the catalytic converter to complete the exhaust system.

Catalytic converter

3-Way catalytic converter construction

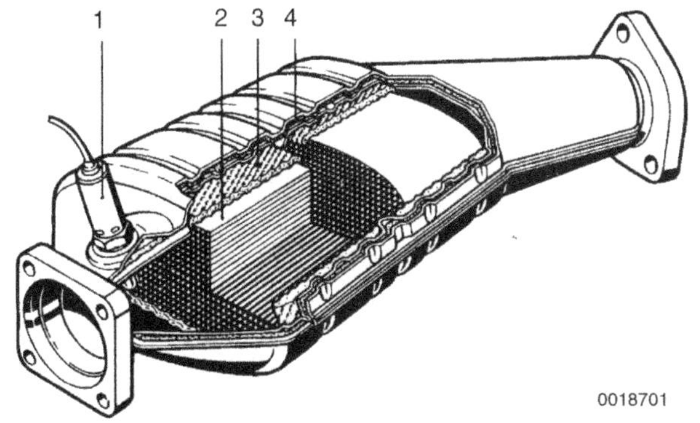

1. Oxygen sensor
2. Ceramic core
3. Wire screen
4. Heat resistant casing

All models covered in this manual are fitted with a catalytic converter. 1978 and 1979 models use a monolith type converter. 1980 and later models with oxygen sensor system use a 3-way type converter.

The simple catalytic converter used in 1978 and 1979 models is used to oxidize unburned hydrocarbons (HC) and to convert carbon monoxide (CO) to carbon dioxide (CO_2).

With the introduction of the oxygen sensor system in 1980 models, a 3-way catalytic converter was installed. This unit oxidizes HC and CO while reducing oxides of nitrogen (NO_x), but only if the air-fuel mixture of the engine is very close to the optimal.

Oxygen sensor

The oxygen sensor (introduced in 1980 models) is mounted in the inlet side of the catalytic converter. The sensor monitors the exhaust gas to provide feedback about combustion efficiency to the oxygen sensor control module. The catalytic converter reduces emissions most efficiently when the percentage of oxygen in the exhaust falls within a narrow range.

Oxygen sensor diagnosis is covered in **240 Fuel Injection**. Oxygen sensor replacement is covered in **030 Maintenance.**

CAUTION—
Some carburettor and fuel injection additives or sprays can damage the oxygen sensor or catalytic converter. Check with an authorized Porsche dealer for approved fuel additives.

NOTE—
A three wire oxygen sensor is shown in the photo. SC models use a single-wire oxygen sensor.

TROUBLESHOOTING

A quick-check for detecting exhaust leaks is to hold a wet shop towel over the tailpipe outlet with the engine running. If any hissing is heard under the car, the system has leaks.

> ***WARNING —***
> *Toxic exhaust gases are colorless and odorless. Do not run the engine in a non-ventilated area. Repair any structural damage to the body or exhaust system immediately.*

Catalytic converter, checking

– Possible indications of faulty catalytic converter:
  - Reduced power
  - Stalling at idle
  - Rattles in the exhaust system
  - High exhaust emissions measured at the tailpipe.

– To visually check catalytic converter, remove muffler and check for evidence of physical damage inside converter.

– Alternatively, tap converter with a block of wood.
  - If rattling noise is heard from inside, replace converter.

– Test efficiency of catalytic converter:
  - Run engine until it and converter are at operating temperature.
  - Using infrared thermometer, measure temperature differential at inlet and outlet of converter.
  - If temperature difference is not more than approximately 55°C (100°F) replace converter.

> ***WARNING —***
> *The catalytic converter operates at a very high temperature and can cause severe burns.*

Tightening torques
- Muffler to catalytic converter (M8) ..... 25 Nm (18 ft-lb)
- Muffler strap bolt (M10)............... 40 Nm (30 ft-lb)

COMPONENT REPLACEMENT

> ***WARNING —***
> - *To avoid injury, wear eye protection and heavy gloves when working on rusty parts of an old exhaust system.*
> - *Do not use flammable workshop chemicals near a hot catalytic converter. They may ignite.*

Use of non-original exhaust system components can change system backpressure and may affect fuel mixture. Thermodynamic effects and pressure affect the way the engine runs. It may be necessary to check and adjust the fuel injection system after installing a non-original manufactured muffler.

NOTE—
- *Use new gaskets and mounting hardware whenever removing and installing exhaust system parts.*
- *When mounting exhaust system components, tighten the mounting nuts in two stages.*

Muffler, replacing

– Raise car and support in a safe manner.

> ***WARNING—***
> *Make sure the car is stable and well supported at all times. Use a professional automotive lift or jack stands designed for the purpose. A floor jack is not adequate support.*

Muffler and muffler support

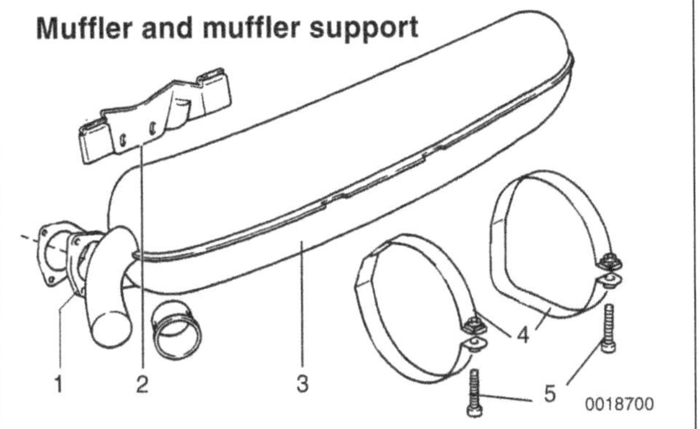

1. Flange to catalytic converter
2. Muffler support
3. Muffler
4. Muffler support straps
5. Muffler strap allen bolts

◄ Remove three M8 bolts and nuts that secure muffler to catalytic converter flange.

– Use penetrating oil to lubricate threads of two M10 allen bolts in muffler straps. Loosen both muffler strap bolts and slide each strap off muffler support while supporting muffler.

– Remove muffler.

– Installation is reverse of removal.
- Use new gaskets and hardware.

Tightening torques
- Muffler to catalytic converter (M8) ..... 25 Nm (18 ft-lb)
- Strap bolt (M10) .... 40 Nm (30 ft-lb)

Heat exchanger, removing and installing

– Raise car and support in a safe manner.

> ***WARNING—***
> *Make sure the car is stable and well supported at all times. Use a professional automotive lift or jack stands designed for the purpose. A floor jack is not adequate support.*

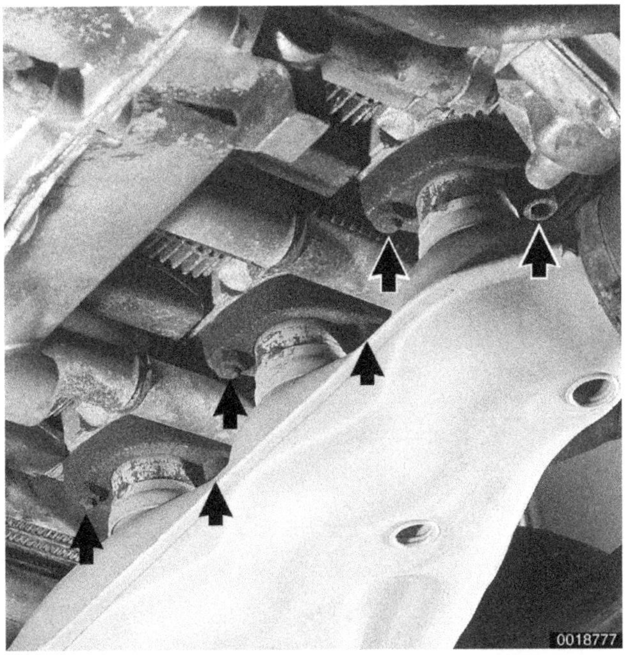

◄ Use penetrating oil to lubricate exhaust header to cylinder head fasteners (**arrows**).

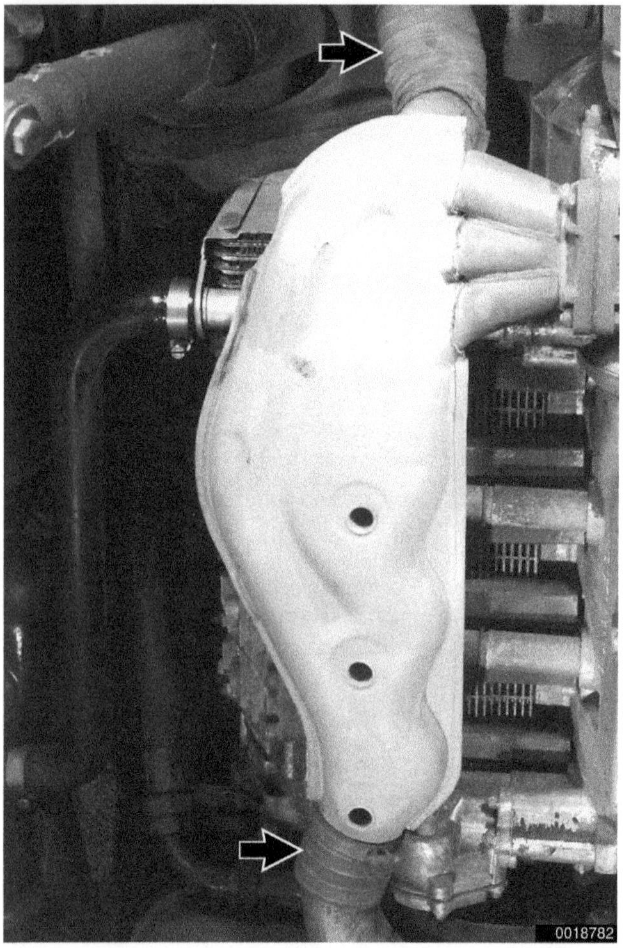

< Detach heater hoses (**arrows**) from heat exchanger.

- Loosen clamps and slide rubber boot back on heater blower cross-over tube.
- Loosen clamp and slide off heater hose that runs between body and heat exchanger.

– Left side: Remove three M8 fasteners connecting heat exchanger to catalytic converter.

– Right side: Remove three M8 fasteners connecting heat exchanger to crossover pipe.

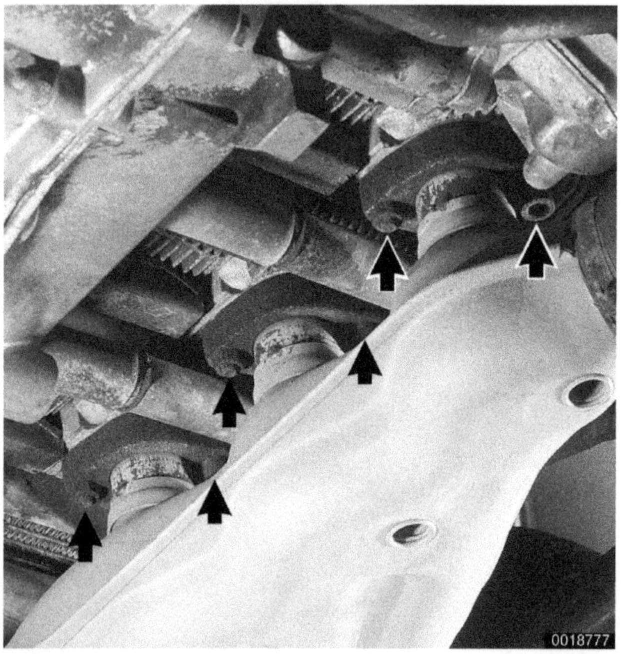

< Remove exhaust header to cylinder head mounting nuts (**arrows**).

- Three fasteners are lock nuts.
- Three are allen-head nuts.

CAUTION —
Support heat exchanger while removing the fasteners.

– Lower heat exchanger down and out.

– Installation is reverse of removal.

- Use new gaskets and hardware.

Tightening torque
- Exhaust runner to cylinder head....... 25 Nm (18 ft-lb)

Catalytic converter assembly

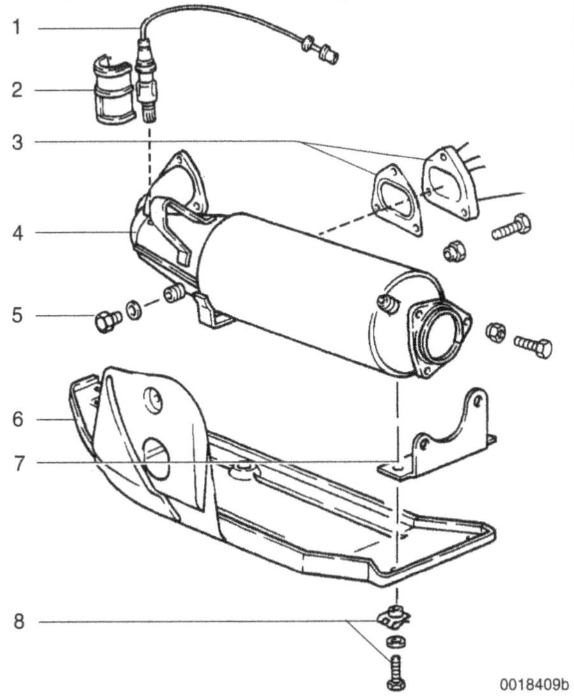

1. Oxygen sensor (1980 - 1983)
2. Oxygen sensor heat shield (1983)
3. Heat exchanger flange and gasket
4. Catalytic converter
5. Exhaust gas analyzer test port
6. Catalytic converter heat shield
7. Heat shield support bracket
8. Heat shield mounting hardware

Catalytic converter, removing and installing

– Raise car and support in a safe manner.

> ***WARNING —***
> *Make sure the car is stable and well supported at all times. Use a professional automotive lift or jack stands designed for the purpose. A floor jack is not adequate support.*

– Remove muffler as described earlier.

◄ Remove catalytic converter heat shield mounting bolts. Remove heat shield.

◄ On 1980 or later model: Open engine compartment and locate oxygen sensor wiring on left side of engine.

- Detach connector (**arrow**).
- Push wire with rubber grommet through engine compartment bulkhead.

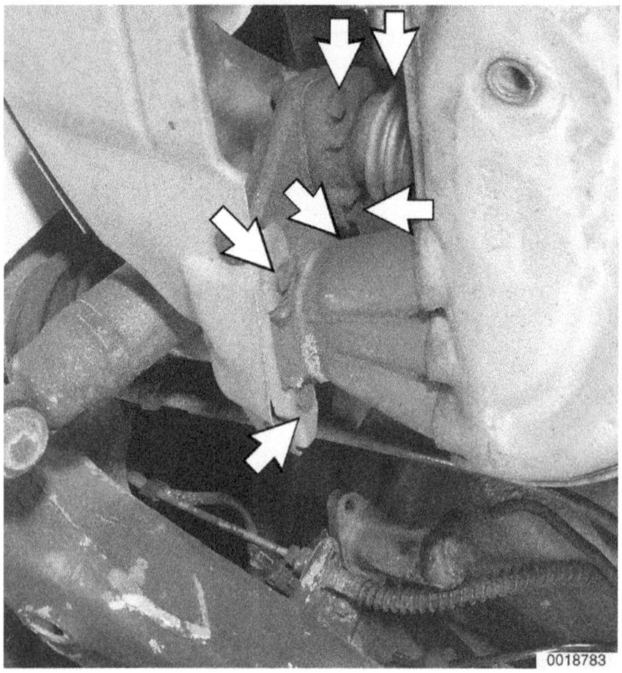

< Remove converter flange fasteners (**arrows**) at left heat exchanger and right crossover pipe.

> ***CAUTION —***
> *Support catalytic converter while removing the fasteners.*

– Remove catalytic converter.

– Installation is reverse of removal.

- Use new gaskets and hardware.

Tightening torque

- Catalytic converter to muffler (M8) . . . . . 25 Nm (18 ft-lb)

270 Battery, Starter, Alternator

TABLES

Special Tools

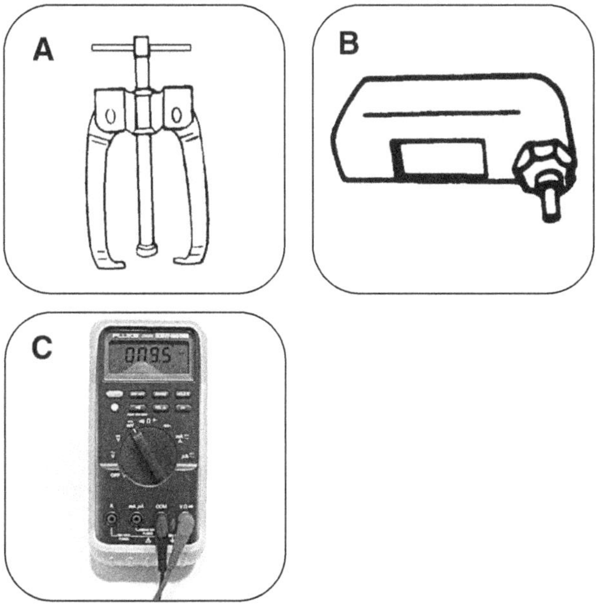

Special tool

A 236P (source: Baum Tools Unlimited) . . Battery terminal puller
B 1080 (source: Baum Tools Unlimited) . Battery electrolyte filler
C Fluke 87 .Automotive digital multimeter

GENERAL

This section covers the battery, starter and alternator components of the electrical system. For additional electrical troubleshooting information that may apply to these components, see **900 Electrical System–General**.

> *CAUTION —*
> *The starter and alternator are not protected by fuses. To prevent accidental shorts that might damage wires and electrical components, the negative (–) battery cable should always be disconnected before working on the starter or alternator.*

Special tools

Some of the test and repair procedures given in this repair group require special tools.

Voltage and polarity

The cars covered by this manual have a 12-volt, direct current (DC), negative ground electrical system. The voltage regulator maintains the voltage in the system at approximately the 12-volt rating of the battery. All circuits are grounded by direct or indirect connection to the negative (–) terminal of the battery.

Battery

The battery is located in the front luggage compartment behind the left headlight.

The battery stores electricity chemically. Lead plates inside the battery interact with a mixture of sulfuric acid and water, called electrolyte, to create a flow of current.

Starter

The starter and its attached solenoid are mounted on the right side of the transmission just above the right drive axle.

Charging system

The charging system consists of a belt-driven alternator and voltage regulator.

NOTE —

- *1978 - 1981 models: The voltage regulator is mounted on the electrical panel to the left of the engine.*
- *1982 - 1983 models: The voltage regulator is mounted to the back of the alternator.*

Maintenance

The following electrical system maintenance is described in **030 Maintenance**.

- Checking battery level.
- Checking alternator V-belt tension.
- Changing alternator V-belt.

Troubleshooting

The cause of a dead battery or an inoperative starter are often related. Before troubleshooting any battery-related problems, check that applicable ground connections are tight and free from corrosion.

Table a lists symptoms of trouble with the battery, starter and charging system, their probable causes, and suggested corrective actions. Most corrective actions are described later in this group.

Table a. Battery, starter and charging system troubleshooting

Sympton	Probable Cause	Corrective Action
Engine turns slowly or not at all, solenoid clicks when starter is operated.	a. Battery cables loose, dirty, or corroded b. Battery discharged c. Body ground strap loose, dirty, or corroded d. Poor connection or no voltage at starter motor terminal 30 e. Starter motor or solenoid faulty	a. Clean or replace cables. See **030 Maintenance.** b. Charge battery, test and replace if necessary. c. Inspect ground strap, clean, tighten and replace if necessary. d. Check connection, test for voltage at starter. e. Test starter.
Battery will not stay charged more than a few days.	a. Short circuit draining battery b. Alternator V-belt loose or damaged c. Battery faulty d. Battery cables loose, dirty, or corroded e. Alternator or voltage regulator faulty	a. Test for static current draw. b. Inspect alternator V-belt. See **030 Maintenance**. c. Load test battery and replace if necessary. d. Clean or replace cables. See **030 Maintenance**. e. Test alternator and regulator.
Battery loses water.	a. Battery being overcharged	a. Test voltage regulator.
Lights dim, light intensity varies with engine speed.	a. Altenator V-belt loose or damaged b. Alternator or voltage regulator faulty c. Body ground strap loose, dirty, or corroded	a. Inspect alternator V-belt. See **030 Maintenance**. b. Test alternator and voltage regulator. Replace as required. c. Inspect ground strap, clean, tighten, and replace as required.

The battery to chassis ground cable connection (**arrow**) in left front of luggage compartment is shown.

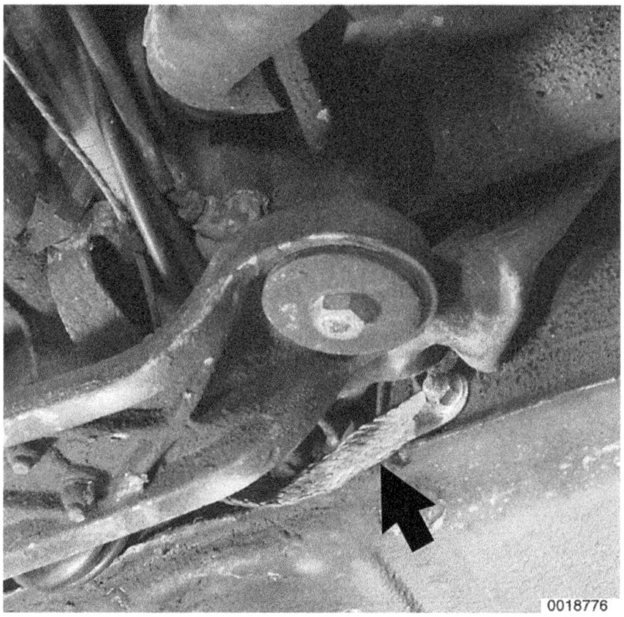

Engine/transmission to body ground strap connection is shown (**arrow**).

BATTERY

> ***WARNING —***
> - *Electrolyte contains sulfuric acid and can cause severe burns.*
>   *–Wear goggles, rubber gloves, and a rubber apron when working with battery electrolyte.*
>   *–If electrolyte is spilled on your skin or clothing, flush the area at once with large quantities of water.*
>   *–If electrolyte gets into your eyes, bathe them with large quantities of clean water for several minutes and call a physician.*
> - *Batteries generate explosive gasses. Keep sparks and open flame away. Do not smoke.*

> ***CAUTION —***
> *Replace a battery if the case is cracked or leaking. Electrolyte can damage the car. If electrolyte is spilled, clean the area with a solution of baking soda and water.*

Low-maintenance batteries have been supplied as original equipment in the cars covered by this manual. These batteries have vented removable filler caps. The battery requires distilled water added periodically to maintain electrolyte level.

The six-cell, 12-volt lead-acid battery capacity is rated by ampere/hours (Ah) and cold cranking amps (CCA). The Ah rating is determined by the average amount of current the battery can deliver over time without dropping below a specified voltage. The CCA is determined by the battery's ability to deliver starting current at 0°F (–18°C).

Temperature affects the efficiency of the battery. The current-producing capacity of a battery at 5°F (–15°C) is only half its capacity at 68°F (20°C), and partly-discharged batteries can freeze due to the higher proportion of water in the electrolyte. A frozen battery produces no current, but can usually be restored when thawed.

Recommended battery type and rating
- Battery type. 12 V negative ground
- Ampere/hours (Ah) 88

Battery testing

> ***CAUTION —***
> - *Never operate the engine with the battery disconnected.*
> - *Prior to disconnecting the battery, read the battery disconnection cautions given on page viii.*

Battery testing determines the state of battery charge. The most common methods are open-circuit and load voltage testing. Batteries that have filler caps can also be tested by checking the specific gravity of the electrolyte. The specific gravity test checks the amount of acid in the electrolyte as an indication of battery charge. Inexpensive specific gravity testers are available at most auto supply stores.

Open-circuit voltage test

An open-circuit voltage test checks battery voltage by connecting an accurate digital voltmeter to the battery posts. Before making any open circuit voltage test on the battery, first load the battery with 15 amps for one minute, for example by turning on the headlights without the engine running. Open-circuit voltage levels and their corresponding percentages of charge are shown in **Table b.**

Table b. Open circuit voltage and battery charge

Open-circuit voltage	State of charge
12.6 volts or more	Fully charged
12.4 volts	75% charged
12.2 volts	50% charged
12.0 volts	25% charged
11.7 volts	Fully discharged

The battery is in satisfactory condition if the open-circuit voltage is at least 12.4 volts. If the open-circuit voltage is at this level or above, but the battery still lacks power for starting, make a load voltage test to determine the battery's service condition. If the open-circuit voltage is below 12.4 volts, recharge the battery. If the battery cannot be recharged to at least 75% it should be replaced.

Battery load voltage test

A battery load tester is required for a load voltage test. The test is made by applying a high resistive load to the battery terminals and then measuring battery voltage. The battery should be fully charged for the most accurate results. The battery cables must be disconnected before making the test. If the voltage is below that listed in **Table c**, the battery should be replaced.

> ***WARNING —***
> *Always wear protective goggles and clothing when performing a load test.*

Table c. Battery load test–minimum voltage (apply 200 amp load for 15 seconds)

Ambient temperature	Voltage
80°F (27°C)	9.6 V
60°F (16°C)	9.5 V
40°F (4°C)	9.3 V
20°F (–7°C)	8.9 V
0°F (-18°C)	8.5 V

Static current draw test

If the battery discharges over time, there may be a constant drain or current draw on the battery. A small static drain on the battery is normal, but a large drain will cause the battery to quickly discharge. Make a static current draw test as the first step when experiencing battery discharge.

– Make sure ignition and all electrical accessories are switched off.

– Disconnect battery negative (–) cable.

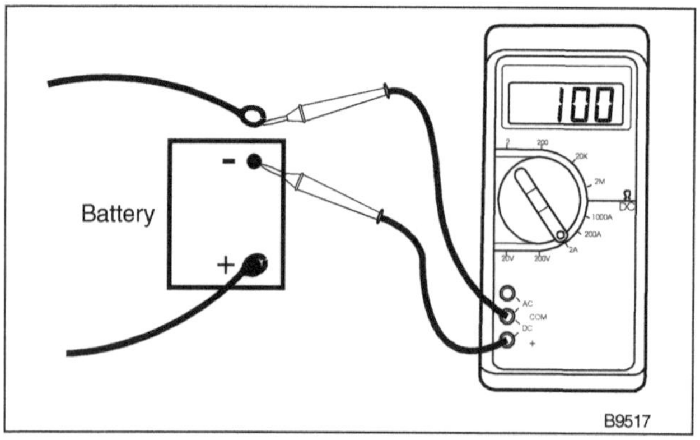

Connect a digital ammeter between negative battery post and negative battery cable to measure current. Wait at least one minute to get an accurate reading.

A range of about 0 to 100 milliampere is normal, depending on the number of accessories that need constant power. A current of 400 milliampere (0.4 amp) or more may indicate a problem.

To determine the circuit or component causing the problem, remove one fuse at a time until the current drops to a normal range.

NOTE —
Large current drains can be detected by using a test light connected in series with the negative cable and negative battery post.

Battery charging

Discharged batteries can be recharged using a battery charger, but a battery can never be charged to a voltage in excess of that which it is capable of producing electro-chemically. Prolonged charging causes gassing that will evaporate the electrolyte to a level that can damage the battery.

Always read and follow the instructions provided by the battery charger's manufactures. Do not use a charger if the instructions are not available. **Table d** lists charging rates and times that should be followed when charging a battery.

Table d. Battery charging specifications

Charging Rate Low-Maintenance Battery	Specific Gravity	Approximate Charging Time
Fast Charge (at 80% to 90% of battery's capacity. Example: 70 to 80 amperes for a 88 ampere-hour battery	1.150 or less 1.150 to 1.175 1.175 to 1.200 1.200 to 1.225	1 hour ¾ hour ½ hour ¼ hour
Slow charge at 10% of battery's capacity. Example: 8.8 ampere for a 88 ampere-hour battery	Above 1.225	Slow charge only to a specific gravity of 1.250 to 1.265

Starting system wiring

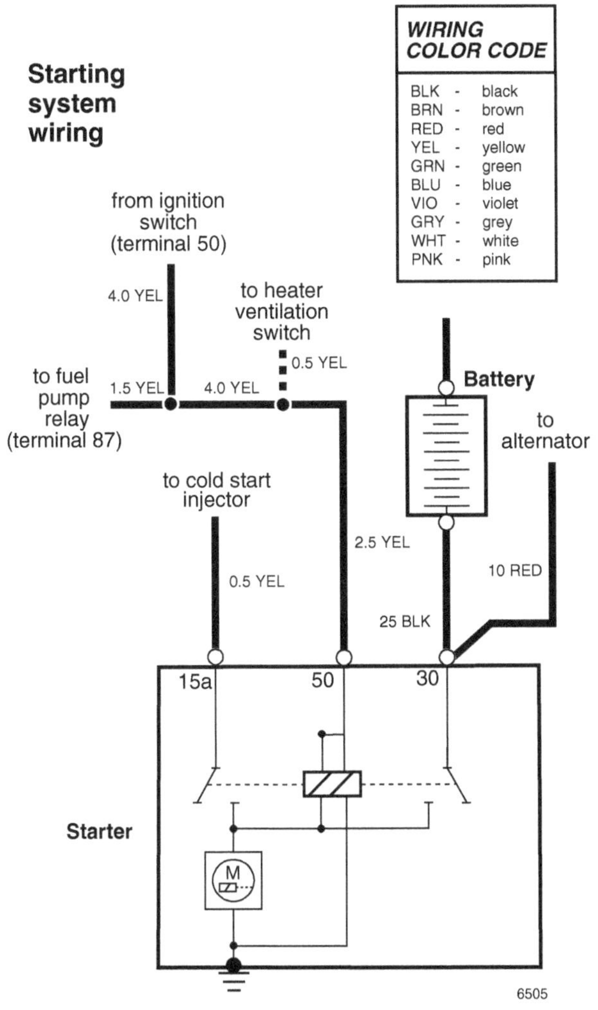

STARTING SYSTEM

When the ignition key is in the START position, battery power energizes a solenoid switch on the starter motor. The solenoid then engages the starter motor. To minimize wear and stress on the drive pinion and ring gear, the solenoid does not switch current to the starter until the drive pinion is fully engaged. Also the drive pinion has an overrunning clutch to prevent the starter from being driven by the engine.

Starting system troubleshooting

The battery and its cables should be in good condition when troubleshooting the starter. If in doubt about the battery, test as described earlier. Starter efficiency is affected by engine oil viscosity. This is especially true in cold weather.

Table e lists symptoms of trouble with the starting system, their probable causes, and suggested corrective actions, most described later in this group.

See **970 Electrical Wiring Diagrams** for a wiring schematic of the starter system that can be used as a guide for troubleshooting.

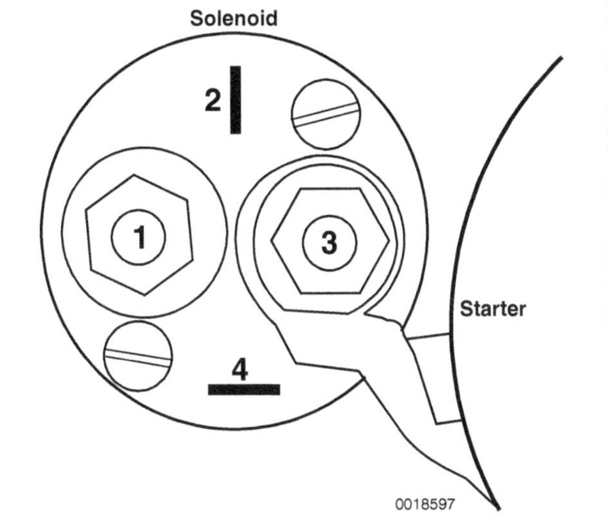

1. Terminal 30 from battery
2. Terminal 50 from ignition switch
3. To starter field winding
4. Terminal 15a to cold start injector

When troubleshooting starting system, refer to terminals in figure.

Table e. Starting system troubleshooting

Symptom	Probable Cause	Corrective Action
Starter does not operate when ignition switch is turned to START	a. Ignition switch or wire leading from ignition switch to solenoid faulty (less than 7.5 volts to solenoid switch) b. Solenoid switch faulty (less than 7.5 volts to starter motor) c. Starter motor fault	a. Test for voltage at terminal 50 of solenoid switch with ignition switch at START. If not at least 7.5 volts, test for voltage at terminal 50 of ignition switch with switch at START. Replace ignition switch (See **480 Steering**) or eliminate open circuit between ignition switch and solenoid switch. b. Test for voltage at field-winding connector with ignition at START. If not at least 7.5 volts, replace solenoid. c. Test for voltage at field-winding connector with ignition at START. If at least 7.5 volts, repair or replace starter motor.
Starter turns slowly or fails to turn engine	a. Dirty, loose, or corroded starter connections b. Dirty, loose, or corroded ground strap between engine/transmission and body c. Starter worn or faulty	a. Remove, clean, and tighten connections. b. Remove and clean or replace ground strap. c. Repair or replace starter.
Starter makes unusual noises, turns erratically, or fails to turn	a. Drive pinion defective b. Starter ring gear damaged	a. Repair or replace starter motor. b. Replace starter ring gear. See **301 Clutch**.
Starter operates, but does not turn engine	a. Starter drive pinion or armature shaft faulty b. Solenoid switch mechanism faulty	a. Repair or replace starter. b. Replace starter solenoid switch.

Starter, removing and installing

- Disconnect negative (–) battery cable and cover battery terminal to keep cable from accidentally contacting terminal.

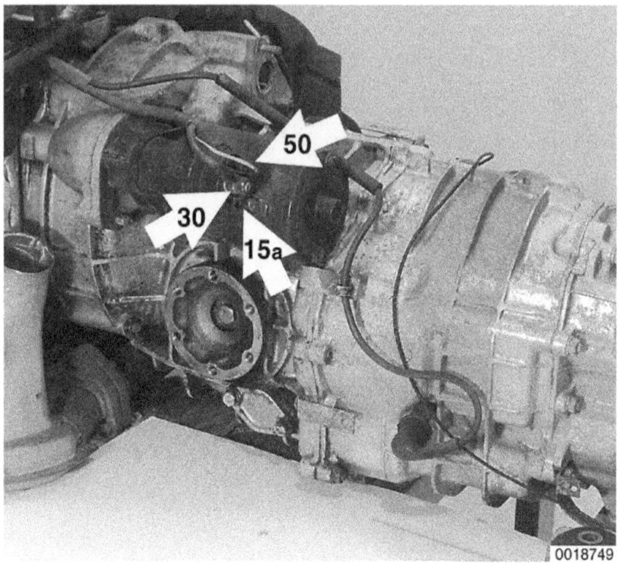

- Remove wires from starter terminals 15, 30 and 50.
- Using 10 mm allen wrench, remove starter mounting nuts.
- Slide starter toward front of car to remove.
- Installation is reverse of removal.

Tightening torques

- Positive cable to starter .... 10 Nm (7.5 ft-lb)
- Starter to transmission or engine ...... 45 Nm (33 ft-lb)

NOTE —

Engine/transmission assembly shown removed to better illustrate starter motor and starter motor wiring.

CHARGING SYSTEM

The charging system provides the current necessary to keep the battery charged and to operate the car's electrical accessories. The system includes a belt-driven alternator to generate the charging current, and a voltage regulator to control the charging current.

> ***CAUTION—***
> - *Disconnect the battery negative (–) cable before disconnecting any wires from the rear of the alternator. Battery voltage is wired directly to the alternator without fuse protection. Reconnect the battery cable after all wires have been safely disconnected and insulated.*
> - *Never operate the engine with the battery disconnected. Never operate the alternator with its output terminal (B+ or 30) disconnected and the other terminals connected.*
> - *To prevent damage to the alternator or regulator when making tests or repairs, make all connections with negative (–) to negative, and positive (+) to positive unless directed otherwise. Even momentary contact with a conductor of the wrong polarity can damage the alternator's diodes.*

The voltage regulator maintains a norminal voltage in the electrical system by feeding excess output from the alternator back to ground. In addition, it regulates the amount of current supplied to the alternator rotor, turning it on and off as needed. If either component is faulty, the result will be an over- or undercharged battery. The brushes wear under normal use, and will eventually need to be replaced.

The brushes contact the alternator's slip rings to supply the current that magnetizes the alternator rotor.

Two versions of charging systems/alternators are used on the cars covered by this manual:

Alternator output
- 1978 - 1981 with external regulator .... 70 amp
- 1982 - 1983 with internal regulator............ 70 amp

NOTE—
Rated output of the alternator when measured at the battery with the engine running can be 10 to 15 amps less than specified. The difference is accounted for by electrical consumers and is a normal condition.

0021051

On 1978 - 1981 models, the charging system uses an external voltage regulator. The voltage regulator is mounted in the left side of the engine compartment (**arrow**).

On 1982-1983 models, the charging system uses an alternator with an internal voltage regulator (**arrow**). The voltage regulator is available as a replacement part.

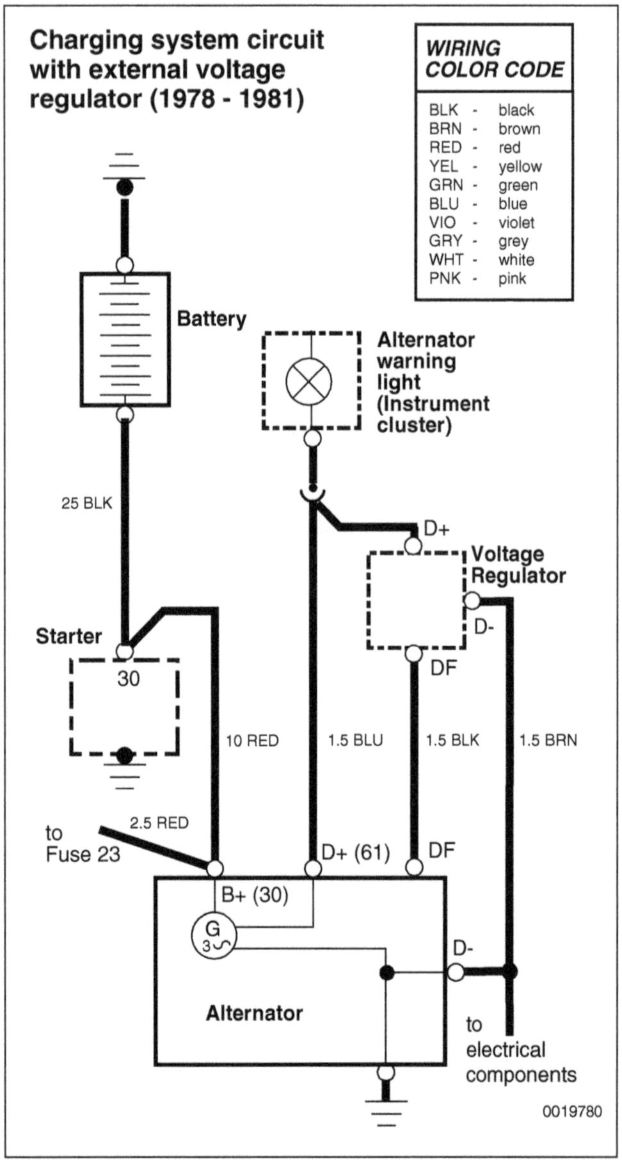

Charging system troubleshooting

Aside from the physical location of the voltage regulator, there are only minor differences in the wiring of the alternator with external regulator and the alternator with built-in regulator. See the accompanying wiring diagrams.

Charging system trouble is indicated by an illuminated alternator warning light on the instrument panel, or by an under- or overcharged battery.

When the engine is running and the alternator is spinning, part of the current the alternator produces energizes its electromagnetic field. When starting, some other current must be provided to initially energize (excite) the field and begin the current generating process. This current is provided by the battery through the alternator warning light in the instrument panel.

NOTE —

On some early 911 SC models, a burned out instrument cluster alternator warning light will prevent the alternator from charging.

As soon as the alternator output equals battery voltage, the warning light goes out. Normally, the warning light should be off when the ignition is off and the engine is stopped. The light should only come on when the ignition is turned on (current to the alternator rotor) and go out again when the engine is started and the battery is being charged.

An undercharged battery is usually associated with starting trouble. Again, make sure that the battery is in good condition and capable of accepting a full charge before blaming the charging system.

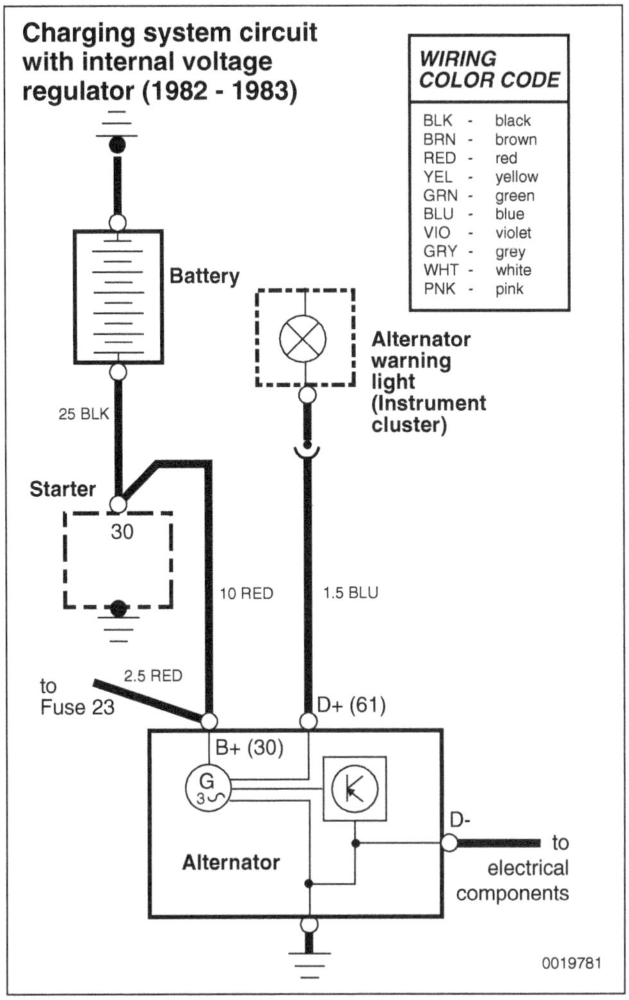

Causes of an undercharged battery may be:

- The simultaneous use of many electrical accessories for long periods of time, especially at low engine speeds
- Leaving accessories or lights in operation with the engine stopped
- Frequent and long periods of starter usage
- Frequent short-trip driving
- Improper alternator V-belt condition or tension
- Broken or frayed charging system wiring
- Corroded connectors at the D+ and B+ terminals of the alternator
- Worn, corroded, or loose battery cable connections at the chassis

Table f lists symptoms of trouble indicated by the alternator warning light, their probable causes, and suggested corrective actions, most described later in this group.

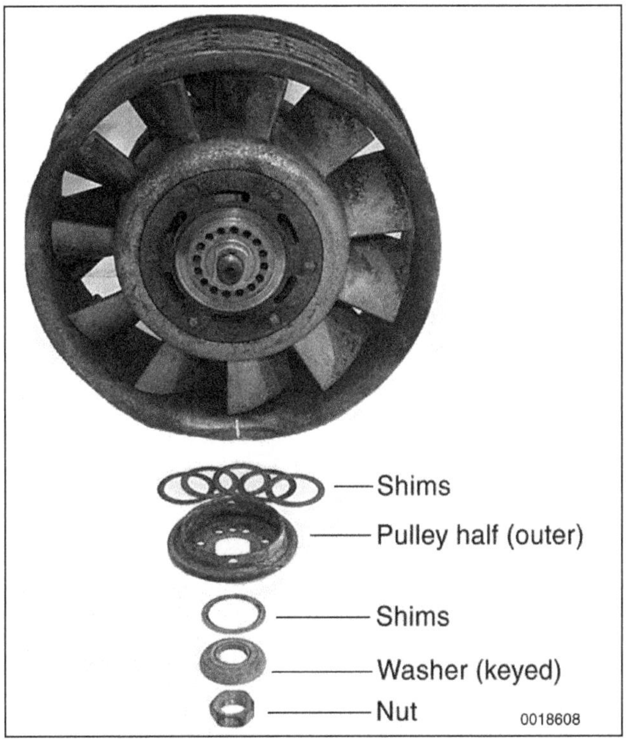

Charging system, testing

The tests described here will help determine the cause of charging system trouble.

< The battery should be fully charged and the alternator V-belt correctly tensioned. Correct belt tension by adding or subtracting V-belt adjusting shims. See **030 Maintenance**.

– The charging system warning light should be operating correctly.

– All electrical connections should be clean and tight.

– Replace wires that are hard and cracked.

– If tests show that the alternator and regulator are operating correctly, but the battery still continually runs down, there may be a short in the electrical system causing a continuous current drain.

– Remember that the clock, radio, and some other components require constant current. See **Static current draw test** earlier in this group.

Table f. Warning light troubleshooting

Symptom	Test and probable cause	Corrective action
Ignition off, engine off, warning light on	*Test:* Disconnect blue wire (D+) from alternator. a. Light goes out: Faulty alternator diodes. b. Light does not go out: short to ground in wiring harness or wiring connector.	a. Repair or replace alternator. b. Repair or replace faulty wiring.
Ignition on, engine off, warning light off	a. Battery fully discharged. b. Bulb burned out. *Test:* Disconnect blue wire (D+) from alternator. With battery connected and ignition on, touch blue wire to ground. c. Light does not come on: Faulty bulb socket, open circuit between socket and terminal 19 of ignition switch, or open circuit between blue wire (D+) on alternator and instrument cluster. d. Light comes on: Loose connection between regulator and alternator or loose connection between brushes and regulator. e. Light comes on, no faults with regulator: Internal alternator faults or faulty regulator.	a. Charge battery. b. Remove and test bulb. Replace faulty bulb. c. Replace instrument cluster bulb holder. Repair wiring to alternator. d. Inspect brushes. Correct loose connections. e. Repair or replace alternator or voltage regulator.
Engine running at any speed, warning light stays on	a. Loose or broken alternator V-belt. *Test:* Disconnect and insulate battery negative (–) terminal then disconnect red wire (B+) from alternator. Do not short wire. Using a voltmeter, test between red wire and ground with battery terminal reconnected. b. No voltage to alternator: Open circuit between red wire on alternator and starter or between starter and positive (+) pole. c. Exciter diodes burned out. d. Faulty regulator or faulty alternator windings. e. High voltage drop between red alternator positive (+) wire and starter due to broken, loose, or corroded wires.	a. Replace or adjust V-belt. b. Repair wire or connections between alternator and starter. c. Repair or replace alternator. d. Test charging system and replace faulty components as needed. e. Repair wires or connectors.

A general test of charging system output can be made with an accurate digital voltmeter. Such a test is given below. The most accurate testing is done using an alternator and regulator tester that applies a high-current load to the alternator. Conclusive tests using this equipment can be made inexpensively by an authorized Porsche dealer or other qualified shop.

Alternator and regulator, checking

The procedure given below assumes that the battery is in good condition and fully charged.

- Attach digital voltmeter to negative (–) and positive (+) battery terminals.
  - Make sure tester is connected to clean areas of terminals.
  - Set voltmeter to DCV scale.
- Start engine and run it at about 1,500 rpm with all electrical accessories turned off. Measure voltage at battery.
- A reading much higher than 14.2 volts most likely indicates a faulty voltage regulator.
- A reading below 13.5 volts means that the battery is not adequately charged. This may be due to:
  - Faulty V-belt
  - Faulty voltage regulator
  - Faulty alternator.

– To determine which component is faulty:
  - Visually check V-belt. See **030 Maintenance**.
  - Keep engine running at about 1,500 rpm and turn on all electrical accessories.
  - Check voltage across battery terminals.
  - A reading that is the same as the first most likely indicates a faulty voltage regulator.
  - A reading that is lower most likely indicates a faulty alternator.

0018607

Noisy alternator

< Alternator mechanical noises may be caused by:
- Misalignment between the V-belt and the pulley
- Loose or broken pulley
- Worn bearings
- Bent rotor shaft.

– Check for bad bearings by removing V-belt and rotating alternator pulley by hand. If shaft bearings grinds, alternator should be replaced.

Alternator, replacing

– Disconnect negative (–) battery cable and cover battery terminal to keep cable from accidentally contacting terminal.

> ***CAUTION —***
> *Be sure to read the battery disconnection cautions given on page viii before disconnecting the battery.*

0018055

– Remove A/C drive belt as described in **870 Air Conditioner**.

– Remove V-belt. Reinstall pulley and nut fingertight.

< Remove allen bolt (**A**) from cooling housing strap.

– Pull alternator and fan out (toward rear of car) partially. Also lift engine shroud slightly and unbolt alternator ground strap from top of engine case.

Remove nuts (**arrows**) holding alternator air guide and slide guide back over wiring harness.

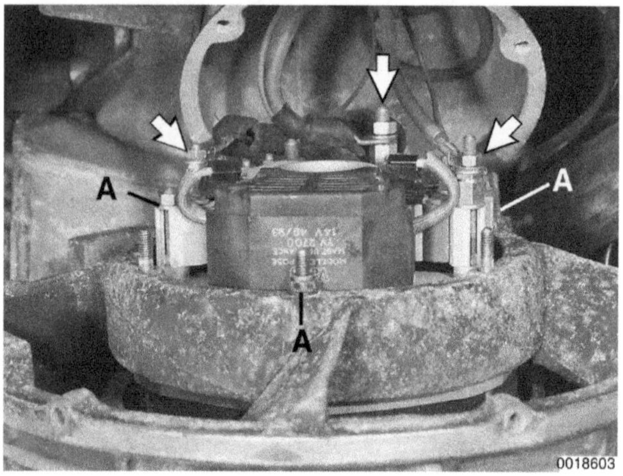

Lift off insulating rubber boots from alternator terminals.

- Detach electrical connections (**arrows**) in back of alternator.
- Remove alternator/cooling fan assembly.
- Working at bench, remove alternator to fan housing mounting nuts (**A**).
- Push alternator out of fan housing.

NOTE —

Photo shows model with internal regulator (1982 - 1983). Model with external regulator varies slightly.

– Pull fan off alternator shaft using a puller.

– Installation is reverse of removal.

Tightening Torques

- Fan housing mounting strap allen bolt . . 25 Nm (18 ft-lb)
- Fan/alternator pulley to alternator shaft . 55 Nm (40 ft-lb)
- Terminal B+ . . . . 6 Nm (53 in-lb)
- Terminal D+ . . . . 3 Nm (27 in-lb)

Voltage regulator, external removing and installing (1978 - 1981)

– Remove cover from electrical panel in left side of engine compartment.

Pull voltage regulator (**arrow**) straight up to remove.

– Installation is reverse of removal.

Brushes, removing and installing 1978 - 1981 (with alternator removed)

Porsche recommends replacing the early style alternator (with external regulator) with an updated alternator with internal regulator. If the brushes are worn on the early alternator, replacement brushes may be available through aftermarket sources.

- Remove alternator as described earlier.
- Remove brush holder from back of alternator.
- Installation is reverse of removal.

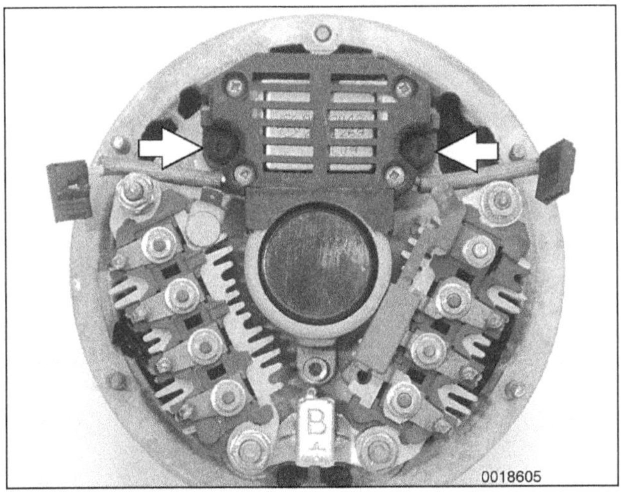

Voltage regulator, internal removing and installing (1982 - 1983) (with alternator removed)

On 1982 - 1983 models, the regulator and brush holder are mounted to the rear of the alternator. To remove the regulator and brushes, remove alternator as described earlier.

< With alternator on a bench remove two screws (**arrows**) holding regulator to alternator housing.

- Slide regulator away from shaft and lift out of alternator housing.
- Installation is reverse of removal.

280 Ignition System

GENERAL

This repair group covers descriptions, troubleshooting and component replacement for the Capacitor Discharge Ignition (CDI) system.

NOTE —
Information on ignition system maintenance, including the prescribed maintenance intervals, can be found in **030 Maintenance.**

Special Tools

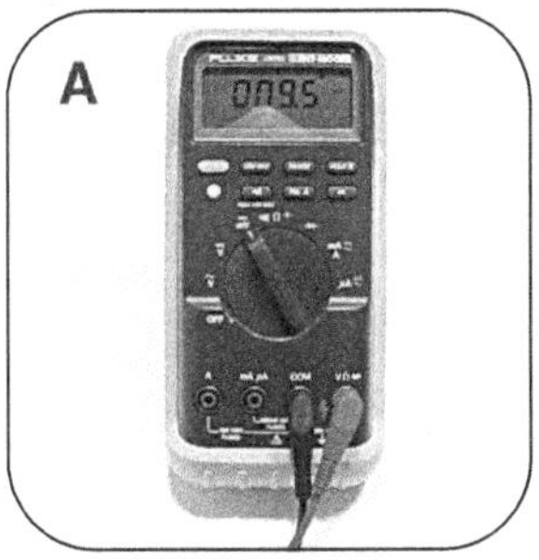

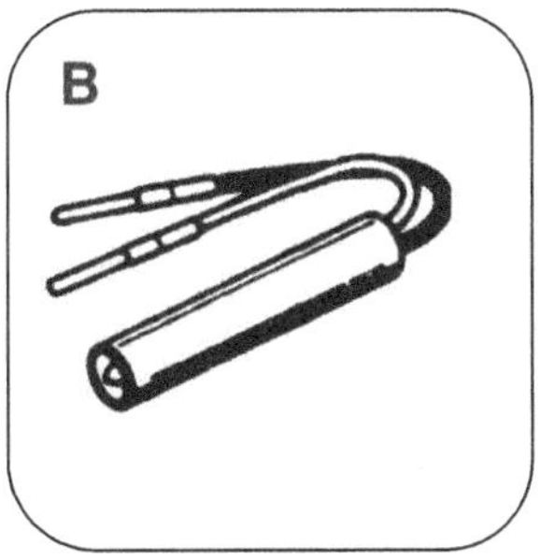

Special tool

- **A** Fluke 87 . . . . . Automotive digital multimeter
- **B** 1115 . . . . . LED tester with thin spade probes (source: Baum Tools Unlimited)
- **C** . . . . . Strobe timing light

Test equipment

◄ Some of the procedures require special tools.

> *CAUTION —*
> *When connecting test equipment be careful that the connectors do not come in contact with a ground. Any type of short circuit can cause damage to the CDI module.*

Capacitor Discharge Ignition (CDI)

The ignition system installed in 911 SC models covered by this manual is called Capacitor Discharge Ignition (CDI). This system was developed by Bosch for high performance applications.

The ignition distributor uses a pulse generator to produce a current pulse which is sent to the CDI module. The CDI module amplifies the pulse to as high as 450 volts before it is sent to the ignition coil.

The centrifugal and vacuum advance and retard operate like a conventional distributor.

Firing order

To determine the start of the firing order, remove the distributor cap and inspect distributor housing for cylinder 1 mark (**arrow**).

NOTE —
The ignition rotor turns in a counterclockwise direction.

– Spark plug wires must be installed so that the spark plugs fire in the correct order.

Ignition firing order
- Porsche 911 SC 1-6-2-4-3-5

Cylinder 1 is at left rear when viewing engine from rear of engine compartment.

Ignition timing, adjusting

Ignition timing should be checked and adjusted (if necessary) at each major maintenance interval or whenever the distributor position has been altered.

NOTE —
Check and adjust timing with engine at operating temperature.

Check ignition timing at the crankshaft pulley using a strobe type timing light:
- **Z1** lined up with scribe mark on lower fan housing is TDC.
- Check for timing specifications in **Table a**.

Table a. timing specifications, 911 SC

Model Year		
1978-79	5° ± 2° BTDC @ 950 rpm, vacuum hose connected	26° ± 2° BTDC @ 6000 rpm, vacuum hose disconnected
1980-83	5° ± 2° BTDC @ 950 rpm, vacuum hose disconnected	21° ± 3° BTDC @ 6000 rpm, vacuum hose disconnected

◄ To adjust ignition timing:

- Loosen distributor housing lock-down fastener (**arrow**).
- Connect timing light according to manufacturer's instructions.
- Start engine and allow it to reach operating temperature.
- Slowly rotate distributor to match timing specifications given in **Table a**.
- Tighten distributor lock-down fastener and recheck timing.

Tightening torque

- Distributor to mounting stud (M8) . . . . . . 18 Nm (14 ft-lb)

NOTE —

Be sure to reattach vacuum hose(s).

Disabling ignition system

To avoid accidental shock and possible personal injury, the ignition system should be disabled when making certain tests such as compression test or starter current draw test.

> ***WARNING —***
>
> *Do not touch or disconnect any of the high tension cables from the coil, distributor cap or spark plugs while the engine is running or being cranked by the starter motor. Fatal voltage are present.*

– Remove cover from electrical panel in left side of engine compartment.

◄ With ignition off, disconnect electrical harness connector at CDI module (**arrow**) located in engine compartment electrical panel.

TROUBLESHOOTING

A complete failure of the ignition system to produce spark at the spark plugs is self-evident. For other problems such as rough idle, misfiring, or poor starting, however, the cause may not be so clear.

Poor driveability may have a variety of causes. The fault may lie with the ignition system, the fuel system, parts of the emission control system, or a combination of the three. Because of the interrelated functions of these systems and their effects on each other, it is often difficult to know where to begin looking for problems. For this reason, effective troubleshooting should always consider these systems in unison, as one major system.

The CDI ignition system contains sensitive electronic components. To protect the system, and for general safety, the warnings and cautions given here should be observed during any ignition system troubleshooting, maintenance or repair work.

> ***WARNING —***
> *Inefficient combustion can cause serious problems in the catalytic converter. The poorly burned mixture can overload the converter with raw fuel, leading to converter overheating or plugging. An overheated catalytic converter can be a fire hazard.*

> ***CAUTION —***
> - *Connect or disconnect ignition system wires, multiple connectors and ignition test equipment leads only while the ignition is off.*
> - *A high impedance digital multimeter should be used when making voltage and resistance tests on the CDI system. Switch multimeter functions or measurement ranges only with the test probes disconnected.*
> - *Do not disconnect terminal 4 (center terminal) from the coil or remove the distributor cap to disable the ignition.*
> - *Do not disconnect battery while engine is running.*
> - *Do not connect a test instrument with a 12-volt supply to terminal 15 (+) of the ignition coil. The voltage backflow can damage the CDI module. In general, make test connections only as specified by Porsche, this manual, or the instrument manufacturer.*
> - *Before operating the starter without starting the engine (e.g., doing a compression test), always disable the ignition, as described earlier in the text.*
> - *Do not quick charge the battery (for boost starting) for longer than one minute, and do not exceed 16.5 volts at the battery with the boosting cables attached. Wait at least one minute before boosting the battery a second time.*
> - *Do not wash engine while it is running, or any time the ignition is switched on.*
> - *Disconnect the battery when doing electric welding on the vehicle or when charging the battery.*
> - *Do not start the engine if the car has been heated above 176°F (80°C) (e.g., in a paint drying booth). Wait for the car to cool to normal temperature.*
> - *Do not connect terminal 1 of the coil to ground as a means of preventing the engine from starting.*

Basic troubleshooting principles

The function of the ignition system is to provide a properly timed spark for efficient combustion.

An engine that has good cranking speed but will not even begin to start may indicate a complete failure of the system to produce spark. Inspect the ignition system visually. Make sure the spark plug wires have not been interchanged. The firing order is given earlier.

If no faults are located, make a basic check to see if spark is being produced as described below under **Ignition system quick-check**. This will be the most important first troubleshooting step. If a strong spark is observed, then the failure to start is due to another cause, such as no fuel being delivered to the engine.

Ignition system quick-check

If the engine does not start, the most fundamental step in troubleshooting the ignition system is to determine whether or not the system is making a spark at the spark plug. If no spark is present, then more detailed testing of the ignition system is necessary.

To check for spark, turn ignition off and remove spark plug wire connector (**arrow**) from a spark plug.

- Connect to known good spark plug.
- Attach jumper wire with alligator ends to spark plug threads and a good engine compartment ground.

> ***WARNING —***
> *Do not hold the spark plug or its connector, even if using insulated pliers. The ignition system is a high-energy system operating in a dangerous voltage range which could prove to be fatal if exposed terminals or live parts are contacted.*

> ***CAUTION —***
> *Any test set-up other than the one described above may lead to damage or inconclusive results.*

– While a helper actuates starter, look and listen for spark in spark plug gap.

- Bright blue spark indicates healthy ignition system.
- Yellow-orange spark is weaker and indicates that, while spark is present and system is functioning, it is not operating at peak efficiency.
- Check condition of ignition system components as described below under **Ignition system visual inspection**.

– Disconnect CDI harness connector from module.

– Turn ignition ON.

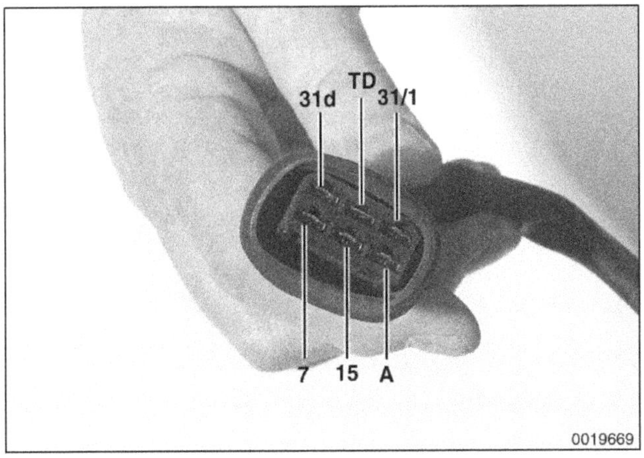

Check for battery voltage at terminal 15 of CDI plug.

> ***CAUTION —***
> - *If ignition system failure is not the problem, the engine may start during this test. Be prepared to turn off the ignition immediately. Running the engine with a spark plug wire disconnected may damage the catalytic converter.*
> - *Before checking the ignition system when there is no spark or a weak spark, make sure that the battery is fully charged.*

– If battery voltage is not present, there is either a fault in wire between terminal 15 and ignition switch, in ignition switch itself, or in wiring from battery to ignition switch.

NOTE —
See **970 Electrical Wiring Diagrams** *for ignition system schematics.*

– If no faults have been detected up to this point but there is still no spark or a weak spark, refer to **Table b** for more troubleshooting information.

– If a strong spark is observed but engine still will not start, the ignition system is most likely OK and not the cause of the no start condition.

Table b. Ignition system troubleshooting

Symptom	Probable cause	Corrective action
No spark or weak spark observed during spark test	Wet or damp distributor cap and/or spark plug wires	Remove cap and wires. Dry and reinstall.
	Faulty wires or connectors (primary circuit)	Inspect and repair wiring as needed.
	Weak or faulty coil	Test and replace as needed.
	Defective spark plug wires	Test and replace as needed.
	Worn or fouled spark plugs	Replace spark plugs. See **030 Maintenance**.

Ignition system visual inspection

The spark plug wires, ignition coil, distributor cap and ignition rotor are subject to wear and electrical breakdown which will impair their ability to deliver a crisply timed and powerful spark. Many of these conditions are most easily detected by a thorough visual inspection. Dirt and moisture on any of these components are also potential causes of poor spark at the spark plugs.

0019608

To inspect distributor cap and ignition rotor, use flat-tipped screwdriver to pry lock clips (**arrow**) away from cap.

- Lift off cap.
- Inspect contacts inside cap and at tip of the rotor for corrosion, wear, or pitting.
- Parts with corroded contacts can be cleaned and reused, but if there is wear, pitting or heavy corrosion, replacements is recommended.
- Center black carbon brush inside the cap should spring back when compressed.

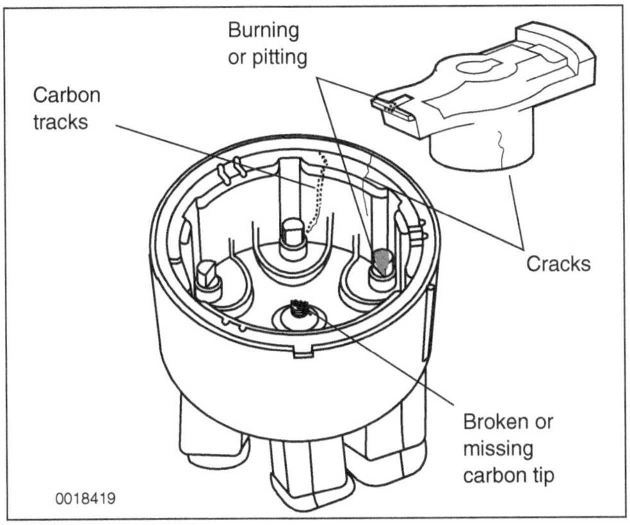

◄ Cracks or carbon tracks in distributor cap may cause shorts to ground.

- Cracks may be fine and difficult to see. Check carefully, especially around the contacts. Carbon tracks are faint lines, usually running between two contacts or to ground, left over from high-voltage arcing.
- A distributor cap that shows any sign of cracks or carbon tracking should be replaced.

— For a quick-check of distributor cap and spark plug wire condition, listen for sound of voltage arcing or watch while engine runs at night.

- In the darkness, the arc of high voltage to ground because of a crack in the cap or poorly insulated wires may be visible as a blue spark.

— The coil should be closely examined for cracks, burns, carbon tracks, or any leaking fluid.

— Inspect all primary wires and connections for any corrosion or damage.

- Clean or repair any faults found. Corroded or loose connections may interfere with ignition function.

NOTE—
A failed CDI control module can sometimes damage the ignition coil. Be sure to check the ignition coil anytime a faulty CDI module is replaced and a no-start condition still exists.

Spark plug wires, testing

— Check resistance of each spark plug end connector as well as wire itself.

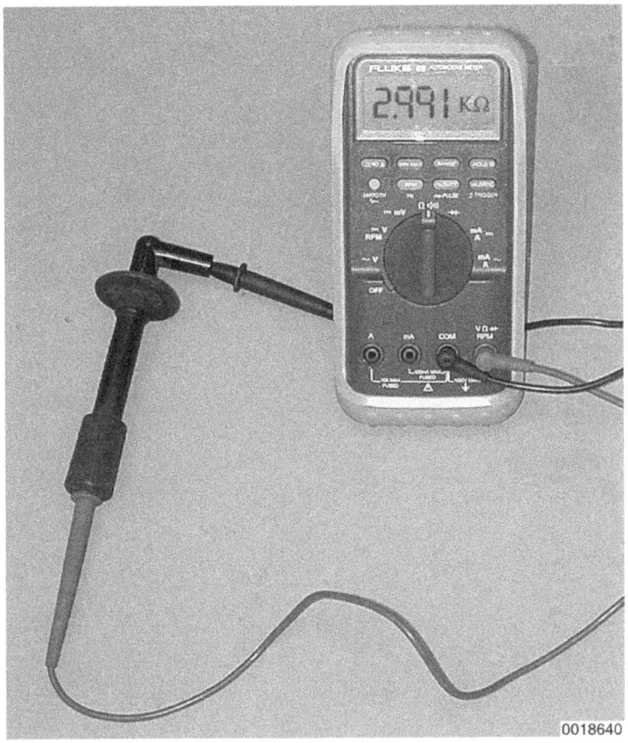

◄ To test connector, use ohmmeter probe on each end of connector.

— Test shielded ends of coil and spark plug wires at distributor cap and coil using same method.

CAUTION—
To avoid damaging the distributor cap, do not wiggle the connectors when removing the spark plug wires. If necessary, twist to loosen. Then, pull straight out from the cap.

- Check for corrosion at connectors.
- If measured resistance is too high, wire assembly should be replaced.

Ignition cable resistance
- Plug connector to distributor cap end . . . . 3 kΩ

NOTE—
Individual spark plug wire end connectors are available as replacement parts from an authorized Porsche dealer.

Distributor cap, testing

◄ To test cap and rotor, unhook the two spring clips (**arrow**) from sides of distributor cap and lift off cap.

- Detach spark plug wires and coil wire.

> ***CAUTION —***
> *To avoid damaging the distributor cap, do not wiggle the connectors when removing the spark plug wires. If necessary, twist to loosen. Then, pull straight out from the cap.*

NOTE —
Be sure to label the spark plug wires and their locations when disconnecting them from the cap.

- Check resistance of cap between tower and its matching contact inside cap. Resistance should be nearly zero ohms.
- If any resistance is detected, replace cap.

Ignition rotor, testing

– After removing cap, pull rotor straight up off drive shaft.

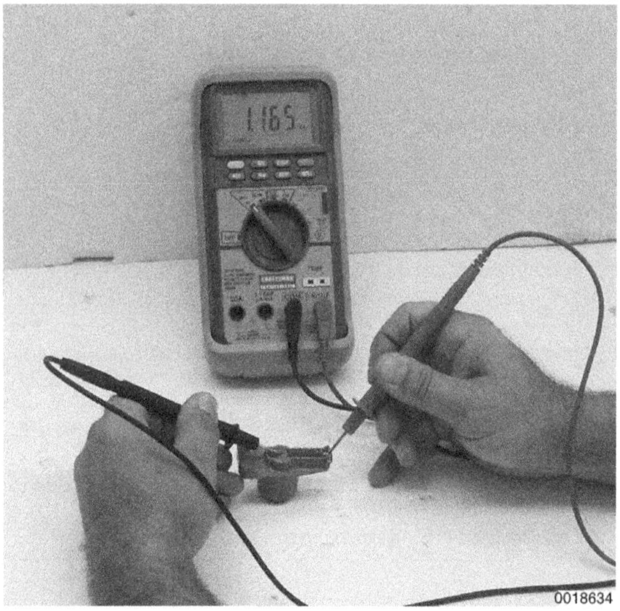

◄ Check rotor resistance with ohmmeter.

- If measured resistance is too high, replace rotor. Breaker-

Ignition rotor resistance

- Standard rotor . . . . 1 kΩ
- Rpm limiter rotor . . . . 5 kΩ

less distributor, testing

– Disconnect harness plug from bottom of CDI module.

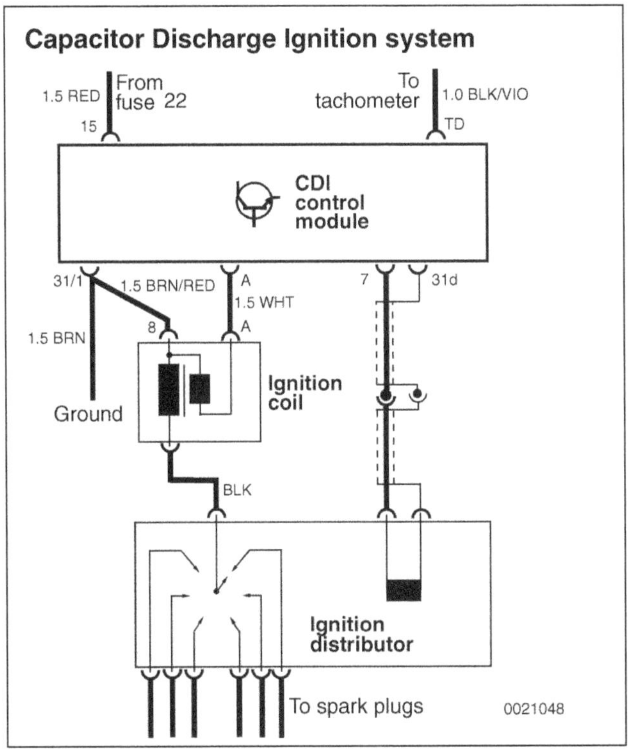

Breakerless distributor, testing

The Capacitor Discharge Ignition (CDI) control module cannot be tested directly. CDI module terminal connections are shown in **Table c**.

Table c. CDI module terminals

Terminal #	Description
31/1	Ground
TD	To tachometer
31d	Distributor pulse wire shielding
A	To positive side of ignition coil
15	From fuse 22 in main fuse box
7	Distributor pulse input

– Remove electrical panel cover in engine compartment.

– Disconnect harness plug from bottom of CDI module.

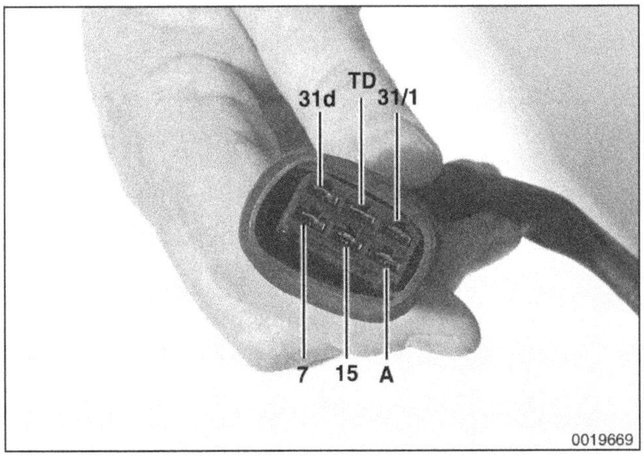

With digital voltmeter, make following tests:

- With ignition ON: Test for battery voltage at CDI module terminal 15. No battery voltage indicates lack of power at fuse 22, or a bad wire.
- With engine CRANKING: Test for voltage (VAC) between terminals 7 and 31d.
- If meter shows no pulse, distributor pulse generator is most likely faulty.
- If meter shows a voltage pulse, CDI control unit or coil are suspect.

NOTE —
To test for defective components, substitute components that are known to be good.

COMPONENT REPLACEMENT

The cap, rotor and spark plug wires all carry high voltage to the spark plugs. Proper engine performance depends on getting the best possible spark at the spark plug. Worn and corroded contacts or poor insulation which allows the spark to short to ground are the primary reasons for replacement of these components.

CDI module, replacing

– Disconnect negative (-)cable battery.

– Remove cover from engine compartment electrical panel.

– Disconnect CDI harness connector.

◄ Remove three 5 mm allen bolts (**arrows**) securing CDI module to panel.

– Installation is reverse of removal.

Distributor cap and ignition rotor, removing and installing

NOTE —

It is recommended that the cap and rotor be replaced as a set during a major maintenance.

◄ To replace cap and rotor, unhook the two spring clips (**arrow**) from sides of distributor cap and lift off cap.

- Pull rotor off distributor shaft.

– If replacing cap and rotor, temporarily leave spark plug wires in old cap.

- Install new rotor.
- Mount new cap on distributor and lock side clips.
- Change over one spark plug wire at a time from old to new cap.

> *CAUTION —*
>
> *To avoid damaging the distributor cap, do not wiggle the connectors when removing the spark plug wires. If necessary, twist to loosen. Then, pull straight out from the cap.*

Spark plug wires, replacing

– When removing spark plug wires, label their position so that they can be reinstalled in the proper places. If the wires get mixed up, see **Firing order** above.

Distributor, removing and installing

◄ Set crankshaft pulley so that **Z1** mark is aligned with scribe mark on lower fan housing.

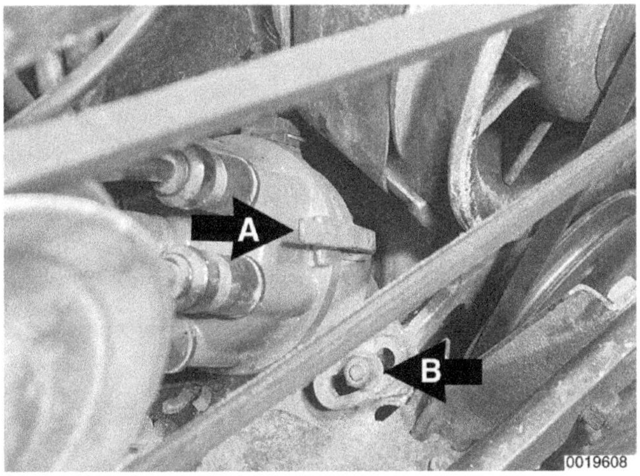

◄ Unhook two spring clips (**A**) from sides of distributor cap and lift off cap.

– Disconnect vacuum hose(s) at distributor.

– Disconnect distributor wiring harness.

– Remove distributor housing lock-down fastener (**B**).

◄ Check that distributor rotor is aligned with scribe mark (**arrow**) on lip of distributor housing.

NOTE —

If distributor rotor is pointing 180° off, rotate crankshaft one full turn and check the rotor again.

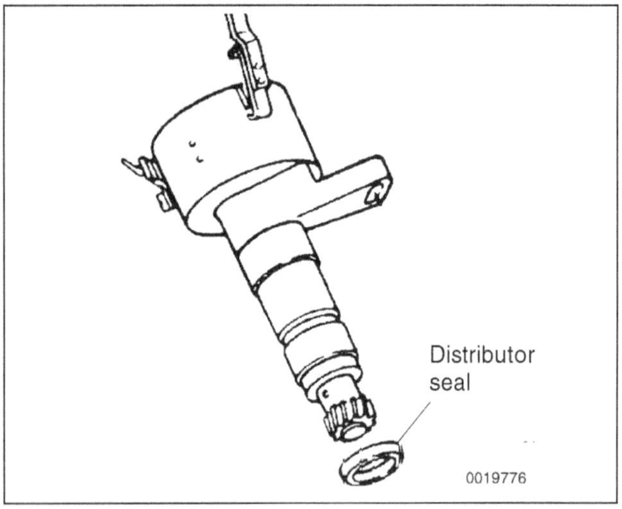

Pull distributor out of crankcase. Inspect O-ring seal on distributor shaft housing and replace if rubber is hard or cracked.

When installing, turn distributor housing so that ignition rotor and housing scribe mark (**arrow**) align.

- Install distributor locking nut and tighten finger tight only.
- Install cap and reconnect wiring harness.
- Using specifications in **Table a,** reset ignition timing.

NOTE —
Check and adjust timing with engine at operating temperature.

- Tighten distributor lock-down fastener and recheck timing.

Tightening torque
- Distributor to crankcase (M8) . . . . 18 Nm (14 ft-lb)

NOTE —
Be sure to reattach vacuum hose(s).

300 Transmission and Clutch–General

GENERAL

This section covers general information about the transmission and clutch. Transmission removal is covered in **101 Engine Removal and Installation**. Refer to the following repair groups for repair information:

- **301 Clutch**
- **340 Manual Transmission–Controls and Case**
- **350 Manual Transmission–Gears and Shafts**
- **390 Differential**

TRANSMISSION

The models covered by this manual are equipped with the 915 transmission with a cable operated clutch system.

Cross-section of 915 transmission

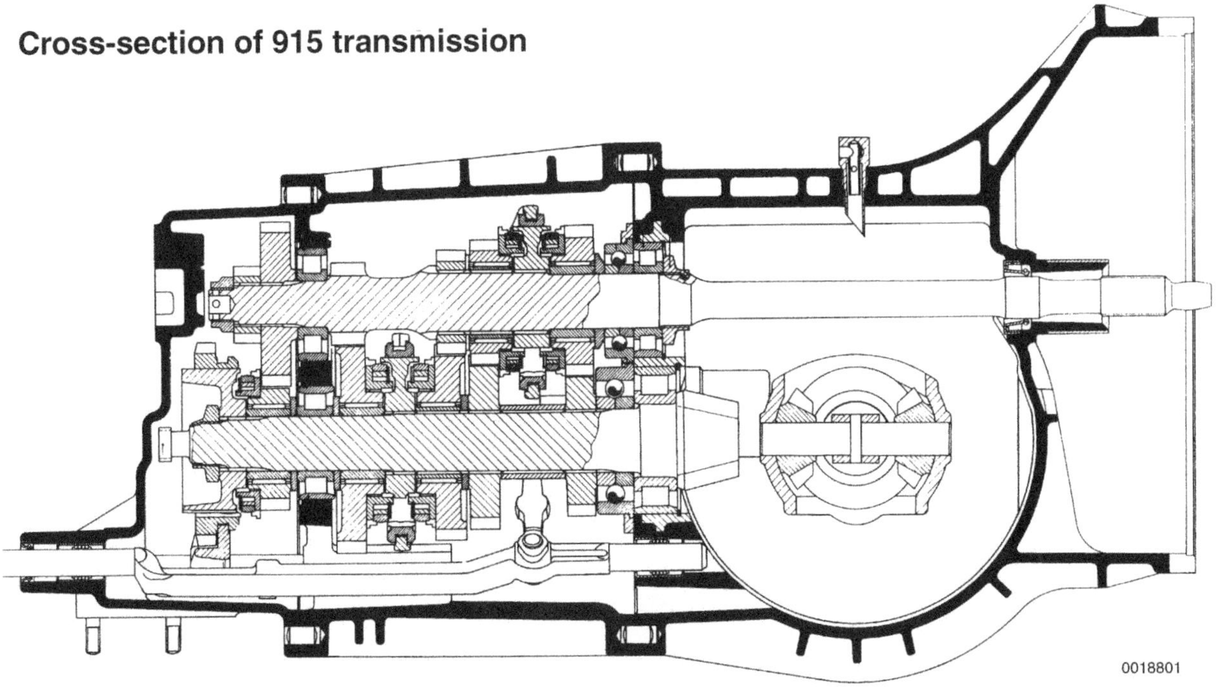

915 transmission speed in gear

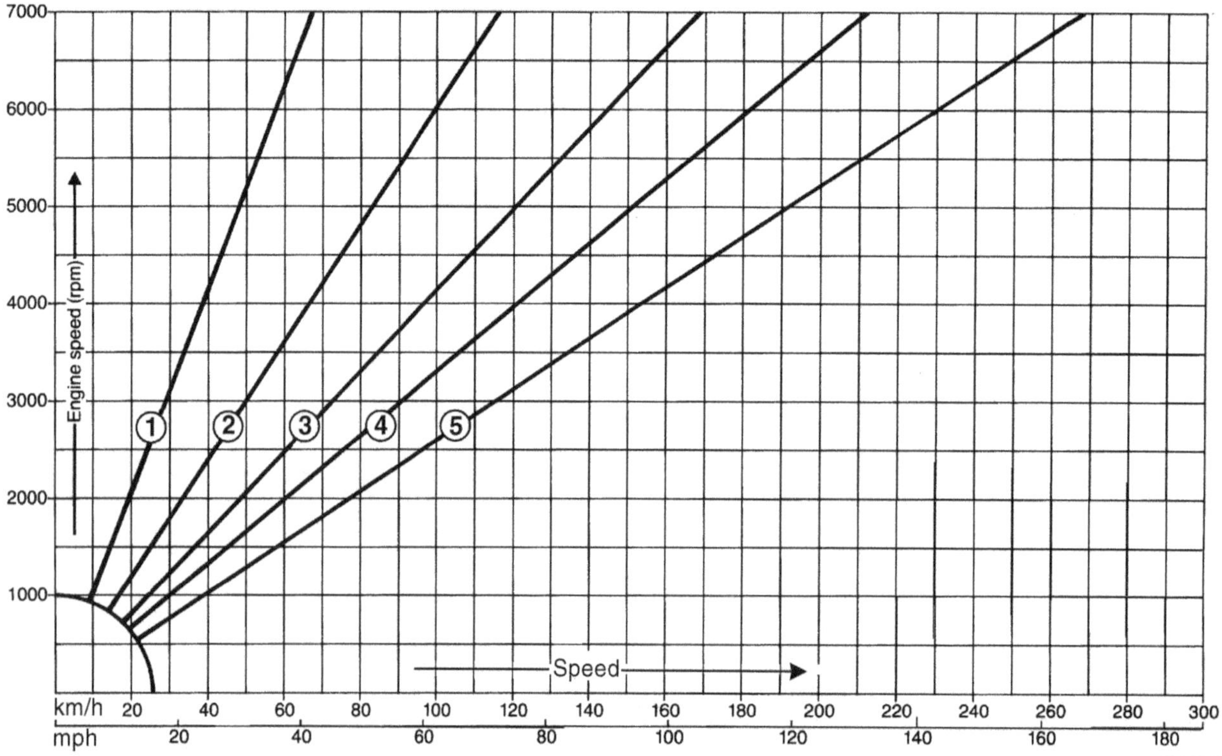

Transmission applications are listed in **Table a**. Transmission gear ratios are in **Table b**.

Table a. Transmission applications

Year	Transmission	Code numbers
1978 - 1979	915/61	71 9 0000 - 10000
1980 - 1983	915/63	74 C 0000 - 10000

Table b. Transmission gear ratios

	915/63	
Gear ratios		
1st	11/35	3.181
2nd	18/32	1.778
3rd	23/29	1.261
4th	26/26	1.000
5th	28/23	0.821
reverse	-	3.325
Final drive		3.8750:1

The transmission case, which contains the transmission gear train, is a lightweight alloy die casting. The transmission gears are of the constant-mesh type. Each forward gear (1st through 5th) is actually a mating pair of gears, and constant-mesh simply means the mating gears are always meshing.

The 915 transmission uses Porsche-designed gear type synchronizers.

Gear selection is made using the gearshift lever in the passenger compartment, which is linked to the transmission by the gear shift linkage. When shifting, the synchronizer matches the speed of the selected gear to the speed of the output shaft. The synchronizer hub then locks the selected gear into rotation with the output shaft. This eases shifting and minimizes gear wear.

Transmission repairs

Repairs fall into two categories: those that can be performed externally, and those that require disassembly of the transmission. Problems that at first appear to be caused by internal faults, such as gear shifting difficulty or noisy operation, can often be corrected externally.

Repairs to the internal parts of the transmission require special tools, knowledge and experience. If you lack the skills and tools, or a suitable workplace for servicing the transmission, we suggest you leave these repairs to an authorized Porsche dealer or other qualified shop. Completely reconditioned transmissions are available from the authorized Porsche dealer or the aftermarket.

Though you may not have the skills and knowledge for doing the actual internal repairs, it is possible to save some of the expense of professional repair by removing and installing the transmission yourself, using the procedures described in **101 Engine Removal and Installation**. It is important to realize, however, that a partially disassembled transmission may be a problem for a mechanic. We strongly advise against taking the transmission apart to begin any repairs that cannot be properly finished.

CLUTCH

The 915 transmission models are equipped with a cable operated clutch release system with a spring dampening clutch disc.

MAINTENANCE

Information on clutch and transmission maintenance and on the prescribed maintenance intervals can be found in **030 Maintenance**.

Fluid capacity

• 915 transmission 1978 - 1983 . . . . 3.0 L (3.2 qts.).

301 Clutch

301

GENERAL

There is one clutch system with two different clutch disc designs used on the 1978 through 1983 911 models. See **Table a**.

Table a. Clutch applications

Model Year	Trans. Code	Operation	Clutch Disc Diameter	Disc design
1978 - 1980	915	Cable	225 mm (8.86 in.)	Rubber damper center
1981-1983	915	Cable	225 mm (8.86 in.)	Spring damper center

Special tools

Some of the procedures in this group require special tools.

Special Tools

A

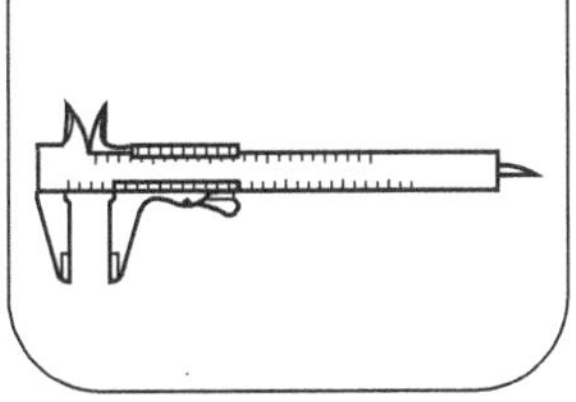

B

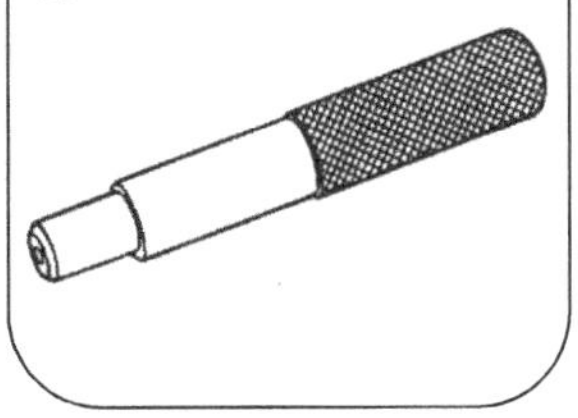

Special tool
A 14 (source: Baum Tools Unlimited) . . . . Vernier caliper
B P 370 . . . . Clutch disc alignment tool (source: Baum Tools Unlimited)

CLUTCH CABLE AND PEDAL CLUSTER

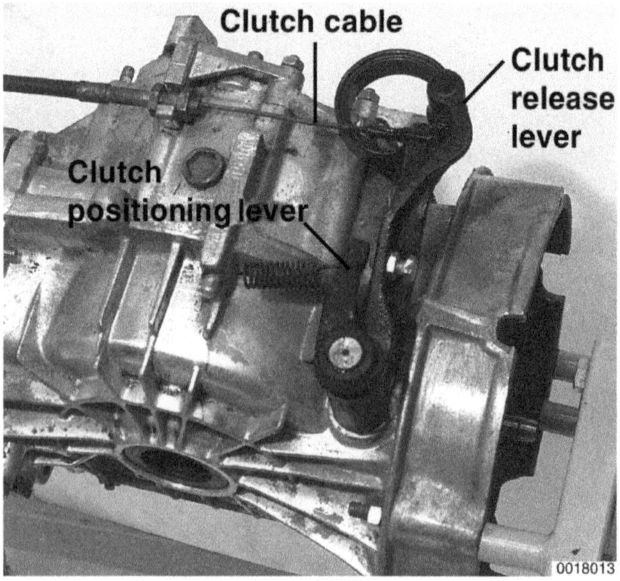

When the clutch pedal is depressed, the clutch cable actuates the clutch release lever via the clutch positioner lever on the transmission.

The positioner lever is splined to the clutch cross-shaft. The clutch release fork is also splined to the cross-shaft in the transmission bellhousing. Turning the cross-shaft pulls the release fork forward and disengages the clutch.

Clutch cable, adjusting

Clutch pedal and cable free play is adjusted at the clutch positioning lever on the transmission.

– Raise car and support in a safe manner.

> ***WARNING —***
> *Make sure the car is stable and well supported at all times. Use a professional automotive lift or jack stands designed for the purpose. A floor jack is not adequate support.*

Working under rear of car, loosen cable housing lock nuts (**arrow**) and disconnect cable from clutch release lever.

Set basic adjustment:

- With cable end disconnected, loosen lock nut and adjust stop bolt until specified gap is obtained between positioning lever and stop bolt.
- Tighten lock nut.

Clutch Cable Basic Adjustment (cable disconnected)

- Gap between clutch positioning lever and stop bolt. . . . . 1.2 mm (0.047 in.)

0018912

< Set cable preload:

- Reattach clutch cable to release lever and cable support on transmission.
- Using cable housing locknuts, remove all slack in cable.
- Tension cable further (at housing locknuts) until specified gap between positioning lever and stop bolt is obtained.

Clutch Cable Preload (cable connected)

- Gap between clutch positioning lever and stop bolt. . . . . 1.0 mm (0.04 in.)

- Tighten cable housing lock nuts. Recheck gap.

NOTE —

If clutch cable housing does not provide the adjustment needed, the clevis at the front of the cable may be adjusted as described under **Clutch cable, replacing**. *Note that the clevis must threaded onto the cable within a range of 17 - 22 mm (0.67 - 0.87 in.).*

Clutch release distance, checking and adjusting

Clutch release distance is limited by an adjustable stop on the floor cover above the pedal cluster. This adjustment should be checked only after adjusting the clutch cable as described earlier. See **Clutch cable, adjusting**.

– Raise car and support in a safe manner.

WARNING —

Make sure the car is stable and well supported at all times. Use a professional automotive lift or jack stands designed for the purpose. A floor jack is not adequate support.

0018815

< Working under transmission, use vernier calipers to measure distance between clutch cable lock nut and end of cable.

– Have helper push clutch pedal to floor stop. Measure distance again.

– Compare difference between two measurements to specification.

Clutch Release Specification

- Clutch release distance (difference measured at transmission) . . . . .25 ± 0.5 mm (0.98 ± 0.02 in.)

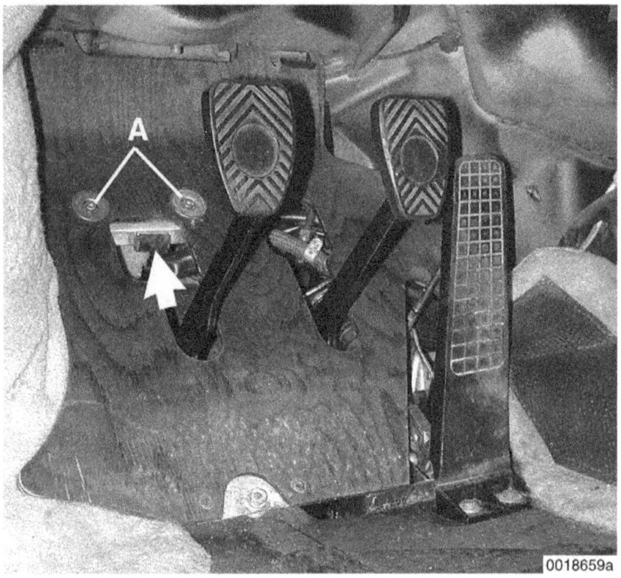

If clutch release distance is out of tolerance, adjust clutch pedal stop (**arrow**) by loosening allen bolts (**A**).

Clutch cable, replacing

– Raise car and support in a safe manner.

> ***WARNING —***
> *Make sure the car is stable and well supported at all times. Use a professional automotive lift or jack stands designed for the purpose. A floor jack is not adequate support.*

Working under rear of car, loosen clutch cable housing locknuts (**A**).

- Detach cable from clutch release lever.

– Working inside car, remove driver's side floor mat.

– Pull accelerator pedal back and disconnect accelerator push rod from pedal.

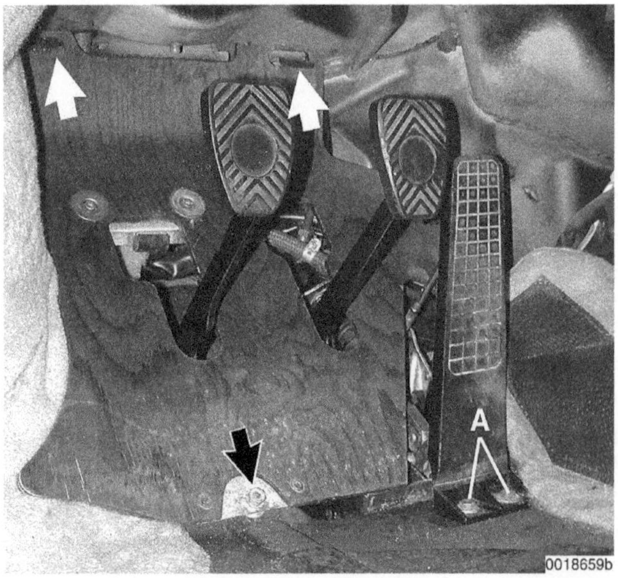

Remove accelerator pedal mounting screws (**A**) and remove pedal.

– Remove pedal cluster cover by removing fastener at bottom and unhooking from above (**arrows**).

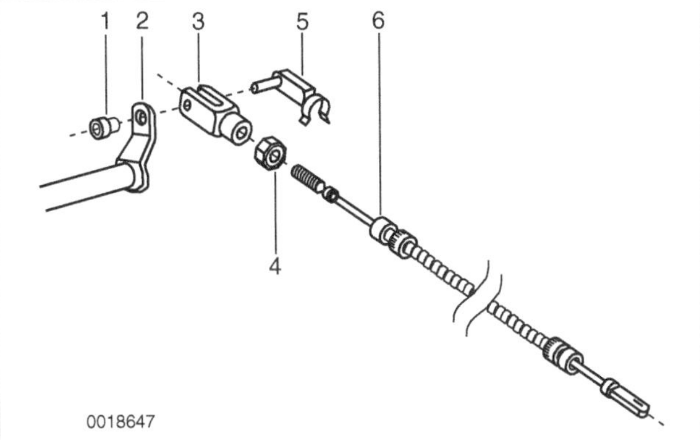

1. Bushing
2. Clutch pedal lever
3. Clevis
4. Clevis lock-nut
5. Clevis clevis pin/locking clip
6. Clutch cable

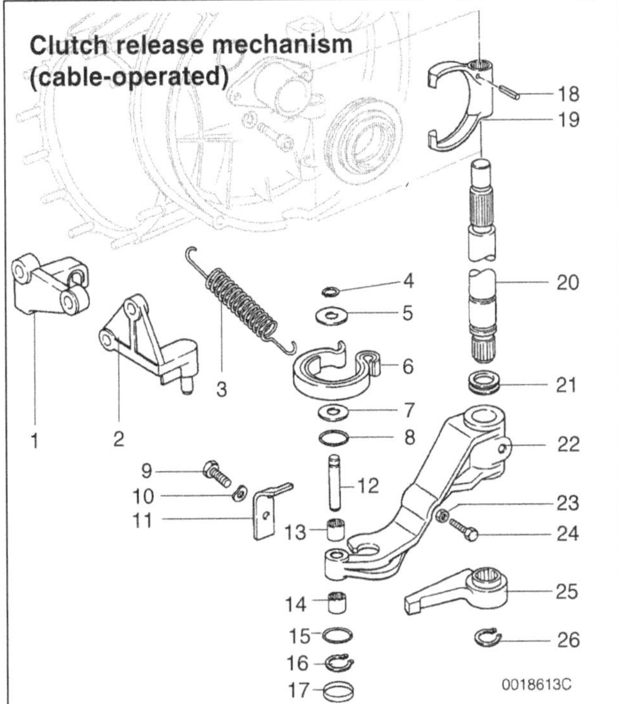

1. Clutch cable anchor
2. Auxiliary spring anchor
3. Positioning lever spring
4. Circlip
5. Washer
6. Auxiliary return spring
7. Washer
8. Seal
9. Bolt
10. Washer
11. Spring anchor
12. Pin
13. Auxiliary spring bushing
14. Auxiliary spring bushing
15. Washer
16. Circlip
17. Plug
18. Dowel pin
19. Clutch release fork
20. Clutch release cross-shaft
21. Seal
22. Clutch release lever
23. Lock-nut
24. Adjusting bolt
25. Clutch positioning lever
26. Circlip

< Locate clutch cable connecting clevis in center tunnel.

- Swivel locking clip upward and pull pin out of clevis.
- Loosen nut behind clevis and remove both clevis and nut.

– Working under rear of car, pull cable out of guide hole in body.

– Install new cable from rear.

– Working at pedal cluster area, install nut and clevis on new cable.

- Adjust clevis so that 17 - 22 mm (0.67 - 0.87 in.) of threaded cable end is screwed into clevis.
- Tighten clevis lock nut.

– Place clevis on pedal lever. Install and lock clevis pin.

– Reinstall pedal cluster cover, accelerator pedal and floor mat.

– Working under rear of car, install clutch cable to transmission.

– Check clutch cable adjustment. See **Clutch cable, adjusting**.

Clutch release lever, lubricating

If the clutch pedal is sticking or not returning all the way, the problem may be in the clutch release lever or auxiliary spring bushings.

> ***CAUTION—***
> *Do not apply lubricant to the pedal bushings to alleviate stiff pedal operation. Most lubricants will cause the bushings to swell.*

– Raise car and support in a safe manner.

> ***WARNING—***
> *Make sure the car is stable and well supported at all times. Use a professional automotive lift or jack stands designed for the purpose. A floor jack is not adequate support.*

< Working under car, disconnect clutch cable from clutch release lever (**22**).

– Disconnect positioning lever spring (**3**).

– Remove circlip holding clutch positioning lever to clutch cross-shaft (**26**).

– Remove positioning lever (**25**).

– Pry release lever (**22**) forward with a large screwdriver.

> ***CAUTION—***
> *Use care when prying release lever forward. The release lever will forcefully snap forward at a certain point.*

– Pull release lever (**22**) off clutch cross-shaft (**20**).

> ***NOTE —***
> *If the release lever will not slide off the clutch cross-shaft, the pedal stiffness problem may have been found. Use heat and back and forth motion to get the lever off.*

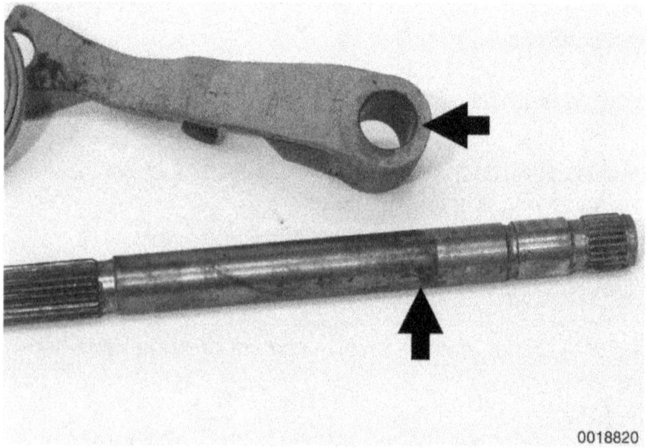

◄ Check cross-shaft and cross-shaft bore (**arrows**) in release lever for rust or corrosion.

- Clean bore and cross-shaft.
- Lubricate cross-shaft and bore, release lever splines, auxiliary spring bushings and clutch cable mount on release lever, using multipurpose grease.

– Replace auxiliary return spring bushings if necessary.

> ***NOTE —***
> *If clutch cross-shaft bore or needle bearings (in transmission bellhousing) are damaged, repairs can only be carried out with the transmission removed form the car. See* **Clutch cross-shaft, removing and installing** *later in this section.*

– Reassembly is reverse of removal.

– Check and adjust clutch cable and pedal throw (clutch release distance). See **Clutch cable, adjusting** and **Clutch release distance, checking and adjusting**.

Pedal cluster, removing and installing

If the pedals malfunction and routine cleaning and maintenance does not relieve the problems, remove and overhaul the pedal cluster as described below.

> ***CAUTION —***
> *Do not apply lubricant to the pedal bushings to alleviate stiff pedal operation. Most lubricants will cause the bushings to swell.*

> ***NOTE —***
> *Road dirt and grime will collect around the pedal cluster area and can eventually cause corrosion and stiff pedal operation. A yearly cleaning in the pedal cluster area will keep it working trouble-free.*

– Raise car and support in a safe manner.

> ***WARNING —***
> *Make sure the car is stable and well supported at all times. Use a professional automotive lift or jack stands designed for the purpose. A floor jack is not adequate support.*

◄ Working under car at transmission, loosen clutch cable mounting nuts (**A**).

- Detach clutch cable from clutch release lever.

– Working inside car, remove driver's floor mat.

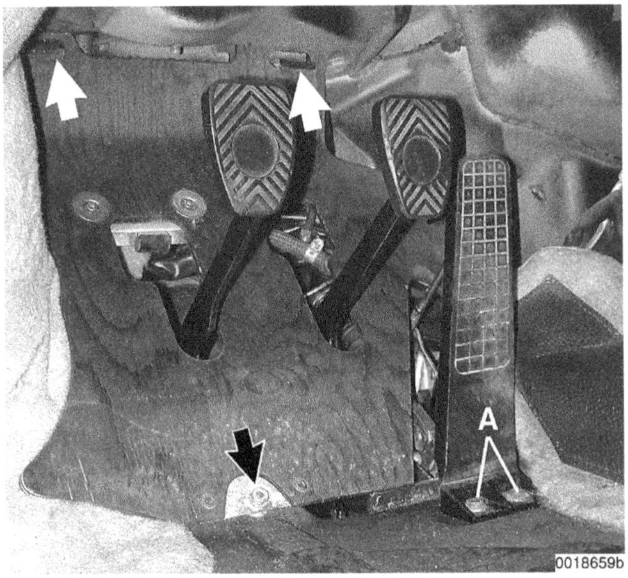

– Pull accelerator pedal back to release accelerator push rod.

< Remove accelerator pedal mounting screws (**A**) and remove pedal.

– Remove pedal cluster cover after removing carpet securing straps at top and fastener at bottom (**arrows**).

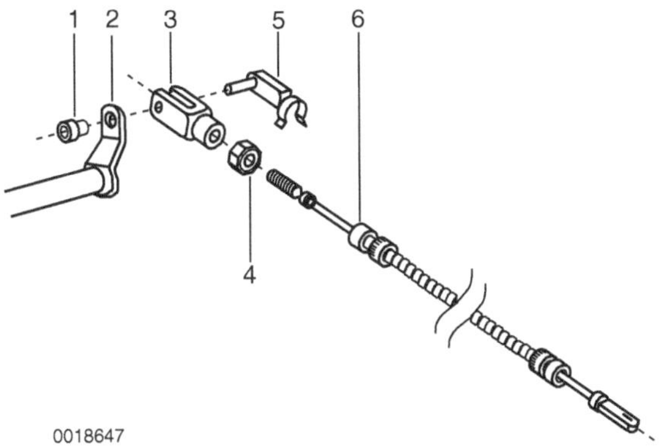

0018647

1. Bushing
2. Clutch pedal lever
3. Clevis
4. Clevis lock-nut
5. Clevis clevis pin and clip
6. Clutch cable

< Locate clutch cable connecting clevis (**arrow**) in center tunnel.

- Remove locking clip and pull pin out of clevis.
- Push clutch cable forward.

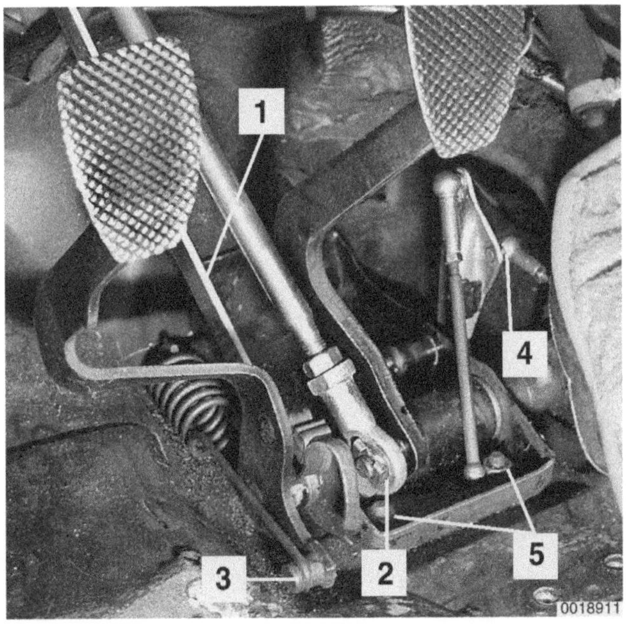

< Disconnect/remove from pedals and/or pedal cluster:

- Vertical brace at left side of pedal cluster (**1**).
- Brake master cylinder pushrod (**2**).
- Return spring at clutch pedal (**3**). Push clutch pedal forward.
- Accelerator cable (**4**) from accelerator bell-crank.
- Pedal cluster mounting nuts (**5**).

Working underneath front of car, remove stone guard mounting bolts (**arrow**). Remove stone guard.

Remove two M8 (8 mm) pedal cluster mounting nuts (**arrows**) on left side of front bulkhead.

– Installation is reverse of removal.

Tightening Torque
- Pedal cluster to floor (M8)............ 17 Nm (12 ft-lb)

Pedal cluster assembly

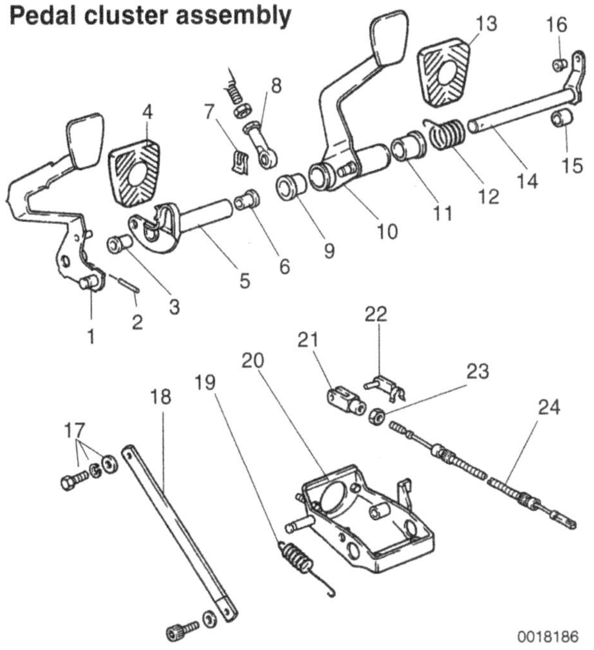

1. Clutch pedal
2. Dowel pin
3. Bushing
4. Pedal pad
5. Brake pedal bearing tube
6. Bushing
7. Locking clip
8. Brake master cylinder pushrod
9. Bushing
10. Brake pedal
11. Bushing
12. Brake pedal return spring
13. Pedal pad
14. Clutch pedal cross-shaft
15. Bushing
16. Bushing
17. Brace mounting hardware
18. Vertical brace
19. Clutch pedal return spring
20. Pedal cluster housing
21. Clutch cable clevis
22. Clutch cable clevis pin
23. Clevis lock-nut
24. Clutch cable

Pedal cluster bushings, replacing

– Remove and clean pedal cluster assembly before starting to disassemble.

– Support pedal cluster in soft jawed vise.

– Drive out dowel pin (**2**) that locks clutch pedal to cross-shaft (**15**). Support shaft to prevent deformation of pedal cluster housing.

NOTE—
Use 4 mm pin punch to drive out dowel pin.

– Remove clutch pedal (**1**) and cross-shaft (**14**).

– Remove brake pedal bearing tube (**5**) from brake pedal base.

– Remove brake pedal and pedal return spring (**12**).

– Remove spring clip from throttle bellcrank mounting shaft on pedal cluster housing (**20**) and slide bellcrank off shaft.

– Remove all bushings and clean all parts. Sand off rust and paint metal parts.

– Install new bushings.

NOTE—
- *Nylon replacement bushings are available from the Porsche dealer parts department.*
- *More durable brass bushings are available from aftermarket sources.*

– Reassembly is reverse of disassembly.

NOTE—
- *Check and replace parts that show any wear.*
- *When assembling, use universal grease.*

CLUTCH ASSEMBLY

Clutch cross-shaft, removing and installing

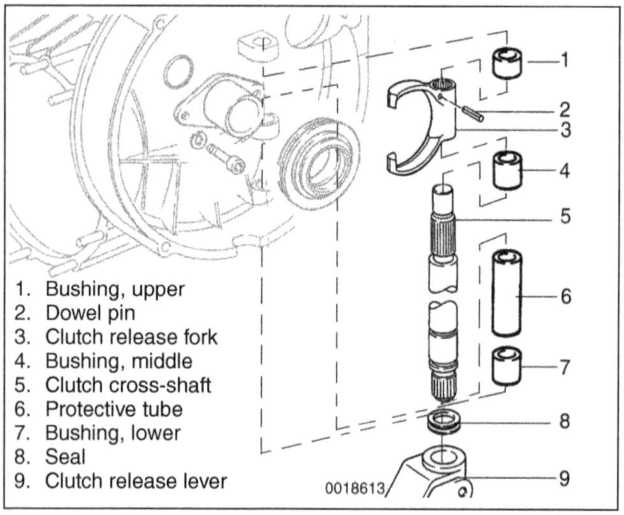

◄ Clutch cross-shaft repairs and upgrades are recommended during a clutch replacement job.

– Remove engine/transmission from car. Separate engine from transmission. See **101 Engine Removal and Installation**.

◄ Drive dowel pin out of clutch fork.

– Pull cross-shaft out of transmission, disengaging from fork.

NOTE —

If the clutch release lever is frozen to the clutch cross-shaft, remove them together by sliding the cross-shaft out of the bottom of the transmission bellhousing.

– Remove release lever from cross-shaft. Clean shaft and release lever.

◄ Check for wear or damage in clutch release fork. Look for cracks in casting at stress points (**arrows**).

– Using a chisel, carefully remove old cross-shaft bushings from transmission housing.

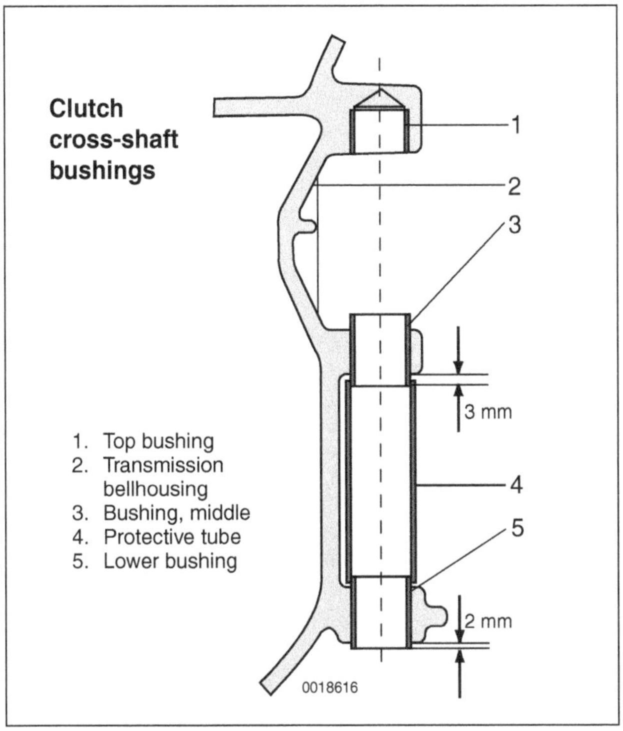

◀ Using suitable drift, tap new top bushing into place.

- Install center and lower bushings, making sure protective tube is in place between them.
- Install clutch fork between top and center shaft supports.
- Install clutch cross-shaft through bushings and clutch fork.

NOTE—
Use a multipurpose grease on cross-shaft bushings and on clutch release lever.

- Tap roll pin through fork to lock it in place on cross-shaft.
- Join transmission to engine and reinstall engine/transmission unit in car. See **101 Engine Removal and Installation**.
- Adjust clutch as described earlier in this section.

Clutch, replacing

The engine and transmission must be removed as a unit and than separated in order to service the clutch. It is normally recommended that the pressure plate and release bearing be replaced when a new clutch disc is installed.

The flywheel and the transmission pilot shaft bearing should be carefully inspected when replacing the clutch. Replacing the clutch disc without replacing other worn components may accelerate clutch disc wear.

- Remove engine/transmission from car. Separate engine from transmission. See **101 Engine Removal and Installation).**
- Insert clutch alignment tool into center of clutch.

◀ Loosen 9 bolts (**arrows**) holding clutch assembly to flywheel.

CAUTION—
To avoid warping the pressure plate, loosen bolts evenly, one turn at a time, until clutch pressure is relieved.

- Once bolts have been loosened, remove all 9 bolts. Remove starter ring gear and clutch.

> ***WARNING —***
> *The clutch disc contains asbestos fibers. Breathing asbestos dust can cause asbestosis. Always wear an approved respirator and protective clothing when handling components containing asbestos. Do not use compressed air, do not grind, heat, weld or sand on or near any asbestos materials. Only approved cleaning equipment should be used to service the clutch disc or areas containing asbestos dust or asbestos fibers.*

– Check components for wear and damage. Repair or replace as needed.

- Check starter ring-gear teeth.
- Check release bearing guide tube.
- Check clutch fork.

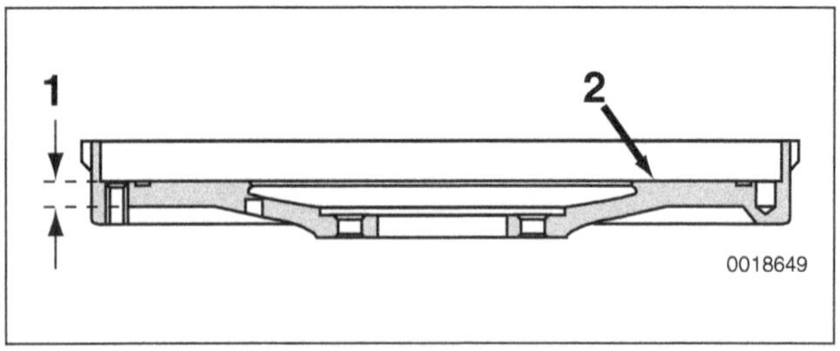

◄ Check flywheel face for damage.

- If damaged or scored, machine to specifications.

1978 - 1983 Flywheel Service Specifications

- Wear limit (**1**). . . . . 8.5 mm (0.335 in.)
- Machining limit . . . . minimum needed for smooth surface
- Max runout (**2**) . . . . 0.1 mm (0.004 in.)

– Check for signs of oil at bottom of transmission or in bell-housing. If found, inspect engine and transmission oil seals.

- To replace transmission input shaft seal, see **340 Manual Transmission–Controls and Case**.
- To replace engine rear main seal (also referred to as the rear crankshaft oil seal), see **130 Crankcase and Crankshaft**.

◄ Check transmission pilot shaft bearing:

- Pilot bearing is lightly pressed into center of flywheel.
- Replace bearing if dry or worn out.
- Lubricate bearing with a multipurpose grease.

– Place new release bearing on pressure plate as described later.

< Using special clutch disc alignment tool P 370, mount clutch assembly to flywheel. Tighten clutch bolts uniformly one turn at a time.

Tightening Torque

- Clutch to flywheel . . . . 25 Nm (18 ft-lb)

– Prior to installing transmission on engine, lubricate transmission input shaft splines with Optimoly® HT or equivalent. Lubricate release bearing guide tube with multipurpose grease containing molybdenum disulfide (MoS_2).

– As transmission slides onto the engine studs, turn clutch cross-shaft so that clutch fork hooks into release bearing.

– Install engine to transmission nuts and torque.

Tightening Torque

- Engine to transmission . . . . 45 Nm (32 ft-lb)

– Install engine/transmission as described in **101 Engine Removal and Installation**.

Clutch release bearing, replacing

– Lay clutch pressure plate with release bearing facing down in a press.

– Lightly press down on pressure plate, using suitable spacers and a steel bar.

Clutch pressure plate and release bearing assembly

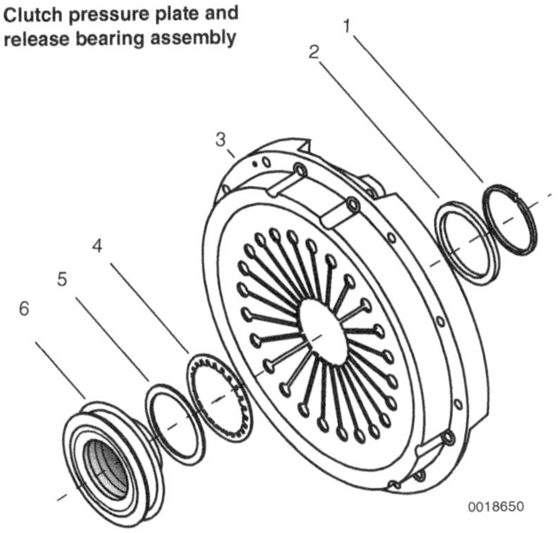

1. Locking ring
2. Thrust washer
3. Pressure plate
4. Spring washer
5. Shim(s)
6. Release bearing

< Remove release bearing locking ring. (Early pressure plate shown. Later models are similar.)

– Remove pressure plate from press and remove release bearing.

NOTE—
Release bearings are filled with a special grease and consequently must not be washed.

– When reinstalling release bearing, place shims and spring washer on first.

– Use press to lightly press down on pressure plate, using suitable spacers and a steel bar.

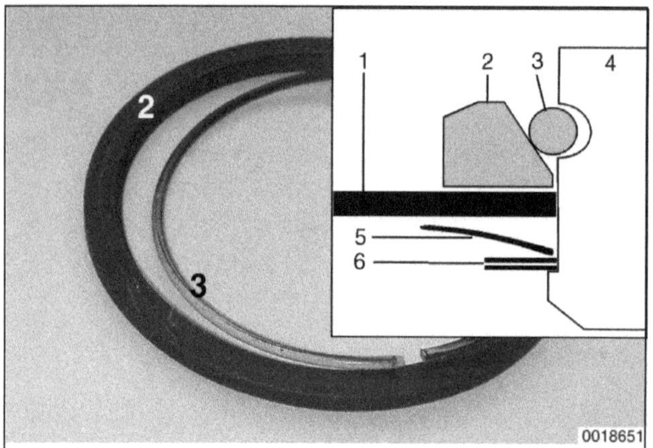

1. Clutch pressure plate	4. Release bearing
2. Thrust washer	5. Spring washer
3. Locking ring	6. Shims

◄ Install release bearing thrust washer (**2**) with beveled inner edge up (toward clutch disc).

- Install locking ring (**3**) inside thrust washer bevel.
- Be sure locking ring snaps into release bearing groove.

– Release pressure on clutch and remove from press.

– Check movement of release bearing in pressure plate. It should be tight but still moveable.

340 Manual Transmission–Controls and Case

TABLES

Special Tools

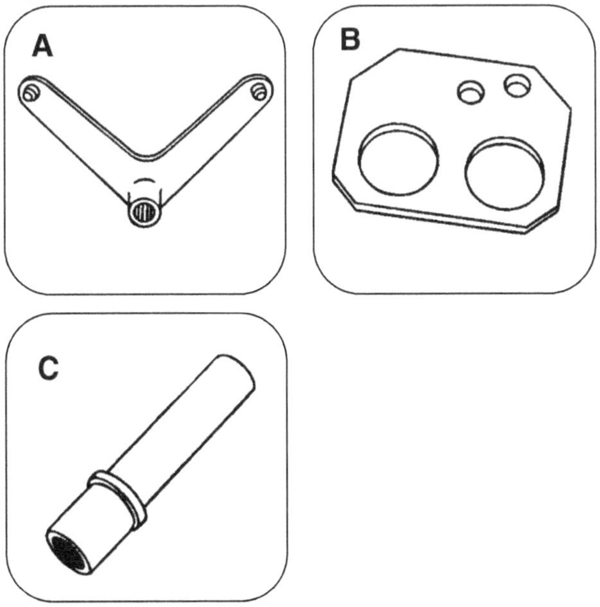

Special tool

A P 37 a. . . . . Input shaft lock (Porsche ordering #000 721 037 10)

B P 260 a. . . . . Shift rod holder (Porsche ordering #000 721 260 10)

C P 381 . . . . Input shaft installing sleeve (Porsche ordering #000 721 381 00)

GENERAL

This repair group covers repairs to the shifter mechanism, transmission fluid and seal service, as well as disassembly and assembly of the transmission down to the input and output shaft assemblies.

Other repair groups dealing with transmission removal and repairs are:

- **101 Engine Removal and Installation**
- **301 Clutch**
- **350 Manual Transmission–Gears and Shafts**
- **390 Differential**

NOTE —
Transmission removal is accomplished by first removing the complete engine and transmission assembly and then separating the transmission from the engine. See **101 Engine Removal and Installation**.

Special tools

Some of the procedures in this group require special tools. If you lack the skills and tools, or a suitable workplace for servicing the transmission, we suggest you leave these repairs to an authorized Porsche dealer or other qualified shop. Completely reconditioned transmissions are available from qualified Porsche repair facilities.

Transmission identification

0018822

The models covered by this manual are equipped with the 915 transmission. The transmission can be identified by a code stamped into the case near the transmission drain plug.

Transmission applications are shown in **Table a**. Gear ratios are shown in **Table**

Table a. Transmission applications

Year, type	Market	Code numbers
1978 - 1979		
915/63	USA, Japan	71 9 0000 - 10000
1980 1983		
915/63	USA, Japan	74 C 0000 - 10000

Table b. Transmission gear ratios

915/63		
Gear ratios		
1st	11/35	3.181
2nd	18/33	1.833
3rd	23/29	1.260
4th	26/26	1.000
5th	28/23	0.821
reverse		3.325
Final drive		3.8750:1

TRANSMISSION FLUID SERVICE

Transmission fluid level, checking

Checking the transmission fluid level involves removing the side drain plug and inserting a finger into the drain plug hole. If the fluid level is up to the bottom of the drain hole (finger wetted by transmission fluid), the level is correct.

NOTE—
The vehicle should be level when checking or filling transmission fluid level.

Transmission fluid, replacing

– Drive vehicle for a few miles to warm transmission.

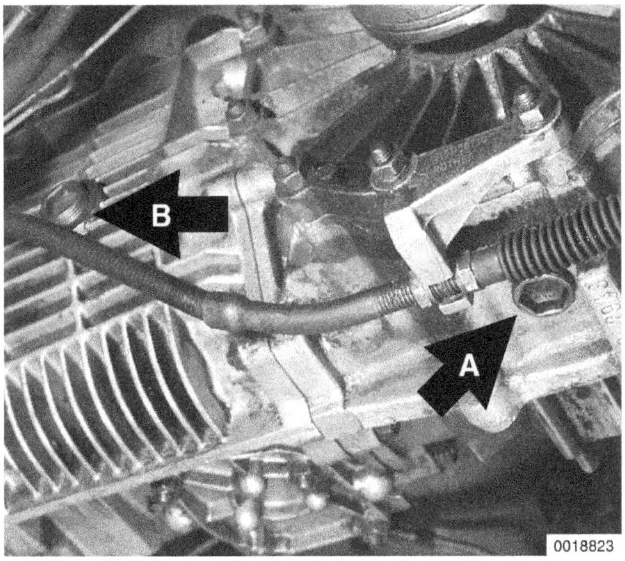

◄ Raise and safely support vehicle to access drain (**A**) and fill (**B**) plugs.

> ***WARNING—***
> *Make sure the car is stable and well supported at all times. Use a professional automotive lift or jack stands designed for the purpose. A floor jack is not adequate support.*

– Place drain pan under transmission. Remove fill plug. Remove drain plug at bottom of transmission. Allow fluid to drain fully.

– Install and torque drain plug.

Tightening Torque
- Drain plug to transmission housing
  915 transmission. . . . . 24 Nm (18 ft-lb)

– Slowly fill transmission until fluid overflows fill hole.

Transmission Fluid
- Type . . . . SAE 75W90
- API classification (alternate type MIL-L2105B). GL5

Transmission Fluid Capacity
- Without oil cooler . . . . 3.0 liter (6.34 pint)
- With oil cooler. . . . . 3.1 liter (6.55 pint)

– Install and torque fluid fill plug.

Tightening Torque
- Fill plug to transmission housing
  915 transmission. . . . . 24 Nm (18 ft-lb)

SHIFTER MECHANISM

Shifter housing, removing and installing

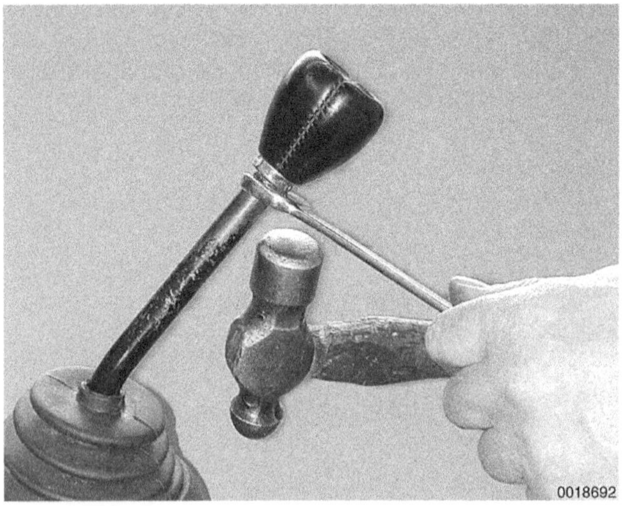

◄ Remove shifter knob by tapping up from below.

NOTE —
A special tool for forcing the shifter knob off the lever can be formed out of thick steel plate cut into a Y-shape. Alternatively, a 15 mm open-end wrench can be used.

– Pull up and remove shift lever boot.

– Remove center console:

- Remove 4 retaining screws on left and right sides of console front cover. Slide cover off center console towards front of car.
- Remove 3 center console retaining screws (2 in front, 1 in rear of console).

– Remove 5 screws at base of shifter housing.

– Remove shifter and housing by pulling straight up and off.

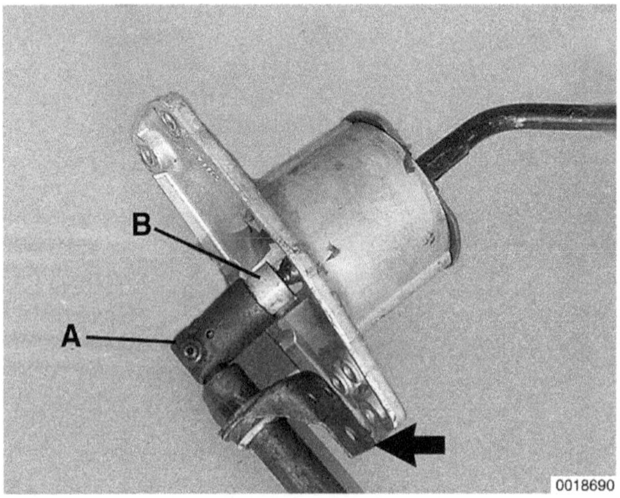

◄ During reinstallation:

- Line up shift rod coupling (**A**) with shift lever ball socket (**B**).
- Position shift rod bushing bracket (**arrow**) in place and start bracket mounting screws before fully lowering shifter housing into tunnel.

Tightening Torques
- Shift rod bushing bracket to housing. . . . .10 Nm (89 in-lb)
- Shifter housing to tunnel . . . . 25 Nm (18 ft-lb)

NOTE —
Check ball socket and replace if worn.

Shifter mechanism, disassembling

– Place removed shift mechanism on its side in a vise.

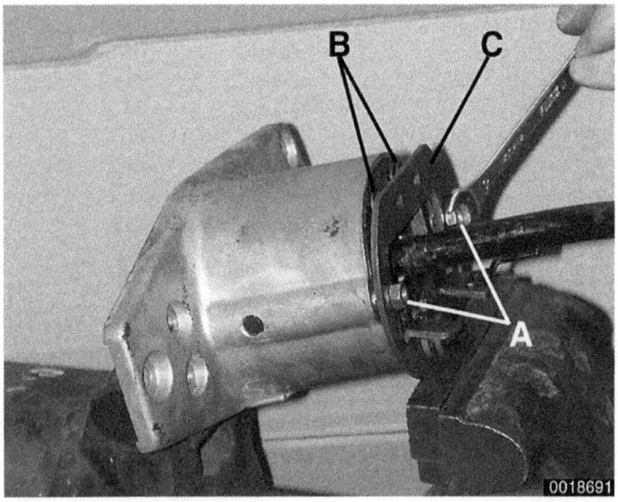

◄ Remove lock pawl plate retaining nuts (**A**).

– Slowly open vise, releasing pawl plate springs (**B**). Remove pawl plate (**C**) and springs.

> ***WARNING —***
> *Be careful when opening vise. The springs under the pawl plate are strong.*

– Remove shifter guide plate through top of housing.

– Remove locking clips from shifter pivot pin and pull pin out of housing.

Shifter mechanism

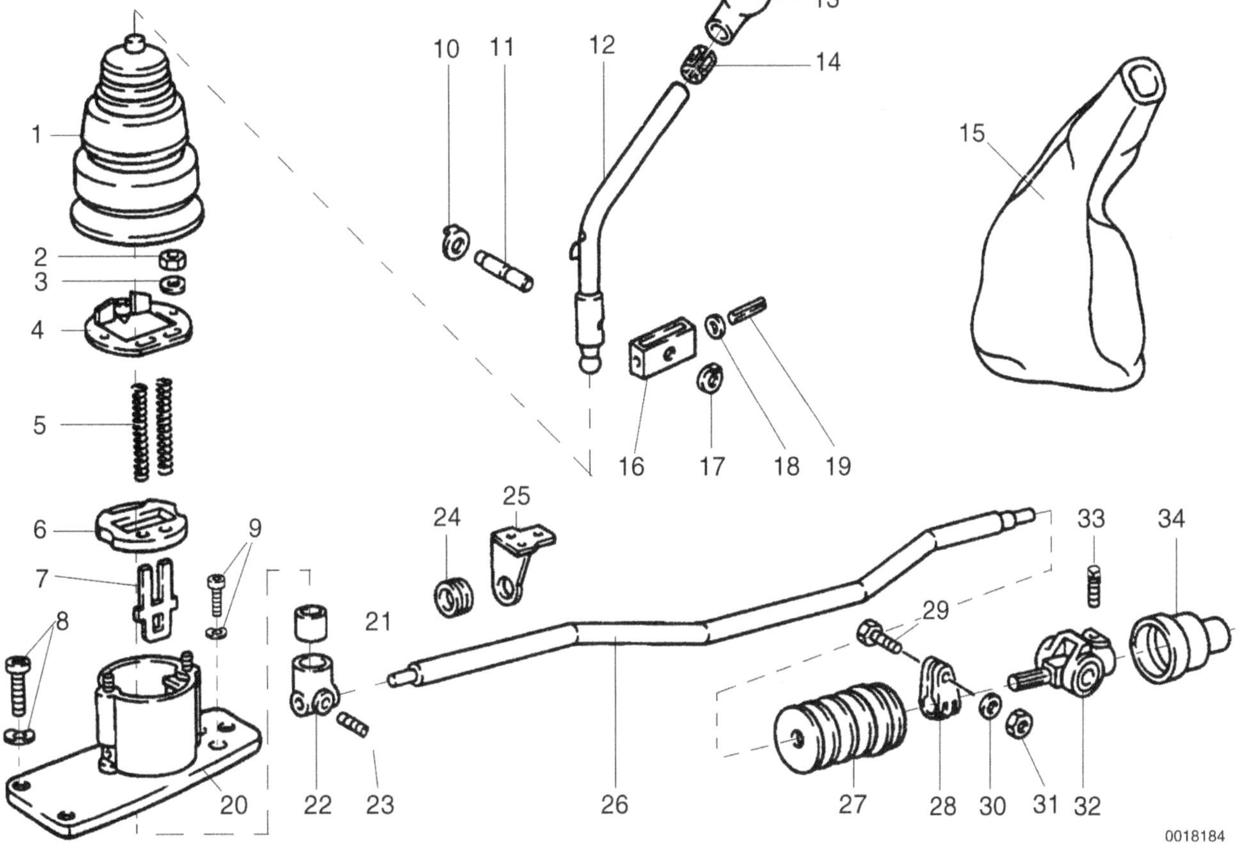

1. Dust boot
2. Nut (M6)
3. Lock washer (M6)
4. Lock pawl plate
5. Springs
6. Stop plate
7. Shifter guide plate
8. M8 bolt and washer (qty. 3)
9. M6 bolt and washer (qty. 2)
10. Locking clip
11. Shifter pivot pin
12. Shift lever
13. Shifter knob
14. Shift knob retaining ring
15. Boot
16. Shift pivot housing
17. Locking clip
18. Spacers
19. Roll pin
20. Shifter housing
21. Ball socket
22. Shift lever coupling
23. Set screw
24. Shift rod bushing
25. Shift rod bushing bracket
26. Shift rod
27. Boot
28. Coupling clamp
29. Clamp bolt (M8)
30. Lock washer (M8)
31. Nut (M8)
32. Shift rod coupling
33. Set screw
34. Transmission selector shaft boot

– Remove gear shift lever from shifter housing.

– Reassembly is reverse of disassembly.

Shift rod bushing, replacing

– Remove shifter mechanism as described earlier.

– Remove set screw (**A**) from shift lever coupling and remove coupling from shift rod.

– Slide shift rod bushing bracket (**B**) off shift rod.

– Snap out old bushing (**C**) and replace with new.

– Installation is reverse of removal.

Shift rod coupling, replacing

- Pull up center floor mat behind front seats. Remove 4 screws (**arrows**) from access panel and remove panel.

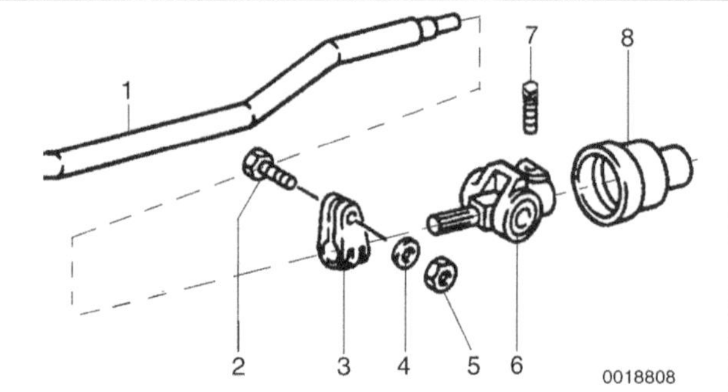

1. Shift rod
2. Pinch bolt
3. Clamp
4. Lock washer
5. Nut
6. Coupling
7. Set screw
8. Selector rod boot

- Remove set screw (**7**) from shift rod coupling to disengage transmission selector shaft.
- Loosen shift rod clamp pinch bolt (**2**).
- Slide coupling forward to remove from transmission selector shaft, then pull back to remove from shift rod.
- Installation is reverse of removal. Adjust as described below.

Tightening Torques
- Coupling clamp to shift rod . . . . 23 Nm (17 ft-lb)
- Set screw to transmission selector shaft. 23 Nm (17 ft-lb)

Shift rod coupling, adjusting

- Access shift rod coupling in rear part of center tunnel as described above.

- Loosen shift rod clamping nut (**A**) on coupling.
- Turn transmission selector shaft counterclockwise as facing rear of car (**arrow 1**).
- With shift lever in neutral, pull shift lever towards driver's seat so that shift lever contacts stop. This will preload the shift rod in a counterclockwise direction (**arrow 2**) as facing rear of car.
- Temporarily tighten shift rod coupling clamp, then test for correct adjustment:
  - Shift transmission into all gears including 5th and reverse. All gears must engage with approximately the same effort and throw.
  - Shift transmission into 5th gear. Push shift coupling boot back. Check transmission selector shaft and coupling for play. Some play must be felt.

– Tighten coupling clamping nut.

Tightening Torques
• Coupling clamp to shift rod . . . . 23 Nm (17 ft-lb)

– Install access cover and floor mat.

TRANSMISSION SERVICE

The transmission case, which contains the transmission and differential gears, is a lightweight alloy die casting. The 915 transmission uses constant-mesh type gears, with Porsche's own design synchronizing system. Each forward gear (1st through 5th) is a mating pair of gears, and constant-mesh simply means the mating gears are always meshing.

This section covers disassembly of the 915 transmission used on 1978 through 1983 models.

NOTE —
For information on removing and installing the transmission, see **101 Engine Removal and Installation**.

Back-up light switch, replacing

– Raise and safely support vehicle.

WARNING —
Make sure car is stable and well supported at all times. Use a professional automotive lift or jack stands designed for the purpose. A floor jack is not adequate support.

– Place oil drain pan under transmission.

0018826

◄ Remove reverse light switch (**arrow**). Install new switch with new sealing washer.

WARNING —
Be sure transmission is cool during this operation. Hot oil can scald.

Reverse Light Switch Location
• 915 transmission. . . . . Transmission end

Tightening Torque
• Reverse light switch to transmission . . . . 35 Nm (25 ft-lb)

– Top up transmission oil as necessary.

Selector shaft seal, replacing

– Raise and safely support vehicle.

> *WARNING —*
> *Make sure the car is stable and well supported at all times. Use a professional automotive lift or jack stands designed for the purpose. A floor jack is not adequate support.*

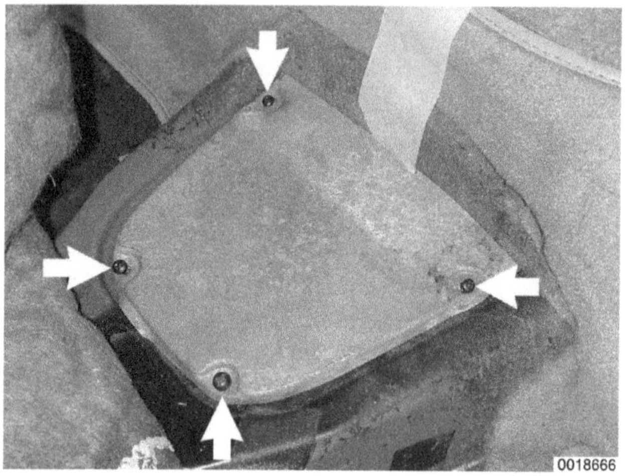

◄ Pull up center floor mat behind front seats. Remove 4 screws (**arrows**) from access panel and remove panel.

– Push back shift coupling protective boot and remove selector shaft set screw.

– Place oil drain pan under transmission.

– Pry seal out of selector shaft housing and install new seal.

> *WARNING —*
> *Be sure transmission is cool during this operation. Hot oil can scald.*

– Top up transmission fluid as necessary.

– Reassemble shift coupling.

Tightening Torques

• Set screw to selector shaft . . . . 23 Nm (17 ft-lb)

Input shaft seal, replacing

– Remove transmission and separate from engine as described in **101 Engine Removal and Installation**.

◄ Remove clutch release bearing guide tube screws (**arrows**).

0018828

⊲ Remove guide tube using a shop made tool and pry bar.

NOTE—

- *A tool to remove the release bearing guide tube can be made using 3/16" piece of steel rod. Bend hooks on both ends of wire to engage holes in guide tube. Lever the guide tube off using a pry bar.*
- *For information on clutch cross-shaft repairs, see* **301 Clutch**.

– Press input shaft seal out through front of guide tube using Porsche special tool P 381 or equivalent.

– Press new seal in through front of guide tube.

– Replace O-ring seal on outside of guide tube.

– Reinstall guide tube and torque screws.

Tightening Torque
- Guide tube to differential housing . . . . 10 Nm (7 ft-lb)

– Reinstall transmission.

Transmission, disassembling

0018829

⊲ Mount transmission on engine stand.

– Drain oil from transmission. Replace drain plug.

Tightening Torque
- Transmission drain plug . . . . 24 Nm (18 ft-lb)

0018826

⊲ Remove back-up light switch and remove switch operating pin from opening in transmission.

< Remove end cover mounting nuts and washers (**arrows**).

- Slide cover off selector shaft.

< Lock transmission input shaft with special tool P 37 a.

< Push 5th - reverse operating sleeve in to engage 5th gear.

NOTE —

With input shaft locking tool installed and transmission in 5th gear, both shafts should be locked in place.

915 transmission components

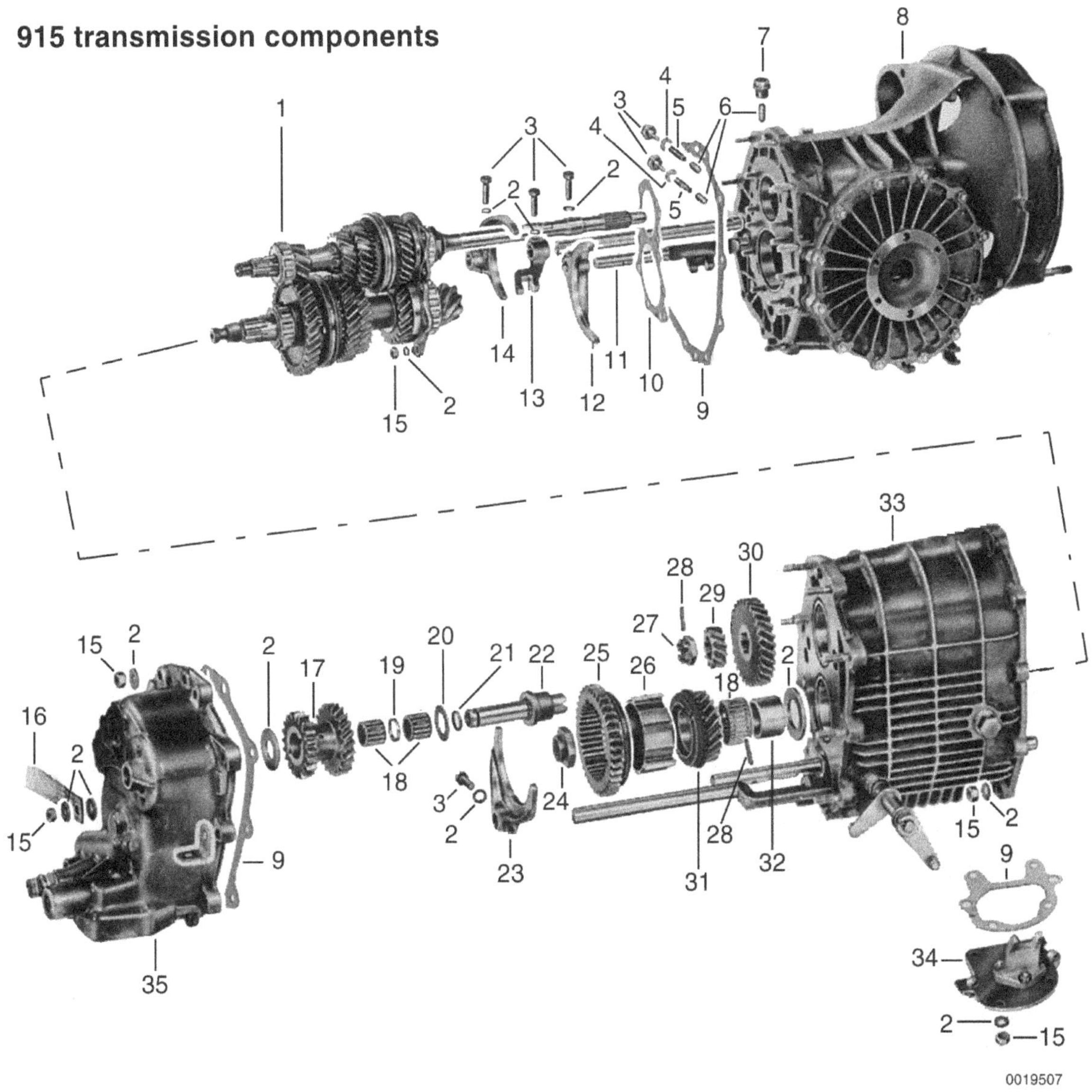

1. Input shaft
2. Washer
3. Bolt
4. Seal
5. Spring
6. Short detent
7. Plug
8. Differential housing
9. Gasket
10. Shim
11. Selector fork rod, 1st/2nd
12. Selector fork, 1st/2nd
13. Shift guide
14. Selector fork, 3rd/4th
15. Nut
16. Ground strap
17. Idler gear, 5th/reverse
18. Needle bearing cage
19. Intermediate ring
20. Thrust needle bearing cage
21. O-ring
22. Shaft, 5th/reverse idler gear
23. Shift fork, 5th/reverse
24. Flange nut
25. Sliding gear, 5th/reverse
26. Guide sleeve
27. Castellated nut
28. Roll pin
29. Gear, reverse
30. Gear, 5th
31. Gear, 5th
32. Bushing
33. Gear housing
34. Cover with guide fork
35. Front transmission cover

◄ Remove roll pin from castle nut on input shaft.

- Remove nuts (**arrows**) from input and output shafts.

– Remove input shaft locking tool (special tool P 37 a).

– Remove reverse light switch operating pin from 5th - reverse gear selector rod.

– Loosen and remove 5th - reverse gear shift fork pinch bolt.

NOTE —

Use a chisel to spread the opening at the shift fork pinch clamp slightly so that the fork moves on the selector rod without binding.

– Side 5th gear operating sleeve and shift fork off selector rod together.

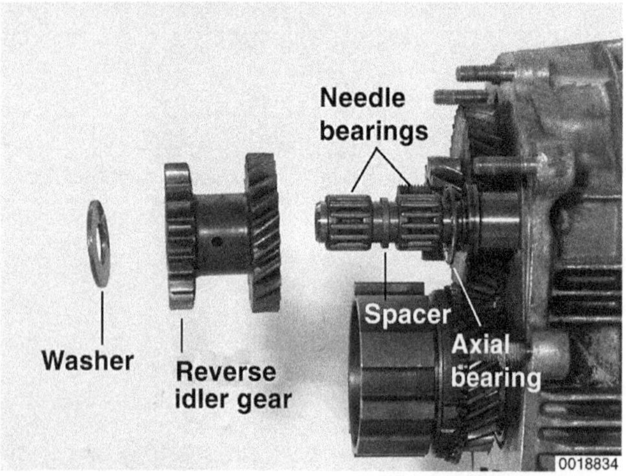

◄ Remove O-ring from end of reverse idler shaft. Remove washer, reverse idler gear (with needle bearings and spacer), and axial bearing.

NOTE —

Keep bearing, spacer, washer and 5th gear together as a set.

◄ Slide 5th - reverse gear operating hub together with 5th gear assembly off output shaft.

0018871

◀ Label and remove needle bearing from output shaft. Carefully pry from behind and remove sleeve and washer from output shaft.

0018872

◀ Remove reverse gear from input shaft. Pull off reverse gear idler shaft together with 5th gear.

0018836

◀ Turn transmission over and remove nuts (**arrows**) holding shift cover in place. Carefully pry cover from transmission case to remove.

- Remove nuts (qty. 10) holding gear housing to differential housing.
- Place 5th gear selector rod in neutral.
- Slide gear housing off with selector shaft.

- Unscrew upper (3rd - 4th gear) shift rod detent plug (**arrow**) from side of differential housing.
- Remove spring and detent from behind plug.

NOTE —
A small magnet can be used to remove the spring and detent.

- Remove pinch bolt (**arrow**) from 1st - 2nd gear shift fork.

NOTE —
Use a chisel to spread the opening at the shift fork pinch clamp slightly so that the fork moves on the selector rod without binding.

0018841

◄ Remove nuts (**arrows**) from bearing flanges on input and output shafts. Left side nuts shown. Remove nuts and lock washers from both sides.

0018842

◄ Slide shafts with gears and forks out of differential housing, leaving 1st - 2nd gear selector rod in transmission case.

NOTE —

*Check the number and thickness of shims (**arrow**) between differential housing and bearing flange. This shimmed surface controls the pinion shaft depth. If no repairs are being carried out on the differential bearing or the differential, the exact shim pack should be reinstalled.*

Transmission, reassembling

– Before assembly clean and check all parts to be sure they are in good operating condition. Lay parts out in order of assembly on a work bench.

◄ Install shims (**arrow**) on differential housing studs.

NOTE —

Unless the pinion bearing is being replaced or repairs have been made to the differential, install the same shims that were removed during disassembly. If the differential bearing is replaced or repairs have been made to the differential, special differential setup procedures must be carried out. Differential rebuild and setup is not covered in this manual.

◄ Insert 1st - 2nd selector rod into lower hole in housing, making sure locking detent is in correct position.

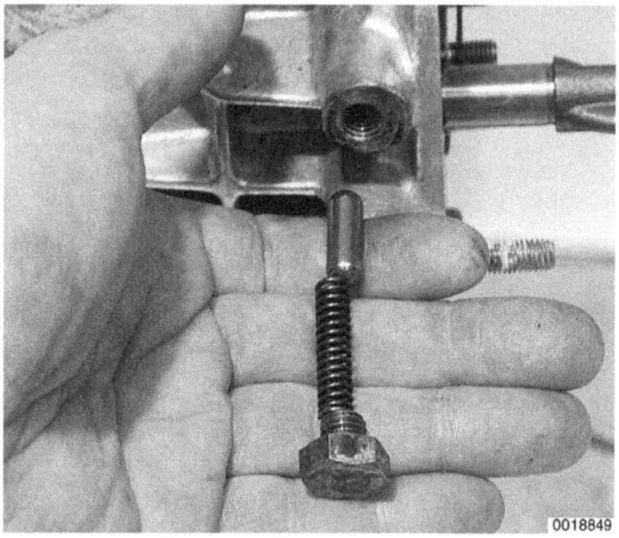

◄ Install 1st - 2nd gear shift rod detent, spring and plug,

Tightening Torque

• Detent plug to transmission case . . . . 17 Nm (12ft-lb)

◀ Place 1st - 2nd gear selector fork on operating sleeve. Insert output shaft into case so that bearing is supported on case.

◀ Mesh input shaft to output shaft and push both shafts into place so that bearing flanges are fully seated.

– Install bearing flange nuts with lock washers and tighten.

Tightening Torque

- Bearing flange to differential housing . . . 25 Nm (18 ft-lb)

– Install and finger tighten 1st - 2nd gear shift fork pinch bolt.

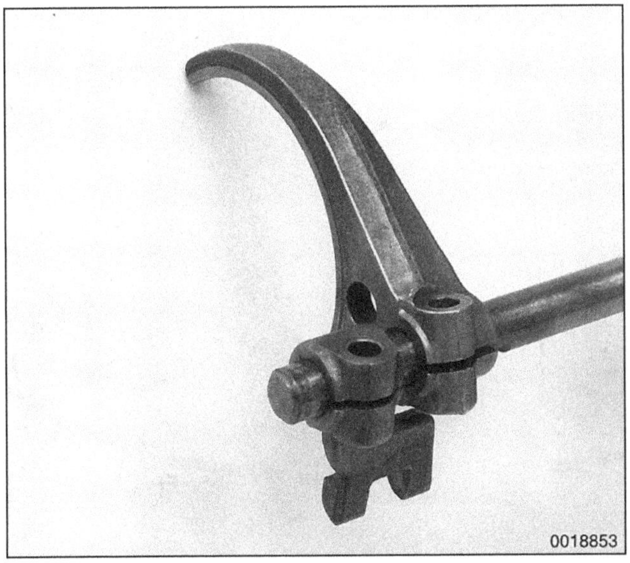

◀ Remove pinch bolts from 3rd - 4th gear shift fork and shift guide. Slide shift fork and shift guide back on selector rod.

Position shift fork onto 3rd - 4th operating sleeve and then slide selector rod into case. Install pinch bolts in forks.

– Insert 3nd - 4th shift rod detent, spring and plug.

Tightening Torque

- Detent plug . . . . 17 Nm (12ft-lb)

Install Porsche special tool P 260 a over shift rods and gear shafts.

Install 5th gear on input shaft with stepped collar facing gearbox. Install reverse gear on input shaft with marking facing away from gearbox.

0018857

Install washer and spacer onto output shaft. Install 5th gear with needle bearing. Install 5th - reverse gear operating sleeve hub and install operating sleeve.

0018882

Lock transmission input shaft with Porsche special tool P 37 a or equivalent.

- Engage 5th gear by sliding 5th - reverse gear operating sleeve forward.

NOTE—
With input shaft locking tool (special tool P 37 a) installed and transmission in 5th gear, both shafts should be locked in place.

– Install and torque input and output shaft nuts.

Tightening Torques
- Input shaft nut . . . . 160 Nm (116 ft-lb)
- Output shaft nut . . . . 250 Nm (181 ft-lb)

– Place 5th gear back in neutral position and remove input shaft lock (special tool P 37 a).

0018859

Turn 1st - 2nd gear selector shaft to left until it stops. Turn it back until unmachined flat inner surface of shift guide is almost vertical.

NOTE—
Do not turn shaft back beyond the center point.

– Set 1st - 2nd gear shift fork so that operating sleeve is precisely centered between 1st and 2nd gear synchronizers. Tighten clamping bolt, making sure shift guide is still in vertical position, as described above.

Tightening Torque

- Shift fork pinch bolt . . . . 24 Nm (18 ft-lb)

< Set 3rd - 4th gear shift fork so that operating sleeve is precisely centered between synchronizers. Install and tighten clamping bolt.

Tightening Torque

- Shift fork pinch bolt . . . . 24 Nm (18 ft-lb)

< Position 3rd - 4th gear shift guide with 2 - 3 mm (0.08 - 0.12 in.) clearance to 1st - 2nd shift guide.

– Tighten 3rd - 4th gear shift guide pinch bolt.

Tightening Torque

- Shift guide pinch bolt . . . . 24 Nm (18 ft-lb)

– Engage 5th gear and lock input shaft using special locking tool P 37 a. Loosen and remove nut from end of output shaft.

– Remove 5th - reverse gear operating sleeve.

– Remove shift fork support plate (Porsche special tool P 260).

– Install new gasket between differential housing and gear housing. Place shifter selector rod in gear housing.

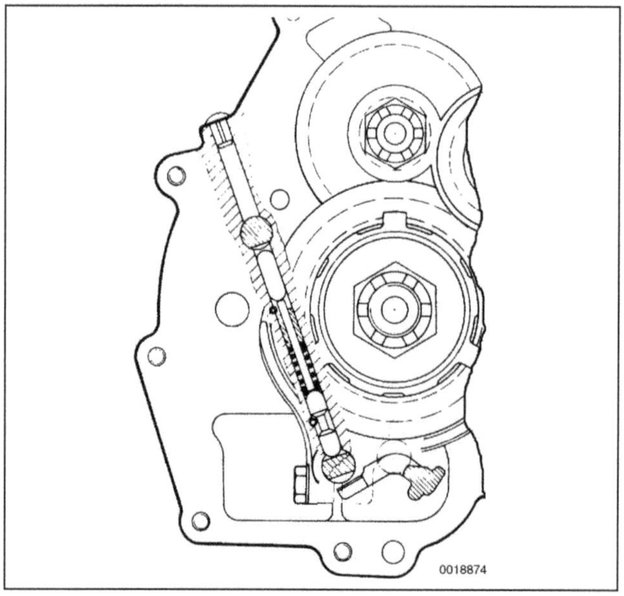

< Install gear housing, making sure detents in housing are correctly positioned in case. Install and tighten housing nuts with washers.

Tightening Torque

- Gear housing to differential housing . . . . 25 Nm (18 ft-lb)

– Install 5th - reverse gear shift rod into housing with detent cutouts up. Position rod in neutral position.

◀ Install 5th gear with reverse idler shaft:

- Place gear in cutout in idler shaft and place shaft with gear into position. Stepped collar on gear must face housing.
- Notch in shaft end must engage locating pin in housing.

◀ Install reverse gear on input shaft.

- Lock input shaft using Porsche special tool P 37 a.
- Install castle nut and tighten.
- Lock nut in place using roll pin (**arrow**).

Tightening torque
- Input shaft nut . . . . 160 Nm (116 ft-lb)

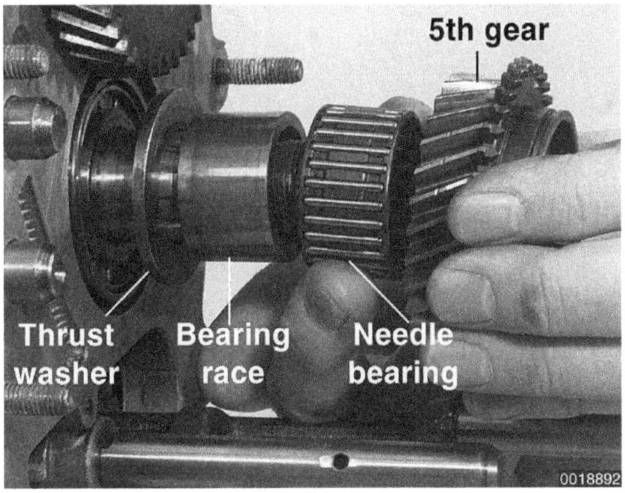

◀ Install thrust washer and 5th gear needle bearing race to output shaft. Install 5th gear with needle bearing.

– Install 5th - reverse gear operating sleeve hub onto output shaft.

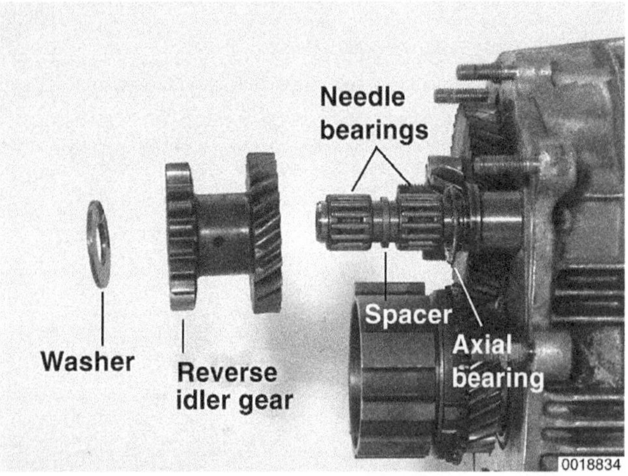

- Unlock input shaft by removing Porsche special tool P 37 a.

< Slide reverse idler gear assembly onto shaft.

- Install sealing O-ring to shaft end.

< Place 5th - reverse gear shift fork on 5th - reverse gear operating sleeve and gear together. Slide operating sleeve and shift fork on hub and selector shaft.

- Install reverse light roll pin into 5th - reverse gear selector rod with pin flush to back side of shaft.

< Lock input shaft with Porsche special tool P 37 a.

- Manually engage 5th gear.
- Install new output shaft nut and tighten.
- Stake nut to shaft (**arrow**),

Tightening Torque
- Output shaft nut . . . . 250 Nm(181 ft-lb)

◄ Push reverse idler against stop and position 5th - reverse shift fork so play between reverse idler gear and reverse operating gear is approximately 1 mm (0.04 in.).

– Tighten shift fork pinch bolt.

Tightening Torques
- 5th gear shift fork pinch bolt . . . . 25 Nm (18 ft-lb)

NOTE—
After tightening shift fork pinch bolt, check the play between reverse operating gear and reverse idler gear by lightly rocking reverse operating gear back and forth. The gears must not contact each other. Readjust shift fork if necessary.

◄ Turn transmission over and install shift cover in place, making sure operating pawl (**arrow**) engages shift rod guides.

Tightening Torques
- Shift cover to transmission housing . . . . 24 Nm (17 ft-lb)

– Turn transmission over and install new O-ring to reverse gear idler shaft.

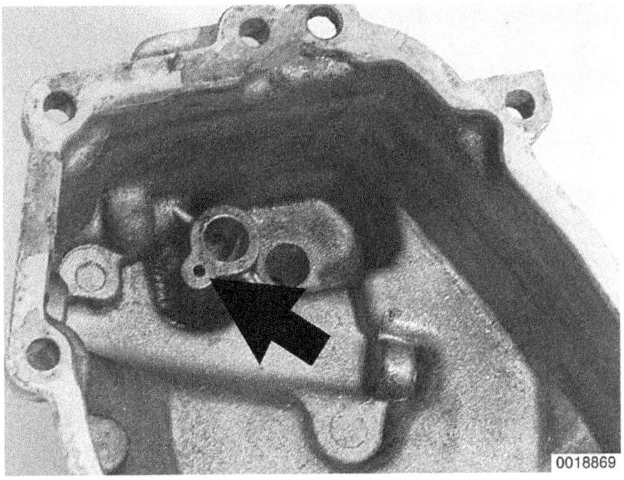

◄ Install new selector shaft oil seal in end cover. Install reverse light switch operating pin into case hole (**arrow**).

– Using new transmission end cover gasket, slide end cover over selector shaft and gears.
  - Install cover nuts and washers and torque.

Tightening Torques
- End cover to gear housing . . . . 25 Nm (18 ft-lb)

– Install new input shaft seal into clutch release guide tube. Install new O-ring to guide tube and reInstall guide tube.

Tightening Torque
- Guide tube to differential housing . . . . 10 Nm (7 ft-lb)

◄ Install reverse light switch with new sealing washer.

Tightening Torque
- Reverse light switch to transmission . . . . 35 Nm (25 ft-lb)

– Reinstall transmission as described in **101 Engine Removal and Installation**. Transmission fluid specification are given below.

Transmission Fluid Specification
- Viscosity . . . . SAE 75W90
- API classification. GL5
- Alternate classification MIL-L2105B

Transmission Fluid Capacity
- Without oil cooler . . . . 3.0 liter (6.34 pt.)
- With oil cooler. 3.1 liter (6.55 pt.)

350 Manual Transmission–Gears and Shafts

Special Tools

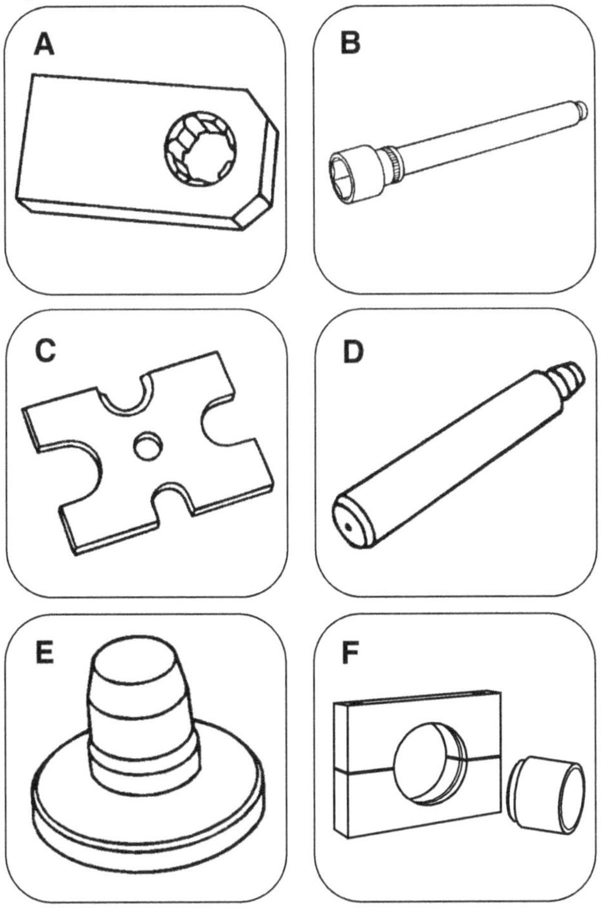

Special tool

- **A** P 355 a . . . . Input shaft mounting tool (source: Baum Tools Unlimited)
- **B** P 252 a . . . . Input shaft collar nut socket (41 mm) (alternate: 9105 Porsche ordering # 000 721 910 50)
- **C** VW 401 or VW 402 . . . . Press support plate
- **D** VW 407 . . . . Input shaft bearing drift
- **E** VW 412 . . . . Input shaft bearing drift
- **F** TL-3100/TL-3150 . . . . Synchronizer gear press tools (source: Automotion)

GENERAL

This section covers the internal repair of the 915 transmission. Other repair groups dealing with transmission removal and repairs are:

- **101 Engine Removal and Installation**
- **301 Clutch**
- **340 Transmission–Controls and Case**
- **390 Differential**

Special tools

Most of the operations and repairs in this section require the use of special tools. If you lack the skills and tools, or a suitable workplace for servicing the transmission, we suggest you leave these repairs to an authorized Porsche dealer or other qualified shop. Completely reconditioned transmissions are available from qualified Porsche repair facilities.

NOTE —
Difficult shifting may be caused by worn out synchronizer teeth, sometimes referred to as "dog" teeth.

INPUT AND OUTPUT SHAFTS

Input shaft, disassembling and assembling

NOTE —
All parts should be installed dry.

– Mount shaft in vise using special tool P 355 a.

– Follow steps in illustration on following page.

Transmission input shaft components

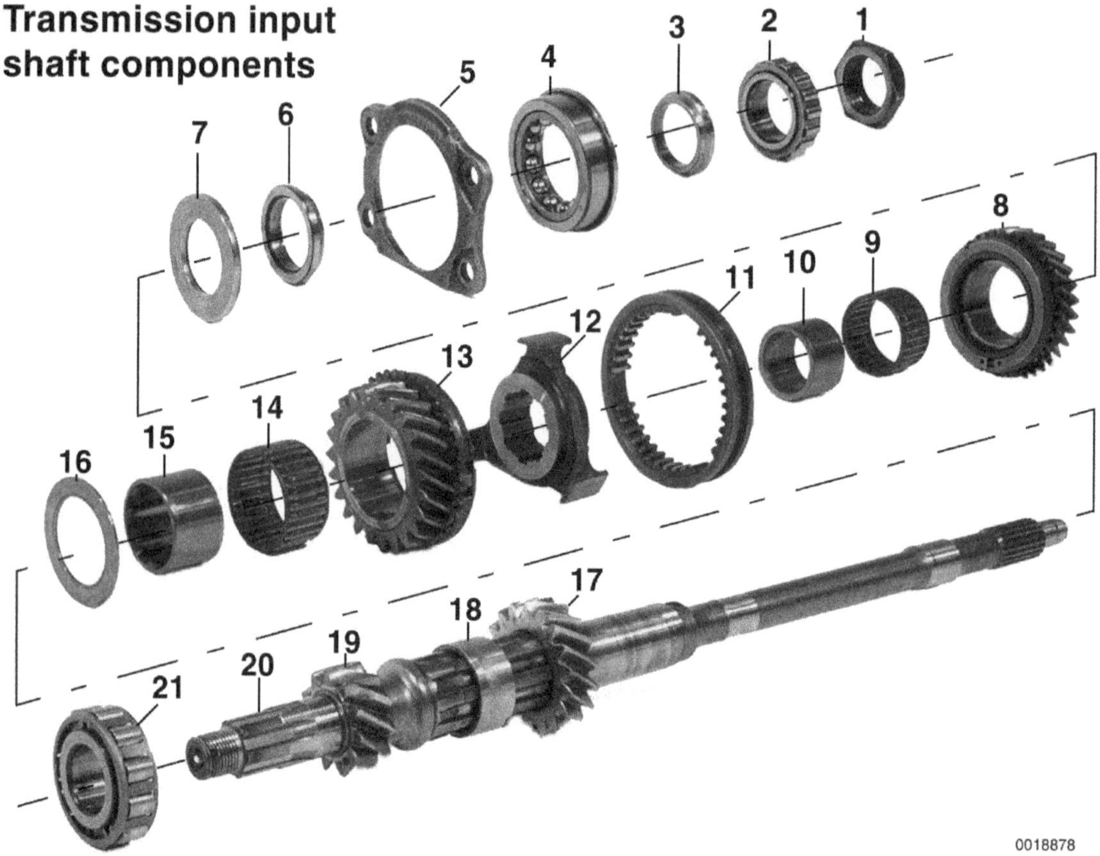

1. Collar nut removal: Use special tool P 252 a (41 mm socket) or equivalent.
   - Installation: Torque to 230 Nm (166 ft-lb).
   - Lock by punching collar. See detail **A**.
2. Roller bearing removal: Press off with special tools VW 401, VW 402, VW 412. When pressing, support under 2nd gear (17).
   - Installation: Heat to approx. 120°C (250°F).
   - Press on with special tools VW 401, VW 454 (press sleeve), VW 407
3. Ball bearing inner race removal: Keep with bearing and other race (**6**) as matched set.
   - Installation: Heat to approx. 120°C (250°F), then press on.
4. Ball bearing
5. Ball bearing flange
6. Ball bearing inner race: same as (**3**)
7. Thrust washer
8. 4th gear: Check synchronizer as described later.
   - Installation: If necessary, replace matched gears in pairs only.
9. Needle bearing: Keep with gear and race.
10. Needle bearing race: Keep with gear and bearing.
11. 3rd - 4th gear operating sleeve: Check for wear. See detail **B**.
12. Operating sleeve hub: Check for wear.
13. 3rd gear: Check synchronizer as described later.
    - Installation: If necessary, replace matched gears in pairs only.
14. Needle bearing: Keep with gear and race.
15. Needle bearing race: Keep with gear and bearing.
16. Thrust washer
17. 2nd gear: Replace matched gears in pairs only.
    - Installation: Manufacturer's marks face 1st gear. See detail **C**.
18. Spacer
19. 1st gear (integral with shaft)
20. Input shaft
21. Roller bearing removal: Press off with special tool VW 407 and suitable pipe.
    - Installation: Heat to approx. 120°C (250°F).
    - Press on with special tool VW 407 and suitable pipe.

– Always check drive shaft runout after tightening collar nut.

- Remove outer bearing races from transmission case and install them on input shaft.
- Place drive shaft bearings on machinist V-blocks.
- Set dial indicator to measure runout at pilot bearing end of shaft.

Input Shaft Runout

- Maximum runout at pilot bearing end . 0.1 mm (0.004 in.)

NOTE—

A shaft with up to 0.3 mm (0.012 in.) runout can be straightened out cold.

Details

A Stake collar nut to input shaft.

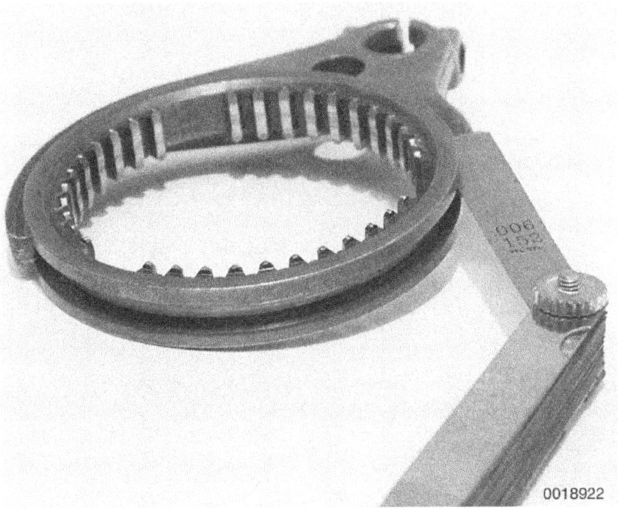

B Check clearance between shift fork and operating sleeve.

Shift Fork Clearance Specification

- Shift fork to operating sleeve. . . . . max 0.5 mm (0.02 in.)

C Install 2nd gear with manufacturer's marks facing 1st gear.

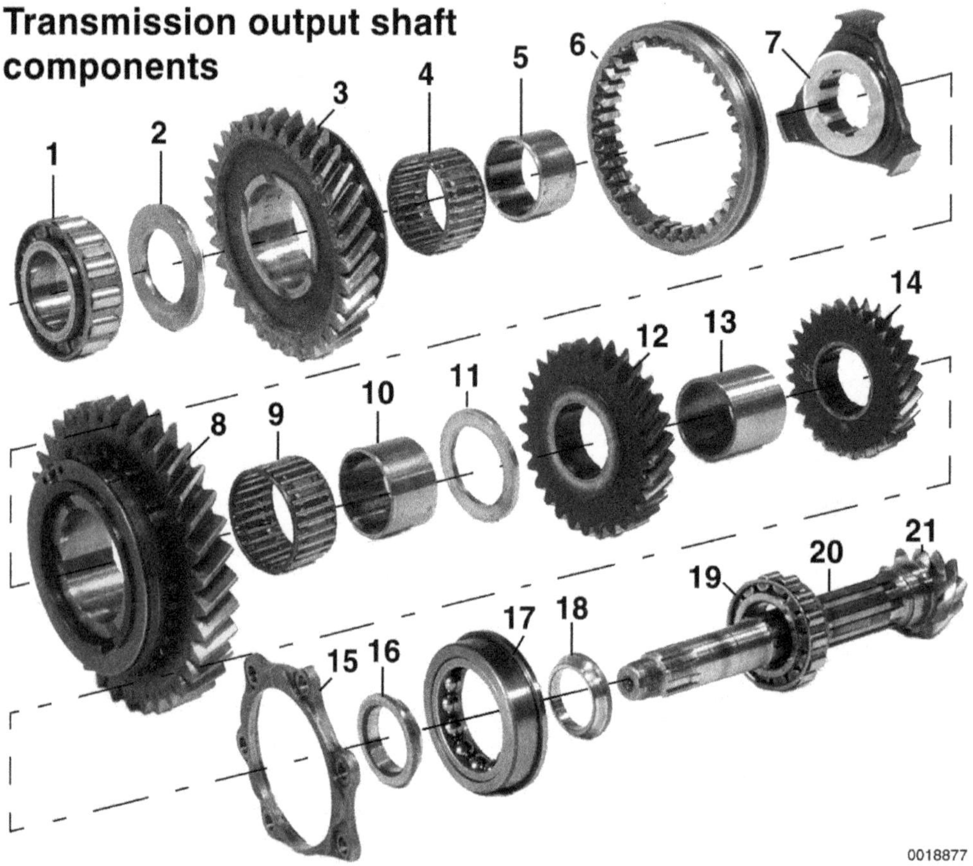

1. Roller bearing removal: Use special tools VW 401, VW 412, P 255 a (press sleeve). Press off by supporting under pinion end roller bearing (**19**).
   - Installation: Heat to approx. 120°C (250°F).
   - Press on with special tools VW 401, VW 412, VW 244 b (installing drift).
2. Thrust washer
3. 1st gear: Check synchronizer as described later.
   - Installation: First gear is matched to input shaft. If necessary, both must be replaced together.
4. Needle bearing: Keep with gear and race.
5. Needle bearing race: Keep with gear and bearing.
6. 1st - 2nd gear operating sleeve: Check for wear. See detail **A**.
   - Installation: Asymmetric teeth and recess face 1st gear. See detail **B**.
7. Operating sleeve hub: Check for wear.
8. 2nd gear: Check synchronizer as described later.
   - Installation: If necessary, replace matched gears in pairs only.
9. Needle bearing: Keep with gear and race.
10. Needle bearing race: Keep with gear and bearing.
11. Thrust washer
12. 3rd gear installation: Replace matched gears in pairs only.
    - Raised collar faces thrust washer. See detail **C**.
13. Spacer
14. 4th gear: Replace matched gears in pairs only.
    - Raised collar faces thrust bearing. See detail **D**.
15. Ball bearing flange
16. Ball bearing inner race removal: Keep with bearing and other race (**18**) as matched set.
    - Installation: Heat to approx. 120°C (250°F), then press on.
17. Ball bearing
18. Ball bearing inner race: same as 16.
19. Roller bearing removal: Use special tools VW 401, VW 412, P 255 a (press sleeve).
    - Install with collar of needle case facing gear set.
    - Installation: Heat to approx. 120°C (250°F).
    - Press on with special tools VW 407 and VW 415 a (press sleeve).
20. Pinion shaft
21. Drive pinion gear must match ring gear in differential.

Output shaft, disassembling and assembling

NOTE—
- *The output shaft is also referred to as the pinion shaft.*
- *All parts should be installed dry.*

– Follow steps in illustration on facing page. Details are covered below.

–Speedometer gear at forward end of shaft (not illustrated):

- Removal: Use puller
- Installation: Heat to approx. 120°C (250°F) and press on.

Details

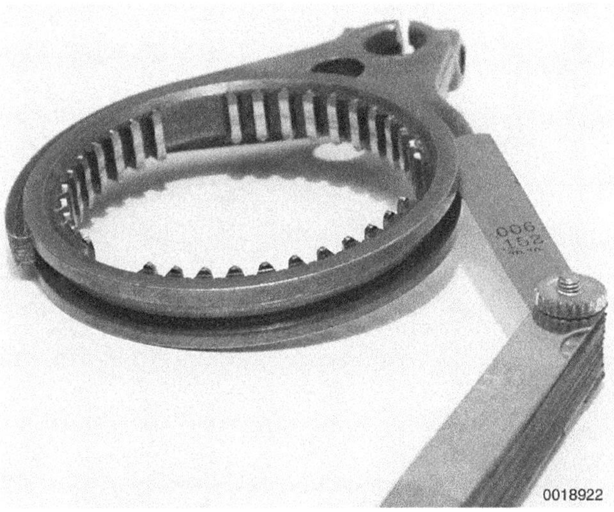
0018922

A Check clearance between shift fork and operating sleeve.

Shift Fork Clearance Specification
- Shift fork to operating sleeve. . . . . max 0.5 mm (0.02 in.)

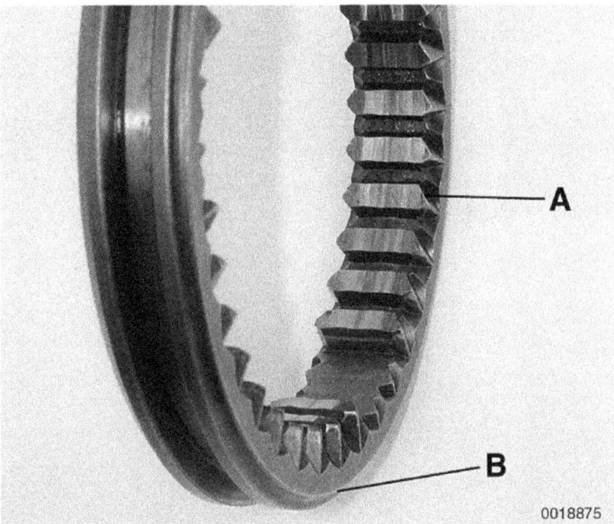

0018875

B Asymmetric teeth (**A**) and recess (**B**) of 1st - 2nd gear operating sleeve must face 1st gear.

C Raised collar (**arrow**) on 3rd gear must face thrust washer.

D Raised collar (**arrow**) on 4th gear must face bearing.

SYNCHRONIZERS

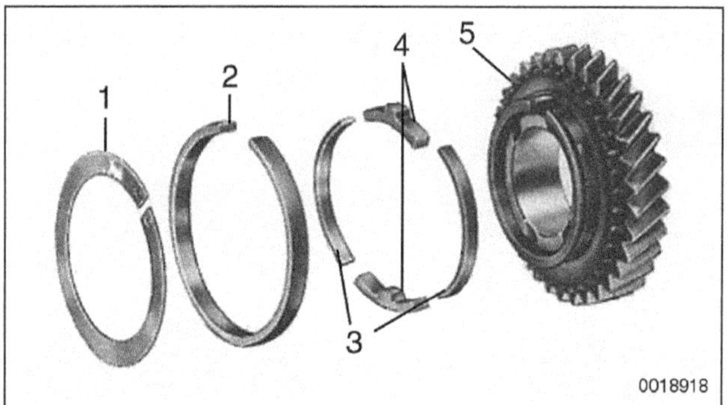

1. Circlip
2. Synchronizer ring
3. Brake band
4. Thrust block
5. 2nd gear with synchronizer ("dog") teeth

The synchronizer mechanism in the 915 transmission is unlike synchronizers in most other transmissions.

Two of the components, the synchronizer ring and the synchronizer ("dog") teeth, take the brunt of wear and tear. They must be carefully checked during transmission overhaul. Replace as necessary.

Synchronizer, checking

0018949

◀ Measure across synchronizer ring with gear and synchronizer fully assembled. If diameter exceeds specifications, replace synchronizer teeth.

Synchronizer Ring Diameter Specification

- 3rd, 4th, 5th gears . . 76.3 ± 0.18 mm (3.004 ± 0.007 in.)
- 1st, 2nd gears. . . . . 86.37 ± 0.17 mm (3.400 ± 0.007 in.)

– Remove circlip. Synchronizer ring should remain coiled inside synchronizer teeth. If not, teeth are worn and should be replaced.

0018927c

◀ Check synchronizer teeth on gear assembly for damage or wear.

NOTE—
The synchronizer teeth can be worn round or chipped.

Synchronizer ring, removing and installing

0018880

◀ Remove circlip holding synchronizer ring in place.

- Remove synchronizer parts and inspect carefully.

CAUTION—
Keep all parts for each synchronizer together.

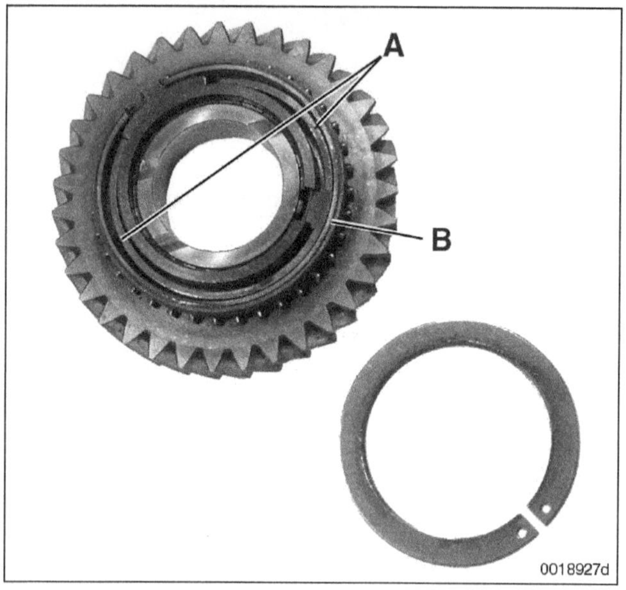

1st gear synchronizer:

- Asymmetric brake bands (**A**).
- Groove in synchronizer ring (**B**).

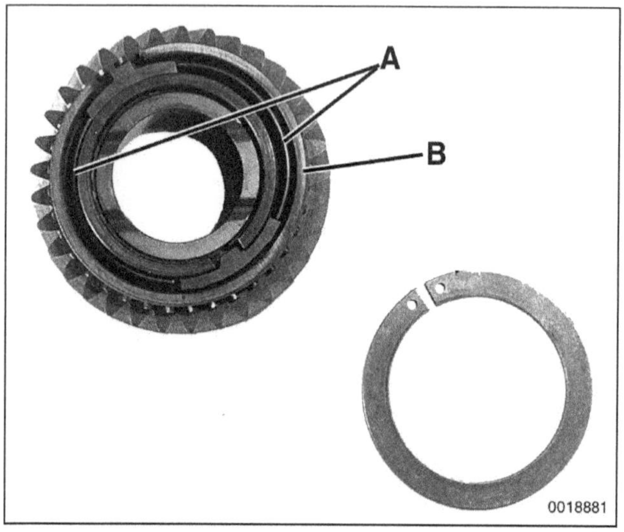

2nd gear synchronizer:

- Symmetrical brake bands (**A**).
- No groove in synchronizer ring (**B**).

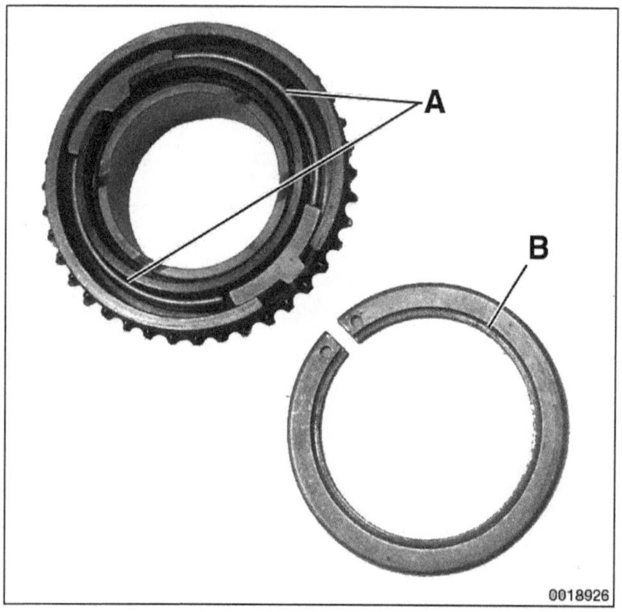

3rd, 4th, 5th gear synchronizers:

- Split brake bands (**A**).

– Install a new synchronizer ring and install circlip.

NOTE —
If the synchronizer ring is reused, install with rough side facing the operating sleeve.

– Install circlip for 3rd, 4th, or 5th gear synchronizer with stepped face (**B**) away from gear.

Synchronizer ("dog") teeth

Synchronizer ("dog") teeth are press fit on each gear. Aftermarket tools and individually available components can be used to replace damaged gear teeth. Consult with the Porsche dealer parts department.

390 Differential

Special Tools

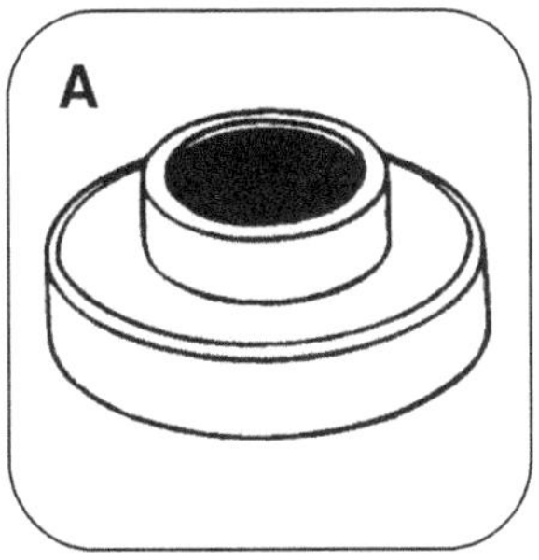

Special tool

A 9252 . . . Seal installer
(Porsche ordering #000 721 925 00)

GENERAL

The Porsche 911 SC transmission assembly houses the forward and reverse gears and the differential, also known as the final drive.

This section covers replacement of the differential flange seals. This does not require disassembly of the differential. Disassembly of the differential requires special tools and knowledge, and is beyond the scope of this manual.

Special tool

The use of a special seal installer is helpful in performing the procedure described below.

Troubleshooting

The source of driveline vibrations and noise can be difficult to pinpoint. Engine, transmission, drive axle or wheel vibrations can be transmitted through the power train mounts to the car body. Noises from the rear of the car may be caused by differential problems or by faulty wheel bearings or constant velocity (CV) joints. Also check that the tires are correctly inflated and are not excessively worn.

To isolate a vibration or noise problem, speed up the engine in the stopped car to the rpm range where the problem occurs. This eliminates the influence of rotating driveshafts and will help indicate if the problem is caused by an engine condition.

Drive the car at the speed where the problem occurs, then shift to different gears to see if the problem changes with road-speed or engine-speed. Road-speed dependent problems usually indicate trouble in the driveline.

Detach the axle shafts at the differential and then operate the engine at the speed and gear position where the problem occurs. If the problem disappears, then it may be in the axle CV joints or wheel bearings.

> ***WARNING —***
> *Make sure the car is stable and well supported at all times. Use a professional automotive lift or jack stands designed for the purpose. A floor jack is not adequate support.*

> ***CAUTION —***
> - *Damage to the differential or axle shaft joints can result if the car is run with the wheels off the ground. The wheels hang below their normal height, putting stress on driveline components. Compress the suspension to normal height with jackstands before running the car.*
> - *Never operate the car with one drive wheel turning at a different speed than the other. Damage to the driveline components can result.*

NOTE —
No-load tests with the car stopped or the drive axles disconnected will only give a general idea of the car's performance under the load of normal operation.

For more information on troubleshooting engine or transmission problems, see **100 Engine–General**, **310 Clutch**, **340 Manual Transmission–Controls and Case**, or **350 Manual Transmission–Gears and Shafts**. For information on troubleshooting rear axle, wheel bearing or constant velocity joint problems, see **420 Rear Suspension, Axle Shafts**.

AXLE FLANGE SEAL

Low oil level caused by leakage due to faulty axle flange oil seals may be the cause of noisy transmission operation. The drive flange oil seals can be replaced while the differential is installed in the car.

Do not mistake leaking CV joints for flange seal leaks. It may be helpful to de-grease the differential to pinpoint the source of the leak prior to replacing seals.

Axle flange seal, removing and installing

NOTE —
Do not remove both differential drive axle flanges at the same time. If both flanges are removed at the same time you may experience difficulty aligning the drive flange splines.

– Raise car and support in a safe manner.

> ***WARNING —***
> *Make sure the car is stable and well supported at all times. Use a professional automotive lift or jack stands designed for the purpose. A floor jack is not adequate support.*

◀ Remove CV joint bolts (**arrows**) from axle flange.

– Screw 2 bolts into threaded holes in flange.

NOTE —
- *On 915 transmission: Use 2 M8x1.25 bolts.*

◀ Counterhold flange using a pry bar between two bolts.
- Remove flange mounting bolt.

– Pull flange out of differential.

– Remove flange seal from differential.

– Install new seal using special tool 9252 or equivalent.

– Reinstall flange and CV joint.

Tightening Torques
- CV joint to flange
  M8 . . . . 42 Nm (30 ft-lb)
  M10 . . . . 83 Nm (61 ft-lb)
- Flange to differential . . . . 44 Nm (32 ft-lb)

400 Suspension, Brakes and Steering–General

TABLES

4

GENERAL

This repair group covers general information for front and rear suspension, the braking system and the steering system. Diagnostic and troubleshooting information on these systems is at the end of the section.

The Porsche 911 SC suspension and steering systems are engineered to provide a compromise between taut, responsive handling and ride comfort. Although the front and rear suspension assemblies are independent subsystems, they work together to achieve Porsche's overall combination of handling and comfort.

The suspension has no lubrication points. All suspension joints are permanently lubricated. To maintain good handling and reliability, carry out periodic inspections as listed in **030 Maintenance**. The main items needing attention are:

- Front wheel bearings, checking, adjusting and repacking
- Wheel nuts, checking tightening torques
- Constant velocity (CV) joint boots, checking for damage
- Tire pressures, checking and correcting

Front suspension and steering

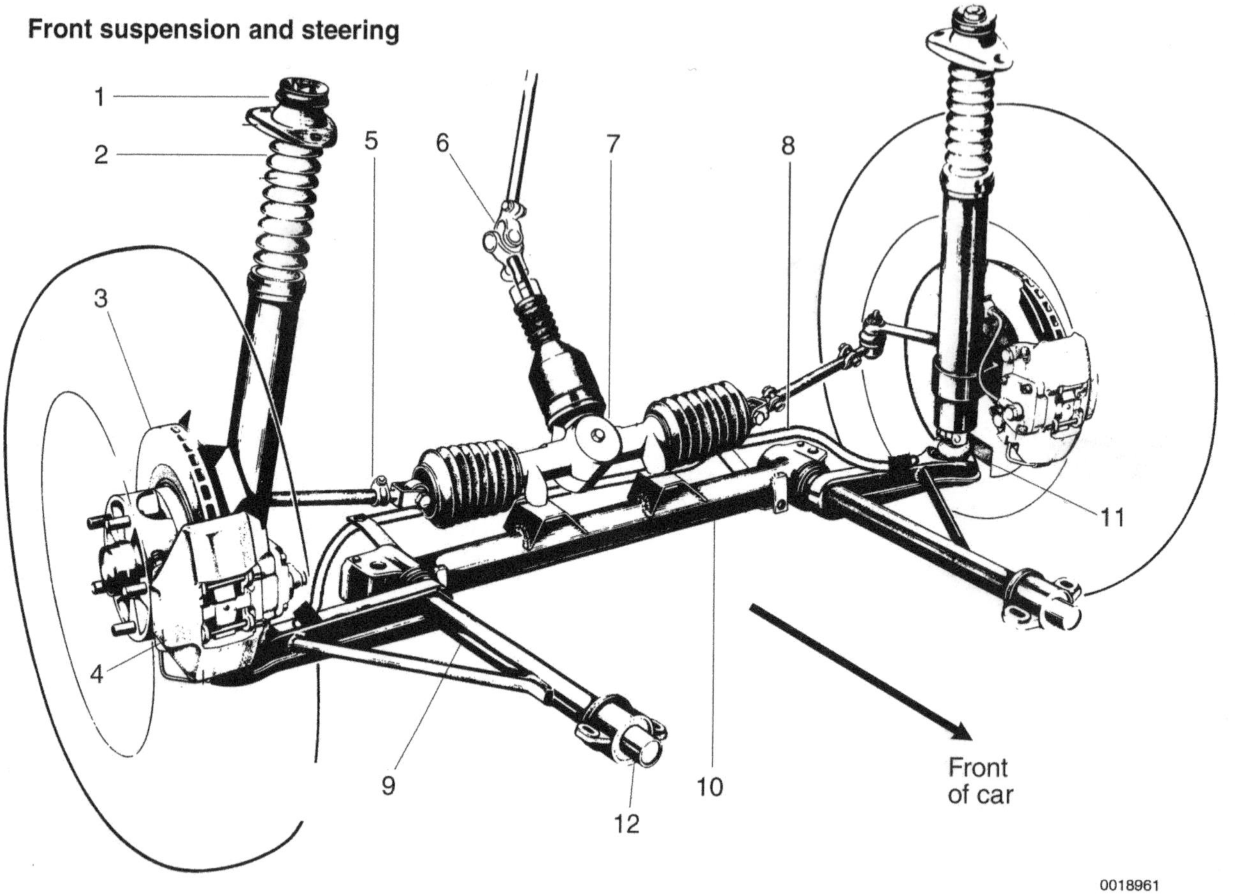

1. Top strut mount
2. Front suspension strut
3. Front brake rotor
4. Front brake caliper
5. Steering tie rod
6. Steering column universal joint
7. Steering rack
8. Stabilizer bar
9. Front control arm
10. Front suspension subframe
11. Ball joint
12. Torsion bar

Front suspension

Each front wheel includes a strut-type shock absorber mounted on a permanently lubricated ball joint. A control arm on each side houses a longitudinal torsion bar and is attached to the suspension subframe.

The torsion bars are adjustable. Suspension travel is limited by a progressive rubber buffer (bump stop) installed on each strut piston.

Cornering control is enhanced by a transversely mounted stabilizer bar connected to the control arms and the rear of the suspension subframe with rubber bushings.

For further information or component replacement procedures for the front suspension, see **401 Front Suspension**.

Rear suspension

The main rear suspension components are the trailing arms, torsions plates, torsion bars, and shock absorbers. Torsion plates connect the wheel bearing housings to the transversely mounted torsion bars. The torsion bars, housed within the rear suspension subframe, are adjustable by means of unequal-numbered inner and outer splines.

The rear wheel bearings are housed in the trailing arms. They are sealed and require no lubrication or adjustment.

Half axle shafts with one constant velocity (CV) joint at each end transmit power from the transmission output flanges to the rear wheels.

For further information or component replacement procedures for the rear suspension, see **420 Rear Suspension, Axle Shafts**.

Rear suspension

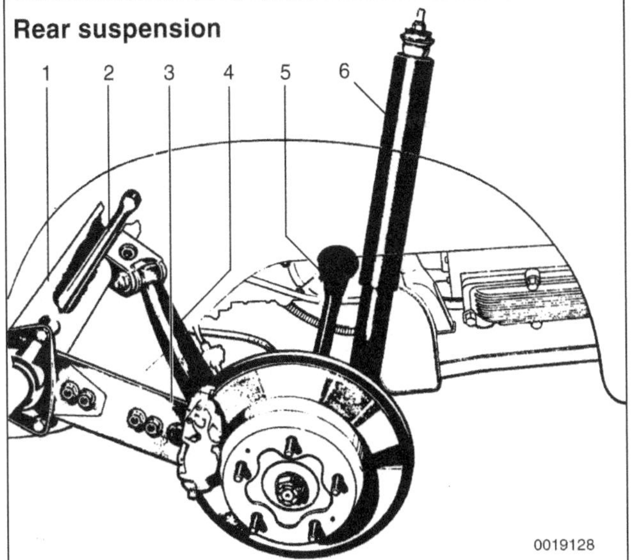

1. Rear suspension subframe (torsion bar carrier)
2. Torsion bar
3. Torsion plate
4. Trailing arm
5. Axle shaft
6. Shock absorber

Table a. Suspension specifications

Component	Front suspension	Rear suspension
Suspension design	Independent wheel suspension with control arm and ball joint	Independent wheel suspension with alloy trailing arms; wheel driven by axle shaft
Track:		
With standard wheels	1361 mm (53.6 in.) with 6J x 15 rims	1367 mm (53.82 in.) with 7J x 15 rims
With optional wheels	1361 mm (53.6 in.) with 6J x 16 rims	1367 mm (53.82 in.) with 7J x 16 rims
Torsion bar design	Longitudinal for each wheel	Transverse for each wheel
Diameter	18.8 mm (0.740 in.)	24.1 mm (0.949 in.)
Shock absorbers:		
Standard	Replaceable hydraulic cartridges	Double-action hydraulic
Optional	Gas-filled cartridges	Gas-filled shock absorber
Stabilizer bar diameter	20 mm (0.787 in.)	18.0 mm (0.709 in.)

Steering rack: Side section

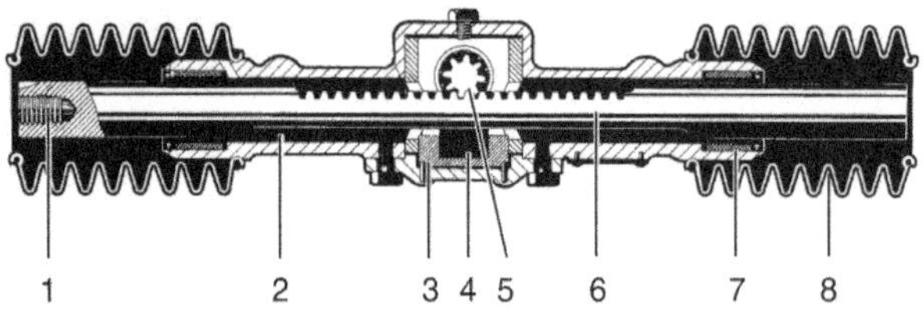

1. Tie rod thread
2. Rack housing
3. Adjusting nut
4. Pressure block
5. Steering pinion
6. Steering rack
7. End bushing
8. Rack boot

Steering Rack: Cross section

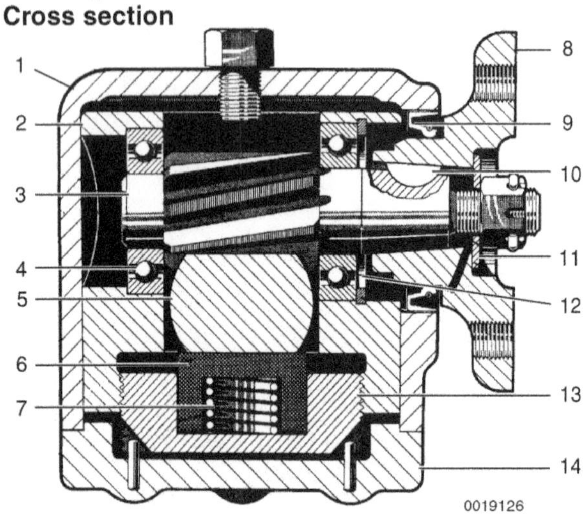

1. Housing
2. Pinion carrier
3. Steering pinion
4. Ball bearing
5. Steering rack
6. Pressure block
7. Spring
8. Steering column flange
9. Oil seal
10. Woodruff key
11. Spacer
12. Lock ring
13. Adjusting nut
14. Base plate

Steering

Rack and pinion steering, used in the Porsche cars, is simple and rugged. There is no power assist to the steering in the cars covered by this manual.

The rack rides in replaceable bushings installed at the ends of the rack housing. The pinion turns in two ball bearings supported in a floating pinion carrier. The rack moves within the confines of the end bushings, pinion gear, a pressure block, spring and adjusting nut. This results in a steering that is very precise with no free play. Torque on the adjusting nut creates a controlled amount of drag in the mechanism.

The steering rack is filled with a special permanent lubricant and needs no further maintenance or lubrication. The protective boots at the tie rod ends of the rack, however, must be inspected periodically and replaced if torn or deteriorated.

For further information or component replacement procedures for the steering system, see **480 Steering**.

Table b. Steering specifications

Characteristic	Specification
Frictional moment of steering (measured at flange of steering column flange, tie rods detached)	80 - 140 Ncm (113 - 198 in-oz.)
Steering ratio	17.78 : 1
Steering wheel turns lock-to-lock	Approx. 3.1
Turning circle diameter	10.70 m (35.1 ft.)

Braking system

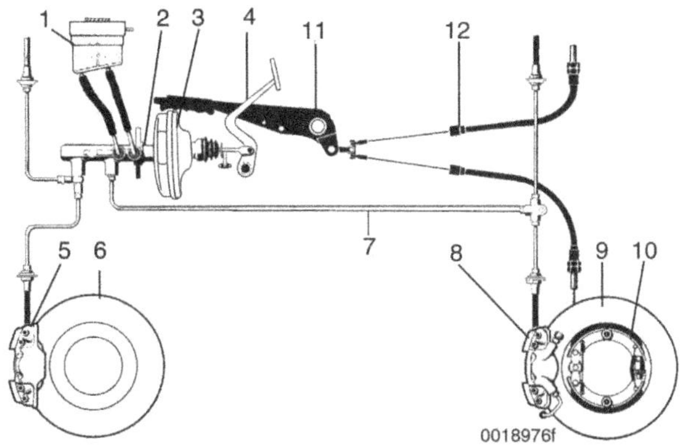

1. Brake fluid reservoir
2. Dual master cylinder
3. Power booster
4. Brake pedal
5. Front brake caliper
6. Front brake rotor
7. Brake fluid line
8. Rear brake caliper
9. Rear brake rotor/drum
10. Parking brake shoes
11. Parking brake handle
12. Parking brake cable

Braking system

Cars covered by this manual are equipped with disc brakes on all four wheels. The brake system is a dual-circuit design, hydraulically actuated by the master cylinder. A brake power booster system is used to reduce pedal effort.

The master cylinder, operated by the brake pedal, creates pressure in the hydraulic system. At the wheels, the hydraulic pressure acts on the calipers, which in turn mechanically apply the brakes. The use of hydraulics makes it possible for the driver to generate high braking forces with a comparatively small amount of effort.

Large pistons in the rear brake calipers improve braking with low pedal effort.

For further information or component replacement procedures for the braking system, see **460 Brakes–Mechanical** or **470 Brakes–Hydraulic**

Table c. Braking system specifications

Description	911 SC
Service brakes	Hydraulic dual circuit, vacuum powered brake booster, vented disc brakes, calipers
Brake booster diameter	17.78 cm (7in.)
Master cylinder:	
Diameter	19.05 mm (.750 in.)
Stroke	18/13 mm (0.748/0.511 in.)
Brake rotor:	
Diameter:	
Front	282.5 mm (11.122 in.)
Rear	290.0 mm (11.417 in.)
Thickness (new):	
Front	20.5 mm (0.807 in.)
Rear	20.0 mm (0.787 in.)
Thickness (wear limit):	
Front	18.5 mm (.728 in.)
Rear	18.0 mm (.708 in.)
Min. thickness after machining	
Front	19.1 mm (0.751 in.)
Rear	18.6 mm (.732 in.)
Max. surface roughness after machining	0.006 mm (0.0002 in.)
Max. thickness variance	0.02 mm (0.0008 in.)
Max. lateral runout	0.1 mm (0.004 in.)
Caliper piston diameter:	
Front	48.0 mm (1.890 in.)
Rear	38.0 mm (1.496 in.)
Brake pad surface area:	
Front (each wheel)	76.0 cm^2 (11.78 sq. in.)
Rear (each wheel)	52.5 cm^2 (8.14 sq. in.)
Total	257.0 cm^2 (39.84 sq. in.)
Brake pad thickness:	
New	10 mm (0.394 in.)
Wear limit	2 mm (0.08 in.)

Table c. Braking system specifications

Description	911 SC
Brake pedal play (brakes bled, engine stopped)	Approx. 10 mm
Parking brake	Cable operated shoes, drum brakes on both rear wheels; drums integral with brake rotors
Brake drum: Diameter (new) Wear limit	 180 mm (7.087 in.) 181 mm (7.126 in.)
Brake shoes: Width Surface area per wheel Thickness Wear limit	 25 mm (0.984 in.) 85 cm^2 (13.175 sq. in.) 4.5 mm (0.177 in.) 2 mm (0.08 in.)

Wheels and tires

The Porsche wheels supplied as original equipment are designed for tubeless, radial-ply tires of a specific size. Replacement tires of non-standard size or construction should be installed only if the tire manufacturer specifies them for your model.

Radial-ply tires are installed as standard equipment on all cars covered by this manual. To retain the car's excellent handling characteristics, it is recommended that the tires only be replaced with those of the same (radial-ply) construction, size and speed rating.

Rim and tire sizes recommended by Porsche are listed in **Table d**.

For further information on wheels and tires, see **440 Wheels, Tires, Alignment**.

Table d. Rim and tire sizes

Model and year	Standard		Option 2	
	Rim size	Tire size	Rim size	Tire size
Front	6 J x 15	185/70 VR 15	6 J x 16	205/55 VR 16
Rear	7 J x 15	215/60 VR 15	7 J x 16	225/50 VR 16

Alignment

Wheel alignment is the precise adjustment of the steering and suspension components to ensure that all wheels are oriented correctly, compared to the other wheels, the chassis, and the direction of travel. Small changes can have a big effect on how the car drives. Proper alignment provides the best compromise between responsiveness, stability, and tire wear.

Alignment specifications differ from model to model. It is important to use the correct alignment specifications for the specific car. Wheel alignment can also be fine-tuned for specific driving conditions. If, for example, the car normally carries only one person or is often heavily loaded, this information may help the alignment technician. Like tire pressure, wheel alignment also has some influence on fuel economy, because of its potential to affect the tires' rolling resistance.

> ***WARNING —***
> *Replacement and/or removal of the trailing arms, torsion bars or torsion plates (rear) disturbs the rear wheel alignment settings. Replacement or adjustment of the front or rear torsion bars or the rear torsion plates also affects vehicle ride height, which directly influences vehicle balance. If ride height is changed, it is strongly recommended that the vehicle balance/weight be checked at all four wheels using suitable wheel-load floor scales. See* **440 Wheels, Tires, Alignment** *for more information.*

TROUBLESHOOTING

This troubleshooting section applies to suspension, steering or braking system problem that affect ride comfort, handling and stability.

Similar problems may be caused by faults in the brakes or steering system. For example, a problem such as consistently pulling to one side may be caused by a bent control arm or by faulty brakes.

Suspension troubleshooting

Stable handling and ride comfort both depend on the integrity of the suspension components. These systems must precisely position the wheels so that the car is stable and controllable, but also allow movement so that the wheels can steer and react to bumps. Any symptom of instability or imprecise road feel may be caused by worn or damaged suspension components.

In addition to inspecting for worn parts, troubleshooting must also consider the condition of tires, wheels, and their alignment. Tire wear and incorrect inflation pressures can dramatically affect handling. Subtle irregularities in wheel alignment angles also affect stability. Mixing different types or sizes of tires, particularly on the same end of the car, can affect alignment and may unbalance a car's handling. See **440 Wheels, Tires, Alignment** for more information.

Table e lists symptoms of suspension problems and their probable causes, and suggests corrective actions.

Table e. Suspension, steering and alignment troubleshooting

Symptom	Probable cause	Corrective action
Pull to one side, wandering	a. Incorrect tire pressure b. Defective/unevenly worn tire c. Incorrect wheel alignment d. Faulty brakes (pulls only when braking) e. Faulty tie rod or tie rod end	a. Check and correct tire pressures. See **030 Maintenance**. b. Inspect tires and replace as needed. See **440 Wheels, Tires, Alignment**. c. Check and adjust wheel alignment. See **440 Wheels, Tires, Alignment**. d. See **460 Brakes–Mechanical** or **470 Brakes–Hydraulic**. e. See **480 Steering**.
Steering heavy, poor return to center	a. Worn front upper strut mount(s) b. Incorrect tire pressure	a. Replace front upper strut mounts. See **401 Front Suspension**. b. Check and correct tire pressures. See **030 Maintenance**.
Vibration or shimmy while driving	a. Severely worn front strut cartridges or rear shock absorbers b. Worn ball joints c. Unbalanced or bent wheels/tires d. Loose wheel lug bolts e. Incorrect tire pressure f. Tire(s) out-of-round (radial runout) g. Faulty tie rod(s) or tie rod end(s) h. Faulty steering universal joint	a. Replace strut cartridges or shock absorbers. See **401 Front Suspension** or **420 Rear Suspension, Axle Shafts**. b. Replace ball joints. c. Balance tires. Check tires for uneven wear patterns. Check wheels for damage. See **440 Wheels, Tires, Alignment**. d. Tighten lug bolts to proper torque. See **440 Wheels, Tires, Alignment**. e. Check and correct tire pressure. See **030 Maintenance**. f. Measure tire radial runout and remount or replace as necessary. g. See **480 Steering**. h. See **480 Steering**.
Poor stability, repeated bouncing after bumps, suspension bottoms out easily	a. Worn (weak) strut cartridges in front b. Worn (weak) shock absorbers in rear	a. Replace strut cartridges. See **401 Front Suspension**. b. Replace rear shocks absorbers. See **420 Rear Suspension, Axle Shafts**.
Suspension noise (especially over bumps)	a. Worn front upper strut mounts. b. Loose front strut cartridge c. Faulty rubber shock absorber buffer (bump stop) d. Worn ball joints. e. Loose or worn stabilizer bar rubber mounts f. Loose front suspension subframe	a. Replace front upper strut mounts. See **401 Front Suspension**. b. Disassemble strut, check for damage. See **401 Front Suspension**. c. Replace rubber buffer. d. Replace ball joints. See **401 Front Suspension**. e. Retorque stabilizer bar rubber mounts. Replace rubber mounts as necessary. f. Check for subframe damage. Tighten mounting bolts.
Uneven ride height	a. Incorrectly adjusted or incorrect torsion bars b. Bent or damaged suspension components c. Sagging torsion bars	a. Measure ride height and adjust or replace torsion bars as required. See **401 Front Suspension** or **420 Rear Suspension, Axle Shafts**. b. Inspect and repair/replace as necessary c. Replace torsion bars in pairs. See **401 Front Suspension** or **420 Rear Suspension, Axle Shafts**.
Scraping, growling or rhythmic knocking from wheels while driving, may be more noticeable when turning	a. Worn wheel bearing(s) b. Faulty brakes c. Worn constant velocity (CV) joint	a. Replace wheel bearings. See **401 Front Suspension** or **420 Rear Suspension, Axle Shafts**. b. Repair brakes. See **460 Brakes–Mechanical** c. Inspect CV joint boots for tears. Replace boots or joints as needed. See **420 Rear Suspension, Axle Shafts**
Steering loose, imprecise	a. Incorrect tire pressure b. Loose steering rack mounting bolts c. Worn tie rod(s) or tie rod end(s) d. Faulty front wheel bearing(s) e. Worn or damaged steering rack	a. Check and correct tire pressure. See **030 Maintenance**. b. Inspect and tighten bolts c. See **480 Steering**. d. Replace front wheel bearing(s). See **401 Front Suspension**. e. See **480 Steering**.

Suspension inspection and testing

Suspension problems can usually be isolated and at least partially diagnosed by careful observation of the symptoms and inspection of the components that are the most likely cause.

Torsion bars support the weight of the car and allow the suspension to travel smoothly over bumps and other road irregularities. A torsion bar that is twisted (loaded) by a bump simply unloads in the other direction with nearly the same force that loaded it.

The function of a shock absorber, or damper, is to moderate the torsion bar action. It slows the bounce and helps the torsion bar return to its normal position. Shock absorbers require no routine maintenance and cannot be serviced.

The best evidence of failing shock absorbers is their behavior in normal driving. Worn shock absorbers will allow extra skittishness over bumps, and a less-controlled and wallowing feel after bumps and in corners. When seriously worn, the shock absorbers present little resistance to torsion bar oscillations. Because they so easily reach the limits of their travel, they may knock when going over bumps.

The most common—though not entirely accurate—test of shock absorber function involves vigorously bouncing each end or corner of the car, and then releasing and observing how quickly the bouncing stops. More than one bounce usually suggests that the shock absorbers are not properly damping the torsion bar action and need to be replaced.

NOTE —
Differences from side to side will affect handling and stability. It is strongly recommended that shock absorber cartridges or strut bearings be replaced in pairs.

Tire wear

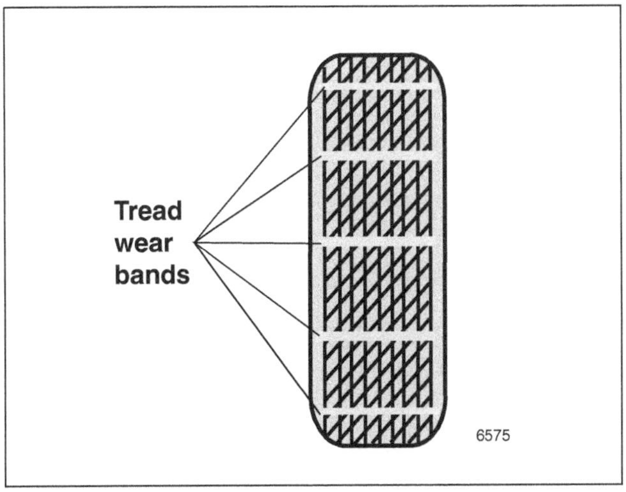

Tire wear is a good indicator of steering and wheel alignment problems. Proper tread wear is difficult to notice, so tires are made with wear-indicator bands that indicate when the tire is nearly worn-out. On an evenly worn tire, these wear-indicator bars will eventually appear as evenly spaced bald "stripes" about ½ in. wide running across the tread surface. The appearance of these tread-wear bands on only one part of the tread indicates uneven wear.

For more information on tire wear, see **440 Wheels, Tires, Alignment**. For information on maintaining proper tire inflation pressures, as well as other general tire maintenance, see **030 Maintenance**.

Isolating pulling symptoms

Consistent pulling to one side in a car driven straight ahead on a level road may be caused either by worn or damaged (bent) suspension components or a faulty tire. In more unusual cases, a brake problem or possibly incorrect torsion bars may be the cause.

To help decide whether the problem is caused by the tires or the suspension, temporarily swap the front wheels (and tires), and then road-test the car. If a tire problem is the cause of the pulling symptom, the problem should switch to the other side of the car when the wheels are switched, and it should now pull to the other side. If the symptom persists, then the problem is probably caused by faulty suspension components.

NOTE —
If no suspension components are found to be faulty, the car may only need a front wheel alignment. See **440 Wheels, Tires, Alignment** *for more information.*

Vibration

Abnormal vibration, if not caused by a mechanical problem, will most often be caused by wheels and tires that are out of balance. Other, less likely, causes of vibration are bent wheels or a tire that has become distorted and out-of-round. These and other tire conditions can be checked by any reputable tire professional.

Before spending the money for balancing, look for an obvious cause of imbalance. Caked-on mud, ice, or snow can dramatically affect wheel balance. Clean the wheels and tires and road-test the car before looking for more serious causes.

Braking system troubleshooting

This heading describes symptoms of trouble with the brakes. The sole function of the brakes is to generate friction to slow or stop the car. Brake problems are usually obvious because they affect the way the car slows and stops. Noise or problems with the car's handling may be caused by the brakes, but may also be caused by faults in the suspension or steering systems. For help in selecting the appropriate repair section, see **Table e** and **Table f**.

Brake diagnostic checks

Component inspection and some general brake system checks can help isolate problems.

Check the brake booster first by pumping the brake pedal approximately 10 times (vacuum-assist systems) or 20 times (hydraulic-assist systems) with the engine off. While holding the pedal down start the engine. The pedal should fall slightly. If not, check for any visible faults before suspecting a faulty brake booster.

Check for leaks in the power booster vacuum line. A faulty check valve or a faulty O-ring between the master cylinder and the vacuum booster will decrease the vacuum assist. Low engine vacuum could also decrease vacuum assist. Test the check valve by removing it from the booster vacuum line and blowing through it. Air should pass through in the direction of the arrow, but not pass through the opposite way.

Brake noise

Occasional groaning or squealing sounds from the disc brakes are usually caused by vibration being transmitted through the brake pads. Brake friction materials that contain little or no asbestos may contribute to brake noise. These noises are normal and rarely indicate a problem.

Although there is no universal solution to the problem of brake noise, proper maintenance and repair can help minimize it. Disc brake caliper assemblies include anti-rattle springs to minimize vibration and noise. Brake pads and calipers should be kept clean and free of corrosion. Always resurface or replace brake rotors when changing brake pads.

Table f lists symptoms of problems commonly associated with the brakes, their probable causes, and suggested corrective actions.

Table f. Brake system troubleshooting

Symptom	Probable cause	Corrective action
Brake squeak or squeal	a. Incorrectly installed brake pads or parking brake shoes b. Brake pad carriers dirty or corroded c. Brake pad anti-rattle springs faulty or missing d. Brake pads heat-glazed or oil-soaked e. Wheel bearings worn (noise most pronounced when turning)	a. Check component installation. See **460 Brakes–Mechanical**. b. Remove brake pads and clean calipers. See **460 Brakes–Mechanical**. c. Install/replace anti-rattle springs. See **460 Brakes–Mechanical**. d. Replace brake pads. Clean rotors. Replace leaking calipers as required. See **460 Brakes–Mechanical** or **470 Brakes–Hydraulic**. e. Replace worn bearings. See **401 Front Suspension** or **420 Rear Suspension, Axle Shafts**.
Pedal goes to floor when braking	a. Brake fluid level low due to system leaks b. Master cylinder faulty	a. Check fluid level and inspect hydraulic system for signs of leakage. Fill and bleed system. See **470 Brakes–Hydraulic**. b. Replace master cylinder. See **470 Brakes–Hydraulic**.
Low pedal even after system bleeding	a. Master cylinder faulty	a. Replace master cylinder. See **470 Brakes–Hydraulic**.
Pedal feels spongy or brakes work only after pedal is pumped	a. Air in brake fluid b. Master cylinder faulty (internal return spring weak) c. Leaking line or hose unions	a. Bleed system. See **470 Brakes–Hydraulic**. b. Replace master cylinder. See **470 Brakes–Hydraulic**. c. Repair or replace lines and hoses. Bleed system. See **470 Brakes–Hydraulic**.

Table f. Brake system troubleshooting

Symptom	Probable cause	Corrective action
Excessive braking effort	a. Brake pads wet b. Brake pads heat-glazed or oil-soaked c. Vacuum booster or vacuum hose connections to booster faulty	a. Use light pedal pressure to dry pads while driving. b. Replace brake pads. Clean rotors. Replace leaking calipers. See **460 Brakes–Mechanical** or **470 Brakes–Hydraulic**. c. Inspect vacuum lines. Test vacuum booster and replace as required. Test vacuum check-valve for one-way air flow. See **470 Brakes–Hydraulic**.
Brakes pulsate, chatter, or grab	a. Warped brake rotors b. Brake pads worn c. Brake pads heat-glazed or oil-soaked	a. Resurface or replace rotors. See **460 Brakes–Mechanical**. b. Replace brake pads. See **460 Brakes–Mechanical**. c. Replace brake pads. Clean rotors. Replace leaking calipers. See **460 Brakes–Mechanical** or **470 Brakes–Hydraulic**.
Uneven braking, car pulls to one side, rear brakes lock	a. Incorrect tire pressures or worn tires b. Brake pads on one side of car heat-glazed or oil-soaked c. Caliper or brake pads binding d. Worn suspension components	a. Inspect tire condition. Check and correct tire pressures. See **030 Maintenance**. b. Replace brake pads. Clean rotors. Replace leaking calipers. See **460 Brakes–Mechanical** or **470 Brakes–Hydraulic**. c. Clean and recondition brakes. See **460 Brakes–Mechanical** or **470 Brakes–Hydraulic**. d. Inspect for worn or damaged suspension components. See **401 Front Suspension** or **420 Rear Suspension, Axle Shafts**.
Brakes drag, bind, or overheat	a. Brake caliper or brake pads binding b. Master cylinder faulty	a. Clean or replace caliper. See **470 Brakes–Hydraulic**. b. Replace master cylinder. See **470 Brakes–Hydraulic**.

401 Front Suspension

401

Special Tools

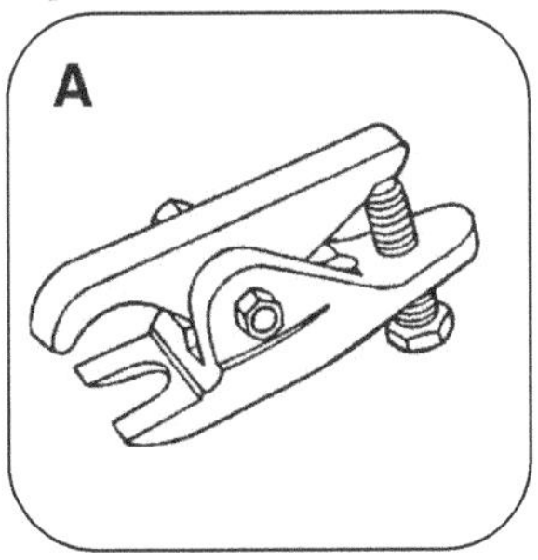

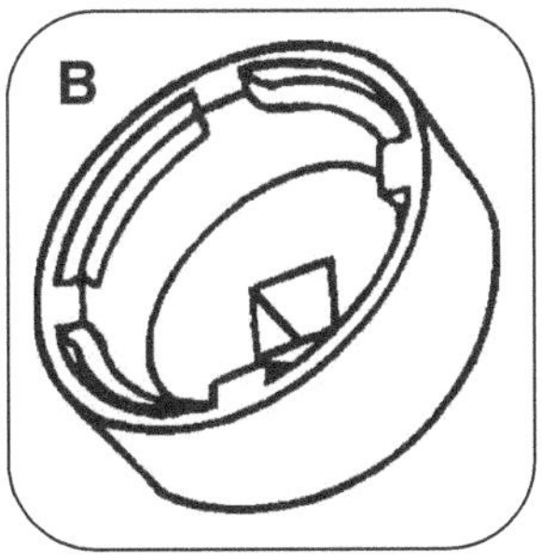

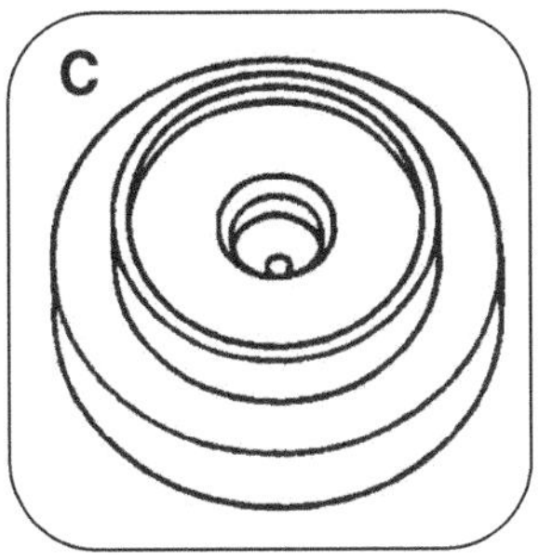

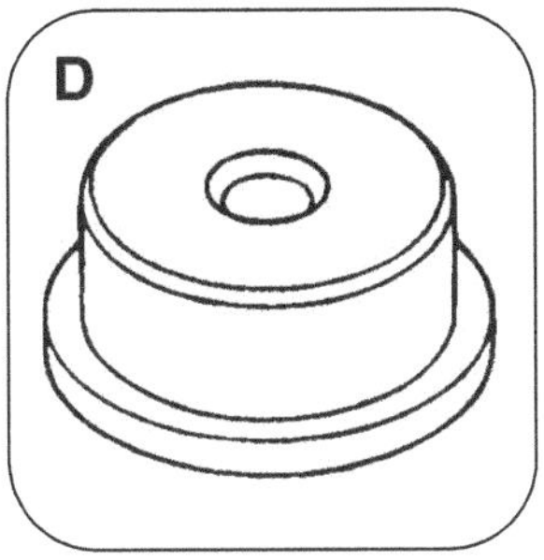

Special tool

- **A** VW 267 a . . . . Tie-rod end ball joint separator (Source: Baum Tools Unlimited)
- **B** P 280 b . . . . Ball joint removal spanner (Porsche ordering #000 721 280 20)
- **C** VW 433 . . . . Front wheel grease seal installer (Source: Baum Tools Unlimited)
- **D** VW 447h/447i . . . . . . Front wheel outer bearing race installers (Source: Baum Tools Unlimited)

GENERAL

This repair group covers replacement procedures for front suspension components. For specifications, diagnosis and troubleshooting information see **400 Suspension, Brakes and Steering–General**. For information on setting ride height and alignment see **440 Wheels, Tires, Alignment**. For information on steering see **480 Steering**.

Special tools

Some of the procedures require special tools.

Front suspension

The Porsche 911 SC is equipped with independent front suspension. Torsion bars are used at all four wheels. The front end uses strut type shock absorbers.

Each front suspension strut assembly includes a replaceable shock absorber cartridge inside a tubular strut housing. The lower end of each strut includes the stub axle for the front wheel assembly.

The front torsion bars are mounted inside the control arms. The spring tension on the front torsion bars is easily adjustable.

The front control arms are mounted to the floor pan. Ball joints are used to connect the control arms to the strut assemblies.

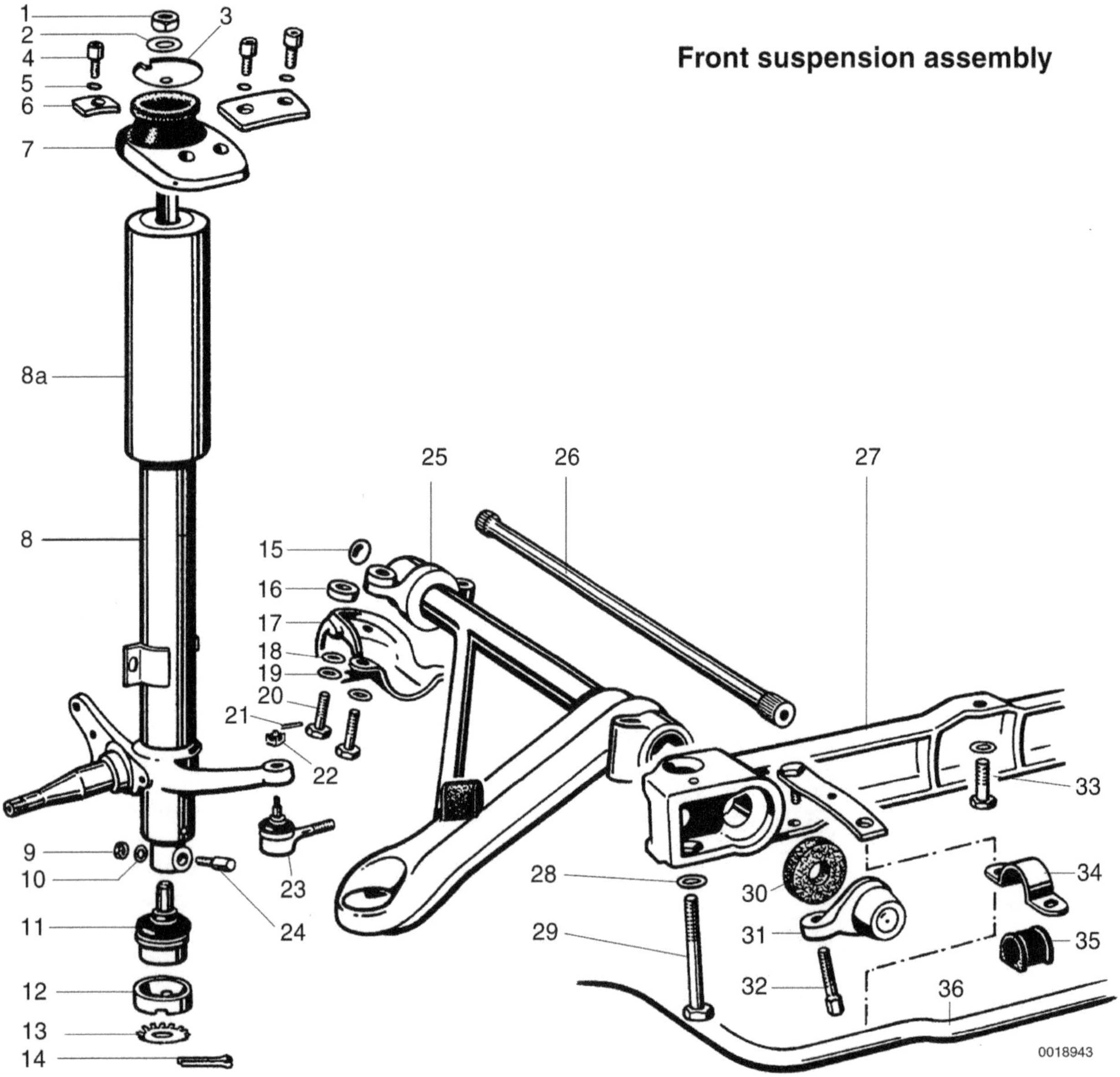

1. Nut (M14 x 1.5)
   tighten to 80 Nm (59 ft-lb)
2. Locking tab: always replace
3. Plate
4. Bolt (M10 x 20 mm)
   tighten to 47 Nm (34 ft-lb)
5. Lock washer (M10)
6. Support plate
7. Top strut mount
8. Strut

8a. Strut dust sleeve

9. Lock nut
   tighten to 22 Nm (16 ft-lb)
10. Washer
11. Ball joint
12. Strut cap
    tighten to 250 Nm (184 ft-lb)
13. Lock plate
14. Cotter pin: always replace
15. Cover
16. Shim
17. Control arm front mounting bracket
18. Washer (M10)
19. Lock washer (M10): always replace
20. Bolt (M10 x 35 mm)
    tighten to 47 Nm (35 ft-lb)
21. Cotter pin (2 x 25): always replace
22. Castle nut (M10 x 1)
    tighten to 45 Nm (33 ft-lb)
23. Tie-rod end
24. Wedge bolt: always replace
25. Control arm
26. Torsion bar
27. Crossmember subframe
28. Lock washer (M12): always replace
29. Bolt (M12 x 1.5 x 85 mm)
    tighten to 90 Nm (66 ft-lb)
30. Foam seal
31. Torsion bar retaining cap
32. Torsion bar adjusting bolt
33. Bolt (M10)
    tighten to 47 Nm (35 ft-lb)
34. Stabilizer bar bracket
35. Stabilizer bar bushing
36. Stabilizer bar

Suspension travel is limited by rubber bump stops on the top of each strut assembly.

A stabilizer bar is mounted to the front control arms and subframe to help reduce body roll when cornering.

When doing any front suspension maintenance or repair, observe the following general cautions.

CAUTION—

- *Do not install bolts and nuts coated with undercoating wax, as correct tightening torque cannot be assured. Always clean the threads with solvent before installation, or install new parts.*
- *Do not attempt to weld or straighten the suspension struts, control arms or subframe. Replace damaged parts.*
- *Do not reuse self-locking nuts or cotter pins. These parts are designed to be used only once.*

FRONT SUSPENSION STRUTS

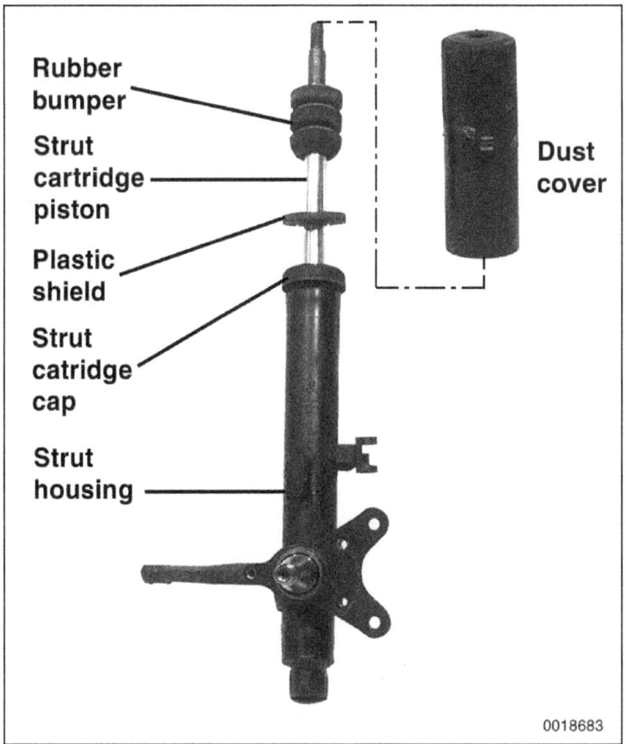

The main components of the front strut assembly are the strut housing, the strut cartridge and the top strut mount. Each of these main components can be replaced separately. The strut housing itself should be replaced if it is damaged.

A variety of struts by different manufacturers, including Boge® and Bilstein®, have been used in Porsche cars. Aftermarket manufacturers, including Koni® and Tokiko®, also offer replacement strut cartridges. The strut assembly must be removed from the car and disassembled, as described later, to replace the internal damper cartridge.

Consult with the Porsche dealer parts department or aftermarket supplier when deciding what replacement parts to use.

NOTE—

Any differences in suspension components from side to side will affect handling and stability. It is strongly recommended that struts, strut cartridges or top strut mounts always be replaced in pairs and at the same time.

Front suspension strut, removing and installing

NOTE —

- *Shock absorbers should be stored upright. If they are stored horizontally or upside-down for an extended period, they should be placed upright at room temperature with shock absorber rods fully extended for at least 24 hours prior to installation.*
- *The non-gas strut will remain in any desired position when removed. The gas-pressure version will have the damper rod all the way out when removed.*
- *If the upper strut mount is not removed with the strut the wheel alignment will not be disturbed.*

— Loosen front wheel lug nuts. Raise front of car and support it securely.

> ***WARNING —***
> *Make sure the car is stable and well supported at all times. Use a professional automotive lift or jack stands designed for the purpose. A floor jack is not adequate support.*

— Remove front wheel.

— Press brake pedal down about 1 in. and hold in place with a pedal stop.

NOTE —
Holding the brake pedal down will prevent brake fluid from dripping out of open brake lines while the work is being performed.

— Disconnect rubber brake hose from metal brake line near bottom of strut. Plug open lines.

> ***CAUTION —***
> *Always counterhold fitting before loosening or tightening brake lines and hoses.*

— Remove caliper mounting bolts and remove caliper. See **470 Brakes–Hydraulic**.

— Remove wheel bearing dust cover.

0018966

< Loosen clamping screw (6 mm allen wrench) in wheel bearing lock nut. Unscrew and remove lock nut from spindle.

— Pull brake rotor with bearing hub and wheel bearings straight off steering spindle.

— Remove cotter pin from tie rod nut and remove castle nut.

— Using special tool (VW 267 a) or equivalent, separate tie rod from strut steering arm.

◀ Remove ball joint wedge bolt (**arrow**).

– Carefully pry out or tap on control arm to push ball joint down out of strut housing.

◀ Working in luggage compartment, bend back strut mounting nut locking tab. Remove strut piston nut from upper strut mount.

– Remove strut assembly from below. If necessary, use a punch to help push strut piston out of upper mount.

– Installation is reverse of removal.

- Always replace ball joint wedge bolt. When installing wedge bolt, line up flat of bolt with cutout in ball joint.
- Repack and adjust wheel bearing as described later.
- Bleed brake hydraulic system. See **470 Brakes–Hydraulic**.
- Check and adjust wheel alignment. See **440 Wheels, Tires, Alignment**.

Tightening torques

- Ball joint wedge bolt (always replace) . . . 22 Nm (16 ft-lb)
- Brake line to brake hose . . . . 12 Nm (9 ft-lb)
- Caliper to strut housing . . . . 70 Nm (51 ft-lb)
- Strut to upper mount (use new locking tab) . . . . 80 Nm (58 ft-lb)
- Tie rod castle nut to tie rod end . . . . 45 Nm (33 ft-lb)
- Wheel bearing lock nut clamp screw (M8) . . . . 15 Nm (11 ft-lb)
- Wheel to wheel hub . . . . 130 Nm (96 ft-lb)

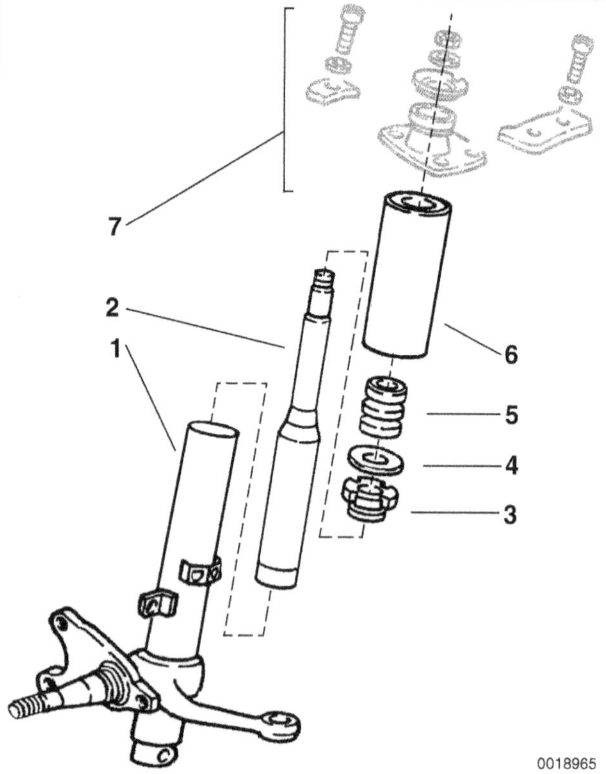

1. Strut housing
2. Strut cartridge (insert)
3. Strut cartridge cap, tighten to 120 + 20 Nm (87 + 14 ft-lb)
4. Plastic shield
5. Rubber bumper
6. Dust cover
7. Upper strut mount assembly

Front strut cartridge, removing and installing

Remove front suspension strut as described earlier. Remove dust cover and rubber bumper from strut piston rod.

– Clamp housing in vise or other strut holding device.

– Using strut nut wrench, remove strut cartridge cap.

- Pull cartridge out of strut housing.
- Empty out oil in strut housing.

– Reassembly is reverse of disassembly.

Tightening torque

- Strut cartridge cap to strut housing . . . . 120 +20 Nm (87 + 14 ft-lb)

Upper strut mount, replacing

– Loosen front wheel lug nuts. Raise front of car and support it securely.

WARNING —

Make sure the car is stable and well supported at all times. Use a professional automotive lift or jack stands designed for the purpose. A floor jack is not adequate support.

– Remove front wheel.

Working in luggage compartment, bend back strut piston mounting nut locking tab and remove nut from upper strut mount.

- Remove 8 mm allen bolts (**arrows**) holding strut mount to body.

NOTE —

Clean off sealing compound around strut mount plates and mark their locations on the inner fender. This will help in reassembling the front suspension close to its original position.

– Using a long punch, push strut piston rod down out of mount and remove mount from below.

– Installation is reverse of removal.

NOTE —

Use a long M6 stud to reach through the strut mount assembly and into the threads at the top of the strut cartridge piston. Then the piston can be pulled up through the mount.

– Reset wheel alignment. See **440 Wheels, Tires, Alignment**.

Upper strut mount assembly

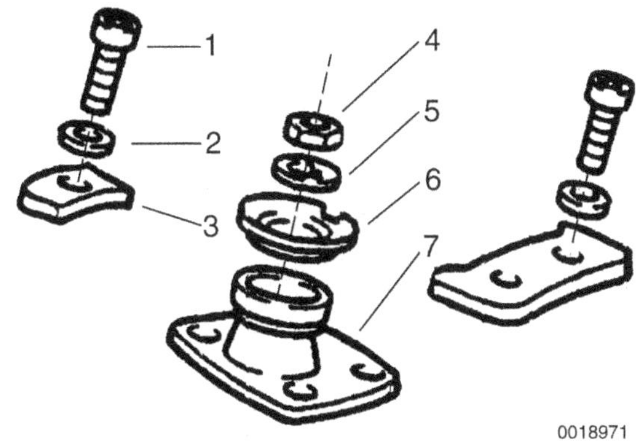

1. Bolt (M10 x 20 mm) tighten to 47 Nm (34 ft-lb)
2. Lock washer (M10)
3. Mounting plate
4. Strut top mounting nut (M10 x 1.5) tighten to 80 Nm (59 ft-lb)
5. Locking tab (always replace)
6. Plate
7. Upper mount

Tightening torques

- Brake line to brake hose . . . . 12 Nm (9 ft-lb)
- Strut to upper mount (use new locking tab) . . . . 80 Nm (58 ft-lb)
- Upper strut mount to body . . . . 47 Nm (34 ft-lb)
- Wheel to wheel hub . . . . 130 Nm (96 ft-lb)

BALL JOINTS, CONTROL ARMS AND TORSION BARS

Ball joint, removing and installing

– Loosen front wheel lug nuts. Raise front of car and support it securely.

> ***WARNING —***
> *Make sure the car is stable and well supported at all times. Use a professional automotive lift or jack stands designed for the purpose. A floor jack is not adequate support.*

– Remove front wheel.

◄ Remove ball joint wedge bolt (**arrow**).

– Carefully pry out or tap on control arm to push ball joint down out of strut housing.

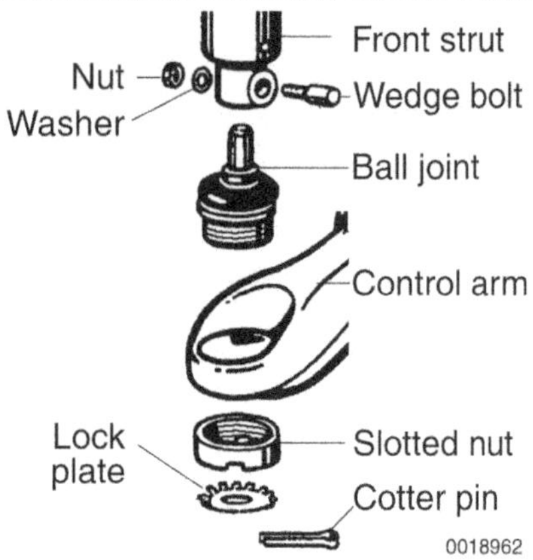

< Remove cotter pin from ball joint lock plate. Remove slotted nut from ball joint using Porsche special tool no. P 280 b.

– Remove ball joint from control arm.

– Installation is reverse of removal.

NOTE —
- *Always replace ball joint wedge bolt.*
- *When installing wedge bolt, line up flat of bolt with cutout in ball joint.*

Tightening torques
- Ball joint slotted nut to ball joint . . . . . 250 Nm (184 ft-lb)
- Ball joint wedge bolt (always replace). . . 22 Nm (16 ft-lb)
- Wheel to wheel hub . . . . 130 Nm (96 ft-lb)

Front control arm, removing and installing

– Loosen front wheel lug nuts. Raise front of car and support it securely.

WARNING —
Make sure the car is stable and well supported at all times. Use a professional automotive lift or jack stands designed for the purpose. A floor jack is not adequate support.

– Remove front wheels.

– Remove cotter pin from ball joint lock plate. Remove slotted round nut from ball joint using Porsche special tool no. P 280 b.

< Working underneath front of car, remove stone guard mounting bolts (**arrow**). Remove stone guard.

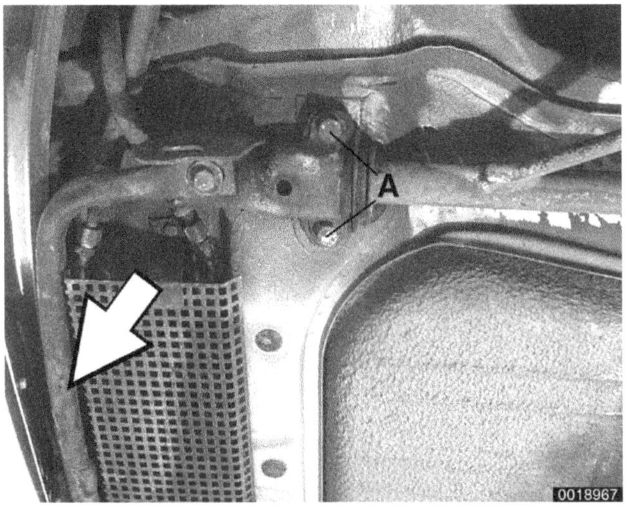

◀ Unbolt and remove A/C condenser protection bar (**arrow**).

– Remove front control arm mounting bracket bolts (**A**).

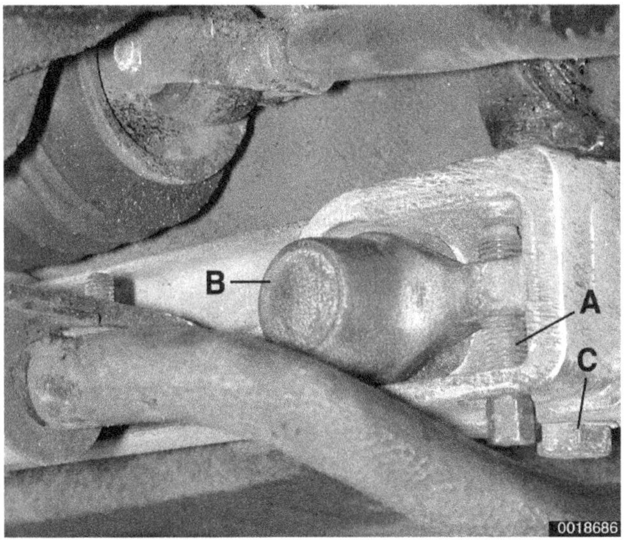

◀ Remove torsion bar adjustment screw (**A**) and retaining cap (**B**).

NOTE —

Record number of threads on adjustment screw while removing. This will help in reassembly.

– Remove single bolt (**C**) at rear of crossmember subframe and pull control arm forward out of subframe.

– With a rubber hammer tap control arm off stabilizer bar.

– Installation is reverse of removal.

NOTE —

- *During installation, use Contifix® or equivalent rubber lubricant when sliding stabilizer bar into control arm stabilizer bushings.*
- *Lubricate torsion bar adjustment screw with multipurpose MoS_2 grease. Install according to number of threads counted during removal.*
- *Be sure to reset front ride height as described in* **440 Tires, Wheels, Alignment**.

Tightening torques

- Ball joint slotted nut to ball joint ..... 250 Nm (184 ft-lb)
- Control arm to body:
  Front bracket (M10).................. 47 Nm (35 ft-lb)
  Crossmember subframe (M12)........ 90 Nm (66 ft-lb)
- Wheel to wheel hub .... 130 Nm (96 ft-lb)

Front torsion bar, removing and installing

> ***WARNING —***
> *Replacement and/or removal of the torsion bar affects vehicle ride height, which directly influences vehicle stability. If ride height is changed, it is strongly recommended that the vehicle balance/weight be checked at all four wheels using suitable wheel-load floor scales. See* **440 Wheels, Tires, Alignment** *for more information.*

> ***CAUTION —***
> *Torsion bars are pretensioned during production. Right and left bars cannot be exchanged. Bars are marked with a R or L on the ends.*

– Raise front of car and support it securely.

> ***WARNING —***
> *Make sure the car is stable and well supported at all times. Use a professional automotive lift or jack stands designed for the purpose. A floor jack is not adequate support.*

◀ Working underneath front of car, remove stone guard mounting bolts (**arrow**). Remove stone guard.

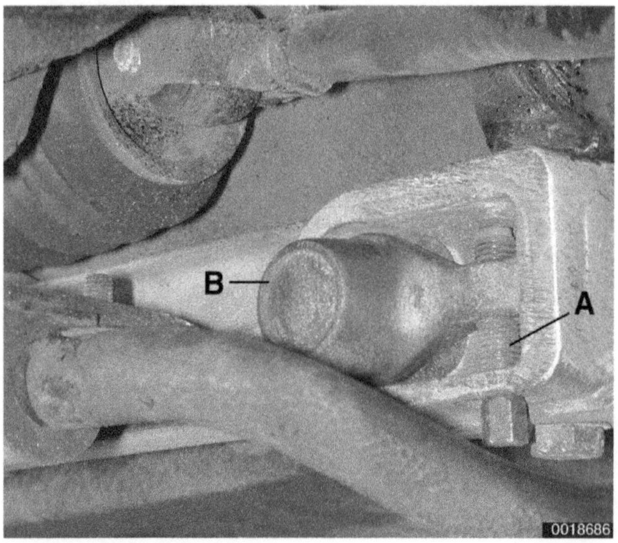

◀ Remove torsion bar adjustment screw (**A**) and retaining cap (**B**).

> ***NOTE —***
> *Record number of threads on adjustment screw while removing. This will help in reassembly.*

– Pull torsion bar out toward rear of car.

> ***NOTE —***
> *Pry stabilizer bar down slightly to clear torsion bar as it is being removed.*

– Installation is reverse of removal.

NOTE —
- *Lubricate torsion bar adjustment screw with multipurpose MoS_2 grease. Install according to number of threads counted before.*
- *Be sure to check front ride height as described in* **440 Tires, Wheels, Alignment**.

Tightening torques
- Stone guard/stabilizer rear bracket to body (M10 bolt) . . . . 47 Nm (35 ft-lb)
- Stone guard/stabilizer rear bracket to crossmember subframe (M10 nut) . . . 28 Nm (21 ft-lb)

WHEEL BEARINGS

The front wheel bearings are a tapered-roller design. Periodic repacking and adjustment is part of the normal maintenance routine for the car. See **030 Maintenance**.

When replacing wheel bearings, press equipment should be used to install the bearing races into the bearing hub.

Front wheel bearing play, adjusting

– Loosen front wheel lug nuts. Raise front of car and support it securely.

> ***WARNING —***
> *Make sure the car is stable and well supported at all times. Use a professional automotive lift or jack stands designed for the purpose. A floor jack is not adequate support.*

– Remove front wheel.

– Carefully pry off grease cap in center of bearing hub.

0018966

< Loosen 6 mm allen clamp screw in wheel bearing lock nut. Loosen bearing lock nut approximately ¼ turn.

Front wheel bearing assembly

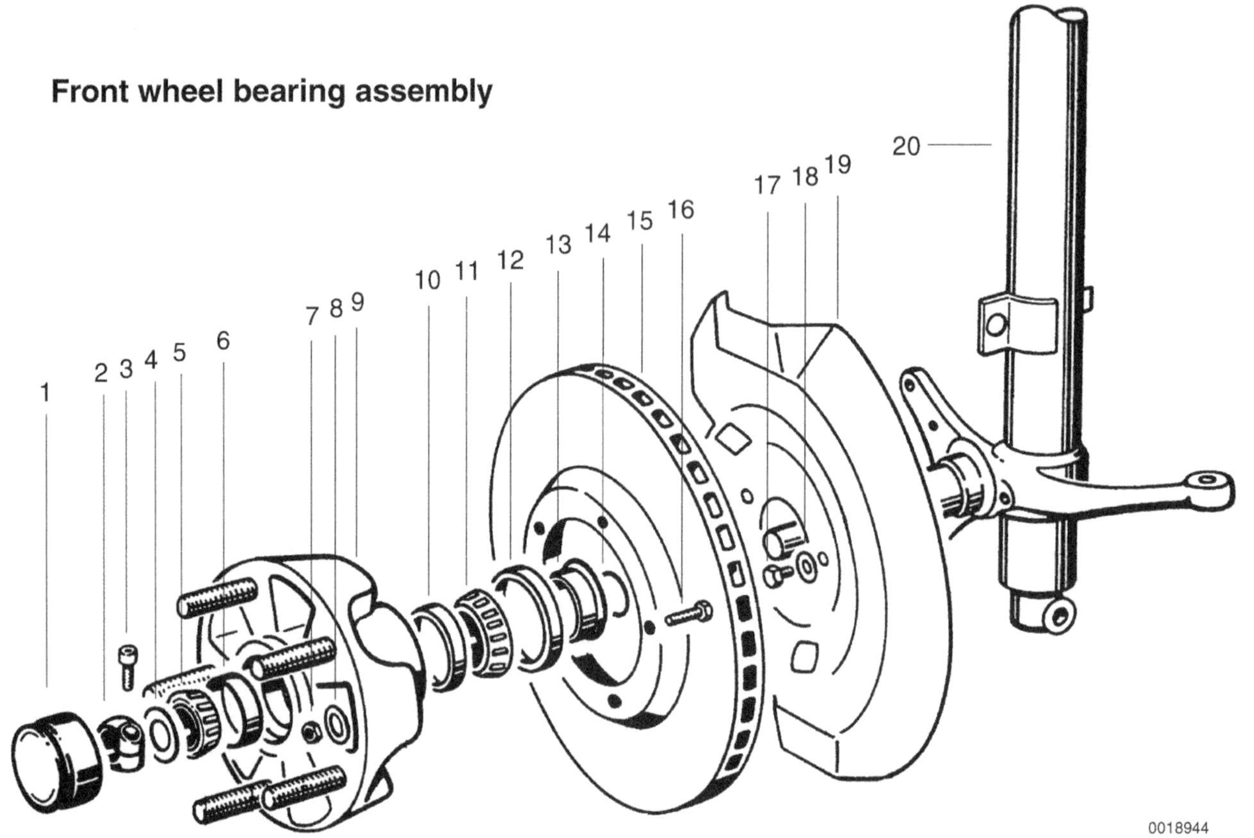

0018944

1. Grease cap
2. Wheel bearing lock nut
3. Clamping allen bolt
   tighten to 15 Nm (11 ft-lb)
4. Thrust washer
5. Outer wheel bearing
6. Outer race, outer wheel bearing
7. Nut (M8)
   tighten to 23 Nm (17 Nm)
8. Washer (M8)
9. Bearing hub
10. Outer race, inner wheel bearing
11. Inner wheel bearing
12. Grease seal
13. Spacer
14. O-ring
15. Brake rotor (disc)
16. Bolt (M8)
17. Bolt (M8)
    tighten to 25 Nm (18 ft-lb)
18. Washer (M8)
19. Backing plate
20. Front strut housing

0018966a

◄ While rotating brake disc and bearing hub, tighten lock nut slowly in small increments until thrust washer can just barely be moved back and forth with a light push on the tip of a screwdriver (do not pry or twist screwdriver).

– Tighten clamping screw to specifications. Recheck tightness of thrust washer.

Tightening torque
- Wheel bearing lock nut clamp screw (M8) .... 15 Nm (11 ft-lb)

– Install grease cap. Reinstall road wheel.

Tightening torque
- Wheel to wheel hub .... 130 Nm (96 ft-lb)

Front wheel bearings, replacing

– Loosen front wheel lug nuts. Raise front of car and support it securely.

> ***WARNING —***
> *Make sure the car is stable and well supported at all times. Use a professional automotive lift or jack stands designed for the purpose. A floor jack is not adequate support.*

– Remove front wheel.

– Press brake pedal down about 1 in. and hold in place with a pedal stop.

NOTE —
Holding the brake pedal down will prevent brake fluid from dripping out of open brake lines while the work is being performed.

– Disconnect rubber brake hose from line near bottom of strut. Plug open lines.

> ***CAUTION —***
> *Always counterhold brake line fittings when loosening or tightening.*

– Remove caliper mounting bolts and remove caliper. See **470 Brakes–Hydraulic**.

– Carefully pry off grease cap in center of bearing hub.

0018966

◄ Loosen 6 mm allen clamp screw in wheel bearing lock nut. Remove lock nut.

– Slide brake rotor and bearing hub straight off axle.

NOTE —
If necessary, rock the brake rotor side-to-side to aid removal of the thrust washer and outer wheel bearing.

Remove grease seal and inner wheel bearing from bearing hub.

Using drift, drive wheel bearing outer races from bearing hub.

- Note cutouts (**arrows**) in bearing hub provided for this purpose.

– Heat bearing hub to 120° - 150°C (250° - 300°F). Press new races into hub.

- Use special tool VW 447h to press outer bearing race.
- Use special tool VW 447i to press inner bearing race.

– Pack wheel bearings with lithium-based wheel bearing grease. Lightly coat bearing races and hub with grease.

> ***CAUTION —***
> *Proper packing of wheel bearings is essential for long bearing service life. Use a wheel bearing packing tool if available.*

Front wheel bearings

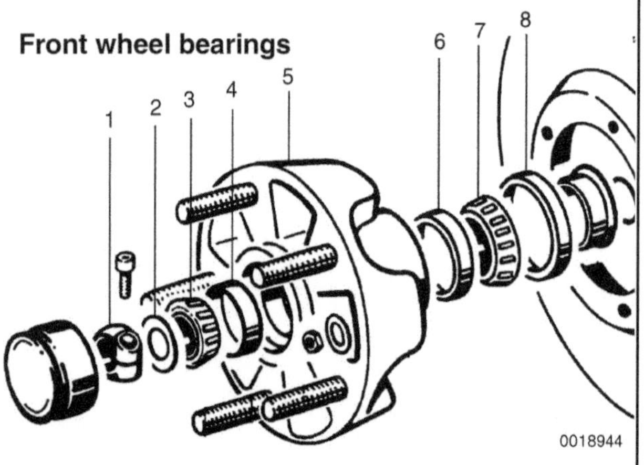

1. Bearing lock nut
2. Thrust washer
3. Outer bearing
4. Inner race, outer bearing
5. Bearing hub
6. Inner race, inner bearing
7. Inner bearing
8. Grease seal

Place inner wheel bearing in race and press in new inner seal using special tool VW 433.

– Apply grease to gap between sealing lips on inner seal.

– Slide bearing hub with rotor and bearings onto stub axle. Use caution to avoid damaging grease seal.

– Install outer wheel bearing, thrust washer and wheel bearing lock nut. Tighten lock nut while rotating brake rotor.

– Adjust wheel bearing play as described earlier. Tighten wheel bearing lock nut clamp screw to specifications. Recheck tightness of thrust washer.

Tightening torque

- Wheel bearing lock nut clamp screw (M8) . . . . 15 Nm (11 ft-lb)

– Fill grease cap with 10 grams of grease and install.

– Remainder of installation is reverse of removal.

Tightening torque
- Brake caliper to strut housing (M12).... 70 Nm (52 ft-lb)
- Wheel to wheel hub .... 130 Nm (96 ft-lb)

STABILIZER BAR

Front stabilizer bar, removing and installing

– Raise car and support in a safe manner.

> ***CAUTION—***
> *Make sure the car is stable and well supported at all times. Use a professional automotive lift or jack stands designed for the purpose. A floor jack is not adequate support.*

◀ Working underneath front of car, remove stone guard mounting fasteners (**arrow**). Remove stone guard.

◀ Loosen stabilizer rear bracket bolt (**A**) on left and right sides.

– Pry stabilizer bracket off crossmember subframe.

– Pull stabilizer backward, first from one control arm, then the other, to remove from car.

– Disassemble stabilizer bracket and inspect rubber bushings for wear. Also inspect bushings in left and right control arm stabilizer brackets. Replace as necessary.

> ***NOTE—***
> *During installation, use Contifix® or equivalent rubber lubricant when sliding stabilizer bar into control arm stabilizer bushings.*

Tightening torques
- Stabilizer bracket to crossmember subframe bracket (M8) .... 25 Nm (18 ft-lb)
- Stone guard/ rear stabilizer bracket to body (M10 bolt).................. 47 Nm (35 ft-lb)
- Stone guard/rear stabilizer bracket to crossmember subframe (M10 nut) ... 28 Nm (21 ft-lb)

SUSPENSION SUBFRAME

Front suspension subframe assembly

The forged light alloy crossmember subframe can be accessed from under the car by removing the stone guard in back of the fuel tank. Details of the subframe assembly are shown in the illustration.

NOTE —

- *Lubricate torsion bar adjustment screw with multipurpose MoS_2 grease.*
- *Be sure to reset front ride height as described in* **440 Tires, Wheels, Alignment**.

Tightening torques

- Steering rack to subframe (M10 x 20 mm) .... 47 Nm (35 ft-lb)
- Stone guard/ stabilizer rear bracket to body (M10 bolt)................... 47 Nm (35 ft-lb)
- Stone guard/ stabilizer rear bracket to crossmember subframe (M10 nut) ... 28 Nm (21 ft-lb)
- Subframe/control arm to body (M12 x 1.5)................. 90 Nm (66 ft-lb)

Crossmember subframe assembly

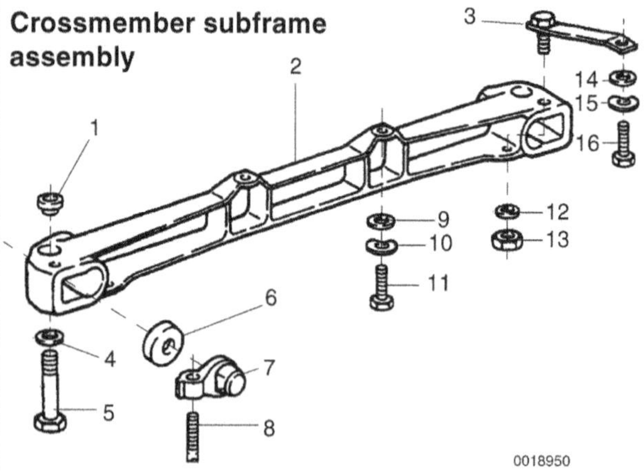

1. Adapter
2. Subframe
3. Stabilizer bar rear mounting bracket
4. Spring washer
5. Bolt (M12 x 1.5) tighten to 90 Nm (66 ft-lb)
6. Foam seal
7. Torsion bar retaining cap
8. Torsion bar adjusting screw (lubricate with MoS_2)
9. Washer
10. Lock washer
11. Steering rack mounting bolt (M10 x 20 mm) tighten to 47 Nm (35 ft-lb)
12. Washer
13. Nut (M10) tighten to 28 Nm (21 ft-lb)
14. Washer
15. Lock washer
16. Bolt (M10) tighten to 47 Nm (35 ft-lb)

420 Rear Suspension, Axle Shafts

TABLE

Special Tools

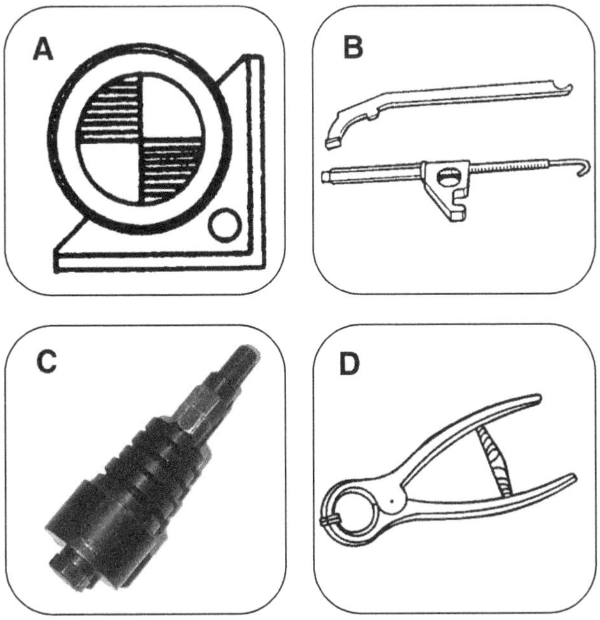

Special tool

- **A** 261/1 . . . . Protractor/level (Source: Baum Tools Unlimited)
- **B** P 289 a. . . . . Rear torsion bar preloading tool (Porsche ordering #000721 289 10)
- **C** B 90 . . . . Wheel bearing extractor/installer (Source: Baum Tools Unlimited)
- **D** VW 161 a . . . . Circlip pliers (Source: Baum Tools Unlimited)

GENERAL

This repair group covers replacement procedures for rear suspension components. For specifications, diagnosis and troubleshooting information see **400 Suspension, Brakes and Steering–General**. For information on setting ride height and alignment see **440 Wheels, Tires, Alignment**. For information on transmission output flange seal repairs, see **390 Differential**.

Special tools

Some procedures found in this repair group require the use of special tools.

Rear suspension

The Porsche 911 SC is equipped with independent rear suspension. Torsion bars are used at all four wheels. The rear end uses conventional shock absorbers.

The rear shock absorbers are conventional hydraulic or gas-pressure dampers. Rear torsion bars are mounted transversely inside the rear suspension subframe (torsion bar carrier). Torsion plates attach to the ends of torsion bars and the trailing arms. Each trailing arm has a wheel-bearing hub and carries a sealed wheel bearing.

A stabilizer bar is mounted to the rear trailing arms to reduce body roll.

NOTE —
The suspension subframe (torsion bar carrier) is welded into the body and is not a replaceable part under normal circumstances.

When doing any rear suspension maintenance or repair, observe the following cautions.

> ***CAUTION —***
> - *Do not install bolts and nuts coated with undercoating wax, as correct tightening torque cannot be assured. Always clean the threads with solvent before installation, or install new parts.*
> - *Do not attempt to weld or straighten the torsion plates, trailing arms or subframe. Replace damaged parts.*
> - *Do not reuse self-locking nuts or cotter pins. These parts are designed to be used only once.*

REAR SHOCK ABSORBERS

The rear shock absorbers are conventional dampers.

Rear shock absorber assembly

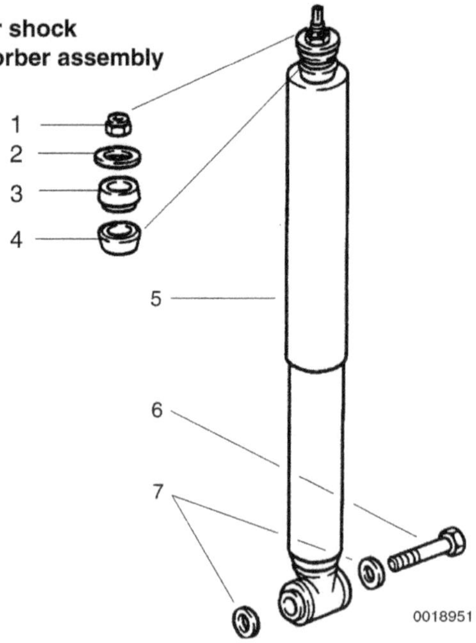

1. Lock nut (M10 x 1)
   - tighten to 25 Nm (18 ft-lb)
2. Washer
3. Rubber bumper
4. Rubber bumper
5. Shock absorber
6. Lower mounting bolt (M14 x 1.5)
   - tighten to 125 Nm (92 ft-lb)
7. Washer

Rear shock absorber, removing and installing

With weight of vehicle on wheels, remove lower shock bolt.

Working in rear of engine compartment, remove shock absorber top mounting nut (**arrow**).

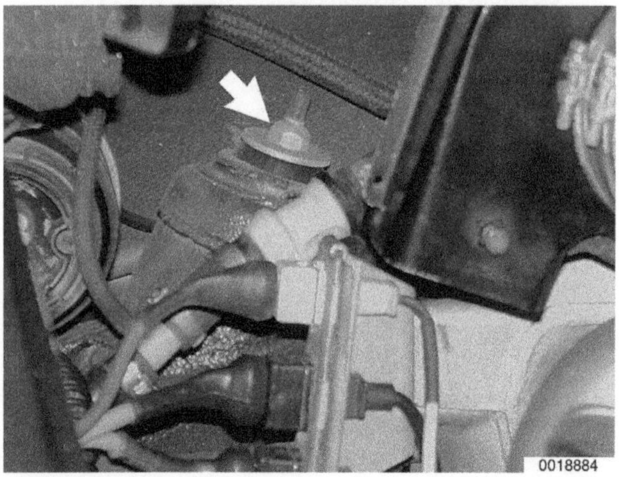

NOTE —
To access the right side top mounting nut, remove air filter cover and, if necessary, housing. See **030 Maintenance**.

- Remove shock absorber.
- Installation is reverse of removal.

Tightening torques
- Rear shock absorber to trailing arm . . . 125 Nm (92 ft-lb)
- Rear shock absorber to body (M10) . . . . 25 Nm (18 ft-lb)

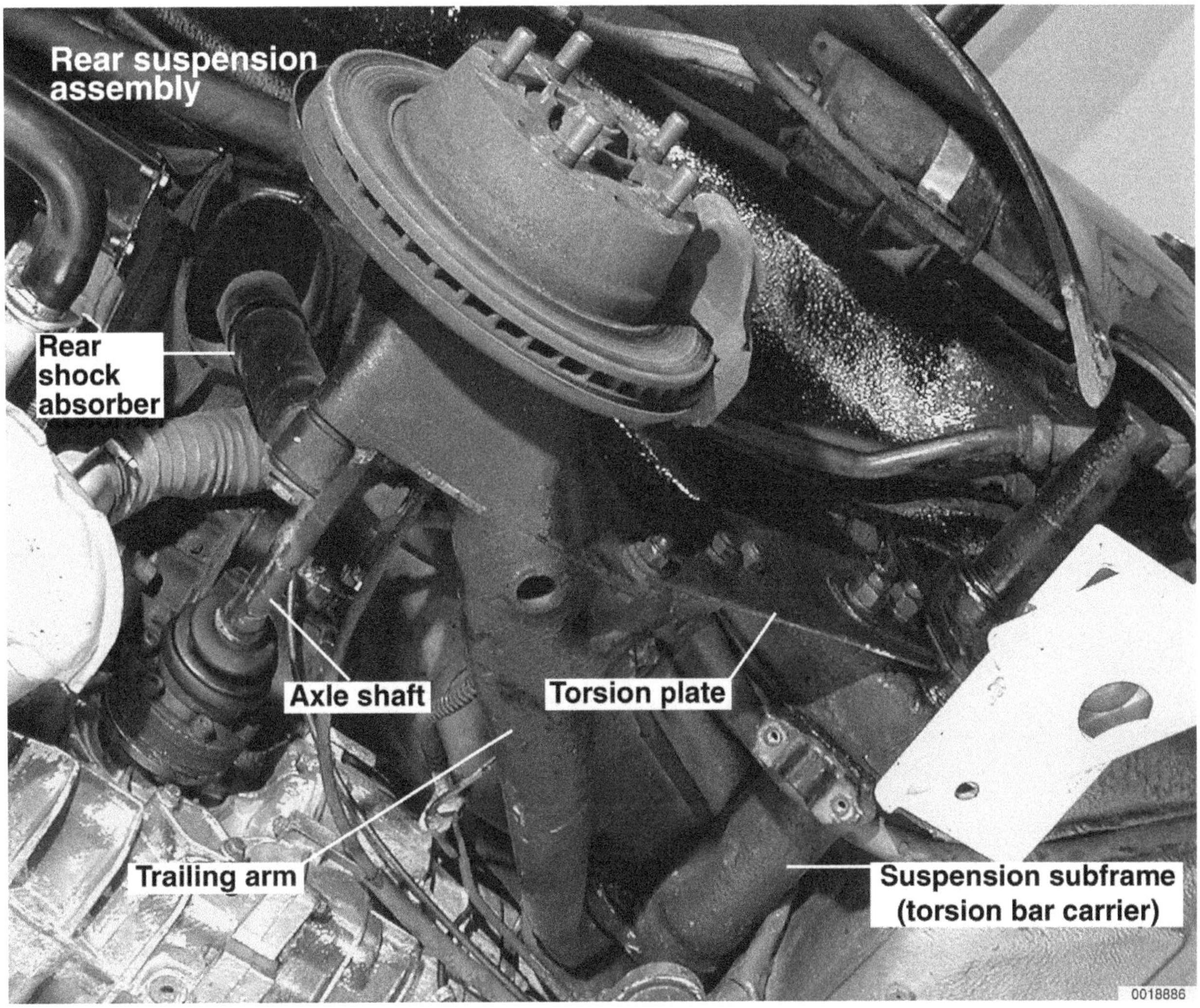

TRAILING ARMS, TORSION PLATES AND TORSION BARS

> ***WARNING —***
> *Replacement and/or removal of the trailing arm, torsion plate or torsion bar disturbs the rear wheel alignment settings. Replacement or adjustment of the torsion plate or torsion bar also affects vehicle ride height, which directly influences vehicle stability. If ride height is changed, it is strongly recommended that the vehicle balance/weight be checked at all four wheels using suitable wheel-load floor scales. See* **440 Wheels, Tires, Alignment** *for more information.*

Rear trailing arm, removing and installing

– With car on ground and parking brake on, loosen rear wheel nuts. Raise car and support securely.

> ***WARNING —***
> *Make sure the car is stable and well supported at all times. Use a professional automotive lift or jack stands designed for the purpose. A floor jack is not adequate support.*

Rear suspension assembly

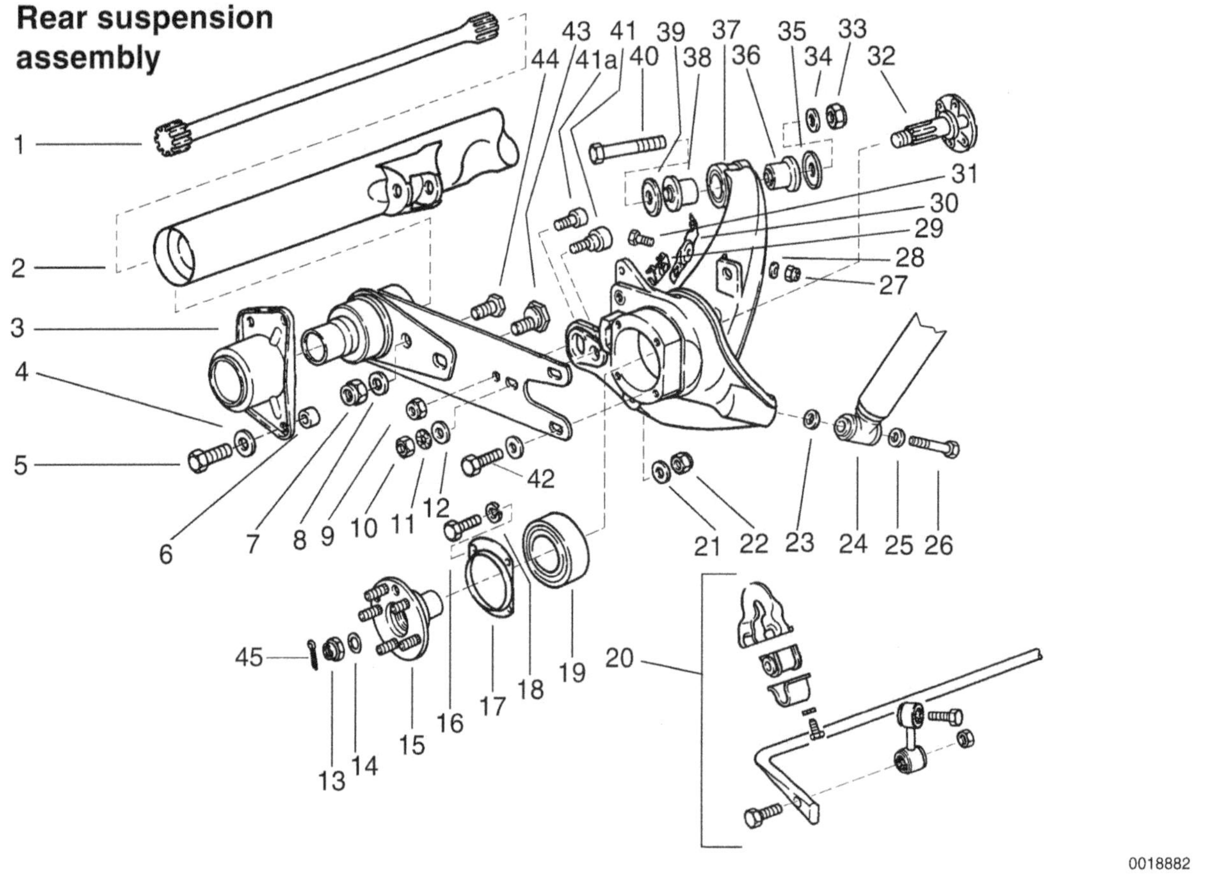

1. Torsion bar
2. Suspension subframe (torsion bar carrier)
3. Torsion bar cover
4. Washer (M10)
5. Bolt (M10)
   - tighten to 47 Nm (34 ft-lb)
6. Spacer
7. Nut (M16)
   - tighten to 245 Nm (177 ft-lb)
8. Washer (M16)
9. Nut (M12)
   - tighten to 85 Nm (61 ft-lb)
10. Nut (M12)
    - tighten to 120 Nm (87 ft-lb)
11. Lock washer (M12)
12. Flat washer (M12)
13. Stub axle nut, M20 castellated nut:
    - tighten to 300-320 Nm (217-231 ft-lb)
14. Washer
15. Wheel hub
16. Bolt (M10)
    - tighten to 47 Nm (34 ft-lb)
17. Wheel bearing cover
18. Washer (M10)
19. Wheel bearing
20. Stabilizer bar assembly
21. Washer (M10)
22. Nut (M12)
    - tighten to 120 Nm (87 ft-lb)
23. Washer (M14)
24. Shock absorber
25. Washer (M14)
26. Bolt (M14)
    - tighten to 125 Nm (92 ft-lb)
27. Nut
28. Washer
29. Clip
30. Clip
31. Bolt
32. Axle shaft
33. Nut (M14)
    - tighten to 100 Nm (72 ft-lb)
34. Washer (M14)
35. Plate
36. Bushing
37. Trailing arm
38. Bushing
39. Plate
40. Bolt (M14)
    - tighten to 100 Nm (72 ft-lb)
41. Eccentric (M12)
    - tighten to 85 Nm (61 ft-lb)

41a. Eccentric (M12)
   - tighten to 85 Nm (61 ft-lb)

42. Bolt (M12)
    - tighten to 125 Nm (90 ft-lb)
43. Eccentric (M16)
    - tighten to 245 Nm (177 ft-lb)
44. Bolt (M16)
    - tighten to 245 Nm (177 ft-lb)
45. Cotter pin

– Remove cotter pin from castellated stub axle nut.

– Temporarily reinstall wheel and lower car to ground. Apply parking brake and shift transmission into reverse.

– Loosen stub axle nut (32 mm socket wrench) through opening in center of wheel.

– Release parking brake and shift car into neutral.

– Raise rear of car and support it securely.

> ***WARNING —***
> - *Make sure the car is stable and well supported at all times. Use a professional automotive lift or jack stands designed for the purpose. A floor jack is not adequate support.*
> - *When working on torsion plates and torsion bars, the car should be no higher than 2 feet off the ground.*

– Remove rear wheel. Remove stub axle nut.

– Press brake pedal down about 1 in. and hold pedal in place with a pedal stop.

NOTE —
This will prevent brake fluid from dripping out of open brake lines while the work is being performed.

0018989

◄ Disconnect brake line from brake caliper. Plug hydraulic lines.

> ***CAUTION —***
> *When working with hydraulic brake lines and hoses be sure to counterhold fitting before loosening or tightening.*

– Remove brake caliper mounting bolts and remove caliper. See **470 Brakes–Hydraulic**.

Remove brake rotor retaining screws (**arrows**) and pull off brake rotor.

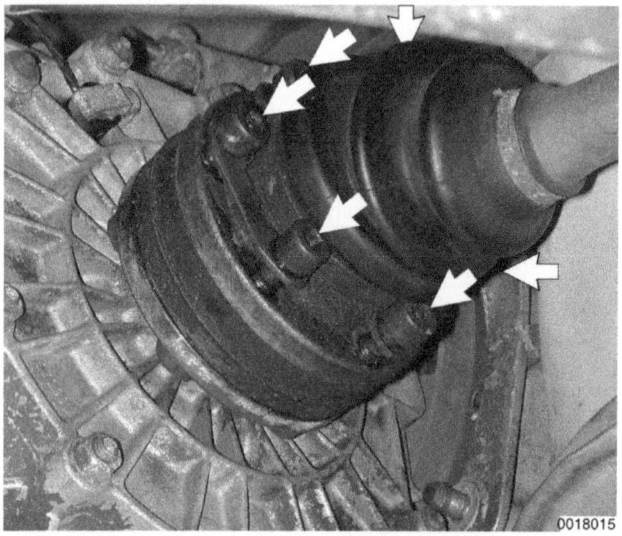

Unbolt CV joint fasteners (**arrows**) from transmission output flange.

- Lower axle shaft. Drive stub axle out of trailing arm.

Working at parking brake shoes pivot, remove cotter pin and castellated nut (**arrow**).

- Pull parking brake cable out of rear of trailing arm.

Disconnect stabilizer bar link bolt (**arrow**) at trailing arm.

Drive out rear wheel hub using Porsche special tool P 297 a or equivalent drift.

Lower car close to ground level. Use floor jack under trailing arm to relieve suspension load on shock absorber. (Brake backing plate shown removed for clarity.)

- Remove rear shock absorber lower mounting bolt from trailing arm (**arrow**).

NOTE —

Porsche special tool P 289 a may be used to hold rear suspension in a compressed position while working on the rear shock absorber bolt.

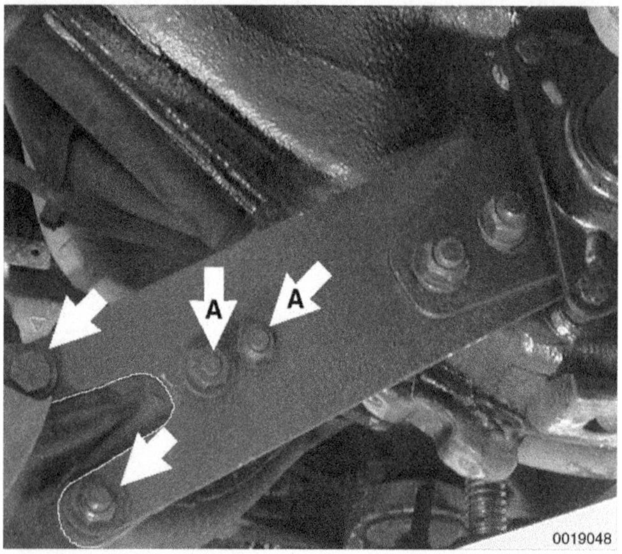

Mark torsion plate position to trailing arm by scribing outline around torsion plate. Remove torsion plate fasteners (**arrows**) and detach trailing arm from torsion plate.

NOTE —

*The torsion plate fastener holes are elongated. Mark relative positions of torsion plate, trailing arm and eccentric bolts (**A**) in order to reassemble suspension close to original alignment setting.*

– Raise car.

Working underneath car, remove trailing arm mounting bolt (**arrow**) from rear suspension subframe (torsion bar carrier). Remove trailing arm.

– Installation is reverse of removal.

- Bleed brakes as described in **470 Brakes–Hydraulic**.
- Reset rear wheel alignment as described in **440 Wheels, Tires, Alignment**.

Tightening torques

- Brake caliper to trailing arm (M12) . . . . . 60 Nm (43 ft-lb)
- Brake hose to brake line . . . . 12 Nm (9 ft-lb)
- Brake rotor to wheel hub (M6) . . . . 5 Nm (44 in-lb)
- CV joint to transmission flange
  M8 bolt . . . . 42 Nm (30 ft-lb)
  M10 bolt . . . . 83 Nm (60 ft-lb)
- Rear shock absorber
  to trailing arm (M14) 125 Nm (92 ft-lb)
- Stabilizer link to trailing arm (M12) . . . . . 85 Nm (61 ft-lb)
- Stub axle shaft to hub
  M20 castellated nut. . . . . . 300 - 320 Nm (217 - 231 ft-lb)
- Trailing arm to suspension subframe
  (torsion bar carrier)(M14) . . . . 100 Nm (72 ft-lb)
- Torsion plate to trailing arm (eccentric) . 85 Nm (61 ft-lb)
- Torsion plate to trailing arm (M12) . . . . 120 Nm (87 ft-lb)
- Wheel to wheel hub . . . . 130 Nm (96 ft-lb)

Rear torsion bar, removing and installing

WARNING —

*If torsion bars are replaced or reset, it is strongly recommended that the vehicle balance/weight be checked at all four wheels using suitable wheel-load floor scales. See **440 Wheels, Tires, Alignment** for more information.*

– Loosen rear wheel lug nuts. Raise rear of car and support it securely.

> ***WARNING —***
> - *Make sure the car is stable and well supported at all times. Use a professional automotive lift or jack stands designed for the purpose. A floor jack is not adequate support.*
> - *When working on torsion plates and torsion bars, the car should be no higher than 2 feet off the ground.*

– Remove rear wheel.

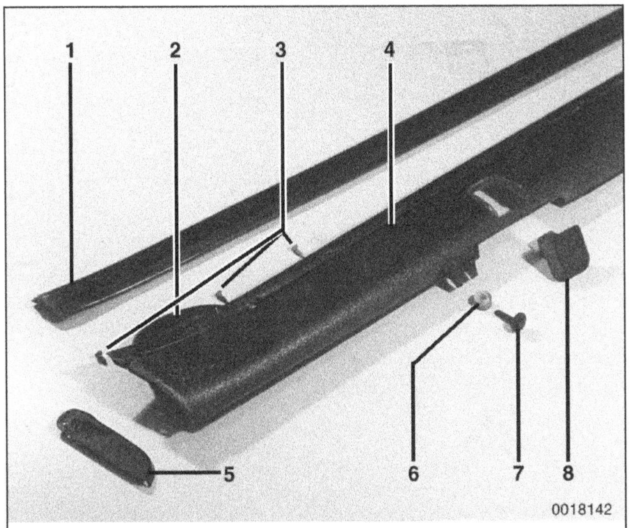

1. Rubber protector
2. Torsion bar access plug
3. Rocker panel mounting screws
4. Rocker panel
5. Rocker panel end cover
6. Plastic spacer
7. Rocker panel bottom mounting screw
8. Jack hole cover

◄ Remove rocker panel.

- Remove rear rocker panel end cap screws and remove cap.
- Working from one end of rocker panel, peel off rubber rocker panel protector.
- Remove rocker panel torsion bar access plug from fender. Plug is retained to body from behind using nut and bar clamp.

◄ Lower car close to ground level. Use floor jack under trailing arm to relieve suspension load on shock absorber. (Brake backing plate shown removed for clarity.)

- Remove rear shock absorber lower mounting bolt from trailing arm (**arrow**).

NOTE —
Porsche special tool P 289 a may be used to hold rear suspension in a compressed position while working on the rear shock absorber bolt.

– Lower and remove floor jack.

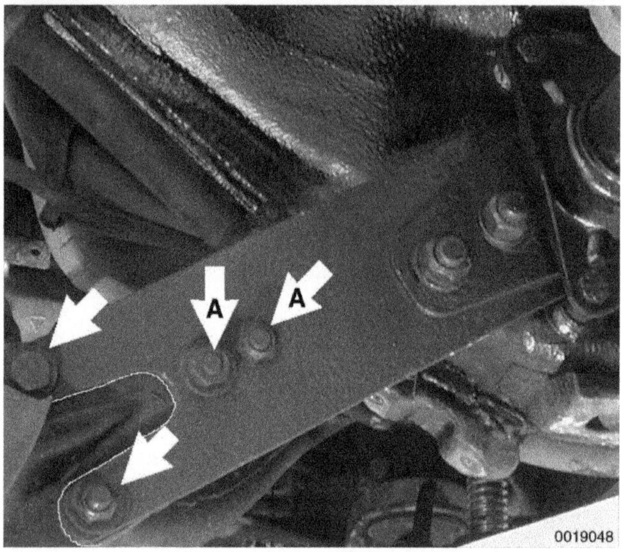

◀ Mark torsion plate position to trailing arm by scribing outline around torsion plate. Remove torsion plate fasteners (**arrows**) and detach trailing arm from torsion plate.

NOTE —

*The torsion plate fastener holes are elongated. Mark relative positions of torsion plate, trailing arm and eccentric bolts (**A**) in order to reassemble suspension close to original alignment setting.*

◀ With car still close to ground, support torsion plate from below using floor jack. Raise torsion plate just until plate comes off lower stop (**arrow**) on torsion bar cover.

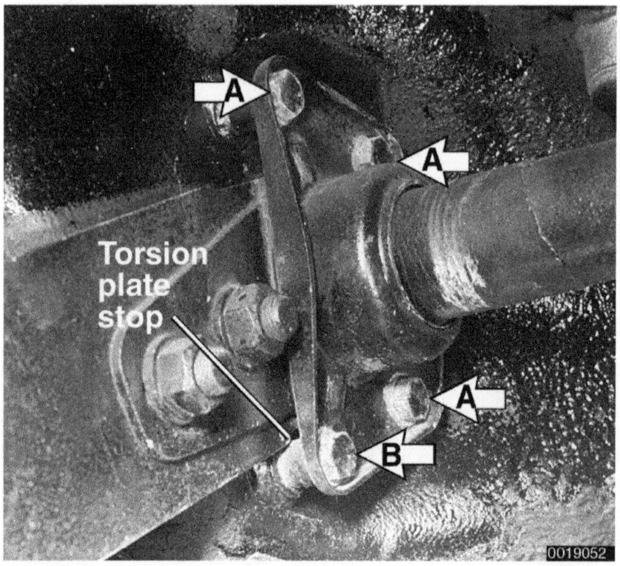

◀ Loosen 3 torsion bar cover bolts (**A**) and remove stop bolt and sleeve (**B**).

– Slowly lower floor jack to relieve torsion bar tension on torsion plate.

0019053

< Raise vehicle on lift. Using spirit level in special protractor tool 261/1 or equivalent, measure inclination of vehicle body at door sill (measurement **A**).

0019057

< Measure inclination of torsion plate (measurement **B**).

– Combine measurements **A** and **B** (add or subtract body inclination measurement) to find value of torsion plate inclination in relation to vehicle body. Record measurement.

Example	
Body inclination (**A**)	–2°
Measured torsion plate inclination (**B**)	34°
Actual torsion plate inclination	**32°**

– Remove torsion bar cover mounting bolts and remove torsion bar cover by prying gently off body.

CAUTION—
When backing off the torsion plate, check that the plate does not hang up on the body mounting boss.

– Remove torsion bar.

NOTE—
Do not interchange rear torsion bars. Bars are marked on the ends "R" and "L" for right and left.

– Prior to installation, carefully clean and check all parts for signs of damage. Replace as needed.

- Check torsion plate for corrosion, chipped paint, scrapes.
- Check torsion bar for cracks.

– Coat torsion bar with multipurpose grease and install.

NOTE—
The inner and outer ends of rear torsion bars have a different number of teeth. Be sure to install the bar with the "L" or the "R" on the end facing out.

- Coat rubber seals on torsion bar cover with a rubber lubricant such as Contifix® or Capella®.

◄ Temporarily install 3 bolts (**arrows**) without spacers to hold cover in place. Leave out lower stop bolt.

0019053

◄ Using special protractor tool 261/1 or equivalent, measure inclination of vehicle body at door sill (measurement **A**).

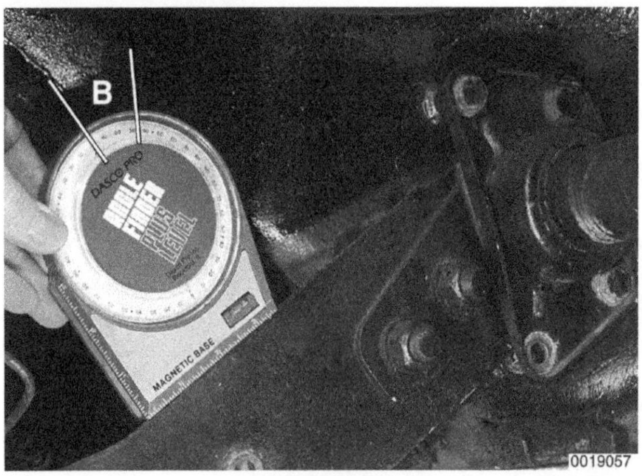

◄ Measure inclination of torsion plate (measurement **B**).

- Combine measurements **A** and **B** (add or subtract body inclination measurement) to find value of torsion plate inclination in relation to vehicle body inclination.

Example	
Body inclination (**A**)	–2°
Measured torsion plate inclination (**B**)	34°
Actual torsion plate inclination	**32°**

Table a. Rear torsion bar specifications

Model	Torsion bar diameter	Torsion bar splines inside/ outside	Torsion plate inclination (see text)
1978 - 1983 SC			
With gas struts	24.1 mm	40/44	40°
Without gas struts	24.1 mm	40/44	39°
Euro-specs	24.1 mm	40/44	34°

– If torsion plate inclination is incorrect:

- Remove torsion bar cover fasteners and slide cover with torsion plate off body.
- Reposition torsion bar in torsion plate and reinstall cover using 3 fasteners.
- Recheck inclination and readjust as necessary.

NOTE—
It may be necessary to try several different positions for inner and outer teeth on the torsion bar to achieve the correct setting.

0019050

Once torsion bar inclination is correct, use floor jack to raise torsion plate just enough to install lower stop bolt with spacer (**arrow**).

- Tighten all four torsion bar cover mounting bolts.

Tightening torque
- Torsion bar cover to body (M10) . . . . 47 Nm (34 ft-lb)

– Remainder of installation is reverse of removal.

- Reset rear wheel alignment and check ride height as described in **440 Wheels, Tires, Alignment**.

WARNING—
Replacement and/or removal of the torsion plate or torsion bar disturbs the rear wheel alignment settings and also affects vehicle ride height, which directly influences vehicle stability. If ride height is changed, it is strongly recommended that the vehicle balance/weight be checked at all four wheels using suitable wheel-load floor scales. See **440 Wheels, Tires, Alignment** *for more information.*

Tightening torques
- Rear shock to trailing arm (M14) . . . . . 125 Nm (92 ft-lb)
- Torsion plate to trailing arm (eccentric) . 85 Nm (61 ft-lb)
- Torsion plate to trailing arm (M12 bolt) . 120 Nm (87 ft-lb)
- Wheel to wheel hub . . . . 130 Nm (96 ft-lb)

Rear torsion bar, setting

WARNING—
If torsion bars are replaced or reset, it is strongly recommended that the vehicle balance/weight be checked at all four wheels using suitable wheel-load floor scales. See **440 Wheels, Tires, Alignment** *for more information.*

The initial position of each rear torsion bar is set by measuring torsion plate inclination in its unloaded position.

0019043

With car close to ground level, use floor jack under trailing arm to relieve suspension load on shock absorber. (Brake backing plate shown removed for clarity.)

- Remove rear shock absorber lower mounting bolt from trailing arm (**arrow**).

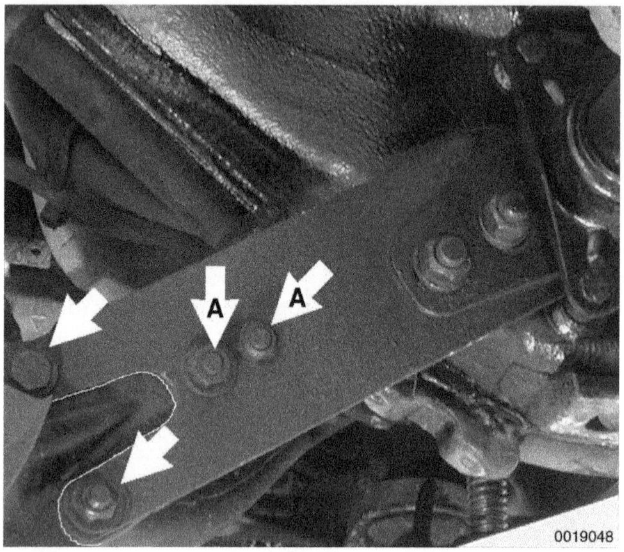

NOTE —

Porsche special tool P 289 a may be used to hold rear suspension in a compressed position while working on the rear shock absorber bolt.

– Lower and remove floor jack.

< Mark torsion plate position to trailing arm by scribing outline around torsion plate. Remove torsion plate fasteners (**arrows**) and detach trailing arm from torsion plate.

NOTE —

*The torsion plate fastener holes are elongated. Mark relative positions of torsion plate, trailing arm and eccentric bolts (**A**) in order to reassemble suspension close to original alignment setting.*

< With car still close to ground, support torsion plate from below using floor jack. Raise torsion plate just until plate comes off lower stop (**arrow**) on torsion bar cover.

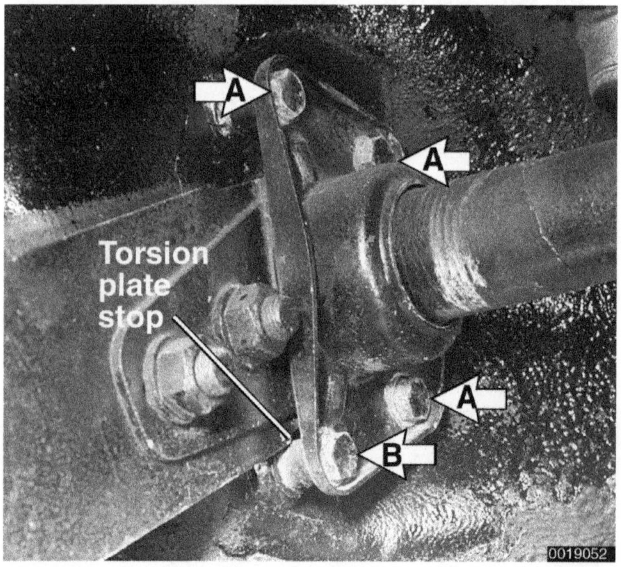

< Loosen 3 torsion bar cover bolts (**A**) and remove stop bolt and sleeve (**B**).

– Slowly lower and remove floor jack to unload torsion bar tension.

Using spirit level in special protractor tool 261/1 or equivalent, measure inclination of vehicle body at door sill (measurement **A**).

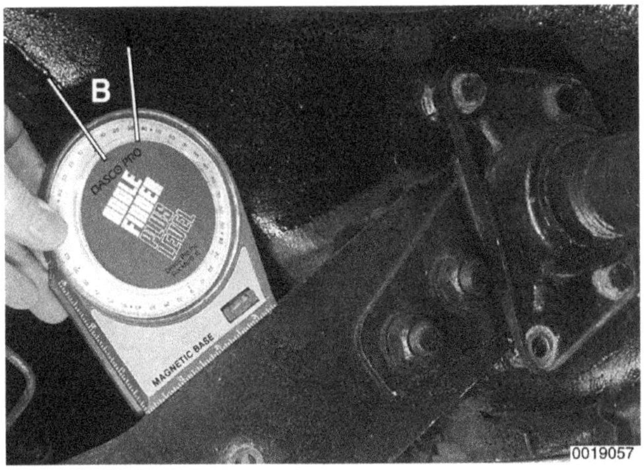

Measure inclination of torsion plate (measurement **B**).

– Combine measurements **A** and **B** (add or subtract body inclination measurement) to find value of torsion plate inclination in relation to vehicle body. Record measurement.

Example	
Body inclination (**A**)	–2°
Measured torsion plate inclination (**B**)	34°
Actual torsion plate inclination	**32°**

NOTE —
There are unequal numbers of splined teeth on the inner and outer ends of each torsion bar. This allows for small incremental changes to be made to the torsion plate inclination.

Torsion bar specifications

Model	Torsion bar diameter	Torsion bar splines inside/ outside	Torsion plate inclination (see text)
1978 - 1983 SC			
With gas struts	24.1 mm	40/44	40°
Without gas struts	24.1 mm	40/44	39°
Euro-specs	24.1 mm	40/44	34°

– If torsion plate inclination is incorrect:

- Remove torsion bar cover fasteners and slide cover with torsion plate off body.
- Reposition torsion bar in torsion plate and reinstall cover using 3 fasteners.
- Recheck inclination and readjust as necessary.

NOTE —
It may be necessary to try several different positions for inner and outer teeth on the torsion bar to achieve the correct setting.

- Once torsion bar inclination is correct, use floor jack to raise trailing arm/torsion plate just enough to install lower stop bolt with spacer.
  - Tighten all four torsion bar cover mounting bolts.

Tightening torque
- Torsion bar cover to body (M10) . . . . 47 Nm (34 ft-lb)

- Remainder of installation is reverse of removal.
  - Reset rear wheel alignment and check ride height as described in **440 Wheels, Tires, Alignment**.

WARNING —
Replacement and/or removal of the torsion plate or torsion bar disturbs the rear wheel alignment settings and also affects vehicle ride height, which directly influences vehicle stability. If ride height is changed, it is strongly recommended that the vehicle balance/weight be checked at all four wheels using suitable wheel-load floor scales. See **440 Wheels, Tires, Alignment** *for more information.*

Tightening torques
- Rear shock to trailing arm (M14) . . . . . 125 Nm (92 ft-lb)
- Torsion plate to trailing arm (eccentric) . 85 Nm (61 ft-lb)
- Torsion plate to trailing arm (M12 bolt) . 120 Nm (87 ft-lb)
- Wheel to wheel hub . . . . 130 Nm (96 ft-lb)

WHEEL BEARINGS

NOTE —
Rear wheel bearings are serviced by replacement only, as they are usually damaged when removed.

The rear wheel bearings are permanently sealed and lubricated and do not require any routine maintenance. Removing and installing the bearing on the car requires a special puller. If the necessary tools are not available, this work is best left to an authorized Porsche dealer or other qualified repair shop.

Check the wheel bearing with the wheel off the ground and the car properly supported. Spin the wheel and listen for any roughness. Alternately, apply pressure to either side of the wheel while checking for play. If any faults are found and no suspension components are found to be at fault, the wheel bearing should be replaced.

Rear wheel bearing, removing and installing

NOTE —
Wheel bearing replacement can be performed by removing the rear trailing arm as described earlier and pressing the bearing out on the bench. The following procedure describes utilizing a special bearing removal and installation tool that eliminates the need to remove the trailing arm from the car.

– With car on ground and parking brake on, loosen rear wheel nuts. Raise car and support securely.

> **WARNING —**
> *Make sure the car is stable and well supported at all times. Use a professional automotive lift or jack stands designed for the purpose. A floor jack is not adequate support.*

– Remove cotter pin from castellated stub axle nut.

– Temporarily reinstall wheel and lower car to ground. Apply parking brake and shift transmission into reverse.

– Loosen stub axle nut (32 mm socket wrench) through opening in center of wheel.

– Release parking brake and shift car into neutral.

– Raise rear of car and support securely.

> **WARNING —**
> *Make sure the car is stable and well supported at all times. Use a professional automotive lift or jack stands designed for the purpose. A floor jack is not adequate support.*

– Remove rear wheel. Remove stub axle nut.

– Press brake pedal down about 1 in. and hold pedal in place with a pedal stop.

NOTE —
This will prevent brake fluid from dripping out of open brake lines while the work is being performed.

0018989

< Disconnect brake line from brake caliper. Plug hydraulic lines.

> **CAUTION —**
> *When working with hydraulic brake lines and hoses be sure to counterhold fitting before loosening or tightening.*

– Remove brake caliper mounting bolts and remove caliper. See **470 Brakes–Hydraulic**.

Remove brake rotor retaining screws (**arrows**) and pull off brake rotor.

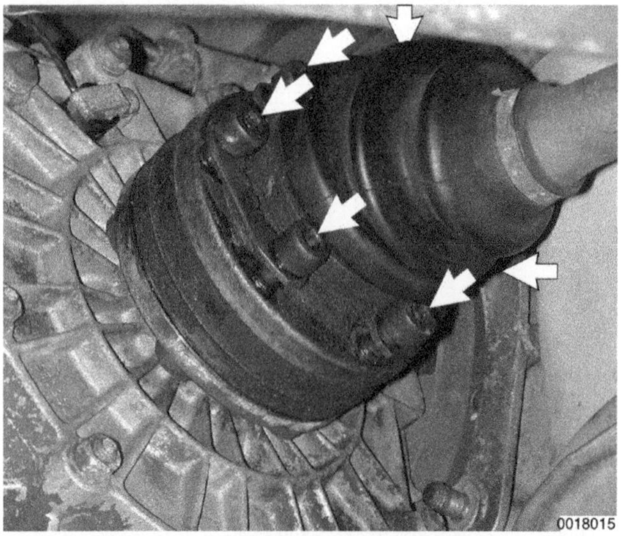

Unbolt CV joint fasteners (**arrows**) from transmission output flange.

- Lower axle shaft. Drive stub axle out of trailing arm.

Working at parking brake shoes pivot, remove cotter pin and castellated nut (**arrow**).

- Pull parking brake cable out of rear of trailing arm.

Drive out rear wheel hub using Porsche special tool P 297 a or equivalent.

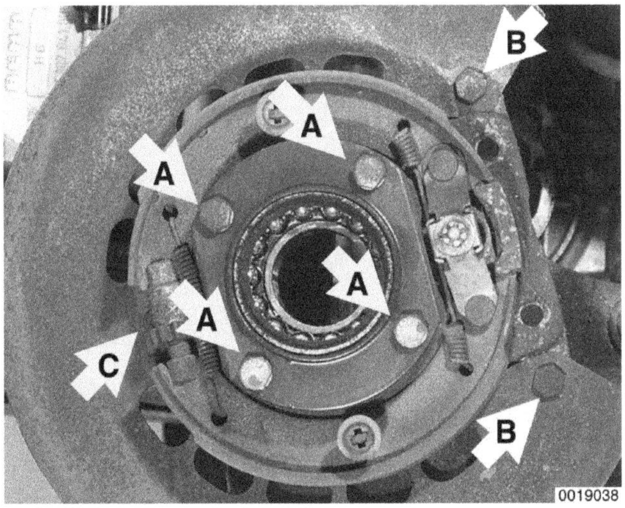

Remove wheel bearing cover plate bolts (**A**). Remove plate with parking brake shoe assembly.

- Remove brake rotor backing plate bolts (**B**, **C**). Remove backing plate.

NOTE —

*Rear backing plate bolt (**C**) is mounted on the backside of the backing plate with threads pointing outward.*

Lower car close to ground level. Use floor jack under trailing arm to relieve suspension load on shock absorber.

- Remove rear shock absorber lower mounting bolt from trailing arm (**arrow**).

NOTE —

Porsche special tool P 289 a may be used to hold rear suspension in a compressed position while working on the rear shock absorber bolt.

Use wheel bearing tool to remove old bearing and install new bearing following tool manufacturer's instructions.

– Remainder of installation is reverse of removal.

- Bleed brakes as described in **470 Brakes–Hydraulic**.

Tightening torques

- Brake caliper to trailing arm (M12) . . . . . 60 Nm (43 ft-lb)
- Brake hose to brake line . . . . 12 Nm (9 ft-lb)
- Brake rotor to wheel hub (M6) . . . . 5 Nm (44 in-lb)
- CV joint to transmission flange
  M8 bolt . . . . 42 Nm (30 ft-lb)
  M10 bolt . . . . 83 Nm (60 ft-lb)
- Rear shock absorber
  to trailing arm (M14) 125 Nm (92 ft-lb)
- Stub axle shaft to hub
  M20 castellated nut. . . . . . 300 - 320 Nm (217 - 231 ft-lb)
- Wheel bearing cover plate
  to bearing housing (M8) . . . . 25 Nm (18 ft-lb)
- Wheel to wheel hub . . . . 130 Nm (96 ft-lb)

STABILIZER BAR

The rear stabilizer bar attachment points at the body have been upgraded. The new stronger braces must be welded to the underbody. This work is best left to an authorized Porsche dealer or other qualified repair shop.

Rear stabilizer bar

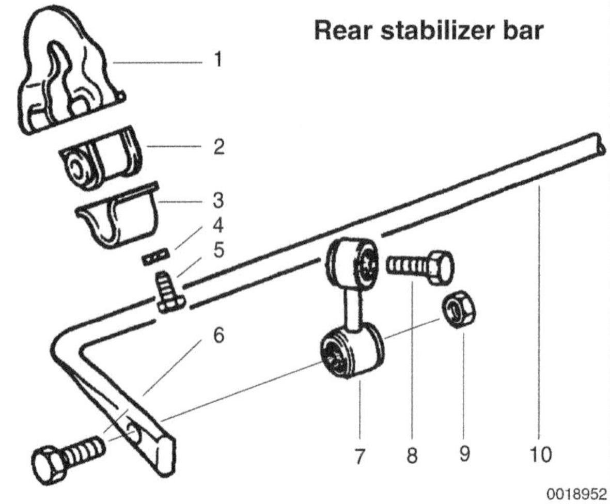

1. Attaching bracket (welded to body)
2. Rubber mount
3. Mounting bracket
4. Washer (M8)
5. Bolt (M8) - tighten to 25 Nm (18 ft-lb)
6. Bolt (M12 x 1.5 x 60), - tighten to 85 Nm (61 ft-lb)
7. Stabilizer bar link
8. Bolt (M12 x 1.5 x 70), - tighten to 85 Nm (61 ft-lb)
9. Nut (M12)
10. Stabilizer bar

Axle shaft assembly

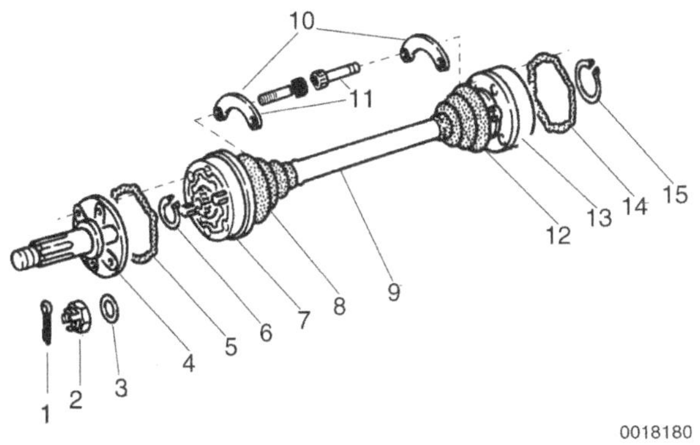

1. Cotter pin
2. Stub axle nut (M20),
   - tighten to 300-320 Nm (217-231 ft-lb)
3. Washer
4. Stub axle
5. Gasket
6. Circlip
7. Outer CV joint
8. Outer CV boot
9. Axle shaft
10. Plate
11. CV joint bolt—M8:
    - tighten to 42 Nm (30 ft-lb)
    CV joint bolt—M10:
    - tighten to 83 Nm (60 ft-lb)
12. Inner CV boot
13. Inner CV joint
14. 15. Gasket
15. 16. Circlip

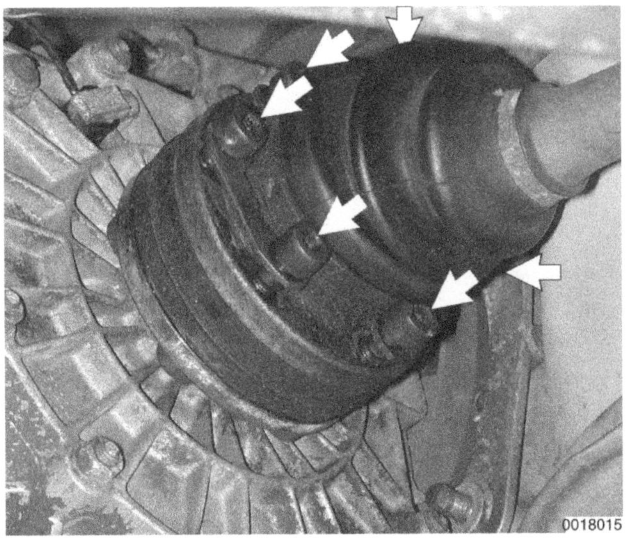

AXLE SHAFTS

In a car with independent rear suspension the drive axles must be able to move with the suspension while delivering power to the wheels. To accomplish this, a constant velocity (CV) joint is located at each end of each axle shaft.

A CV joint is similar in function to the more familiar universal joint. The CV joint allows for the smooth flow of power to the drive wheels even though the final drive, axles and wheels are not lined up with each other. The advantage of the more complicated CV joint is that it operates more smoothly and can accept more radical drive axle angles than a universal joint.

CV joint boots should be inspected periodically to see that they are in good condition. A damaged boot will let in dirt, which quickly destroys the joint.

Axle shaft assembly, removing and installing

– Raise rear of car and support it securely.

> ***WARNING —***
> *Make sure the car is stable and well supported at all times. Use a professional automotive lift or jack stands designed for the purpose. A floor jack is not adequate support.*

Unbolt inner CV joint fasteners (**arrows**) from transmission output flange.

- Push inner CV joint away from output flange.
- Unbolt outer CV joint from stub axle flange.
- Lower axle shaft.

– Installation is reverse of removal.

Tightening torque

- CV joint to drive or stub axle flange:
  M8 bolt . . . . 42 Nm (30 ft-lb)
  M10 bolt . . . . 83 Nm (60 ft-lb)

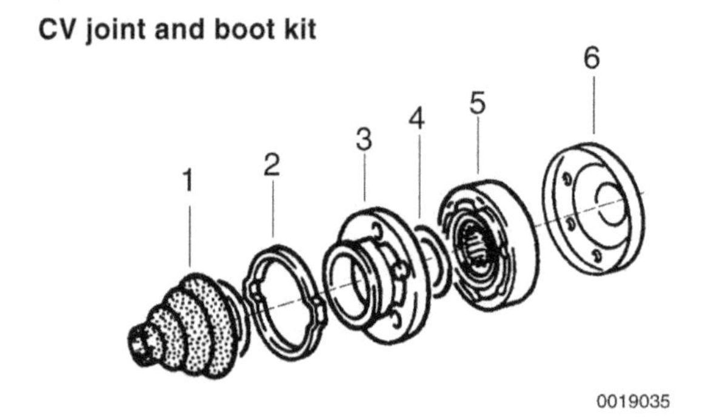

1. Small clamp
2. CV joint boot
3. Large clamp
4. Boot flange
5. Concave washer (install with concave facing CV joint
6. CV joint
7. Inner dust cover

CV joint and boot, removing and installing

NOTE —
On all SC models both the inner and outer CV joints are replaceable separately.

– Remove axle shaft assembly as described earlier and place in shop vice with soft jaws.

< Cut off small and large CV boot clamps.

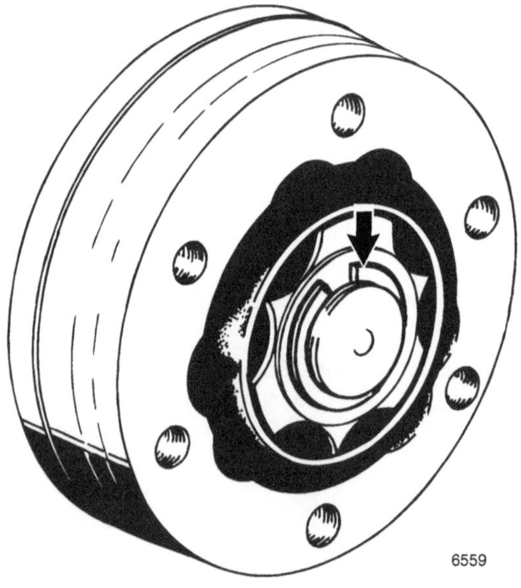

< Use special tool VW 161 a or equivalent to spread retaining circlip (**arrow**) and remove from shaft. Discard circlip.

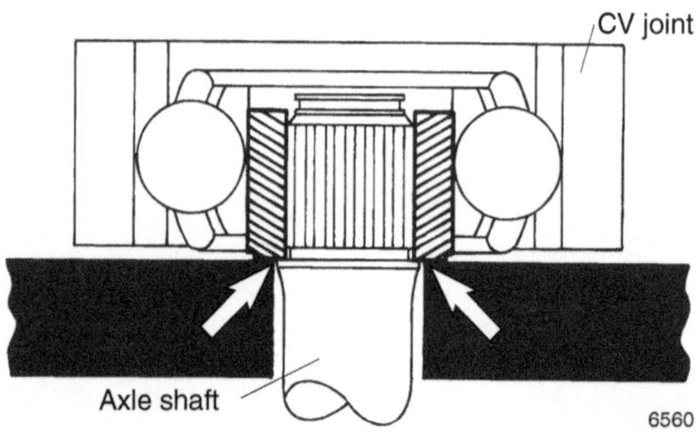

< Support CV joint on inner hub (**arrows**) while pressing off axle shaft.

– Slide boot off shaft.

– Clean all old lubricant off shaft and joint splines.

NOTE —
Inspect the axle shaft and splines for wear or damage. If the shaft is damaged, the shaft and the outer CV joint will have to be replaced as an assembly.

– Inspect CV joint carefully, as described below. If reusing CV joint, clean off old grease and repack with fresh.

– Slide on new boot and secure with new clamps.

NOTE —
A new CV joint boot kit includes new boot, clamps, lubricant and inner CV joint circlip.

– Press inner CV joint on shaft while supporting inner hub of joint. Secure with new circlip.

– Pack CV joint and boot with Optimoly® grease or equivalent.

CV joint lubricant capacity
• Per joint (inner or outer) . . . . 80 gram (2.8 oz.)

– Reinstall in car as described earlier.

CV joint, inspecting

The components of each CV joint are precisely matched during manufacture and cannot be serviced or replaced individually. To inspect a CV joint, clean away the grease and look for galling, pitting and other signs of wear or physical damage. Any of these is cause for replacement. Discoloration due to overheating indicates lack of lubrication. A discolored joint that is otherwise in good condition may be reinstalled if thoroughly cleaned and repacked with the proper amount of new grease.

NOTE —
Polished interior surfaces or visible ball tracks alone are not necessarily cause for replacement.

The balls and grooves allow the hub to move, but the parts should fit snugly and move only with some effort. A joint with obvious free play between inner hub, balls and outer cage should be replaced.

Disassembly of the CV joints is not normally recommended. If the joint must be disassembled for cleaning, mark the relative positions of the inner hub, the ball cage, and the outer housing. Remove the balls one at a time, and keep track of them so that each can be reinstalled in its original position in the hub and cage.

CAUTION —
- *The cage, housing, and balls are precisely matched. When disassembling more than one joint, do not intermix components.*
- *The CV joint should go together firmly but smoothly. Heavy force should not be required. If in doubt, start over and recheck the alignment. A joint that is forced together may lock and not come apart again.*

440 Wheels, Tires, Alignment

TABLES

440

GENERAL

This repair group covers wheels, tires and alignment information. Troubleshooting information is in **400 Suspension, Brakes and Steering**. For detailed information relating to suspension or steering components, refer to the following sections:

- **401 Front Suspension**
- **420 Rear Suspension, Axle Shafts**
- **480 Steering**

Special Tools

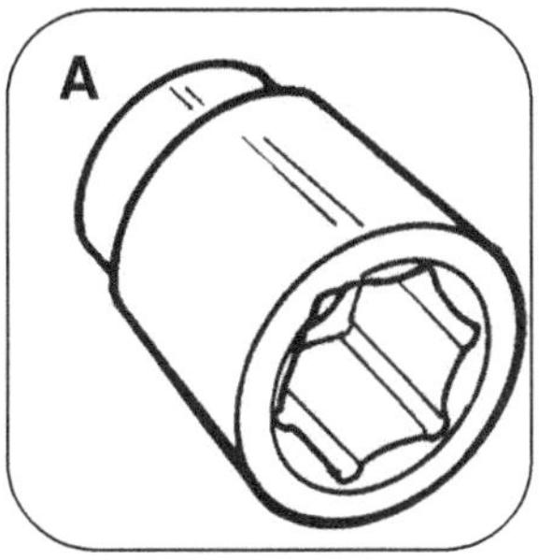

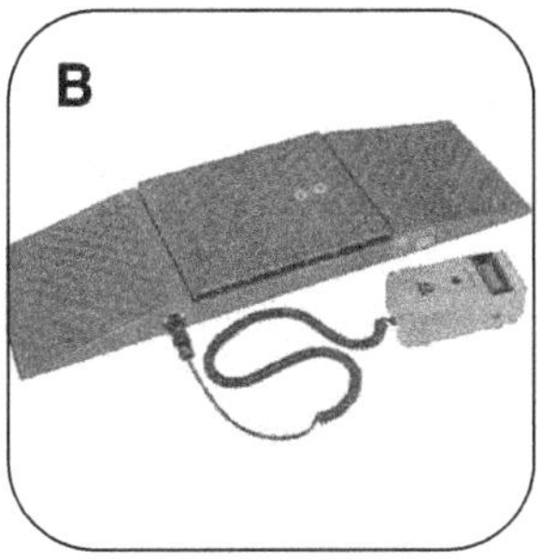

Special tool

A P 300 . . . . . 19 mm wheel bolt socket
(Porsche ordering #000 721 300 00)
B Maha RWA 2 U . . . . . Wheel load scale

Special tools and equipment

Much of the work described in this section requires special tools and experience. If you lack the skills, the tools, or a suitable workplace for tire or alignment work, we suggest you leave such repairs to an authorized Porsche dealer or other qualified repair shop.

Wheels, tires and alignment

Tires greatly influence the handling, stability and cornering performance of a Porsche. In order to gain the best possible performance from a Porsche all four wheels must be aligned correctly. If the torsion bars or torsion bar components have been replaced, the vehicle's corner weights (weight at each wheel) must be checked before an alignment is performed.

For stability and control, wheels and tires must be in good condition and properly aligned. Tire inflation pressures, tire wear and wheel alignment will all influence how the car feels and responds on the road.

Precise wheel alignment can only be accomplished when the tires, the steering, and the suspension are in good condition. Uneven tire wear, different size tires, or worn suspension and steering parts all affect wheel alignment. Reputable wheel alignment technicians will always inspect the suspension and steering for worn parts before an alignment, and will recommend that any necessary repairs be made first.

Correct tire pressure information can be found on an information label (**arrow**) on the rear engine compartment apron and in the owner's manual.

Tire and wheel specifications are given in **Table a**.

Wheels and tires, general information

- Check for tire directional markings (arrow on sidewall) before mounting a tire.
- Always replace tire valves when mounting new tires. Replace stick-on valve stem supports if missing.
- Use only rubber lubricant on the tire beads when mounting tires. Do not use silicone.
- Clean lubricant from bead area before inflating the tire to reduce the possibility of the tire slipping on the wheel.
- Fast acceleration and hard braking should be avoided for the first 150 miles after mounting a new tire.
- Inflate tire to 4 bar (56 psi) to seat tire bead correctly on the wheel. When inflating, tire bead should seat at no greater pressure than 3.3 bar (48 psi) to avoid damage to the tire bead.
- Wheel locks should be placed on the stud closest to the tire valve.
- Lubricate wheel lugs and shoulders with Optimoly® (Porsche part number 000 043 020 00).
- Tighten wheel lug bolts to 130 Nm (96 ft-lbs).
- Check tire pressures only when tires are cold. The tire pressure increases when hot.

Table a. Wheel and tire sizes

Axle	Standard		Optional	
	Wheel size	Tire size	Whee l size	Tire size
Front	6 J x 15	185/70 VR 15	6 J x 16	205/55 VR 16
Rear	7 J x 15	215/60 VR 15	7 J x 16	225/50 VR 16

WHEELS

The Porsche wheels supplied as original equipment are designed for tubeless, radial-ply tires of a specific size. Replacement tires of non-standard size or construction should be installed only if the tire manufacturer specifies them for your make and model.

The wheels are sized according to diameter and width. When fitting a non-standard wheel with different dimensions, there is danger of wheels or tires interfering with the body or with suspension parts. Even if there is no interference, incorrectly sized wheels may place additional loading on wheel bearings or adversely affect steering geometry. For best results, rely on the advice of your Porsche dealer or reputable wheel and tire professionals.

Wheel lug bolts should always be tightened using Porsche special tool P 300 (19 mm socket) and a torque wrench. Too little torque is obviously dangerous as the wheel may come loose, but excessive torque is also a problem. Over-tight wheel lug bolts can warp the hub, causing vibration and other problems affecting stability and braking.

Tightening torque

- Wheel to wheel hub . . . . 130 Nm (96 ft-lb)

0019141

Damaged wheels may also be a source of vibration problems. Alloy wheels are constructed of a soft aluminum-magnesium alloy and can be easily damaged.

Check for a bent wheel by measuring lateral or radial runout at the part of the wheel where the tire bead meets the wheel.

Alloy wheel runout specification

- Maximum allowable runout . . . . 1.0 mm (0.04 in.)

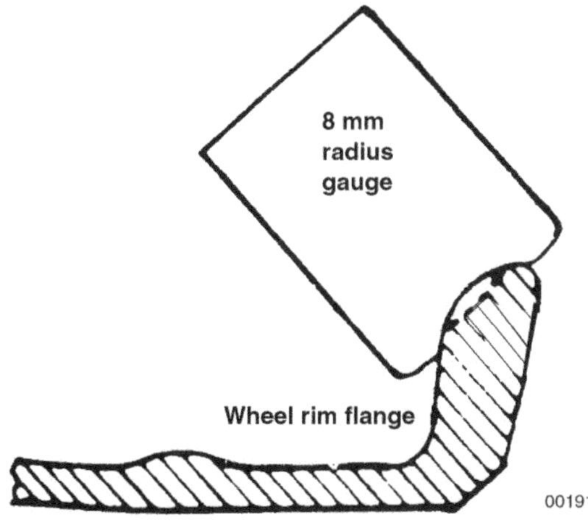

0019142

Check alloy wheel flanges for wear. The inside flange is more subject to wear. Remove sharp edges and burs, then use an 8 mm radius gauge to check for wear.

TIRES

Radial-ply tires are installed as standard equipment on all cars covered by this manual. To retain the car's excellent handling characteristics, it is recommended that the tires only be replaced with those of the same (radial-ply) construction, size and speed rating.

Tire wear

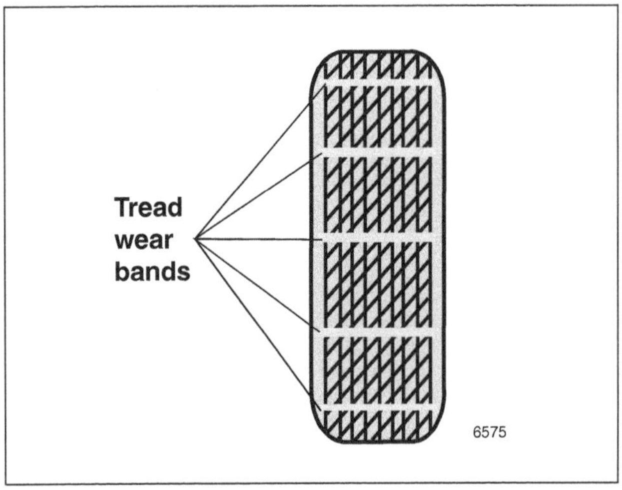

Tire wear is a good indicator of steering and wheel alignment problems. Proper tread wear is difficult to notice, so tires are made with wear-indicator bands that indicate when the tire is nearly worn-out. On an evenly worn tire, these wear-indicator bars will eventually appear as evenly spaced bald "stripes" about ½ in. wide running across the tread surface. The appearance of these tread-wear bands on only one part of the tread indicates uneven wear.

Uneven tire wear is usually caused by improper tire inflation pressures or misalignment. Other kinds of uneven tread wear may be the result of faulty components or improper alignment. The following figures illustrate the influence of these factors on tire wear. (Drawing courtesy of Goodyear Tire and Rubber Co.)

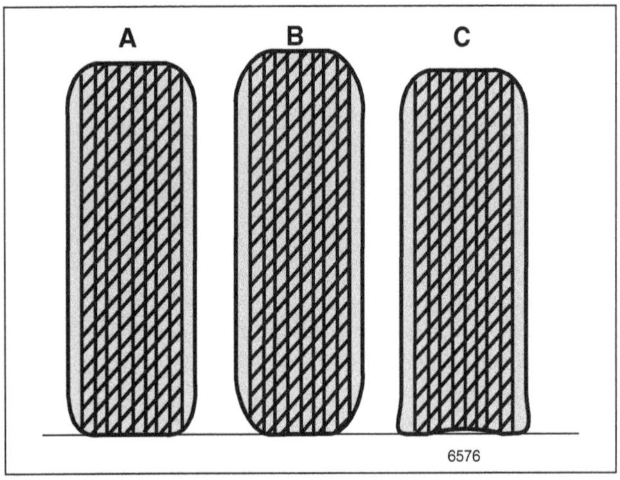

Effects of tire inflation pressure on wear pattern. Condition (**A**) is normal. Overinflation (**B**) causes increased wear in center of tire tread. Underinflation (**C**) causes increased wear on outer edges of tire tread. (Drawings courtesy of Goodyear Tire and Rubber Co.)

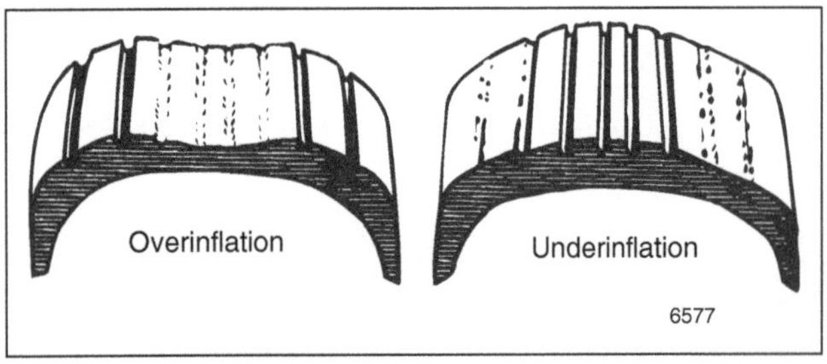

Examples of uneven tread wear caused by overinflated and underinflated tires. (Drawings courtesy of Goodyear Tire and Rubber Co.)

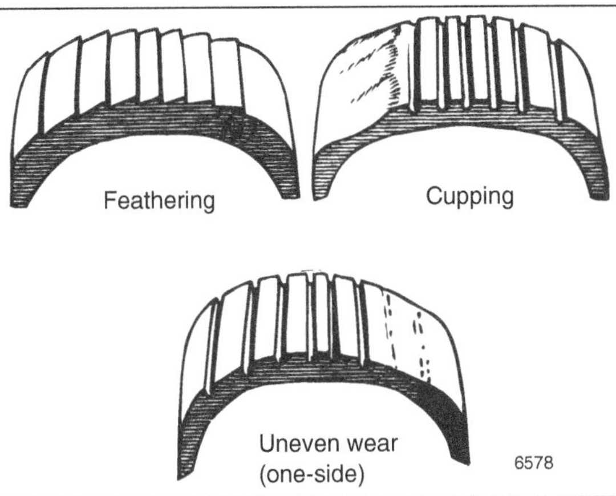

Examples of uneven tread wear caused by misalignment and/or worn or damaged components:

- Feathering suggests incorrect alignment.
- Uneven wear on one side suggests incorrect alignment or repeated hard cornering.
- Cupping results from vibration, often caused by unbalanced tire or faulty shock absorber. (Drawings courtesy of Goodyear Tire and Rubber Co.)

Tire wear that is uneven across the tread—one side worn more than the other, or unusual wear of individual tread ribs—indicates an alignment problem, and perhaps worn suspension or steering components. Cupping or scalloping—wear that is uneven around the circumference of the tire—is a telltale sign of an unbalanced tire or a worn-out shock absorber.

Winter tires

Though inferior to regular tires for dry-road wear and handling, winter (mud and snow) tires offer a big improvement on snow-covered winter roads. Studded winter tires improve traction on icy surfaces, but may be more dangerous than non-studded tires on dry roads. They should be used only if icy conditions predominate. Check local laws. Studded tires are restricted or completely prohibited in many states.

CAUTION —

If winter tires are to be installed on only two of the four wheels, they should be installed on the rear driving wheels. To avoid creating dangerous handling characteristics, they must be of the same type (radial-ply construction) as those on the front of the car.

Tire storage

- The rubber in tires is sensitive to environmental conditions and foreign substances. Ozone, fuels, greases and oil all can have a destructive effect on tires.
- Store tires indoors where the temperature is between 14° F and 68° F and the air is dry.
- Wash tires before storing them.
- Keep tires out of sunlight or fluorescent lighting.
- Store tires on wheels with full tire pressure.
- Do not hang tires on hooks or stack them on top of each other.

WHEEL ALIGNMENT

WARNING —
- *Replacement and/or removal of front or rear suspension components, such as front or rear torsion bars or rear torsion plates, disturbs the wheel alignment settings. Replacement or adjustment of the torsion bars, rear torsion plates or rear trailing arms can affect vehicle ride height, which directly influences vehicle stability, chassis balance and even braking at all wheels. Vehicle balance/weight must be checked at all four wheels, and ride height must be reset. See* **Extended alignment** *later in this group.*

- *While performing alignment procedures, make sure the car is stable and well supported at all times. Use a professional automotive lift or jack stands designed for the purpose. A floor jack is not adequate support.*

Depending on the circumstances, there are two procedures for wheel alignment:

- **Basic alignment** is required:
  –when replacing suspension components which cause no alteration in ride height
  –In case of excessive or uneven tire wear
  –After a new vehicle has "settled"
  –In case of handling complaints

- **Extended alignment** is required:
  –If suspension replacement components alter vehicle height
  –In case of incorrect vehicle height

Basic alignment requires the use of alignment analyzer equipment. Minor resetting of alignment specifications is permitted.

Extended alignment requires checking and resetting vehicle ride height and corner weights with the use of wheel load scales and then performing the basic alignment.

Basic alignment specification for the front and rear wheels are in **Table b** and **Table c**.

Pre-alignment checklist (repair groups in bold)
- Fuel tank full
- Tool kit and spare tire in luggage compartment
- Ball joint and suspension parts checked. See **401**.
- Wheel bearing play set and OK. See **401**.
- Tire pressures set to specifications.
- Steering wheel and steering rack in center position. See **480**.
- Check and, if necessary, reset front and rear ride height. See **Extended alignment** in this group.

Basic alignment

– Always start a four wheel alignment procedure by beginning at the rear wheels. Make sure the car is loaded with a normal load. See below.

Normal load

- Each front seat .... 68 kg (150 lb.)
- Center of rear seat .... 68 kg (150 lb.)
- Luggage compartment .... 21 kg (46 lb.)
- Fuel tank .... full

Table b. Front wheel alignment

	Specification	Max. deviation
Toe angle (total)	+15' ± 5'	
Toe difference angle (inside wheel @ 20° lock)	0° to + 30"	Can be changed only by replacing steering arm
Camber	0° ± 10'	10'
Caster	6°5' ± 15'	30'
Ride height	108 mm (4.25 in.)	± 5 mm (± 0.02 in.)
Steering wheel lock to lock	Approx. 3 turns	
Turning circle diameter	10.35 m (34 feet)	
Track width (7 x 15 wheel)	1361 mm (53.6 in.)	

Table c. Rear wheel alignment

	Specification	Max. deviation
Toe per wheel	+10' ± 10'	20'
Camber	- 1° ±10'	20'
Ride height	37 mm (1.46 in.)	± 5 mm (± 0.02 in.)
Torsion plate setting US/Canada spec. European spec.	 40° 35°	 0.5° 0.5°
Track width (7 x15 wheel)	1380 mm (54.3 in.)	

Extended alignment

After a basic alignment if it is determined that, despite good tires and correct inflation, the handling characteristics of the vehicle are not correct, or if ride height is incorrect, or if the braking is uneven, wheel load scales must be used to set correct wheel load and ride height.

Follow the equipment manufacturer's instructions for measuring and setting wheel load.

NOTE —

- *Before setting wheel load and vehicle height, bounce vehicle a few times to let the suspension settle into the normal ride height position.*
- *On cars fitted with Bilstein shocks, the "normal ride height position" can vary with the car in the rest position. Pulling the suspension up or pushing it down at the bumper can result in different ride height positions. When checking or adjusting the ride height, manually position the car in the 'mean' or middle position.*

Wheel load, checking

Changing ride height at any wheel changes wheel loading at all wheels. Increasing the initial tension of a torsion bar raises the vehicle at the corresponding wheel and increases wheel load at that wheel. Decreasing the initial tension of a torsion bar lowers the vehicle at the corresponding wheel and decreases wheel load at that wheel.

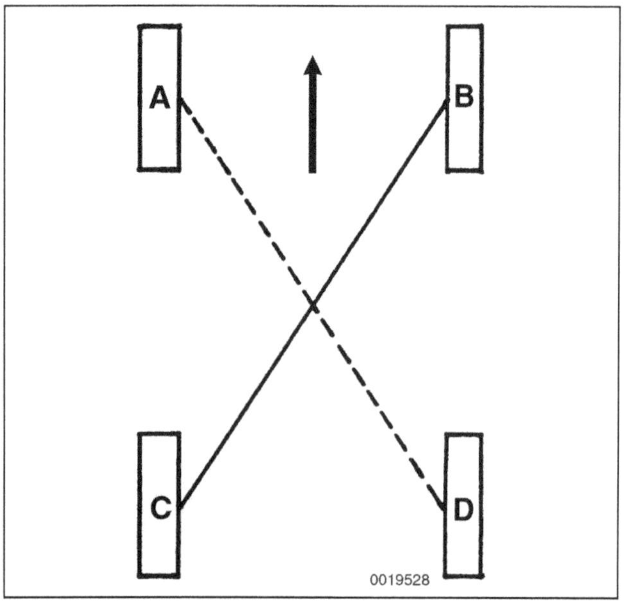

Changing load on one wheel directly affects the wheel at a diagonal from it, while it has the opposite effect on the wheels on the other diagonal.

- For example, increasing load at wheel **A** increases load at wheel **D**.
- Increasing load at wheel **A** decreases load at wheels **B** and **C**.

Wheel load specification (front or rear axle)

- Maximum left-right load difference ...... 20 Kg (44 lbs.)

– Check wheel load using Maha wheel load scales RWA 2 U or equivalent equipment.

– Adjust wheel load by setting front and rear ride height, as described later.

NOTE —

- *Any adjustment of a torsion bar will have a wheel load effect on the other three torsion bars.*
- *Increasing the initial torsion bar tension on one side (raising vehicle) will increase wheel load on that wheel.*
- *Decreasing the initial torsion bar tension on one side (lowering vehicle) will decrease wheel load on that wheel.*

Front ride height, setting

Set front ride height after replacing front suspension components such as control arms or torsion bars. The car must have a full fuel tank and the spare tire must be in luggage compartment for correct adjustment.

– Place car on a level floor.

NOTE —

- *Before setting wheel load and vehicle height, bounce vehicle a few times to let the suspension settle into the normal ride height position.*
- *On cars fitted with Bilstein shocks, the "normal ride height position" can vary with the car in the rest position. Pulling the suspension up or pushing it down at the bumper can result in different ride height positions. When checking or adjusting the ride height, manually position the car in the 'mean' or middle position.*

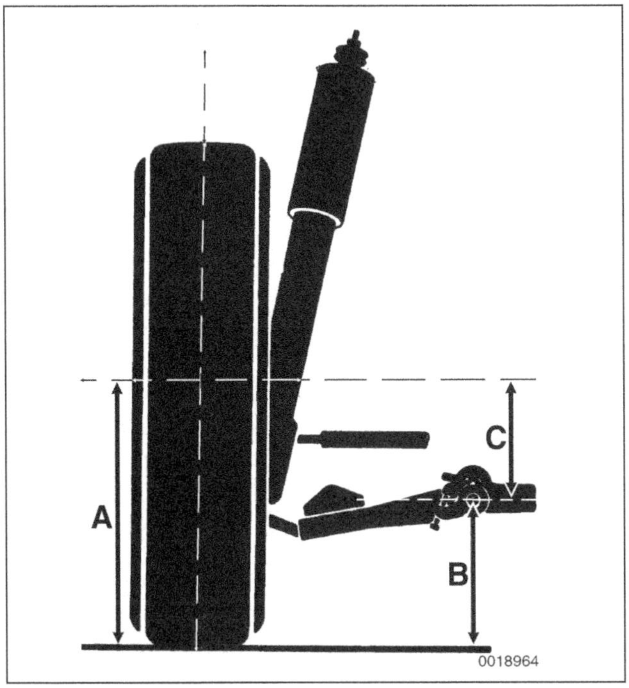

Measure from floor to center of front wheel dust cap to obtain dimension **A**.

- Measure from floor to center of torsion bar adjusting cap to obtain dimension **B**.
- Difference (**A** – **B** = **C**) is ride height.

Front ride height

- Dimension **C**
  US/Canada spec. . . . . 108 ± 5 mm (4.25 ± 0.20 in.)
  European spec.. . . . . 99 ± 5 mm (3.90 ± 0.20 in.)
- Max. difference between left and right . . . 5 mm (0.20 in.)

Adjust to specifications using front torsion bar adjusting screw (**arrow**).

NOTE—

- *Height-adjustment specifications apply to new vehicles. Older vehicles can be up to 10 ± 5 mm (0.4 ± 0.2 in.) lower.*
- *If a vehicle is slightly low in front it can be raised without causing an excessive difference in wheel load. To do this, evenly turn the front torsion bar adjustment screws clockwise. (This will have an effect on the vehicles toe setting.)*

Rear ride height, setting

Set rear ride height after replacing rear suspension components such as trailing arms, torsion plates or torsion bars. Initial setting of rear torsion bars is covered in **420 Rear Suspension, Axle Shafts**.

– Place car on a level floor.

NOTE —

- *Before setting wheel load and vehicle height, bounce vehicle a few times to let the suspension settle into the normal ride height position.*
- *On cars fitted with Bilstein shocks, the "normal ride height position" can vary with the car in the rest position. Pulling the suspension up or pushing it down at the bumper can result in different ride height positions. When checking or adjusting the ride height, manually position the car in the 'mean' or middle position.*

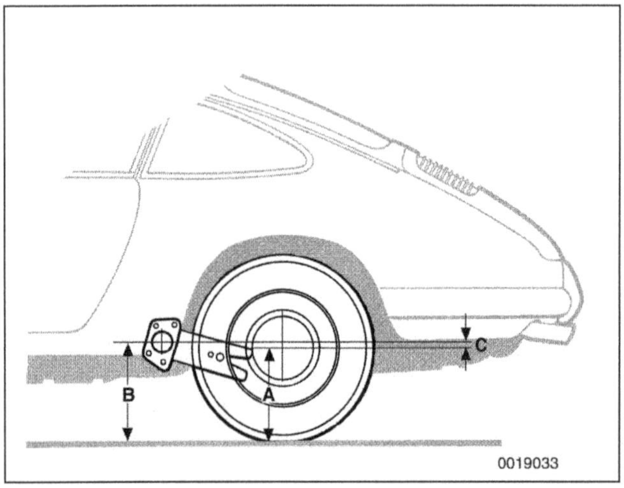

Measure from floor to center of rear wheel dust cap to obtain dimension **A**.

- Measure from floor to center of torsion bar cover to obtain dimension **B**
- Difference (**B - A = C**) is ride height.

Adjust to specifications using height-adjusting eccentric.

Rear ride height
- Dimension **C**
  US/Canada spec. . . . . 37 mm ± 5 mm (1.46 ± 0.2 in.)
  European spec.. . . . . 16 mm ± 5 mm (0.63 ± 0.2 in.)
- Max. difference between left and right . . . . 8 mm (0.3 in.)

NOTE —

- *Height-adjustment specifications apply to new vehicles. Older vehicles can be up to 10 ± 5 mm (0.4 ± 0.2 in.) lower.*
- *The two-piece torsion plate is set at the factory so that the vehicle can be raised with the height adjustment eccentric, but not lowered.*

Tightening torque
- Clamping or eccentric bolt . . . . 245 Nm (181 ft-lb)

Rear wheel camber and toe, adjusting

> ***WARNING —***
> *While performing alignment procedures, make sure the car is stable and well supported at all times. Use a professional automotive lift or jack stands designed for the purpose. A floor jack is not adequate support.*

> ***CAUTION —***
> *Changing camber at one rear wheel will cause a change in the load at that wheel and consequently at all other wheels. Be sure to check vehicle corner loads as a final check. Changing camber at one rear wheel will also affect the toe setting at that wheel.*

NOTE —
A change in the rear toe by adjusting the toe eccentric will have very little effect on camber, wheel load or ride height.

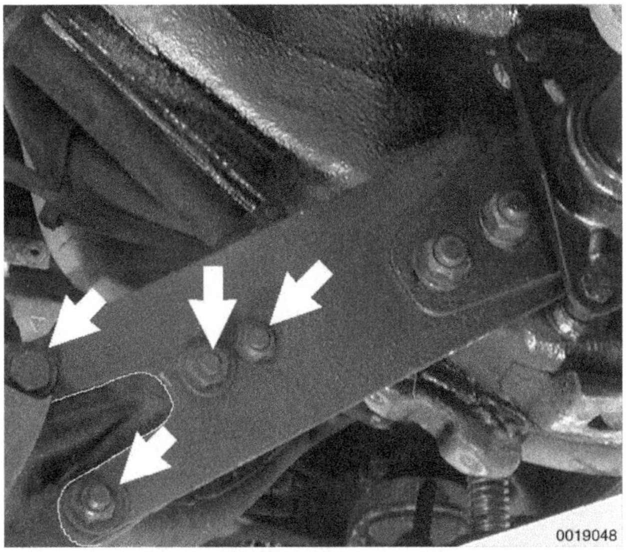

◄ Mark torsion plate position to trailing arm by scribing outline around torsion plate.

- Loosen torsion plate fasteners (**arrows**).

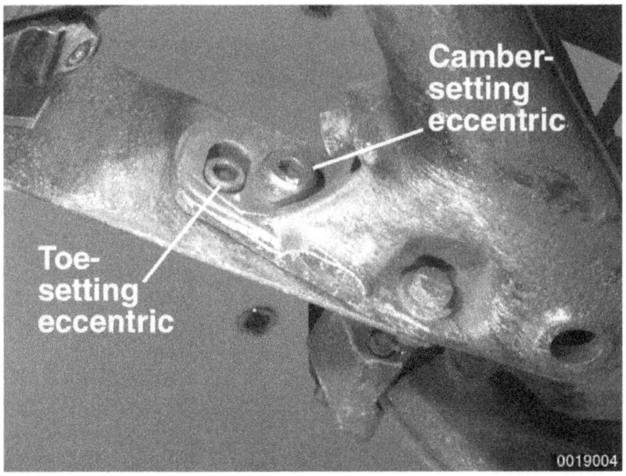

◄ Set rear wheel camber and toe using an alignment analyzer.

Tightening torque
- Torsion plate to trailing arm (eccentric). . 85 Nm (63 ft-lb)
- Torsion plate to trailing arm (M12) . . . . 120 Nm (89 ft-lb)

Front wheel camber and caster, adjusting

◄ Working in luggage compartment, clean sealing compound away from front upper strut mounts and bolts (**arrows**).

– Loosen strut mount bolts slightly.

◄ Set front wheel camber and caster using an alignment analyzer.

Tightening torque
- Upper strut mount to body . . . . 47 Nm (35 ft-lb)

NOTE —
Reseal upper strut mounts with sealing compound.

Front wheel toe, adjusting

– Check to see if steering rack is in its center position when steering wheel is centered:

- Starting from center position, turn steering wheel in one direction to stop, counting turns.
- Turn to stop in opposite direction while counting turns.
- If necessary, remove and center steering wheel at half the number of turns from stop to stop.

◄ Loosen 22 mm inner and outer tie rod locking nuts (**arrows**).

- Adjust tie rod by turning center tube to change length.
- To keep steering wheel centered, adjust both tie rods equal amounts.

– When final toe adjustments have been made tighten tie rod locking nuts.

NOTE —
The toe difference angle cannot be adjusted. If the difference is excessive, steering or suspension parts should be inspected for damage or deformation.

460 Brakes–Mechanical

TABLES

Special Tools

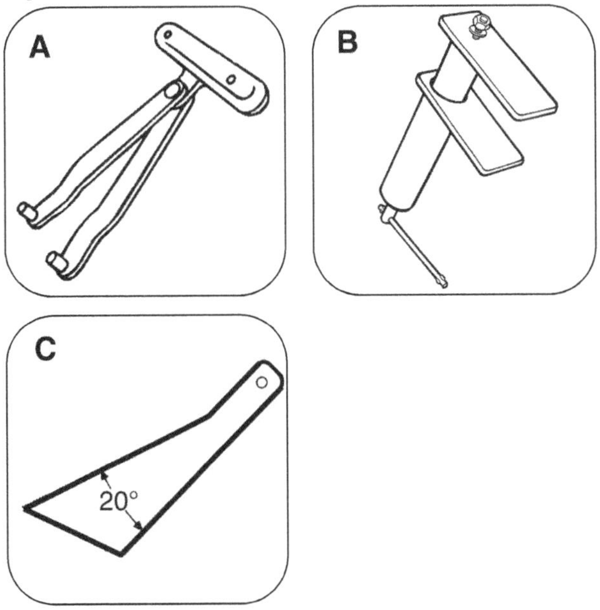

Special tool

A P 86 .... Brake pad removal tool
(Porsche ordering #000 721 086 00)
B 1023/4 .... Caliper piston retractor
(source: Baum Tools Unlimited)
C P 84 .... Caliper piston 20° gauge
(source: Baum Tools Unlimited)

GENERAL

This repair group covers replacement procedures for mechanical brake components including brake pads, brake rotors and parking brake components.

For general specifications, diagnosis and troubleshooting information see **400 Suspension, Brakes and Steering–General**. Repairs to the brake master cylinder, power booster, calipers and associated components are covered in **470 Brakes–Hydraulic**. Pedal cluster repairs are covered in **301 Clutch**.

Special tools

Some procedures require special tools.

> ***WARNING —***
> - *Brake friction materials such as brake linings or pads may contain asbestos fibers which can lead to illnesses. Do not create dust by grinding, sanding, or cleaning the pads with compressed air. Avoid breathing any asbestos fibers or dust.*
> - *Brake fluid is poisonous, highly corrosive and dangerous to the environment. Wear safety glasses and rubber gloves when working with brake fluid. Do not siphon brake fluid with your mouth. Immediately clean away any fluid spilled on painted surfaces and wash with water. Dispose of brake fluid properly.*
> - *Do not reuse self-locking nuts, or fasteners. They are designed to be used only once and may fail if reused. Always replace them with new self-locking nuts.*

Braking system

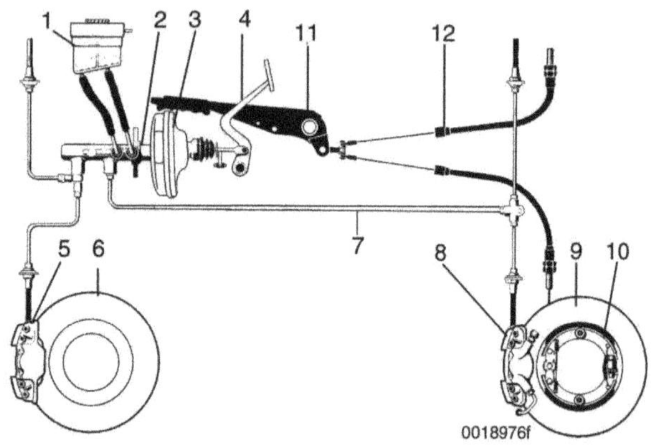

1. Brake fluid reservoir
2. Dual master cylinder
3. Power booster
4. Brake pedal
5. Front brake caliper
6. Front brake rotor
7. Brake fluid line
8. Rear brake caliper
9. Rear brake rotor/drum
10. Parking brake shoes
11. Parking brake handle
12. Parking brake cables

Braking system

Porsche 911 SC models are equipped with power-assisted four-wheel disc brakes.

A brake booster power-assists a dual-circuit master cylinder. A vacuum boost system is used in all models.

Dual-piston calipers act on vented front and rear rotors.

The parking brake is mechanically operated via dual cables. The parking brake system is independent of the rear disc brake system. The drum-style parking brake shoes act on a small drum, which is integral with the rear brake rotor.

Brake noise

Occasional groaning or squealing sounds from the disc brakes are usually caused by vibration being transmitted through the brake pads. Brake friction materials that contain little or no asbestos may contribute to brake noise. These noises are normal and rarely indicate a problem.

Although there is no single solution to brake noise problems, proper maintenance and repair can help minimize noise. Disc brake caliper assemblies include anti-rattle springs to minimize vibration and noise. Brake pads and calipers should be kept clean and free of corrosion. Always resurface or replace brake rotors when changing brake pads.

Troubleshooting

Brake performance is mainly affected by three things: the level and condition of the brake fluid, the system's ability to create and maintain hydraulic pressure, and the condition of the friction components.

Air in the brake fluid will make the brake pedal feel spongy during braking or will increase the brake pedal force required to stop. Fluid contaminated by moisture or dirt can corrode the system. Inspect the brake fluid inside the reservoir. If it is dirty or murky, or is over a year old, the fluid should be replaced.

Visually check the hydraulic system starting at the master cylinder. To check the function of the master cylinder hold the brake pedal down hard with the car stopped and the engine running. The pedal should feel solid and stay solid. If the pedal slowly falls to the floor, either the master cylinder is leaking internally, or fluid is escaping from other points in the system. If no leaks can be found, the master cylinder is faulty and should be replaced. Check all brake fluid lines and couplings for leaks, kinks, chafing or corrosion.

Check the brake booster by pumping the brake pedal approximately 10 times with the engine off. Then hold the pedal down and start the engine. The pedal should fall slightly. If not, check for any visible faults before suspecting a faulty brake booster, and check the non-return valve for one-way flow.

NOTE—
- *The brake booster non-return valve is in the vacuum hose at the left side manifold, in the engine compartment.*
- *See* **400 Suspension, Brakes and Steering–General** *for a brake troubleshooting table.*
- *See* **470 Brakes–Hydraulic** *for repair information for hydraulic brake components or the brake power-booster.*

Worn or contaminated brake pads will cause poor braking performance. Oil-contaminated or glazed pads cannot produce as much friction, and stopping distances will increase. Inspect the rotors for glazing, discoloration, scoring or oiliness.

Steering wheel vibration while braking at speed is often caused by warped brake rotors, but can also be caused by worn suspension parts.

Maintenance

– Porsche specifies the maintenance steps below to be carried out at particular time or mileage intervals for proper maintenance of the brakes.

- Check brake fluid. See **030 Maintenance**.
- Inspect brake hoses, lines and wheel calipers for leaks.
- Check brake pads and shoes for wear.
- Replace brake fluid. See **470 Brakes–Hydraulic**.

BRAKE PADS AND ROTORS

Brake pads can be replaced without disconnecting the brake fluid line from the caliper or having to bleed the brakes.

Brake rotor removal and replacement requires that the caliper be disconnected from the brake line in order to avoid bending the metal line. The rear rotors can be replaced without dissembling the wheel hub and bearing, but front rotor replacement requires that the wheel bearing hub be removed with the rotor and disassembled on a work bench.

CAUTION—
- *Always machine or replace brake rotors in pairs.*
- *Always replace brake pads in sets (one set = one axle).*

Brake pads, inspecting

– Raise car and remove wheels.

> ***WARNING —***
> *Make sure the car is firmly supported on jack stands designed for the purpose. Place the jack stands beneath a structural chassis point. Do not place jack stands under suspension parts.*

◄ Visually inspect brake pads for thickness and compare to specifications. Replace as necessary.

Brake pad thickness specification

- Wear limit **A** (front or back) . . . . 2 mm (0.08 in.)

Brake pads, replacing

> ***CAUTION —***
> *To prevent brake fluid overflow when caliper pistons are pushed back, siphon some fluid out of brake master cylinder reservoir.*

◄ This procedure is applicable to both front and rear brake pads.

– Raise car and remove wheels.

> ***WARNING —***
> *Make sure the car is firmly supported on jack stands designed for the purpose. Place the jack stands beneath a structural chassis point. Do not place jack stands under suspension parts.*

Front brake assembly

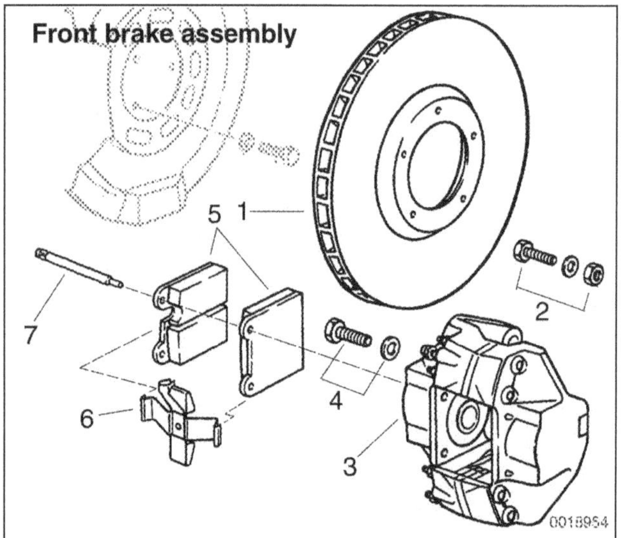

1. Brake rotor
2. Brake rotor to bearing hub fasteners (M8)
   - tighten to 23 Nm (17 ft-lb)
3. Brake caliper
4. Brake caliper to strut housing fastener (M12)
   - tighten to 70 Nm (52 ft-lb)
5. Brake pads
6. Anti-rattle cross spring
7. Retaining pin

- Use drift to drive brake pad retaining pins out while depressing anti-rattle cross-spring.
- Remove anti-rattle cross spring.

- Remove brake pads from caliper with Porsche special tool P 86 or equivalent.

NOTE—
If brake pads are to be reused, mark original position.

- Push pistons back into caliper body with Porsche special tool P 83 or equivalent (Baum tool 1023/4 shown).

NOTE—
If the special tool in not available, a hard wood block can be used to push the piston back.

- Use coarse file or wire brush to clean brake pad contact surfaces in calipers.

◀ Inspect caliper piston dust boots (**arrows**). Replace caliper with damaged boots.

NOTE—
Brake caliper rebuilding procedures are beyond the scope of this manual. Remanufactured calipers are available from the Porsche dealer parts department, as well as a variety of aftermarket sources.

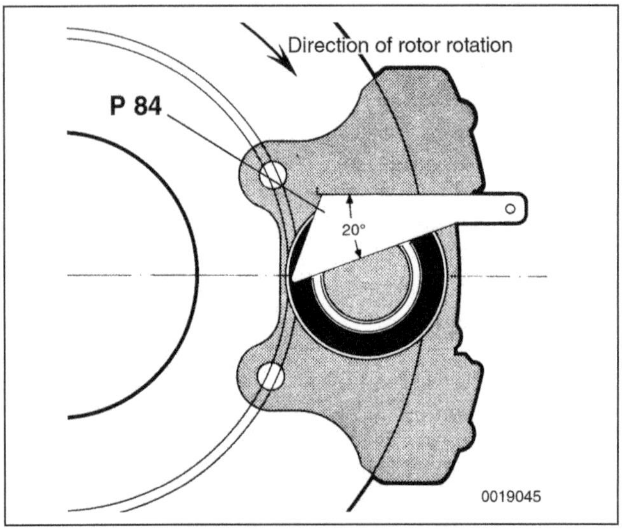

◀ Set caliper piston cutout alignment using Porsche special tool P 84.

– Check brake rotor for wear as described later. Replace or reface if necessary.

– Install new brake pads in caliper.

NOTE—
Apply thin coat of Optimoly® HT or equivalent anti-seize paste to sliding edges of pads.

– Install anti-rattle cross spring and retaining pins.

– Remainder of installation is reverse of removal.
  - Before driving car after brake work has been performed, depress brake pedal several times to seat pads against brake disc.
  - Recheck brake fluid level.

Tightening torque
- Wheel to wheel hub . . . . 130 Nm (96 ft-lb)

Brake rotor, checking

Check the brake rotors for wear any time the brakes are serviced. Rotors that are scored with sharp ridges, warped, worn irregularly or cracked should be replaced.

Brake rotors should always be resurfaced in pairs, with an equal amount of material removed from both sides of each rotor. **Table a** lists the dimensions for resurfacing rotors. Rotors which fail to meet these requirements should be replaced.

NOTE—
On original equipment rotors, the minimum thickness dimension can be found stamped into the rotor's hub.

Table a. Brake rotor reconditioning

	Front	Rear
Thickness when new	20.5 mm (0.807 in.)	20.0 mm (0.787 in.)
Minimum thickness after machining	19.1 mm (0.751 in.)	18.6 mm (0.732 in.)
Wear limit (minimum permissible thickness)	18.6 mm (0.732 in.)	18.0 mm (0.708 in.)
Thickness tolerance (difference between any two measurements)	0.02 mm (0.0008 in.)	0.02 mm (0.0008 in.)
Maximum lateral runout	0.1 mm (0.004 in.)	0.1 mm (0.004 in.)

Use a micrometer to measure rotor thickness at eight to ten positions on rotor braking surface. Rotor runout should be measured using a dial indicator setup.

If a low speed shimmy goes away when the brakes are released, excessive rotor runout is probably the cause of the shimmy. If rotor runout is within limits and the shimmy still persists, check the front wheel bearings as described in **401 Front Suspension**.

Brake rotors can be resurfaced by most automotive machine shops.

Front brake rotor, removing and installing

CAUTION—
Always replace or machine brake rotors in pairs.

– Press brake pedal down about 1 in. and hold in place with a pedal stop.

NOTE—
This will prevent brake fluid from dripping out of open brake lines while the work is being performed.

– Raise car and remove front wheels.

WARNING—
Make sure the car is firmly supported on jack stands designed for the purpose. Place the jack stands beneath a structural chassis point. Do not place jack stands under suspension parts.

– Disconnect rubber brake hose from metal brake line near bottom of strut. Plug hydraulic lines.

CAUTION—
When working with hydraulic brake lines and hoses be sure to counterhold fitting before loosening or tightening.

– Remove caliper mounting bolts and remove caliper. See **470 Brakes–Hydraulic**.

– Remove wheel bearing dust cover.

0018966

◄ Loosen 6 mm allen clamping screw in wheel bearing lock nut and unscrew and remove lock nut from spindle.

– Pull brake rotor with bearing hub and wheel bearings straight off steering spindle.

0018987

◄ Remove 8 mm bolts (**arrows**) mounting brake rotor to hub.

- Tap hub out of rotor with a soft hammer.

0018988

◄ Clean rotor mounting surface on hub with a wire brush.

– Install new rotor on hub and torque bolts.

Tightening torque

- Brake rotor to bearing hub (M8) . . . . 23 Nm (17 ft-lb)

– Reinstall hub and rotor on spindle and install outer wheel bearing.

- Adjust wheel bearing as described in **401 Front Suspension**.
- Install bearing dust cap.

Tightening torque
- Wheel bearing lock nut clamp screw (M8) . . . . 15 Nm (11 ft-lb))

– Remainder of assembly is reverse of disassembly.
  - Bleed brakes as described in **470 Brakes–Hydraulic**.

Tightening torques
- Brake caliper bleeder screw (7 mm) . . . . . 3 Nm (27 in-lb)
- Brake line to brake hose . . . . 12 Nm (9 ft-lb)
- Caliper to strut housing. . . . . 70 Nm (52 ft-lb)
- Wheel to wheel hub . . . . 130 Nm (96 ft-lb)

Rear brake rotor, removing and installing

WARNING —
Always replace or machine brake rotors in pairs.

– Press brake pedal down about 1 in. and hold in place with a pedal stop.

NOTE —
This will prevent brake fluid from dripping out of open brake lines while the work is being performed.

– Raise car and remove rear wheels.

WARNING —
Make sure the car is firmly supported on jack stands designed for the purpose. Place the jack stands beneath a structural chassis point. Do not place jack stands under suspension parts.

< Disconnect metal brake line from brake caliper. Plug hydraulic fittings.

– Remove brake caliper mounting bolts and remove caliper. See **470 Brakes–Hydraulic**.

Remove two screws (**arrows**) mounting brake rotor to hub and remove rotor.

NOTE —
If rotor hangs up and will not slide off, back off parking brake adjuster. See **Parking brake shoes, adjusting**, *later in this section.*

– Clean hub face and install new rotor.

– Remainder of assembly is reverse of disassembly.
  - Adjust parking brake shoes as necessary.
  - Bleed brakes as described in **470 Brakes–Hydraulic**.

Tightening torques
- Brake caliper bleeder screw (7 mm) . . . . . 3 Nm (27 in-lb)
- Brake rotor to hub screws. . . . . 5 Nm (3.5 ft-lb))
- Caliper to trailing arm . . . . 60 Nm (44 ft-lb))
- Hydraulic brake line to caliper. . . . . 14 Nm (10 ft-lb)
- Wheel to wheel hub . . . . 130 Nm (96 ft-lb)

Parking brake assembly

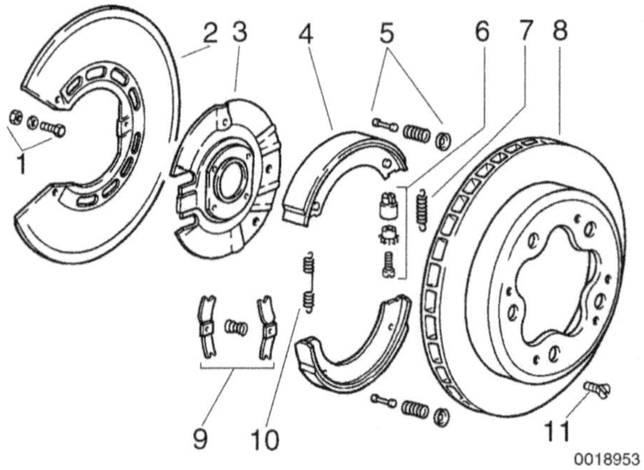

1. Bolt, washer, nut (M8)
   - tighten to 25 Nm (18 ft-lb)
2. Backing plate
3. Brake shoes carrier
4. Brake shoe
5. Shoe hold-down assembly
6. Star-wheel adjuster assembly (Right side shown. Left side is installed upside down from illustration.)
7. Return spring
8. Brake rotor/drum
9. Shoe expander assembly
10. Return spring
11. Brake rotor retaining screw
    - tighten to 5 Nm (3.6 ft-lb)

PARKING BRAKE

The parking brake uses a small brake drum inside the rear rotor. The parking brake shoe assembly is fitted in the drum. The mechanical parking brake operates only on the rear wheels and is independent of the main hydraulic brake system.

Each of the two parking brake cables is connected to an expander assembly on each of the brake shoe assemblies. When the parking brake is applied, the brake shoes are forced out against the brake drum.

Parking brake shoes, adjusting

There is no automatic adjusting mechanism for the parking brake. The parking brake shoes need periodic adjustment to compensate for wear. The parking brake should also be adjusted if a cable, brake rotor, or parking brake shoes are replaced. Check to see that the parking brake functions properly whenever the cables have been detached for rear brake or rear suspension service.

– Raise car and remove rear wheels.

WARNING —
Make sure the car is firmly supported on jack stands designed for the purpose. Place the jack stands beneath a structural chassis point. Do not place jack stands under suspension parts.

NOTE —
- *Parking brake should be fully released for this procedure.*
- *Push brake pads away from brake rotor so that rear rotor turns freely.*

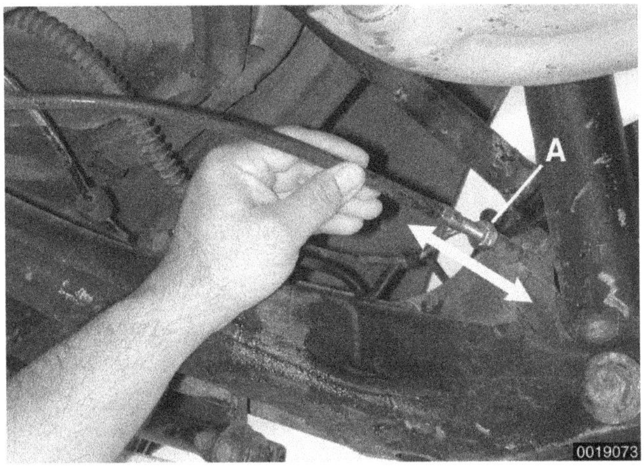

◄ Working underneath car, check for axial (in and out) play in brake cable at trailing arm.

- There should be slight movement in brake cable.
- Adjust locking nuts (**A**) as necessary.
- Repeat on other side.

– Turn brake rotor until brake shoe adjuster (star wheel) is visible through adjustment/inspection hole.

◄ Use a screwdriver to turn adjuster until brake rotor can no longer be turned by hand.

- Loosen adjuster by 4 clicks and check rotor for free movement.
- Loosen more if necessary.
- Adjust other side.

NOTE —
On both wheels, the screwdriver handle must be pulled backward (toward back of car) to tighten and forward to loosen adjuster. The two star wheel assemblies, though identical, are installed upside down from each other on the two sides of the car.

– With parking brake lever pulled up two teeth, both rear wheels should be difficult to turn.

– Release parking brake lever and check wheels. Both should turn freely.

Parking brake cable, adjusting

Parking brake cable adjustment is necessary if:

- Parking brake lever needs to be pulled up more than 4 clicks (max. 5 clicks) to lock rear wheels.
- One wheel stops before the other when parking brake is applied.

NOTE —
If parking cable was replaced, pull up and release the parking brake handle several times to stretch and set the cable.

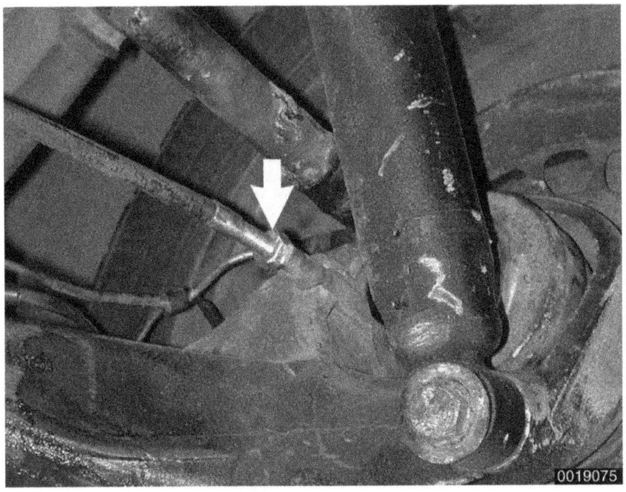

◄ Working underneath rear of car and with parking brake fully released, adjust cables at locking nuts (**arrow**) so that equal amounts of thread show on each cable.

- Check to be sure there is slight axial (in and out) play in both cables at the trailing arms.
- Check to be sure both wheels turn freely with parking brake released.

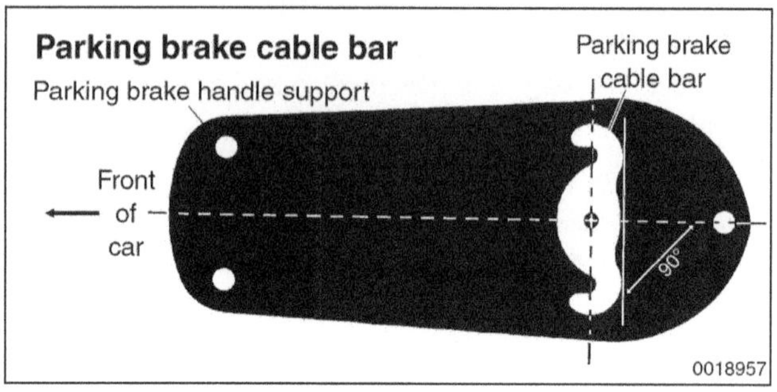

Access parking brake handle assembly as described later under **Parking brake cable, replacing**.

- Check to make sure cable bar is exactly transverse (90° angle) to car.
- Adjust cables as necessary.

Parking brake cable, replacing

– Press brake pedal down about 1 in. and hold in place with a pedal stop.

NOTE—
This will prevent brake fluid from dripping out of open brake lines while the work is being performed.

– Raise car and remove rear wheel.

> ***WARNING—***
> *Make sure the car is firmly supported on jack stands designed for the purpose. Place the jack stands beneath a structural chassis point. Do not place jack stands under suspension parts.*

Disconnect metal brake line from brake caliper. Plug hydraulic fittings.

– Remove brake caliper mounting bolts and remove caliper. See **470 Brakes–Hydraulic**.

Remove two screws (**arrows**) mounting brake rotor to hub and remove rotor.

NOTE—
If rotor hangs up and will not slide off, back off parking brake adjuster as described earlier.

◄ Remove cotter pin (**arrow**) from castellated nut on end of brake cable and remove nut.

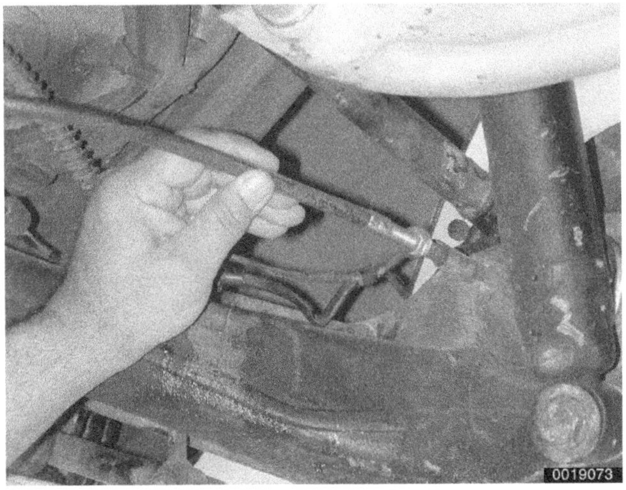

◄ Working underneath car, pull parking brake cable out of brake backing plate and trailing arm.

- Remove passenger seat to gain access to heater controls and parking brake handle assembly.
- Remove access panel from right side of heater control.

◄ On cars with automatic heating, disconnect heater control linkage rod (**arrow**).

- Remove heater control trim screws from floor. Lift heater control and lay aside.
- Pull carpet up around parking brake handle.

Remove bolts (**arrows**) holding parking brake handle support to floor.

Parking brake handle components

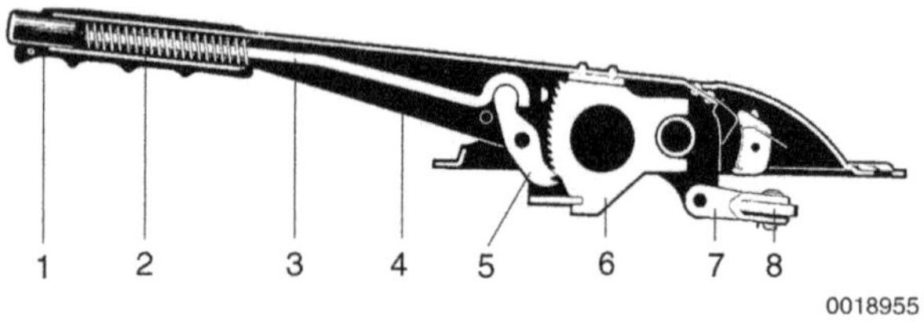

1. Release button	5. Pawl
2. Spring	6. Toothed sector
3. Push rod	7. Link
4. Parking brake handle	8. Cable bar

Lift up parking brake handle mechanism and disconnect link at base of handle by removing clevis pin from cable bar.

– Detach cables from cable bar.

– Pull out old cable and replace with new.

NOTE —

With the old cable removed, blow out cable conduit in body with compressed air.

– Attach new cable eye to cable bar at handle mechanism.
 - Reinstall handle mechanism.

– Push cable through trailing arm and backing plate.
 - Install castellated nut.
 - Use new cotter pin.

Parking brake cable attachment at rear wheel

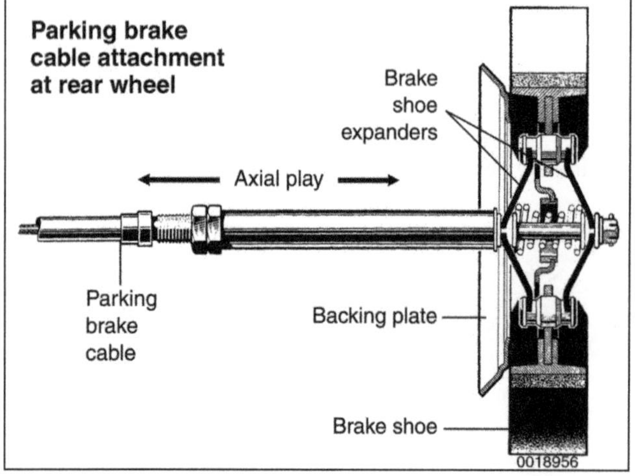

Check brake cable housing axial play. If play is found, adjust nuts on cable until slight axial play is felt. Lock adjusting nuts together.

– Reinstall brake rotor and caliper.

Tightening torques

- Brake caliper to trailing arm (M12) . . . . . 60 Nm (43 ft-lb)
- Brake rotor to hub (M6). . . . . 5 Nm (3.5 ft-lb)
- Hydraulic brake line to caliper. . . . . 14 Nm (10 ft-lb)

– Adjust parking brake shoes adjuster as described earlier.

– Pull parking brake handle up and release 4 or 5 times to stretch and set new cable.

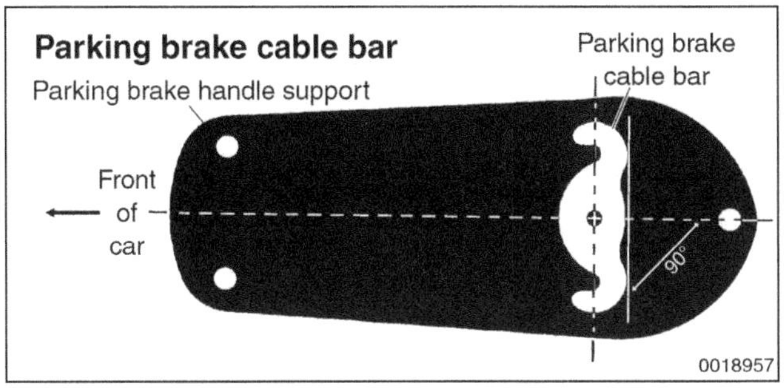

◀ Working inside car, check to make sure cable bar below parking brake handle is exactly transverse (90° angle) to car.

- Adjust cables as necessary.

– Raise parking brake handle 2 clicks. With lever in this position both rear wheels should be difficult to turn.

– Bleed brakes as described in **470 Brakes–Hydraulic**.

– Install rear wheel.

Tightening torque

- Wheel to wheel hub . . . . 130 Nm (96 ft-lb)

– Replace rug over parking handle housing. Reinstall heater controls and passenger seat.

Parking brake shoes, replacing

– Press brake pedal down about 1 in. and hold in place with a pedal stop.

NOTE —
This will prevent brake fluid from dripping out of open brake lines while the work is being performed.

– Raise car and remove rear wheel.

> ***WARNING —***
> *Make sure the car is firmly supported on jack stands designed for the purpose. Place the jack stands beneath a structural chassis point. Do not place jack stands under suspension parts.*

◀ Disconnect metal brake line from brake caliper. Plug hydraulic fittings.

– Remove brake caliper mounting bolts and remove caliper. See **470 Brakes–Hydraulic**.

◀ Remove two screws (**arrows**) mounting brake rotor to hub and remove rotor.

NOTE —

If rotor hangs up and will not slide off, back off parking brake adjuster as described earlier.

◀ Remove cotter pin (**arrow**) from castellated nut on end of brake cable. Remove nut.

– Remove shoe expander and spring.

Parking brake assembly

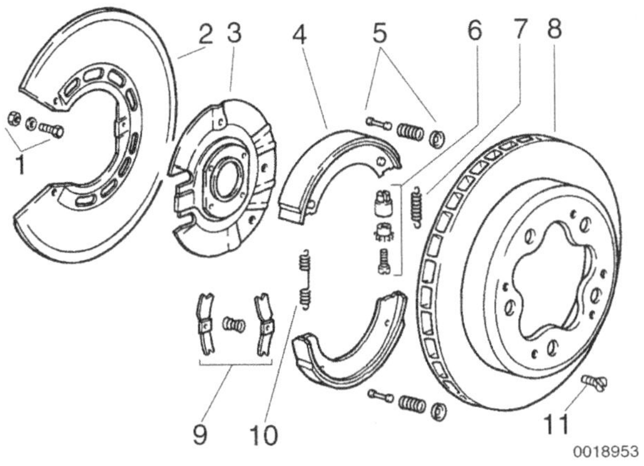

1. Bolt, washer, nut (M8)
   - tighten to 25 Nm (18 ft-lb)
2. Backing plate
3. Brake shoes carrier
4. Brake shoe
5. Shoe hold-down assembly
6. Star-wheel adjuster assembly (Right side shown. Left side is installed upside down from illustration.)
7. Return spring
8. Brake rotor/drum
9. Shoe expander assembly
10. Return spring
11. Brake rotor retaining screw
    - tighten to 5 Nm (3.6 ft-lb)

◀ Remove brake shoe hold-down springs and cups.

– Remove brake shoe adjuster (star wheel) and rear return spring.

NOTE —

Take note of brake shoe adjuster orientation:
–The left side adjuster slotted head points ***up****.*
–The right side adjuster slotted head points ***down****.*

– Pull shoes forward and remove with front return spring.

– Clean and inspect all parts. Replace as necessary.

- Replace oily, worn or cracked shoes.
- Lubricate adjuster to make sure it is functioning properly.

– Installation is reverse of removal.

- Adjust parking brake as described earlier.
- Bleed brakes as described in **470 Brakes–Hydraulic**.

Tightening torques

- Brake caliper to trailing arm (M12) . . . . . 60 Nm (43 ft-lb)
- Brake rotor to hub (M6). . . . . 5 Nm (3.5 ft-lb)
- Hydraulic brake line to caliper. . . . . 14 Nm (10 ft-lb)
- Wheel to wheel hub . . . . 130 Nm (96 ft-lb)

470 Brakes–Hydraulic

Special Tools

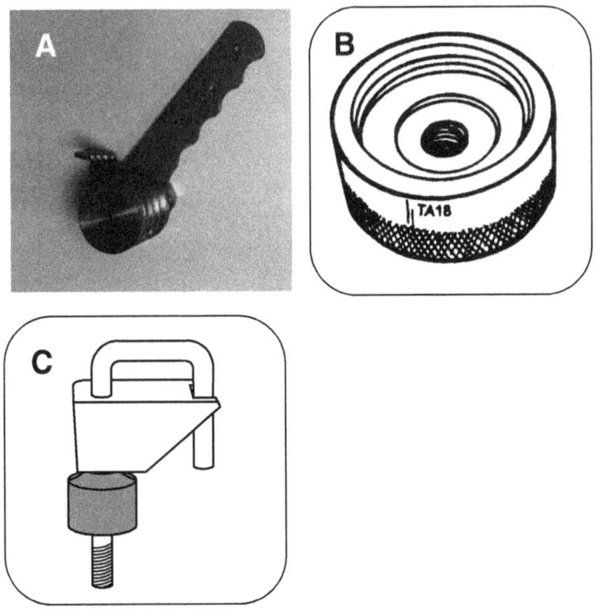

Special tool no.

A . . . Tubing bender
B TA18. . . . ATE brake master cylinder pressure bleeder adapter
C 647 b . . . Flexible line clamp tool
(source: Baum Tools Unlimited)

GENERAL

This repair group covers brake bleeding as well as replacement procedures for hydraulic brake components, including brake master cylinder, brake power booster and brake calipers.

For general specifications, diagnosis and troubleshooting information see **400 Suspension, Brakes and Steering–General**. Repairs to the brake pads and rotors or to the parking brake system are covered in **460 Brakes–Mechanical**. Pedal cluster repairs are covered in **301 Clutch**.

Special tools

Some procedures require special tools.

> ***WARNING —***
> - *Brake friction materials such as brake linings or pads contain abrasive fibers which can lead to illnesses. Do not create dust by grinding, sanding, or cleaning the pads with compressed air. Avoid breathing any brake friction fibers or dust.*
> - *Brake fluid is poisonous, highly corrosive and dangerous to the environment. Wear safety glasses and rubber gloves when working with brake fluid. Do not siphon brake fluid with your mouth. Immediately clean away any fluid spilled on painted surfaces and wash with water. Dispose of brake fluid properly.*
> - *Do not reuse self-locking nuts, bolts or fasteners. They are designed to be used only once and may fail if reused. Always replace them with new self-locking nuts.*

470

Braking system

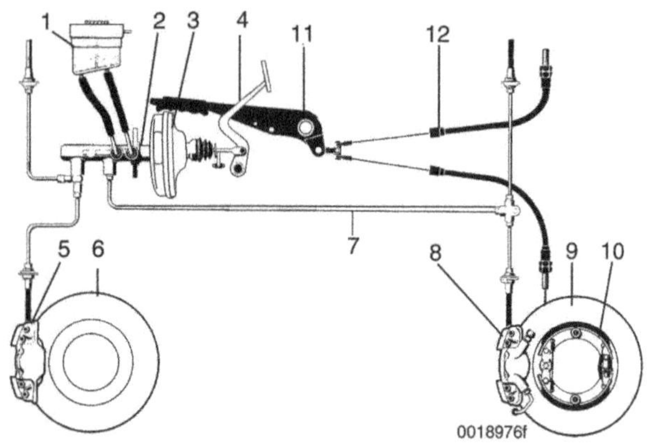

1. Brake fluid reservoir
2. Dual master cylinder
3. Power booster
4. Brake pedal
5. Front brake caliper
6. Front brake rotor
7. Brake fluid line
8. Rear brake caliper
9. Rear brake rotor/drum
10. Parking brake shoes
11. Parking brake handle
12. Parking brake cables

Braking system

All Porsche 911 SC models are equipped with power-assisted four-wheel disc brakes. A brake booster power-assists a dual-circuit master cylinder. A vacuum boost system is used in all models.

Dual-piston calipers act on vented front rotors and rear rotors. The dual-drum parking brake is integrated with the rear brake rotors.

Replacing brake fluid

Porsche recommends that the brake fluid be replaced at least once every two years. This is due to the fact that brake fluid readily absorbs moisture. Moisture in the brake fluid can adversely affect braking performance and may also damage the system, leading to costly repairs.

Replace the brake fluid using one of the procedures described later to expel the old fluid. Remove the filter/strainer from the brake fluid reservoir and clean it in new, unused brake fluid. Using new, unused brake fluid, pump at least 1 pint (500 cc) of brake fluid through each caliper to completely flush the system and expel the old fluid. Then refill the reservoir and bleed the brakes as described above. See **030 Maintenance** for brake fluid specifications. Use only Porsche-recommended brake fluid.

Flushing brake system

Do not rely on flushing alone to clean a system contaminated with dirt or corrosion. The flushing procedure may actually force dirt in the lines into the calipers. To do the job thoroughly, the system must be disassembled and the parts individually cleaned. Use only brake fluid to flush the lines. Alcohol must not be used since it will encourage the accumulation of water in the system.

When preparing to bleed the brakes, be sure to read the following warnings and cautions.

WARNING —
When working underneath the car, make sure the car is firmly supported on jack stands designed for the purpose. Place the jack stands beneath a structural chassis point. Do not place jack stands under suspension parts.

CAUTION —
- *Brake fluid is highly corrosive and dangerous to the environment. Dispose of it properly.*
- *When adding or replacing brake fluid always add new brake fluid from an unopened container.*
- *It is important to bleed the entire brake system whenever any part of the brake hydraulic system has been opened.*

Brake lines

Straight and some pre-shaped brake lines are available as replacements. To replace a brake line, first remove the old line from the car and measure its length. using a new straight line of the same length, bend it to match the pattern of the old line. Each bend radius should be no smaller than about 15 mm (9/16) in.). Check that the protective coating on the brake line is not damaged after bending.

Make sure the new brake line has sufficient clearance from the body and suspension when installed, to allow for suspension bottoming and steering travel.

Brake hoses must be checked for rubber deterioration, cracking or other damage and replaced as necessary.

Tightening torques

- Brake line to brake component . . 10 - 16 Nm (7 - 11 ft-lb)
- Brake line to brake hose . . . . . . 13 - 16 Nm (10 - 12 ft-lb)

CAUTION —
The use of a tubing bender is highly recommended for shaping replacement brake lines. Bending brake lines by hand can cause kinks that weaken the line and restrict fluid flow. Inexpensive tubing benders are available at most auto parts stores.

Troubleshooting

Brake performance is mainly affected by three things: the level and condition of the brake fluid, the system's ability to create and maintain hydraulic pressure, and the condition of the friction components.

Air in the brake fluid will make the brake pedal feel spongy during braking or will increase the brake pedal force required to stop. Fluid contaminated by moisture or dirt can corrode the system. Inspect the brake fluid inside the reservoir. If it is dirty or murky, or is over a year old, the fluid should be replaced.

Visually check the hydraulic system starting at the master cylinder. To check the function of the master cylinder hold the brake pedal down hard with the car stopped and the engine running. The pedal should feel solid and stay solid. If the pedal slowly falls to the floor, either the master cylinder is leaking internally, or fluid is escaping from other points in the system. Check all brake fluid lines and couplings for leaks, kinks, chafing or corrosion. If no leaks can be found, the master cylinder is faulty and should be replaced.

Check the brake booster by pumping the brake pedal approximately 10 times with the engine off. Then hold the pedal down and start the engine. The pedal should fall slightly. If not, check for any visible faults before suspecting a faulty brake booster, and check the one-way check-valve for air flow.

NOTE —
- *See* **400 Suspension, Brakes and Steering–General** *for a brake troubleshooting table.*
- *See* **470 Brakes–Hydraulic** *for hydraulic brake components or brake power-booster repair information.*

Maintenance

– Porsche specifies the maintenance steps below to be carried out at particular time or mileage intervals for proper maintenance of the brakes.
  - Check brake fluid. See **030 Maintenance**.
  - Inspect brake hoses, lines and wheel calipers for leaks.
  - Check brake pads and shoes for wear.
  - Replace brake fluid (described in this section).

BLEEDING BRAKES

Brake bleeding is usually done for one of two reasons; either to replace old brake fluid as part of routine maintenance or to expel trapped air in the system that resulted from opening the brake hydraulic system during repairs.

Brake system bleeding can be done with a pressure bleeder, a vacuum bleeder or manually. Pressure bleeding is recommended as the most efficient method.

When bleeding the brakes, start at the wheel farthest from the master cylinder and progress in the following order:

- Right rear brake
- Left rear brake
- Right front brake
- Left front brake.

Pressure bleeding brakes

– Using clean syringe, remove old brake fluid from reservoir. Clean out reservoir with a lint-free cloth.

BLEEDING BRAKES

◄ Using special tool 647 b or equivalent, clamp off vent line from master cylinder reservoir.

Top off brake fluid and connect pressure bleeder (**arrow**) to reservoir.

– Pressurize system to approximately 1 bar (14.5 psi).

> ***CAUTION —***
> *Do not exceed a pressure of 2 bar (29 psi) when pressure bleeding the brake system. Excessive pressure will damage the brake fluid reservoir.*

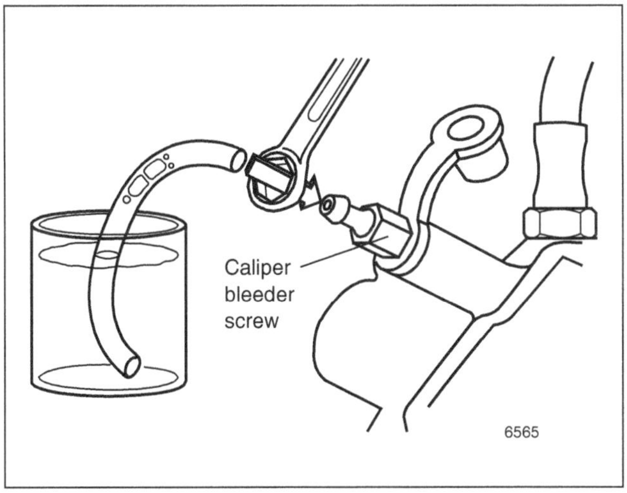

Connect bleeder hose and waste bottle to right rear caliper bleeder screw.

– Have helper hold brake pedal down.

Open bleeder screw (**arrow**). Have helper slowly pump brakes *at least 12 times* with bleeder screw open, holding pedal down on last pump.

- Close bleeder screw and release brake pedal.
- When finished, replace rubber bleeder cap.

Tightening torques

- Bleeder screw (7 mm). . . . . 3 Nm (27 in-lb)

– Refill brake fluid reservoir and proceed to next wheel in order given above.

Vacuum bleeding brakes

– Using clean syringe, remove old brake fluid from reservoir. Clean out reservoir with a lint-free cloth.

– Top off brake fluid and replace reservoir cap.

– Connect vacuum pump to right rear brake bleeder screw. Route discharge hose into waste bottle.

> ***CAUTION —***
> *Vacuum bleeding requires an air-tight seal between the pump and the bleeder screw. Any leakage will result in faulty bleeding.*

– Open bleeder screw and withdraw brake fluid until it is clean and without air bubbles.

NOTE —
Fill the brake fluid reservoir level frequently to prevent it from emptying during brake bleeding.

– Close bleeder screw.

- Replace rubber bleeder cap.

Tightening torques
- Bleeder screw (7 mm). . . . . 3 Nm (27 in-lb)

– Refill brake fluid reservoir and proceed to next wheel in order given above.

Manually bleeding brakes

– Using clean syringe, remove old brake fluid from reservoir. Clean out reservoir with a lint-free cloth. Top off brake fluid and replace reservoir cap.

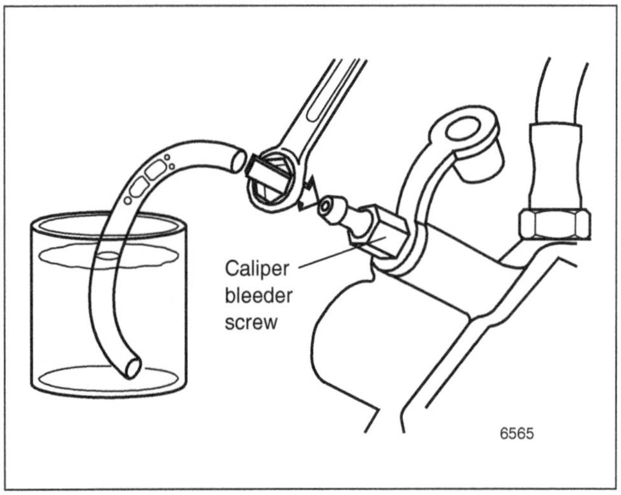

◄ Connect bleeder hose and waste bottle to right rear bleeder screw.

– Have helper pump brakes several times. Then hold pedal down firmly on last pump.

– Open right rear caliper bleeder screw and allow fluid to expel. Close bleeder screw.

– Repeat previous two steps at least 12 times, or until discharged brake fluid is clean and without air bubbles.

NOTE —
Fill the brake fluid reservoir level frequently to prevent it from emptying during brake bleeding.

– Close bleeder screw.

- Replace rubber bleeder cap.

– Refill brake fluid reservoir and proceed to next wheel in order given above.

Tightening torques
- Bleeder screw (7 mm) . . . . 3 Nm (27 in-lb)

BRAKE MASTER CYLINDER AND POWER BOOSTER

The master cylinder is mounted to the rear of the brake booster. The master cylinder, power booster and brake fluid reservoir are mounted in the luggage compartment.

The brake pedal operates the master cylinder via a pedal pushrod that passes through the firewall and operates a reversing lever. The reversing lever in turn operates the master cylinder/brake booster push rod. The master cylinder piston acts on the brake fluid to create pressure in the system.

The master cylinder and brake booster can be removed and installed with ordinary hand tools. Be sure to have an adequate supply of new, unopened brake fluid on hand, as bleeding the brakes, described earlier, will be necessary after installation.

Although master cylinder parts and rebuilding kits are available for most models, replacing the master cylinder as a complete unit is preferable. Whether or not it can be successfully rebuilt depends on its internal condition, which can only be determined after it is disassembled.

The brake master cylinder and power booster are removed and installed as one unit. Once removed from the car and placed on a work bench, the two components can be separated and replaced as necessary.

Brake master cylinder/power booster, removing and installing

– Remove driver's side floor mats and pedal cluster cover.

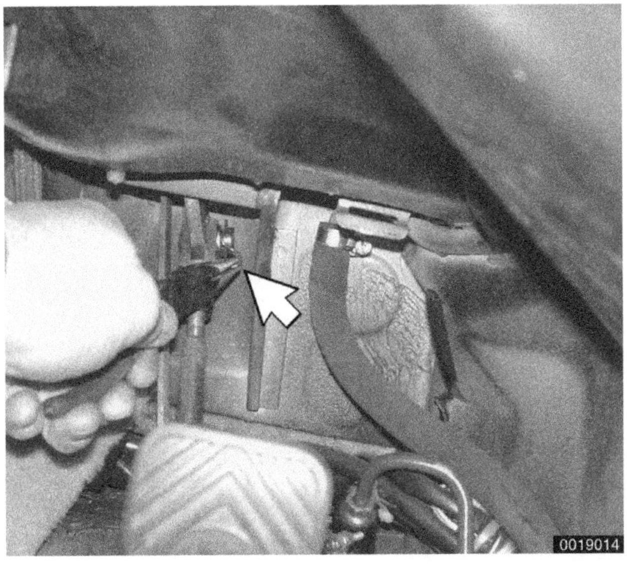

< Working above and forward of pedal cluster, remove locking clip (**arrow**) from brake pedal push rod and pull out pin.

Working above pedal cluster, remove master cylinder mounting bolt (**arrow**) above driver's knees.

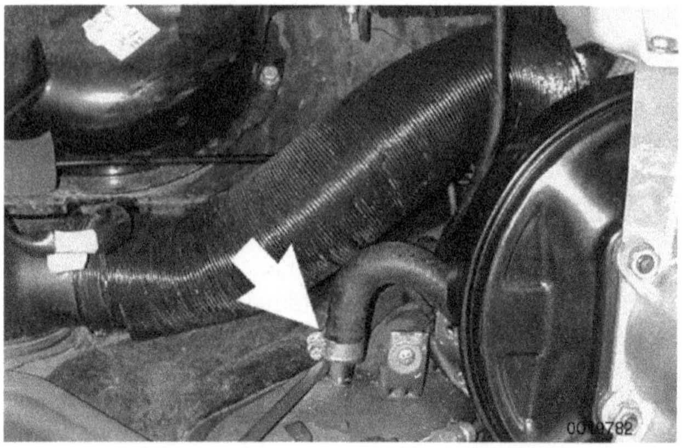

Working in luggage compartment, disconnect power booster vacuum hose at connection in luggage compartment floor (**arrow**).

Detach and remove air duct hoses (**arrows**) behind master cylinder/power booster assembly.

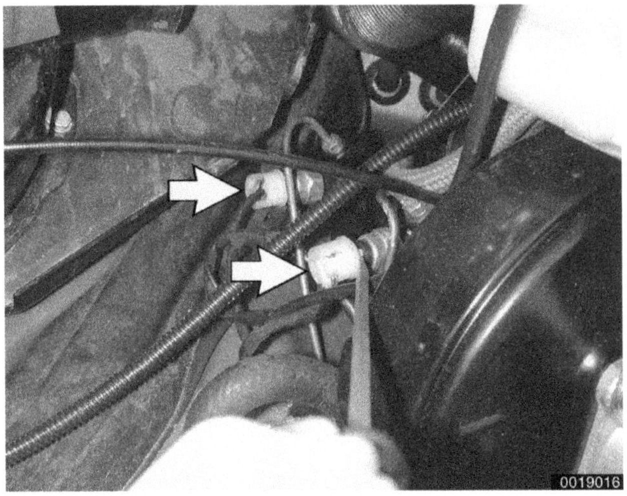

Disconnect both stop light switch electrical harness connectors (**arrows**) at master cylinder.

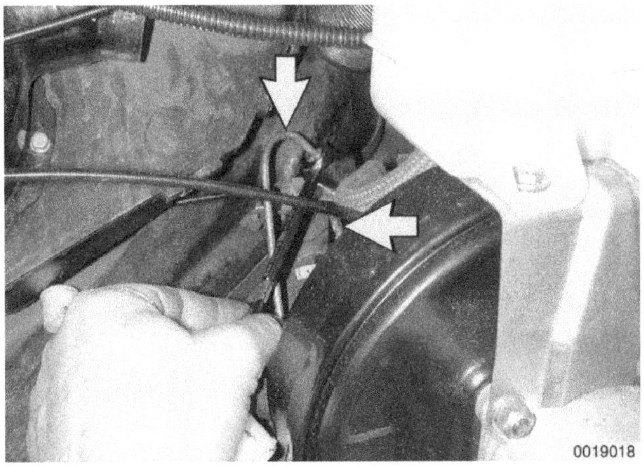

Loosen and remove brake lines (**arrows**) from master cylinder.

– Drain master cylinder reservoir using a syringe.
  - Detach master cylinder feed lines.
  - Remove fluid reservoir and set aside.

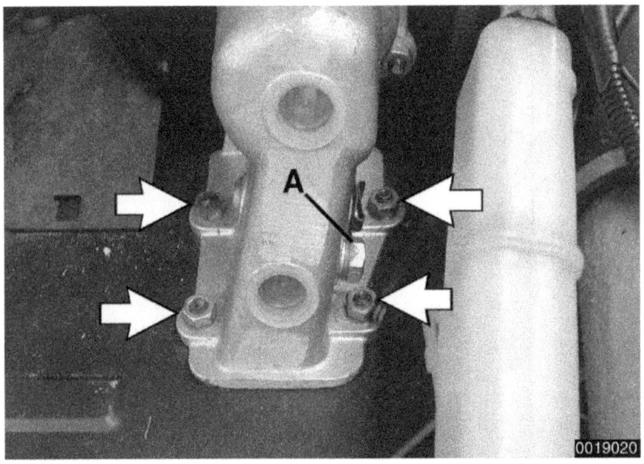

Remove pedal cluster brace bolt (**A**) from side of power booster mounting bracket.

– Remove power booster bracket mounting nuts (**arrows**). Remove master cylinder and power booster as a unit.

0019022

With booster and master cylinder on work bench, remove master cylinder mounting nuts (**arrows**) and pull cylinder out of power booster.

– Installation is reverse of removal. Always replace the sealing O-ring between the master cylinder and the power booster.

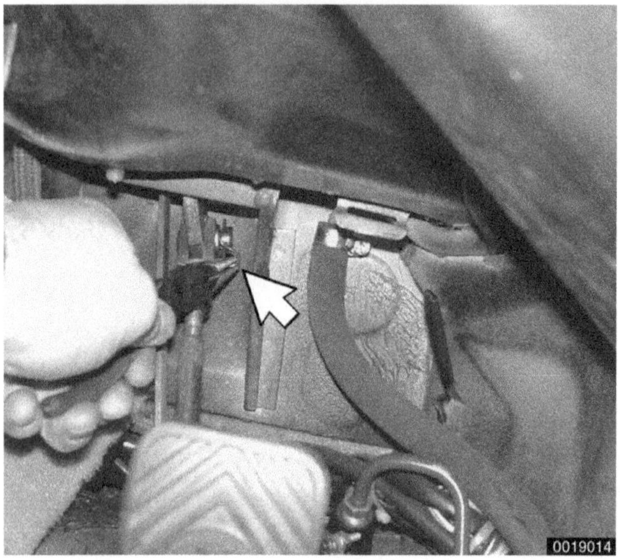
0019014

When attaching brake pedal push rod to reversing lever, be sure to insert pivot pin so that locking clip (**arrow**) can be installed from the right side.

– Bleed complete brake system as described earlier.

Tightening torques

- Bleeder screw (7 mm). . . . . 3 Nm (27 in-lb)
- Brake line to master cylinder. . . . . 14 Nm (10 ft-lb)
- Brake light switch to master cylinder. . . . 15 Nm (11 ft-lb)
- Master cylinder to booster . . . . 25 Nm (18 ft-lb)
- Power booster to luggage compartment floor. . . . . 25 Nm (18 ft-lb)
- Pedal cluster brace to power booster . . . 46 Nm (33 ft-lb)

BRAKE CALIPERS

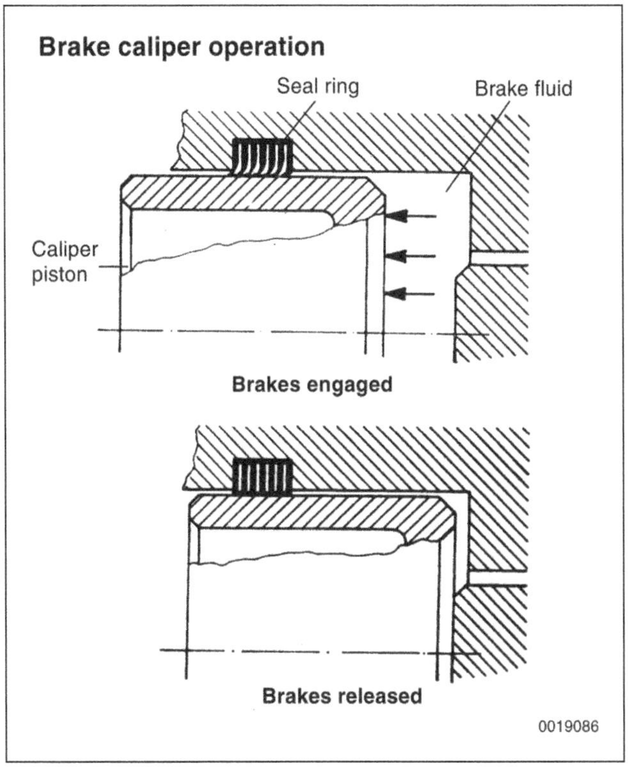

Brake calipers are self-adjusting units. When brake fluid pressure is applied to the caliper piston, the seal ring in the caliper wall deforms slightly, allowing the piston to move out and engage the brakes. When brake fluid pressure is reduced, the seal ring returns to its original shape, pulling the piston back and disengaging the brakes.

As brake pads or rotors wear, the deformation of the seal ring reaches its limit and the piston actually slides out further before regaining a stable position again.

Brake fluid seeping or leaking from around the caliper piston is the result of a failed or damaged piston seal ring. A damaged seal ring is usually caused by corrosion, scoring, or pitting of the piston or caliper bore. The seal can be replaced separately, but a damaged piston will quickly damage the new seal. To remedy a leaking caliper piston seal and avoid future problems, complete replacement of the caliper is recommended.

Brake caliper, removing and installing

This procedure is applicable to both front and rear brake calipers.

NOTE—
It is recommended brake calipers be replaced in pairs.

– Raise car and remove wheels.

> ***WARNING—***
> *Make sure the car is firmly supported on jack stands designed for the purpose. Place the jack stands beneath a structural chassis point. Do not place jack stands under suspension parts.*

– Depress brake pedal approximately 1 inch and hold in place with a pedal depressor.

NOTE—
This will prevent brake fluid from dripping out of open brake lines while the work is being performed.

– Front brakes: Working in wheel housing, disconnect rubber brake hose from metal brake line near bottom of strut. Plug hydraulic lines.

> ***CAUTION —***
> *When working with hydraulic brake lines and hoses, be sure to counterhold fitting before loosening or tightening.*

- Disconnect metal brake line from brake caliper. Plug hydraulic fittings.

- Remove caliper mounting bolts (**arrows**) and remove caliper. (Front caliper shown. Rear is similar).

- Inspect brake rotor for heat damage due to caliper pistons hanging up.

- Installation is reverse of removal.

- Bleed brakes as described earlier.

Tightening torques

- Bleeder screw (7 mm) . . . . 3 Nm (27 in.-lb)
- Brake line to caliper . . . . 14 Nm (10 ft-lb)
- Front caliper to strut . . . . 70 Nm (52 ft-lb
- Rear caliper to arm . . . . 60 Nm (43 ft-lb)
- Wheel to wheel hub . . . . 130 Nm (94 ft-lb)

480 Steering

GENERAL

This repair group covers servicing of the steering system, including steering rack replacement, rack boot and tie rod replacement and repairs to the steering column universal joints.

Troubleshooting information is in **400 Suspension, Brakes and Steering–General**. For information on the suspension or alignment, see the following repair groups:

- **401 Front Suspension**
- **420 Rear Suspension, Axle Shafts**
- **440 Wheels, Tires, Alignment**

Special tools

Some procedures require special tools.

Special Tool

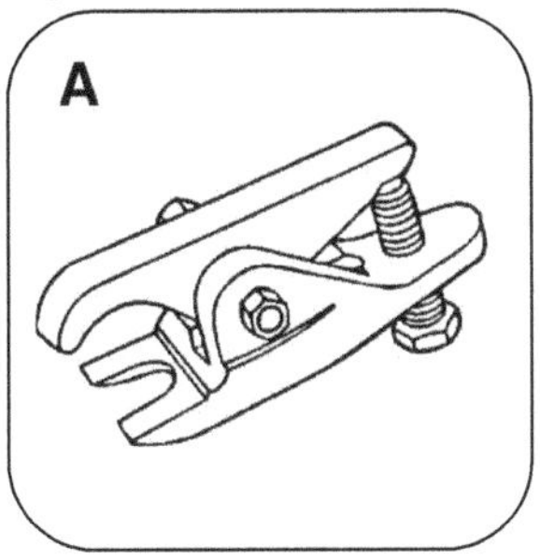

Special tool

A VW 267 a . . . Tie rod removal tool

> ***WARNING —***
> - *Do not reuse self locking nuts. They are designed to be used only once and may fail if reused. Always replace them with new locking nuts.*
> - *Do not install bolts and nuts coated with undercoating wax, as correct tightening torque cannot be assured. Always clean the threads with solvent before installation, or install new parts.*
> - *Do not attempt to weld or straighten any steering components. Always replace damaged parts.*

480

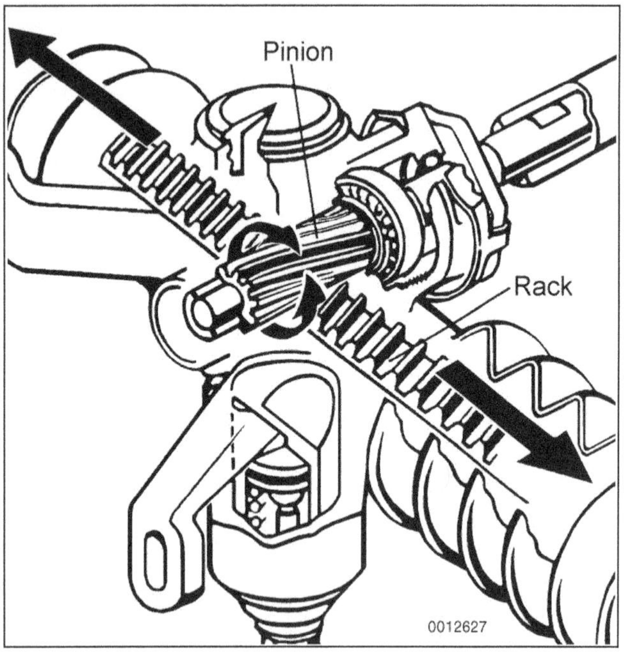

Steering system

The Porsche 911 SC is equipped with manual rack and pinion steering.

The steering wheel and column are connected to the steering rack by a short shaft and two universal joints. Turning the steering wheel turns the rack pinion gear which in turn moves the rack and steering linkage.

Wear and excessive play or clearance anywhere in the steering system will cause sloppy, loose steering. On cars with high mileage, tie rod ends are prone to wear.

If the steering requires too much effort or lacks precise feel, the steering mechanism may not be to blame. Worn suspension parts may also cause or contribute to the symptoms. Before checking for any steering problems, check tires, tire inflation pressures front suspension components.

There are no provisions for lubrication of the steering rack or linkage. In general, the steering system is serviced by the replacement of worn parts.

Table a. Steering specifications

Characteristic	Specification
Frictional moment of steering (measured at flange of steering column flange, tie rods detached)	80 - 140 Ncm (113 - 198 in.-oz)
Steering ratio	17.78 : 1
Steering wheel turns lock-to-lock	Approx. 3
Turning circle diameter	10.95 m (35.93 ft.)

STEERING WHEEL AND COLUMN

NOTE —
Removal and installation of the steering column switches and the ignition switch is covered in **960 Electrical Switches, Interior Lighting**.

Steering wheel, removing and installing

The steering wheel must be removed for access to the steering column switches. The steering wheel is held in place by a 27 mm nut.

– Disconnect negative (–) battery cable.

Pull horn pad straight back to release locking clips.

NOTE —
There are three locking clips, one at top center of pad and one at each lower corner.

0019091

< Disconnect horn wire from horn pad.

– Make sure front wheels are pointed straight ahead. Center steering wheel.

– Remove steering wheel center nut and washer.

– Lightly punch mark or scribe position of steering wheel to steering column shaft.

– Pull steering wheel straight off steering shaft.

– Installation is reverse of removal.
  - Align horn pad with steering wheel locking clips. Then press horn pad down on steering wheel until clips snap into place.
  - Check operation of horn and steering wheel.

Tightening torques

- Steering wheel to column . . . . 50 Nm (36 ft-lb)

Steering wheel, centering

If the steering rack was disturbed while the steering wheel was removed, or if a new unmarked steering wheel is being installed, the steering rack will have to be centered prior to installing the wheel in a straight ahead position.

To center the steering rack, temporarily position the steering wheel on the column and turn until the road wheels point straight ahead. Reposition steering wheel and install lock washer and nut. Back car up a short distance and drive forward a short distance to see if the car is moving in a straight line with the steering wheel centered. Torque steering wheel nut, install horn pad. Road test to check steering wheel position.

Steering column universal joints, removing and installing

Access to the steering column universal joints is through the floor of the luggage compartment. The fresh air plenum and blower must first be removed.

– Disconnect negative (–) battery cable.

– Unsnap luggage compartment floor carpet and lift off.

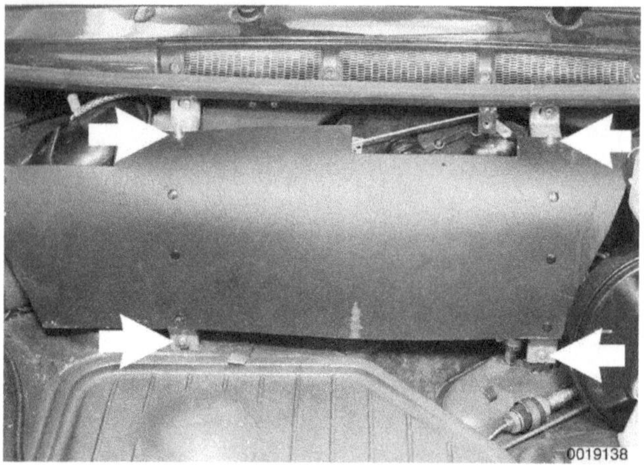

Remove fresh air blower cover retaining fasteners (**arrows**). Remove cover.

Remove screws (**arrows**) holding fresh air vent trim in cowl.

– Remove vent trim and screen.

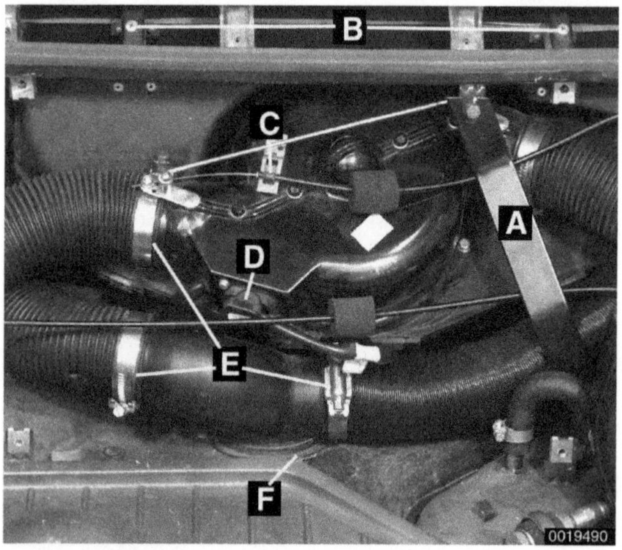

Working inside luggage compartment, detach fresh air plenum and blower connections:

- Remove diagonal brace (**A**).
- Remove screws (**B**) holding fresh air blower housing in place.
- Disconnect heater/defroster cable (**C**) from housing.
- Disconnect fresh air hoses (**D**) from housing.
- Disconnect electrical harness connector (**E**).
- Remove retaining screw (**F**) and pull off large plastic air hose T-fitting.
- Remove blower.

< Remove drain hose (**arrow**) at bottom of fresh air plenum.

– Remove fresh air plenum.

NOTE —
If plenum sticks to cowl, pry carefully to avoid damage to plenum housing.

< Pry off speed nuts (**arrows**) used to hold cover in place over steering shaft. (Brake master cylinder/power booster unit shown removed for clarity.)

- Remove cover.

< Open access door in floor behind fuel tank.

Remove locking nut and bolt (**arrow**) from upper universal joint.

Remove locking nut and bolt (**arrow**) from lower universal joint.

– Remove steering shaft with both universal joints.

NOTE —

*It may be necessary to remove allen bolts (**A**) from steering rack shaft support and to remove bracket.*

Steering universal joints

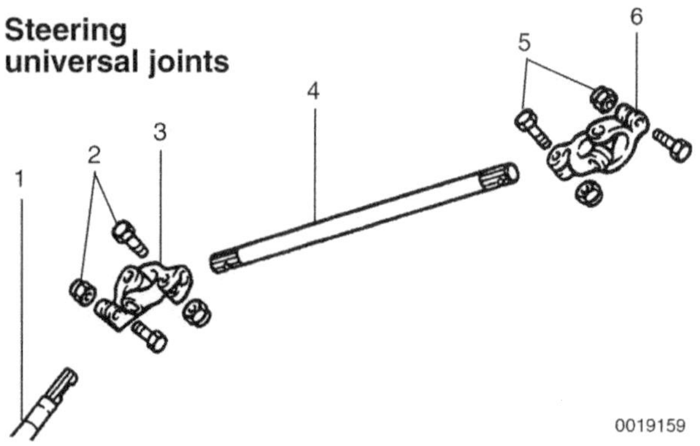

1. Steering rack shaft
2. Universal joint pinch bolt/nut - tighten to 20 Nm (15 ft-lb)
3. Lower universal joint
4. Steering shaft
5. Universal joint pinch bolt/nut - tighten to 20 Nm (15 ft-lb)
6. Upper universal joint

Inspect steering universal joints. Disassemble universal from shaft and replace as necessary.

NOTE —

Always replace universal joint lock nuts.

– Installation is reverse of removal.

Tightening torque

- Steering rack shaft support to body (6 mm allen) . . . . 25 Nm (18 ft-lb)
- Steering universal pinch bolt (M8) . . . . . 20 Nm (15 ft-lb)

Steering column, removing and installing

– Disconnect negative (–) battery cable.

– Working inside luggage compartment, detach upper steering column universal as described earlier.

– Working in passenger compartment, remove steering wheel and steering column switches as described earlier.

< Working under dashboard, remove knee trim panel screw (**arrow**) located above luggage compartment lid release handle.

< Remove left outer A/C vent from dash.

- Remove screw (**A**) from trim panel on left side of dash.
- While pulling on A/C vent, release locking tabs (**arrows**) at top and bottom of vent with a small screwdriver.

< Remove knee trim:

- Working through A/C vent opening, remove trim mounting nut and washer from left stud (**A**).
- Working just below heater controls in center of dashboard, remove trim mounting nut and washer from right stud (**B**).

- Use a small screwdriver to carefully pry tachometer out of dashboard.
  - Disconnect wiring from back of tachometer while removing.

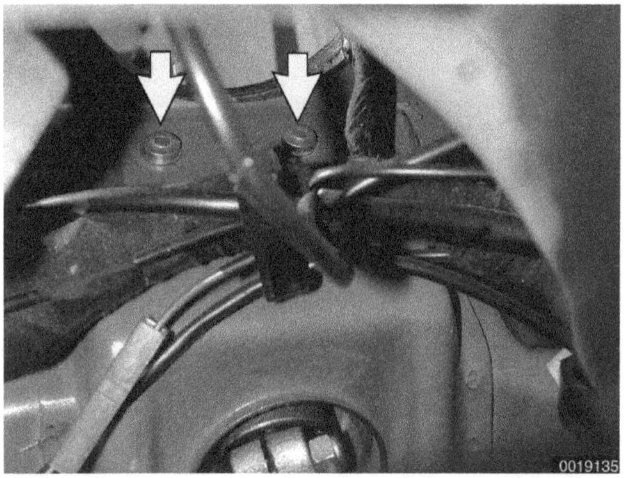

◄ Use a 1/8 in. bit in an angle drill to drill holes in each steering column shearbolt located behind tachometer.

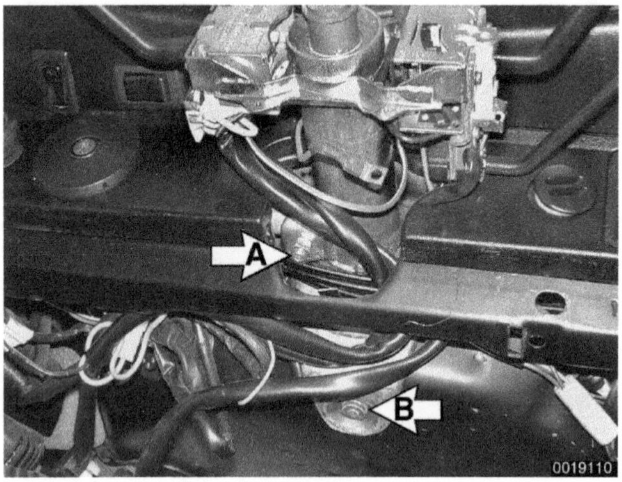

◄ Loosen lock nut and allen bolt (**A**) holding ignition lock arm in column support.

- Drill 1/8 in. hole in shearbolt (**B**) located above driver's knees.
- Use an easy-out to remove all 3 shearbolts.
- Remove column by sliding it to right to release steering lock.
- Installation is reverse of removal.

NOTE—
Always use shearbolts to attach steering lock to steering column.

Tightening torques
- Steering shaft universal joint pinch bolt . 20 Nm (15 ft-lb)
- Steering wheel to column . . . . 50 Nm (36 ft-lb)

Ignition lock, removing and installing

– Disconnect negative (–) battery cable.

< Working under dashboard, remove knee trim panel screw (**arrow**) located above luggage compartment lid release handle.

< Remove knee trim:

- Working from below, remove trim mounting nut and washer from left stud (**A**).
- Working just below heater controls in center of dashboard, remove trim mounting nut and washer from right stud (**B**).

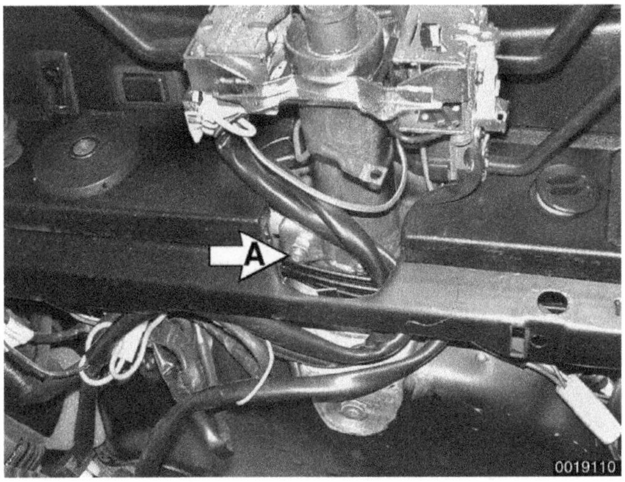

< Loosen lock nut and allen bolt (**A**) holding ignition lock arm in column support.

- Working from rear of ignition lock, disconnect harness connector from ignition switch.
- Remove ignition switch rubber trim.
- Drill out shear bolts (**arrows**) mounting lock to dashboard. Use a screw extractor to remove shear bolts from ignition lock.
- Push ignition lock through dash and remove from behind.
- Installation is reverse of removal. Use new shear bolts and tighten until bolt heads shear off.

STEERING RACK

The illustration shows the basic components of the rack-and-pinion steering gear.

The steering gear and linkage require no maintenance other than alignment and periodic inspection for worn components.

Steering rack assembly

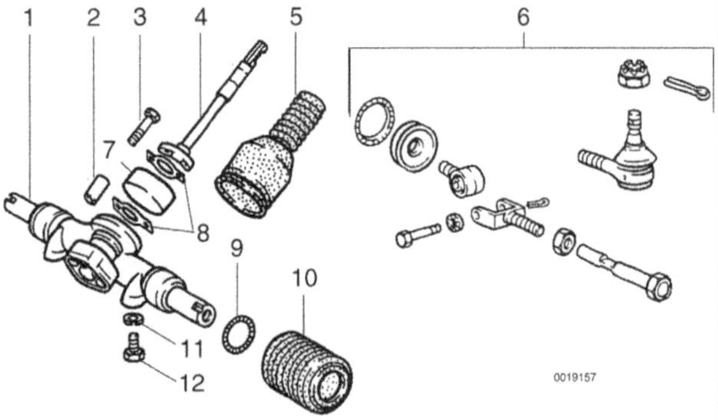

1. Steering rack
2. Spacer bushing
3. Bolt (M8)
   - tighten to 25 Nm (18 ft-lb)
4. Steering shaft
5. Bellows
6. Tie rod assembly
7. Steering shaft coupling
8. Lock plate
9. Spring clamp
10. Rack boot
11. Washer (M10)
12. Bolt (M10)
   - tighten to 47 Nm (35 ft-lb)

Steering rack, removing and installing

- Raise car and remove front wheels.

> ***WARNING —***
> *Make sure the car is firmly supported on jack stands designed for the purpose. Place the jack stands beneath a structural chassis point. Do not place jack stands under suspension parts.*

- Disconnect negative (–) battery cable.

Working underneath front of car, remove stone guard mounting fasteners (**arrow**). Remove stone guard.

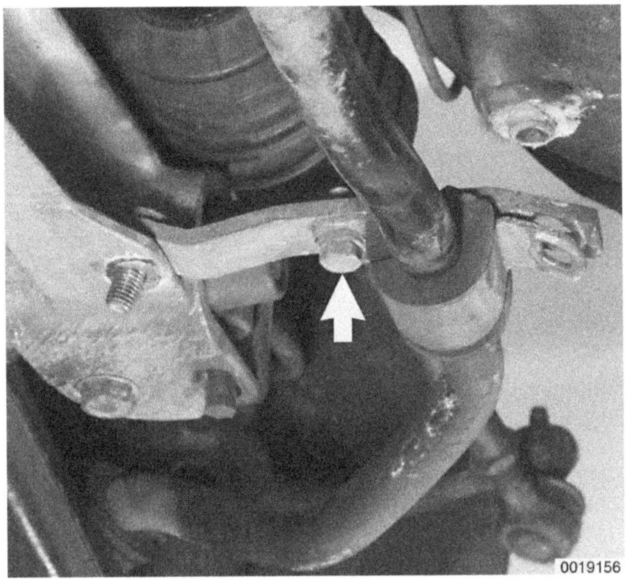

Remove front stabilizer bar mounting clamp bolts (**arrow**). (Right side shown.)

- Remove stabilizer bar.

NOTE —

Remove one end of the stabilizer bar at a time by tapping with a rubber hammer.

Access steering shaft universal joints as described earlier.

- Remove locking nut and bolt (**arrow**) connecting lower universal joint to steering rack.
- Pull universal off steering rack.
- Remove allen bolts (**A**) from steering rack shaft support. Remove bracket.

– Remove cotter pins and castle nuts from both outer tie rod ends.

Use special tool VW 267 a or equivalent to release tie rod from steering arm. Repeat for other side.

– Remove horn and mounting bracket from front suspension subframe.

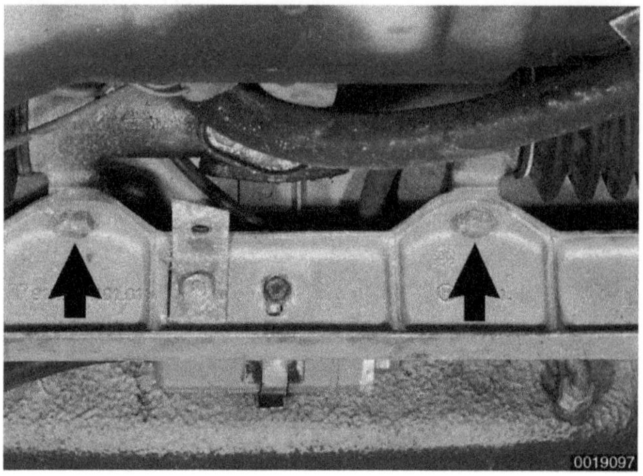

◀ Remove steering rack mounting bolts (**arrows**).

– Remove steering rack, being careful not to damage brake lines and A/C lines.

– Installation is reverse of removal.

NOTE —
Always replace universal joint lock nuts.

Tightening torques

- Steering rack shaft support to body (6 mm allen) .... 25 Nm (18 ft-lb)
- Steering rack to suspension subframe .. 47 Nm (35 ft-lb)
- Steering universal pinch bolt.......... 20 Nm (15 ft-lb)
- Stone guard to body
  M10............................. 45 Nm (33 ft-lb)
  M8.............................. 25 Nm (18 ft-lb)
- Tie rod to steering arm .... 45 Nm (33 ft-lb)
- Wheel to wheel hub .... 130 Nm (96 ft-lb)

Steering rack boot, removing and installing

– Raise car and remove front wheel on side to be repaired.

WARNING —
Make sure the car is firmly supported on jack stands designed for the purpose. Place the jack stands beneath a structural chassis point. Do not place jack stands under suspension parts.

– Remove cotter pin and castle nut from outer tie rod end.

◀ Use special tool VW 267 a or equivalent to release tie rod from steering arm.

Remove spring clamps by sliding over end of boot.

Slide boot off over tie rod.

— Installation is reverse of removal.

Tightening torques

- Tie rod to steering arm . . . . 45 Nm (33 ft-lb)
- Wheel to wheel hub . . . . 130 Nm (94 ft-lb)

Tie rod assembly

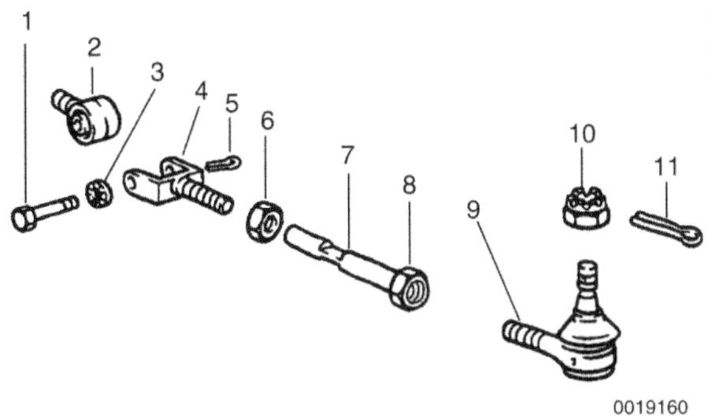

1. Bolt (M10)
   - tighten to 47 Nm (34 ft-lb)
2. Rack end bushing
3. Lock washer (M10)
4. Inner tie rod end yoke
5. Cotter pin
6. Inner tie rod adjuster lock nut (left-hand thread)
   - tighten to 45 Nm (33 ft-lb)
7. Tie rod shaft
8. Outer tie rod adjuster lock nut (right-hand thread)
   - tighten to 45 Nm (33 ft-lb)
9. Outer tie rod
10. Castle nut
    - tighten to 45 Nm (33 ft-lb)
11. Cotter pin

TIE RODS

Each tie rod has two replaceable ends. The outer ball joint end connects to the steering arm with a castle nut. The inner end is a metal yoke that attaches to a rubber bushing on the steering rack end.

It is possible to replace the outer tie rod end, but not the inner, without removing the tie rod from the car

Tie rod, removing and installing

NOTE —
To replace a worn tie rod end, first measure and record the overall length of the tie rod. This will ensure that the alignment will be disturbed as little as possible because of removal and installation.

– Raise car and remove front wheel on side to be repaired.

> ***WARNING —***
> *Make sure the car is firmly supported on jack stands designed for the purpose. Place the jack stands beneath a structural chassis point. Do not place jack stands under suspension parts.*

Loosen inner and outer tie rod adjuster lock nuts (**arrows**).

NOTE —
The inner tie rod adjuster threads are left-handed.

– Remove cotter pin and castle nut from outer tie rod end.

Use special tool VW 267 a or equivalent to release tie rod from steering arm.

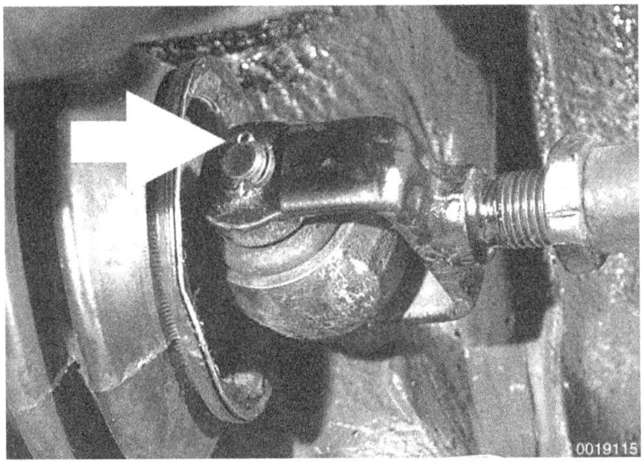

Remove cotter pin from inner tie rod end bolt. Remove bolt.

– Either tie rod end can now be replaced by clamping tie rod shaft in shop vise.

NOTE —
Clean all threads and coat with anti-seize paste.

– Installation is reverse of removal.
- Reset tie rod total length to measurement obtained earlier.
- Have car aligned professionally.

Tightening torques
- Inner tie rod to rack bushing . . . . 47 Nm (34 ft-lb)
- Tie rod adjuster nut to tie rod shaft (M14). . . . . 45 Nm (33 ft-lb)
- Tie rod to steering arm . . . . 45 Nm (33 ft-lb)
- Wheel to wheel hub . . . . 130 Nm (94 ft-lb)

Outer tie rod end, removing and installing

NOTE —
- *Measure and record the overall length of the tie rod. This will ensure that the alignment will be disturbed as little as possible because of removal and installation.*
- *Another technique for maintaining the basic alignment setting is to count the number of turns as the tie rod end is unscrewed and reinstall the new end with exactly the same number of turns.*

– Raise car and remove front wheel on side to be repaired.

WARNING —
Make sure the car is firmly supported on jack stands designed for the purpose. Place the jack stands beneath a structural chassis point. Do not place jack stands under suspension parts.

Loosen outer tie rod lock nut (B).

– Remove cotter pin and castle nut from outer tie rod end.

◀ Use special tool VW 267 a or equivalent to release tie rod from steering arm.

– Unscrew old tie rod end and replace with new.

NOTE —
Clean all threads and coat with anti-seize paste.

– Reassembly is reverse of removal.

- If necessary, loosen inner tie rod adjuster nut and reset tie rod length to measurement obtained earlier.
- Have car aligned professionally.

Tightening torques

- Tie rod adjuster nut to tie rod shaft (M14) . . . . 45 Nm (33 ft-lb)
- Tie rod to steering arm . . . . 45 Nm (33 ft-lb)
- Wheel to wheel hub . . . . 130 Nm (94 ft-lb)

500 Body Assembly–General

GENERAL

The 1978 - 1983 Porsche 911 SC was produced with an all-steel, galvanized body shell. Three body styles were available: Coupe, Targa and Cabriolet.

Targa and Cabriolet models

Both the Targa and Cabriolet models are all-season vehicles with a high level of functionality, excellent interior comfort, and acoustic refinement.

The Cabriolet model, only available in the 1983 model year, uses a manually operated top.

All pillars and frame members are of thin wall sheet metal and have been deep-drawn for maximum strength. This allows for control cables, linkages and heater lines to be accommodated. The outer skin extending from the windshield pillar to the tail end is welded to the floor assembly, the wheel arches and pillars to form a rigid unit.

Front fenders

The front fenders are attached with bolts. The fuel filler flap is located in the left front fender.

Coupe body dimensions

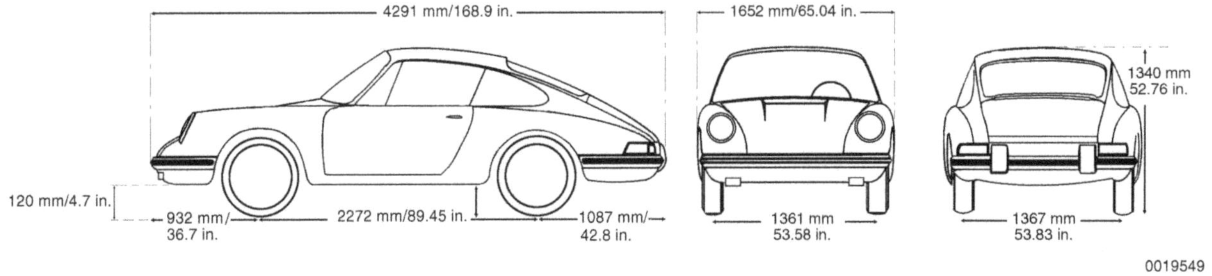

0019549

Bumpers

The bumpers are detachable. They are painted in the same color as the car and have rubber trim strips.

Doors

The doors are fixed to the front pillars of the body with internally fitted hinges.

If equipped with the anti-theft device option, the left door has a built-in alarm lock cylinder (**arrow**).

Lids

The rear lid release handle is located on the drivers door post (B-pillar) and for safety reasons can only be accessed when the door is open. The luggage compartment lid is double-latched. Gas support struts enable the lids to be open in any position. The rear lid covers the engine compartment and carries the air intake grille for the engine.

Seats

Front seats are adjustable backward and forward. The backrest angle can be adjusted all the way back down to the horizontal position. The front seat backrest locks in the upright position.

Luggage compartment

Luggage space is provided in the front compartment under the front lid. The spare wheel, tool kit, battery and fuse/relay panel are also located here. Additional luggage space can be made in the back of the driving compartment if the seat backs of the rear seats are folded forward.

Interior fittings

All interior surfaces are covered with padded leather in many colors and patterns. Perforated leather is used for the headliner. Carpeting covers the interior floor and console as well as the luggage compartment.

For sound proofing and insulation all hollow components are sprayed or painted with rust inhibitor before welding. All surfaces which are not painted in the body color, in particular the floor group with the wheel arch panels and the interior compartment are sprayed with an efficient sound deadening compound which at the same time provides rust protection. For further sound-proofing the passenger and luggage compartments may be covered with bitumen board, bitumen felt, jute felt and cork felt mats.

Heating and air conditioning

Heated air from the engine is blown by the engine compartment blower through heater ducts into the car. The air volume is controlled by either manual levers or an automatic heating control unit mounted in the center console between the seats.

Air conditioning is provided by an A/C system that uses some of the same ducts as the heating system but is otherwise completely separate in its controls.

There is also a fresh air ventilation system for the interior utilizing a separate blower and ducts.

Instruments and controls

Everything in the passenger compartment essential to the driver is logically grouped and easy to reach. All instruments and controls have been arranged ergonomically and are fully integrated into the overall design of the vehicle.

510 Fenders

GENERAL

This repair group covers replacement of the front fenders.

FRONT FENDERS

The front fenders are among the few body panels that are bolted on and can be replaced separately.

For corrosion protection, the galvanized fenders are assembled with a sealing strip between the top mating surfaces to prevent moisture from becoming trapped. This strip should be replaced any time the fenders are removed.

Front fender assembly

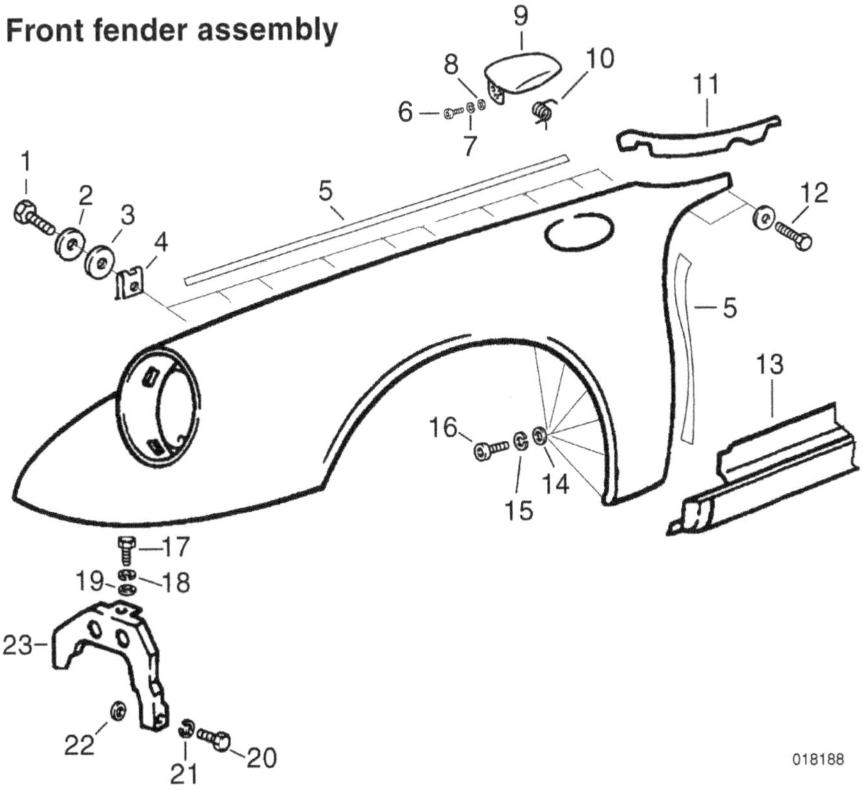

1. Bolt
2. Washer
3. Sealing washer
4. Speed nut
5. Seal
6. Bolt
7. Lock washer
8. Sealing washer
9. Fuel filler door
10. Spring
11. Gasket
12. Bolt
13. Rocker panel
14. Sealing washer
15. Lock washer
16. Bolt
17. Bolt
18. Lock washer
19. Sealing washer
20. Bolt
21. Lock washer
22. Nut
23. Bracket

Fender, removing and installing

- Raise front hood.
- Remove front bumper. See **630 Bumpers.**
- Peel up rubber hood seal on side to be repaired.
- Remove screw securing headlight ring (**arrow**) and remove ring.

- Remove headlight assembly screws (**arrows**). Disconnect electrical harness connector and remove headlight assembly.
- Remove metal wire end from headlight harness plastic plug.
- Withdraw wiring harness from headlight bucket.
- Working inside headlight bucket, remove fender mounting bolt (M8).

- Left fender only:
  - Disconnect cable from fuel tank lid release (**A**).
  - Pry out retaining ring from fuel filler neck sealing bellows (**B**).
  - Remove filler neck bracket bolts (**C**).

– Working at underside of fender, disconnect windshield washer fluid hose from fluid tank.

– Working inside wheel housing, remove bolts (qty. 6) along fender vertical mounting edge.

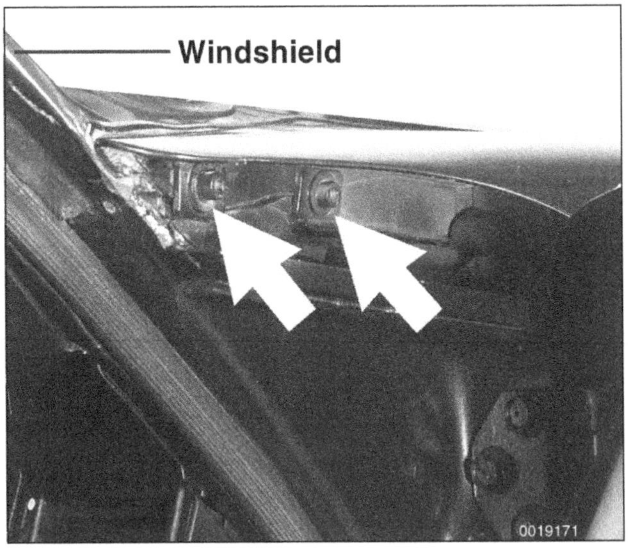

< Open door fully. Remove two screws below windshield (**arrows**).

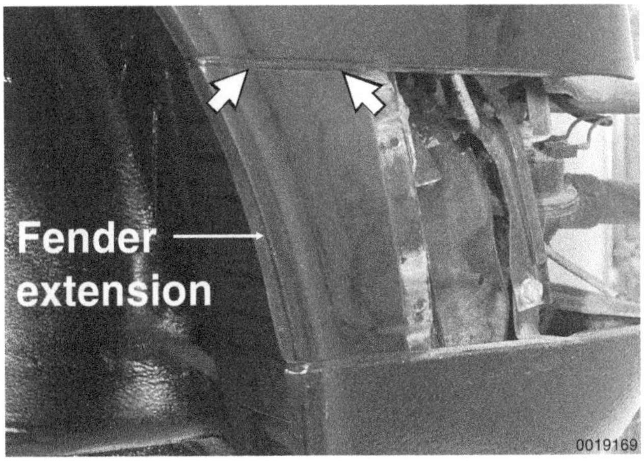

< Working from inside wheel housing, remove upper screws securing fender extension to fender (**arrows**).

– Remove screws securing rubber bumper cover to fender.

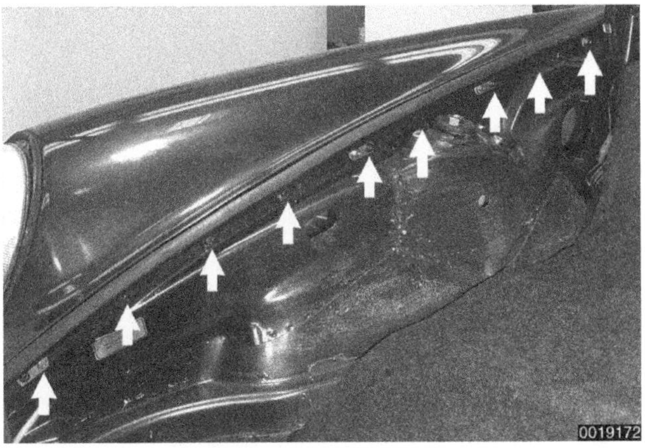

< Working inside luggage compartment, remove mounting bolts along fender edge (**arrows**).

– When removing right side fender, remove upper bolt from oil cooler bracket.

– Carefully separate fender from body and remove.

– When installing new fender, clean off old sealant and protective coating from body mounting surface.

– Replace all gaskets and seals.

– Position fender in place and loosely install all mounting fasteners. Align fender correctly with door and windshield pillar (A-pillar), then tighten bolts.

– Remainder of installation is reverse of removal.

- Align panel so that even gaps exists at all joints.
- Repair any paint damage and paint any exposed metal.

Fender gaps

- Fender to door . . . . 4 mm (0.16 in.)
- Fender to hood . . . . 3 mm (0.12 in.)
- Wheel housing to top edge of fender . . . 17 mm (0.67 in.)

550 Engine Lid, Hood

GENERAL

This repair group covers removal and installation of the hood and engine lid.

FRONT HOOD ASSEMBLY

Front hood assembly

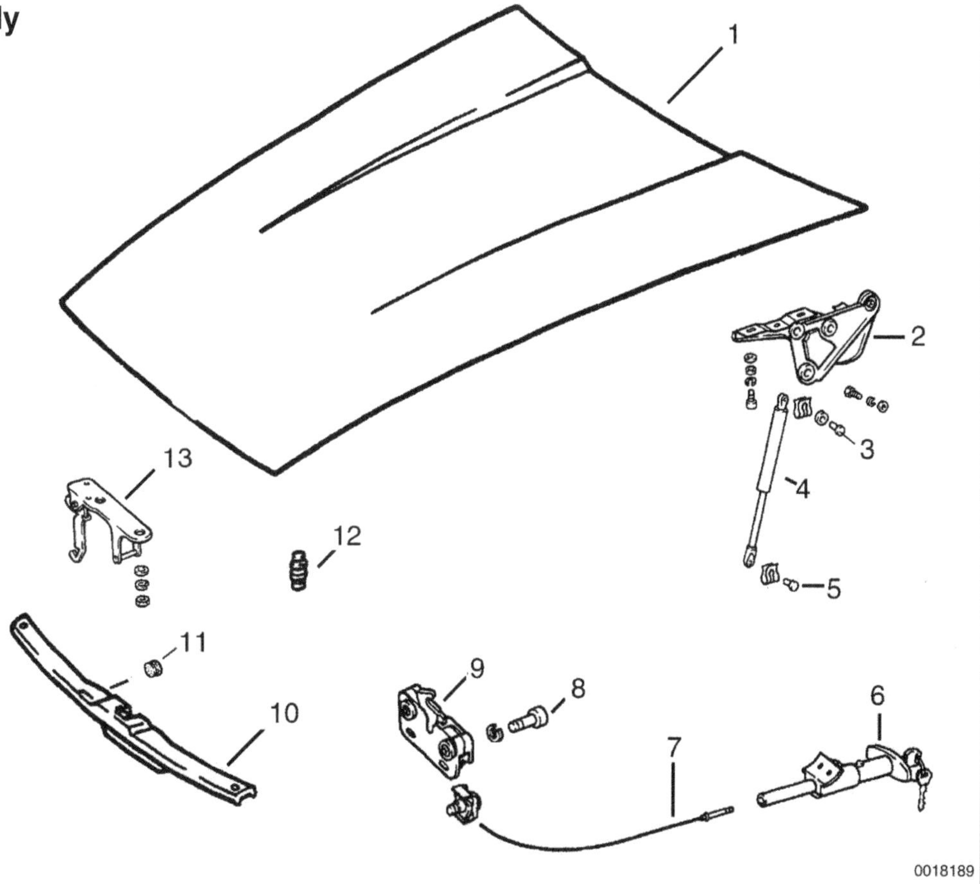

1. Hood
2. Hinge
3. Hinge fasteners
4. Gas support strut
5. Gas support pin and locking clip
6. Locking hood release
7. Hood release cable
8. Bolt
9. Hood lock assembly
10. Hood lock trim
11. Trim plug
12. Rubber stop
13. Hood latch plate

Front hood, removing and installing

- Raise hood.
- Disconnect harness connector for luggage compartment light. Unclip harness from hood hinge. Remove light switch retaining nut from switch.
- Precisely mark hinge location on hood panel.

- Loosen hood mounting bolts on left and right sides (**arrows**).
- With aid of a helper, support hood while removing mounting bolts. Remove hood, guiding electrical harness through hinge.
- Installation is reverse of removal.
  - Repair any paint damage and paint any exposed metal.
  - Check hood alignment as described next.

Hood, aligning

When installing the hood, align hinges as close to the original position as possible.

- Align hood so that gaps are even on sides and rear.

Hood gaps

- Hood to cowl panel . . . . 4 - 5 mm (0.16 - 0.20 in.)
- Side gaps . . . . 4 mm (0.16 in.)

- Set hood to same height as fender joint edge by adjusting threaded rubber stops on either side of hood latch support.

Hood latch, adjusting

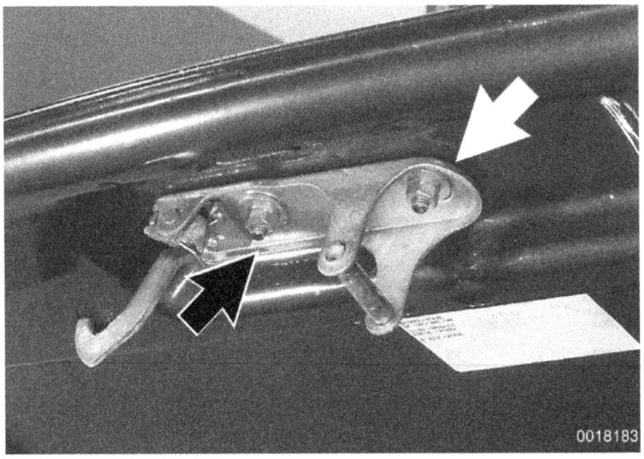

Before adjusting hood latch, be sure the hood is aligned evenly to fenders and cowl panel.

The latch must locate in the center of the lock to close properly. The height at the front of the hood can be altered by sliding the lock assembly up or down on its mounting bolts.

- Loosen latch plate bolts (**arrows**). Tighten bolts finger tight.
- Lower, **but do not latch**, hood so that latch plate pin centers itself in latch.

> ***CAUTION —***
> *Do not let hood lock with latch plate bolts loose.*

- Open hood without altering position of latch plate. Tighten latch plate mounting bolts.

Hood cable, replacing and adjusting

- Raise hood.
- Remove broken cable.
- Install new cable and thread end through latch release arm and into cable clamp (**arrow**).

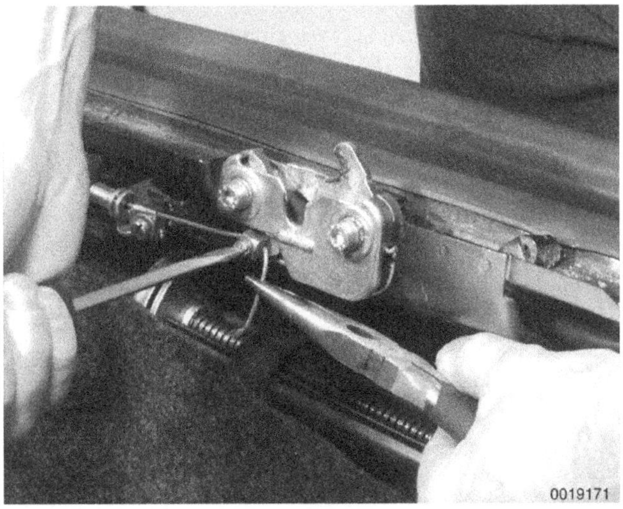

- While holding cable using pliers, tighten cable clamping screw.
- Check for proper hood latch operation before closing hood. While a helper pulls hood release, check that latch releases smoothly and returns to the locked position. If necessary re-adjust cable.

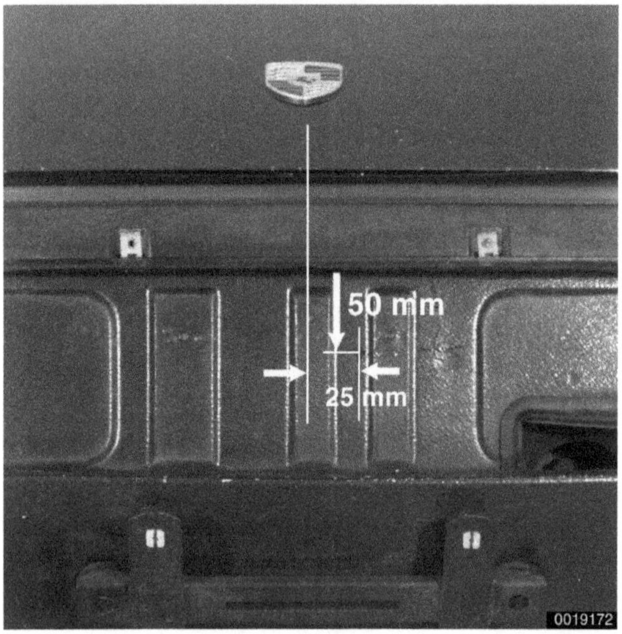

Hood, emergency opening with broken cable

– Remove front bumper assembly. See **630 Bumpers**.

◄ Locate and punch mark position 50 mm (2 in.) down from bottom of bumper trim panel and 25 mm (1 in.) to the left of body center line. Drill an 8 mm (5/16 in.) hole through front body panel.

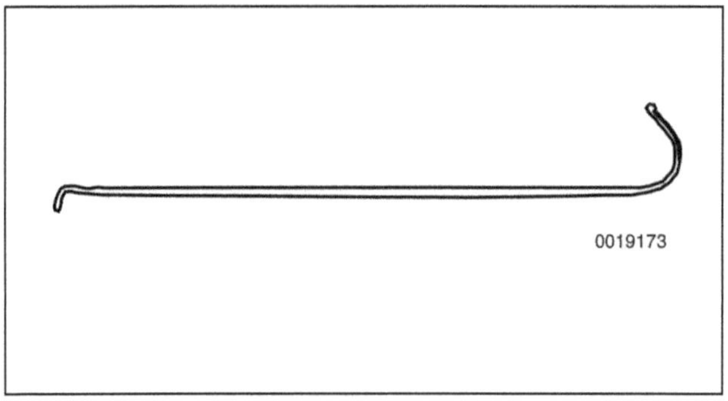

◄ Using a piece of stiff steel wire (6 mm or 1/4 in.) approximately 40 cm (16 in.) long, make a 90° bend 15 mm (0.6 in.) from wire end.

– Insert tool upward through hole in panel to catch hood release. Turn tool to right to release lock.

– Plug drilled hole (use Porsche part no. 999 703 044 50).

– Reinstall front bumper.

Hood hinge, replacing

– Raise hood and support in up position.

– Mark hinge location on hood.

NOTE —
When replacing both hood hinges, do one hinge at a time.

◄ Remove hood mounting bolts (**arrows**).

– Disconnect hood support strut from hinge.

- Remove upper hood strut pin locking clip and remove pin.

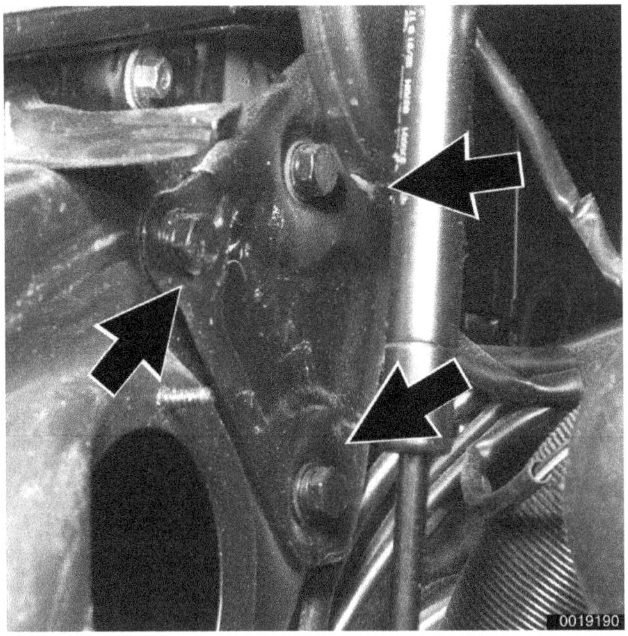
0019190

◄ Remove hinge mounting bolts (**arrows**) and remove hinge.

– Installation is reverse of removal.
  - Repair any paint damage and paint any exposed metal.
  - Check hood alignment as described earlier.

Hood support struts, replacing

When replacing both hood support struts, do only one strut at a time.

– Raise hood and support in up position.

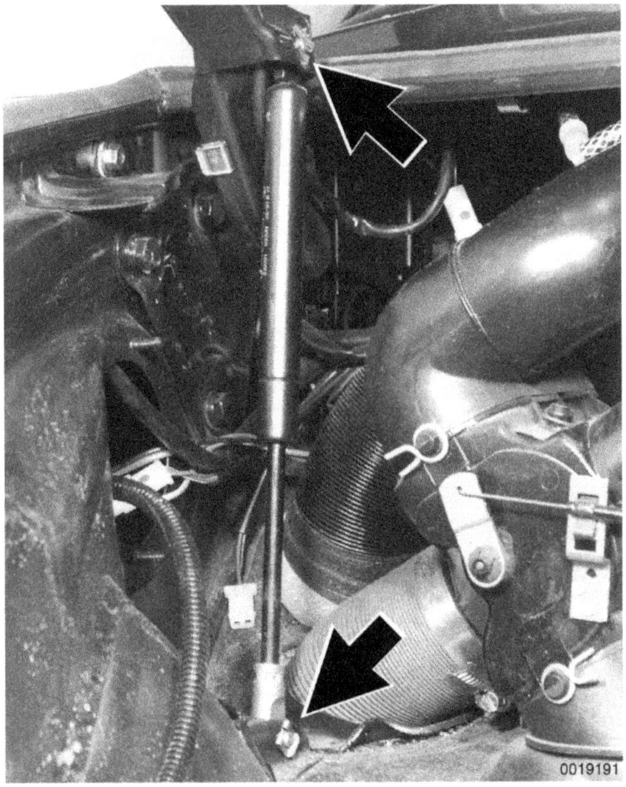
0019191

◄ Remove locking clip and clevis pin from support strut upper and lower mountings (**arrows**).

– Place new strut in position and install lower pin and locking clip.

– While compressing strut slightly, install upper pin and locking clip.

ENGINE LID

Engine lid assembly

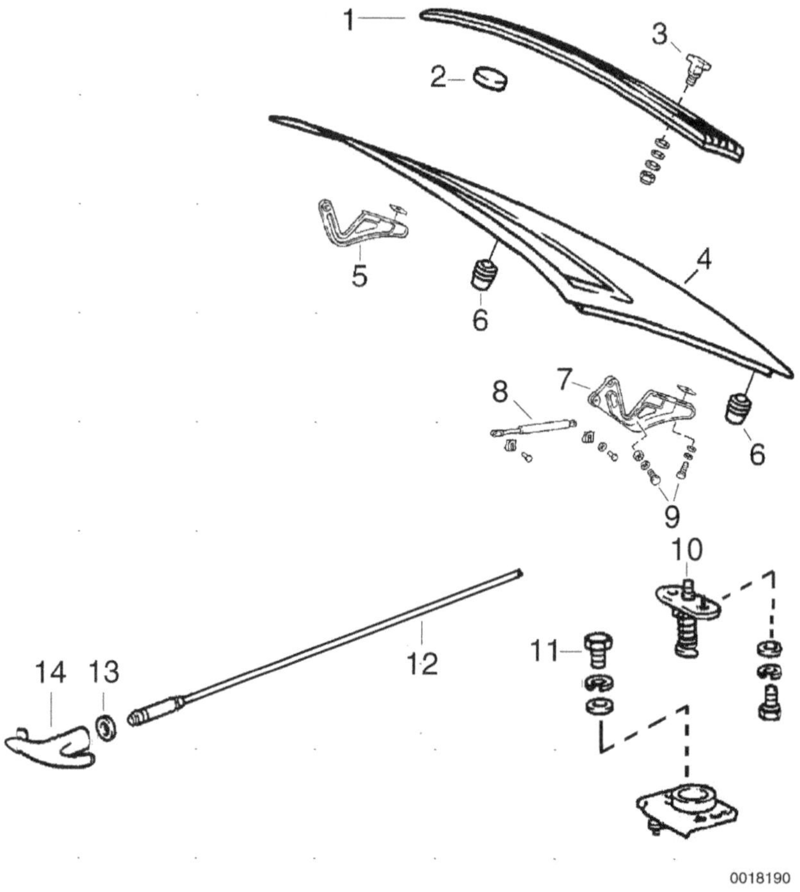

1. Grill
2. Cap
3. Screw
4. Engine lid
5. Right hinge
6. Lid stop
7. Left hinge
8. Gas support pin
9. Hinge bolts
10. Lock
11. Bolt
12. Lid release cable
13. Washer
14. Handle

Engine lid, removing and installing

– Raise engine lid.

◄ Remove 2 screws from A/C condenser center bracket and remove bracket (**arrow**).

> ***CAUTION —***
> *Do not loosen or disconnect A/C refrigerant lines from condenser.*

– Disconnect engine compartment light wiring harness connector and unclip harness from right hinge.

– Mark location of hinges on engine lid.

◄ Loosen engine lid mounting bolts on left (**arrows**) and right sides.

– With aid of helper, support lid and remove hinge mounting bolts, noting position of any shims. Remove lid.

– Installation is reverse of removal. Install shims in same place as originally installed. Align lid to hinges using marks made during removal. Align engine lid so that gaps are even on sides and rear.

Engine lid, aligning

– Open engine lid.

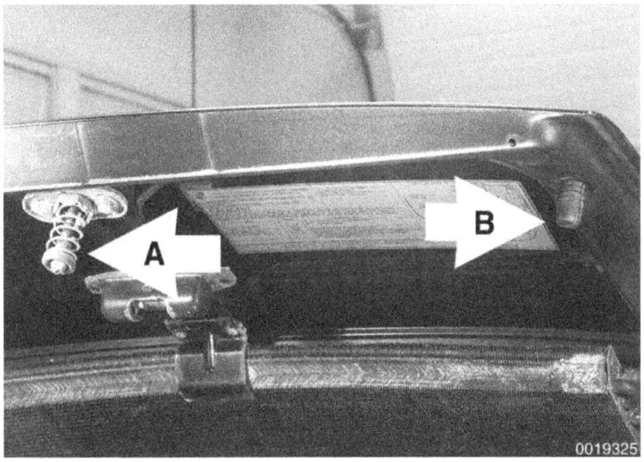

– Loosen engine lid hinge bolts and then tighten finger tight.

– Close lid but do not latch and align lid evenly to opening in body.

Engine lid gaps
- Top . . . . 4 - 5 mm (0.16 - 0.20 in.)
- Sides. . . . . 4 mm (0.16 in.)

– Open engine lid and tighten hinge mounting bolts.

◄ Set engine lid to same height as fender joint edge by adjusting threaded latch pin (**A**) and threaded rubber stops on left and right sides of lid (**B**).

Engine lid grill, removing and installing

– Remove engine lid as described earlier.

NOTE —
Do not disconnect A/C lines from condenser.

◄ Remove grill mounting nuts from rear side of lid and remove grill from front.

– Installation is reverse of removal.

Engine lid strut, removing and installing

- Raise engine lid and support in up position.
- Remove clips and pins from support strut.
- Install pins and clips in new strut.

NOTE —
Models with rear spoiler may have two struts.

570 Doors

GENERAL

This repair group covers door repair information, including removal and installation of the door and the interior door panel.

NOTE—
- *For information on the power windows and door glass replacement, see* **640 Windows**.

Door assembly

1. Door handle
2. Door lock cylinder
3. Base bracket
4. Washer
5. Nut
6. Door
7. Plug
8. Vapor barrier
9. Door check
10. Nut
11. Cover
12. Bushing
13. Shim
14. Hinge
15. Bolt
16. Hinge pin
17. Gasket

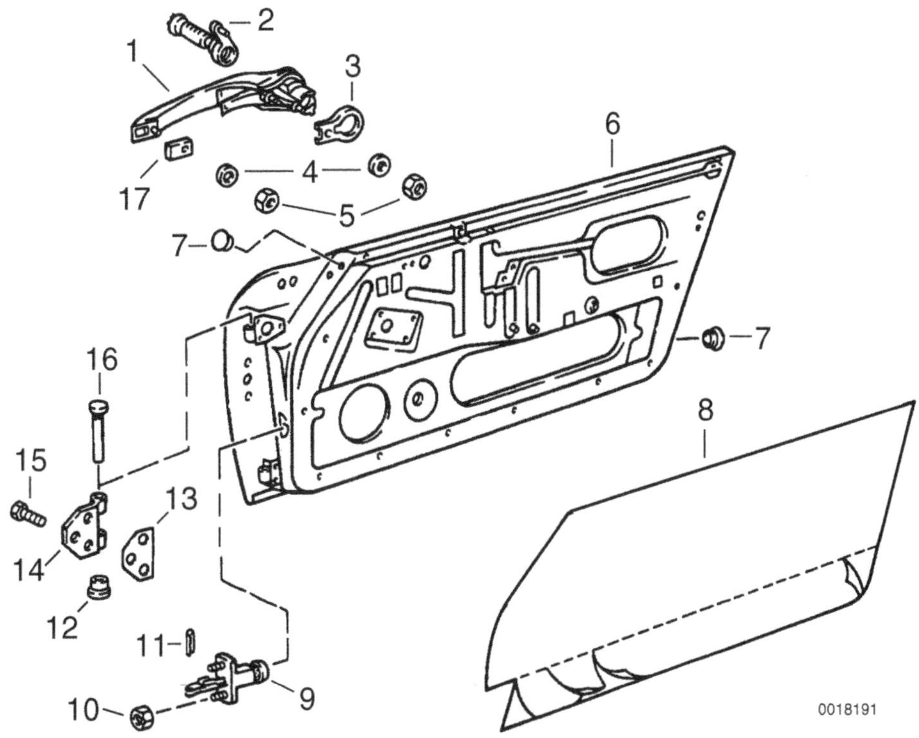

DOORS

Door, removing and installing

- Disconnect battery negative (–) cable.
- Remove door panel. See **Door panel, removing and installing**.
- Label and disconnect wiring from all switches. Arrange wiring so that it can be easily withdrawn through door once door is unbolted.

Table a. Window switch wiring connections

Switch terminal	Left window driver's door switch	Right window driver's door switch	Right window passenger door switch
1	Black	Red / white	Black
2	Green	Red / black	Green
3	Red	Red	Red / white
4	Brown	Brown	Brown
5	Double red	Double red / blue	Red / black

0019263

While supporting door, remove upper and lower hinge bolts on body. Carefully withdraw wiring harnesses when removing door.

- Installation is reverse of removal, noting the following:
  - Align door so that panel gaps are equal on either side, as described later. If necessary adjust hinges as described later.
  - Adjust door striker so that trailing edge of door is slightly higher than front edge.
  - Repair any paint damage and paint any exposed metal.

Door panel, removing and installing

0019197

Remove trim screw and trim piece (**right arrow**) at front door edge. Remove two screws behind trim piece.

- Pry off concealing plug at rear of trim panel and remove screw behind plug (**left arrow**).
- Remove upper trim panel by lifting up and over door lock pin.

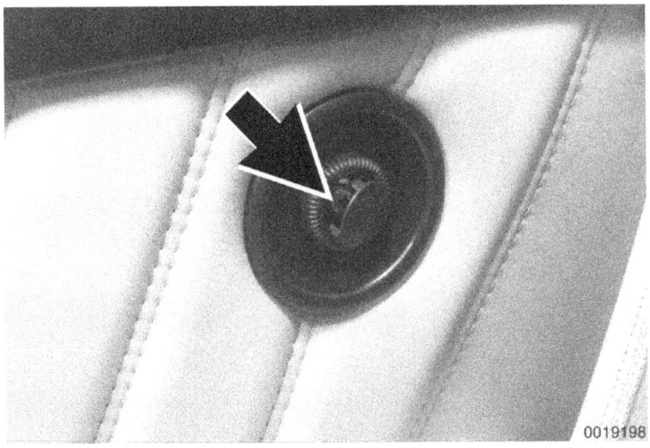

◄ Pry off cover from door lock knob (**arrow**). Remove knob mounting screw and knob.

– Remove lock knob trim by turning counterclockwise.

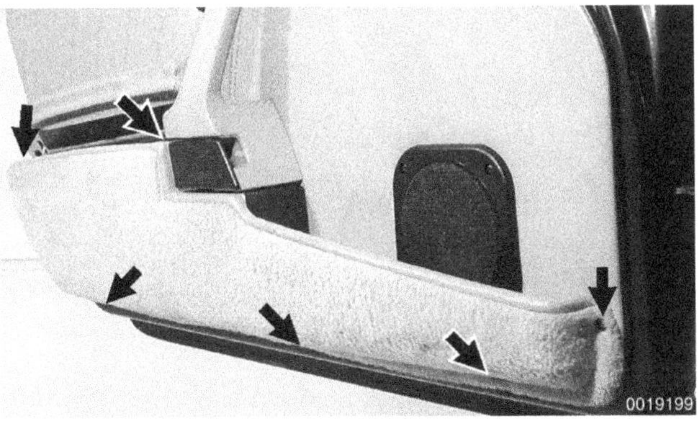

◄ Remove screws from map pocket perimeter (**arrows**) and remove carpeted map pocket.

– Remove two screws from pocket door hinge and remove map pocket door.

– Remove speaker grill screws and remove grill and speaker.

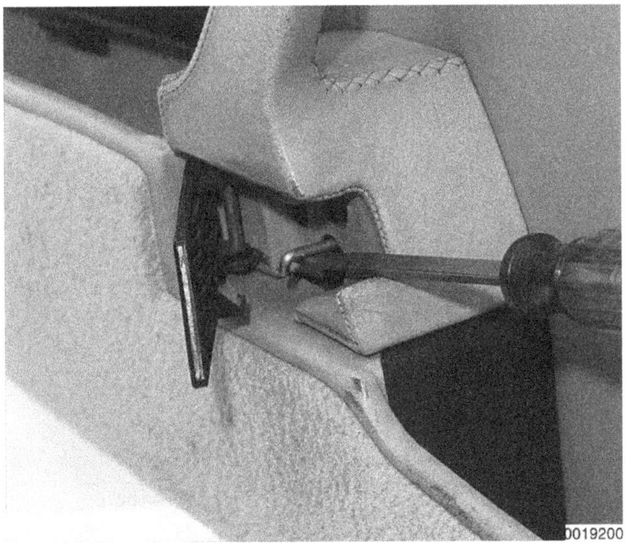

◄ Pull door handle out and pry out actuating rod from handle.

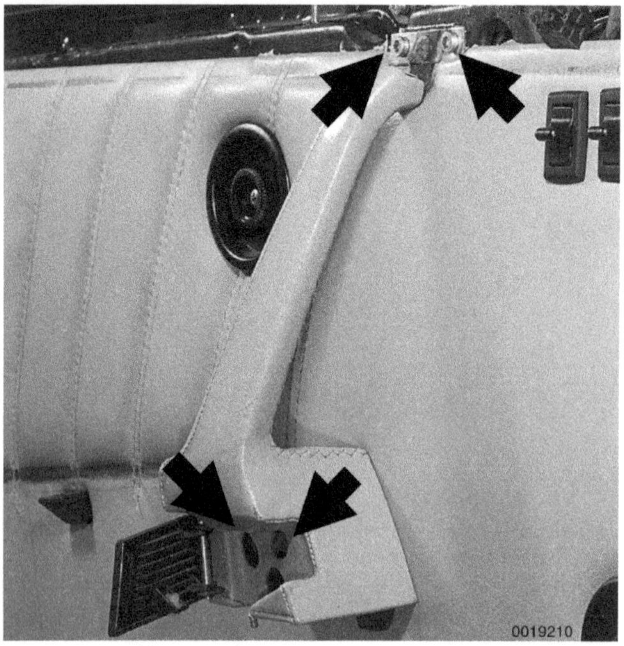

⊲ Remove bolts at bottom and top of door pull (**arrows**).

– Pry out power window switches from door panel.

NOTE —
Label wires on switches before disconnecting.

– Carefully unclip door panel at perimeter and remove.

– Installation is reverse of removal. Check door trim clips for damage. Replace any broken clips before reinstalling door panel.

Door check, removing and installing

– Disconnect battery negative (–) cable.

– Remove interior door panel as described earlier.

⊲ Drive out door check pivot pin (**arrow**).

– Remove door check mounting bolts at door edge (**arrows**). Withdraw door check from inside door.

– Installation is reverse of removal.

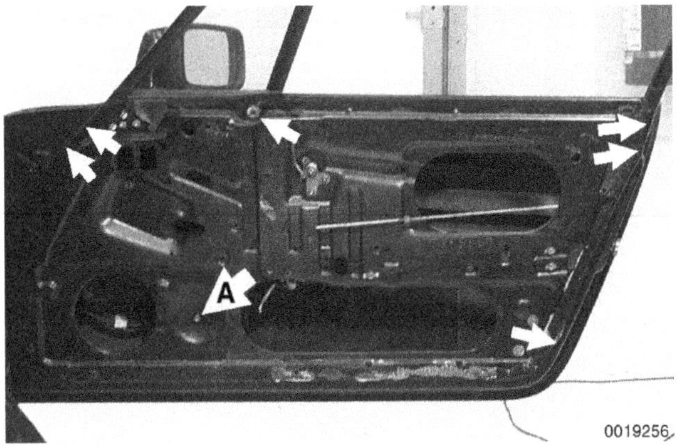

Door window frame, removing and installing

– Remove door panel. See **Door panel, removing and installing**.

– Remove windows glass. See **640 Windows**.

< Remove window frame mounting bolts (**arrows**).

– Working inside door, remove bracket from lower end of window channel (**A**).

– Slide window frame up an out of door.

– Installation is reverse of removal.

- Align marks on window frame with door.
- Check that window seats squarely in opening and frame closes evenly on door gasket.
- Adjust window frame if necessary. The frame tilt (in-out) can be adjusted at rear and middle channel bracket. Frame height (up-down) can be adjusted at two upper bolts at front of door.

Door lock assembly

1. Locking knob
2. Locking rod
3. Lock control rod
4. Support
5. Socket
6. Latch assembly
7. Washer
8. Screw
9. Return spring
10. Concealing plug
11. Screw
12. Washer
13. Rotary Knob
14. Trim cover
15. Screw
16. Locking swivel
17. Socket
18. Washer
19. Gasket
20. Latch release swivel
21. Release cable
22. Support
23. Support
24. Release rod

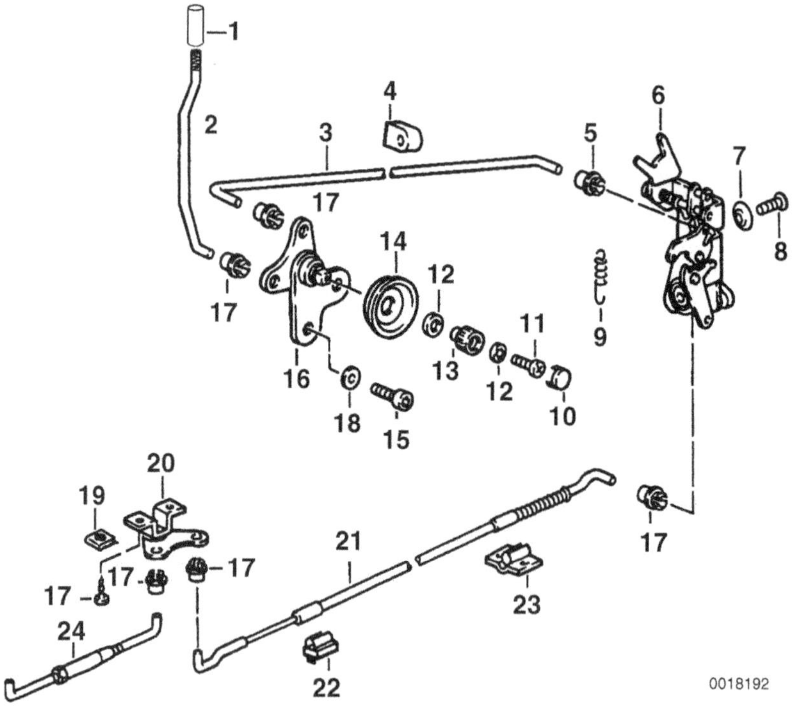

DOOR LOCK MECHANISM

Door latch, removing and installing

– Remove door panel. See **Door panel, removing and installing**.

– Remove window glass. See **640 Windows**.

– Remove window frame. See **Door window frame, removing and installing**.

– Working inside door, disconnect release cable and power door lock control rod from latch assembly.

◀ Disconnect locking rod from locking swivel on door. Remove rod support from door (**arrows**).

Working at door exterior, remove latch bolts (**arrows**) and remove latch from inside door.

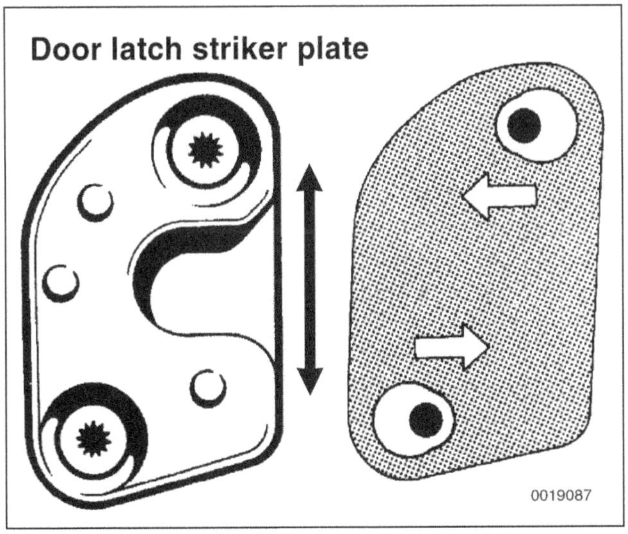

Installation is reverse of removal. If necessary, adjust latch striker position and tighten bolts.

Door outside handle, removing and installing

- Close window completely.
- Remove door inner trim panel. See **Door panel, removing and installing**.
- Disconnect lock wiring harness from connection at bottom of door.

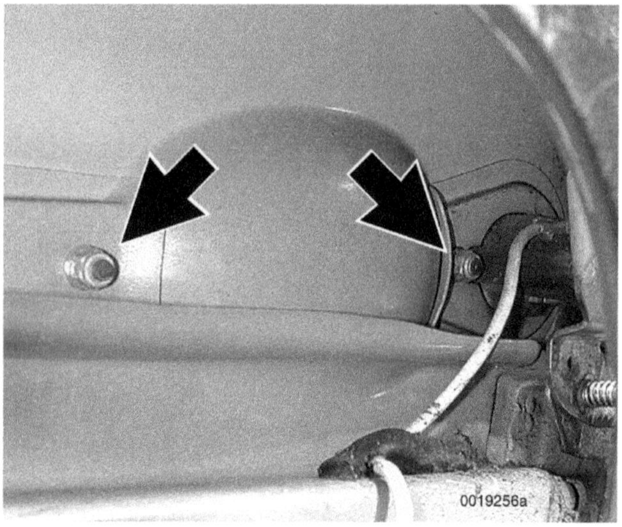

- ◄ Working from inside door, remove door handle securing nuts (**arrows**).
- Carefully pull door handle out of door.
- Installation is reverse of removal. Use new gaskets between door and body when installing handle.

Door striker, replacing

- Open door. Mark position of striker on body.

- ◄ Remove bolts securing striker to body.
- Remove striker and striker shims (where applicable).
- Installation is reverse of removal.

601 Sunroof

GENERAL

The electric sunroof is operated by a motor drive and drive cables. The sunroof panel can be adjusted without removing it from the car. Replacement of the sunroof liner or components such as the cable assembly require that the sunroof panel be removed. The sun roof drive motor is protected from damage by a slip coupling fitted between the motor and the drive.

NOTE —

Be sure to check the front and rear drains in the corners of the sunroof for blockage. If water is entering through the sunroof one of more of these drains are usually plugged.

Sunroof assembly

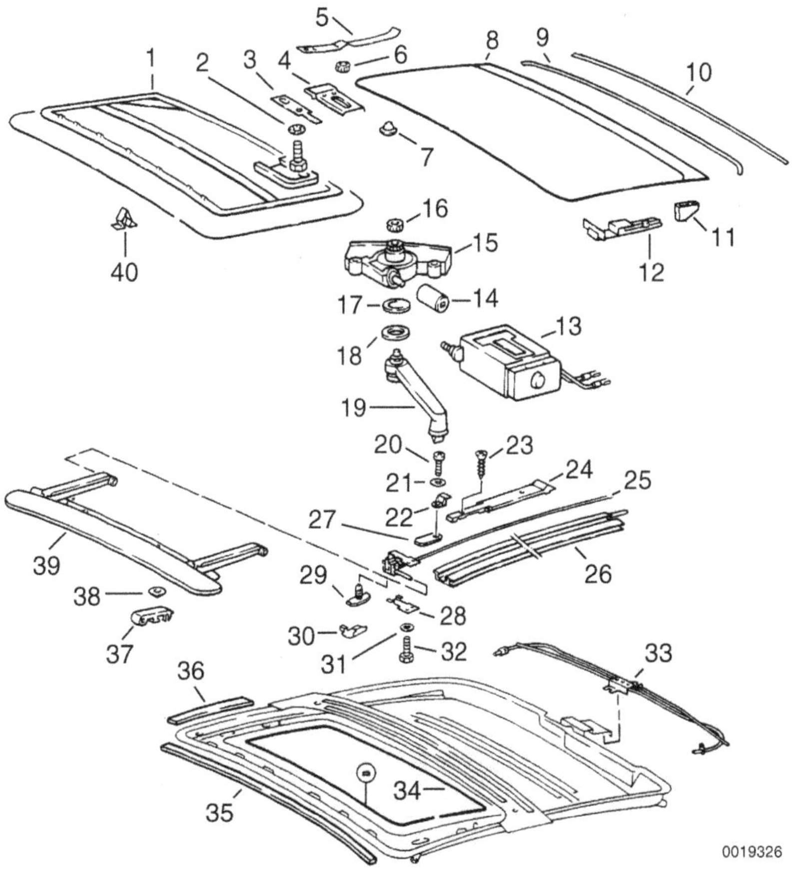

1. Frame
2. Washer
3. Lock plate
4. Guide
5. Leaf spring
6. Knurled nut
7. Plug
8. Cover
9. Velvet strip
10. Sealing rubber
11. Guide shoe
12. Support
13. Motor
14. Clutch
15. Transmission
16. Drive pinion
17. Cap
18. Washer
19. Hand crank
20. Screw
21. Washer
22. Terminal piece
23. Screw
24. Intermediate piece
25. Cable
26. Guide rail
27. Threaded plate
28. Reinforcement
29. Plug
30. Guide piece
31. Lock washer
32. Bolt
33. Cable guide
34. Beading
35. Front seal
36. Side seal
37. Clip
38. Protection cap
39. Wind deflector
40. Clamp

SUNROOF ASSEMBLY

The motor and drive unit are located in the roof just ahead of the rear window. Access to the motor and drive is a zipper opening in the rear headliner.

Sunroof panel, emergency closing

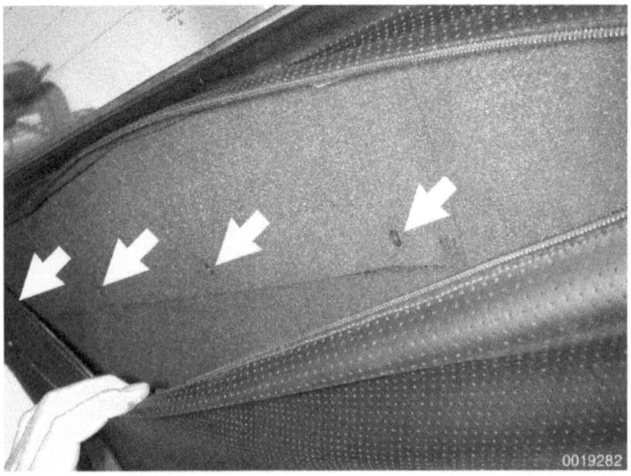

◄ Working in rear headliner, just forward of top of rear window, open zipper in headliner.

– Remove sunroof drive assembly access panel screws (**arrows**). Remove panel.

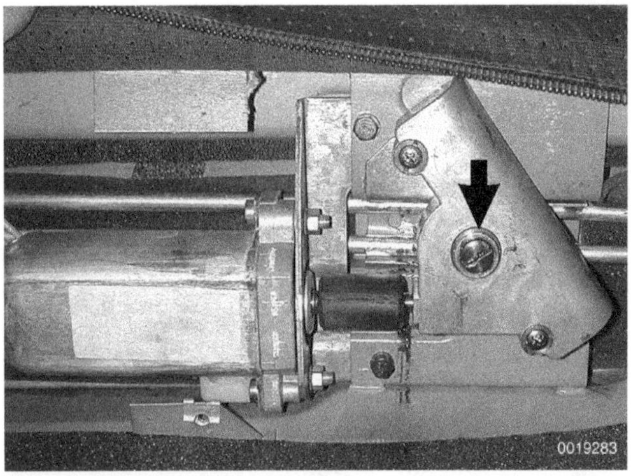

◄ Remove screw from drive unit (**arrow**). Makes sure spacer(s) come off with screw.

NOTE —
The hand crank for closing the sunroof can usually be found in the glove compartment or in the tool kit.

◄ Insert hand crank into drive unit while aligning drive keys in hand crank with drive slots in drive unit. Tighten knurled screw to secure crank to drive unit.

– Turn hand crank to close sunroof.

Sunroof panel, adjusting height

The sunroof panel should be adjusted:

- If the top of the closed sunroof becomes misaligned with the roof of the car;
- If it does not close squarely;
- If there are wind noises at speed;
- If there are water leaks;
- If the sunroof has been removed.

◄ Open sunroof approximately 100 mm (4").

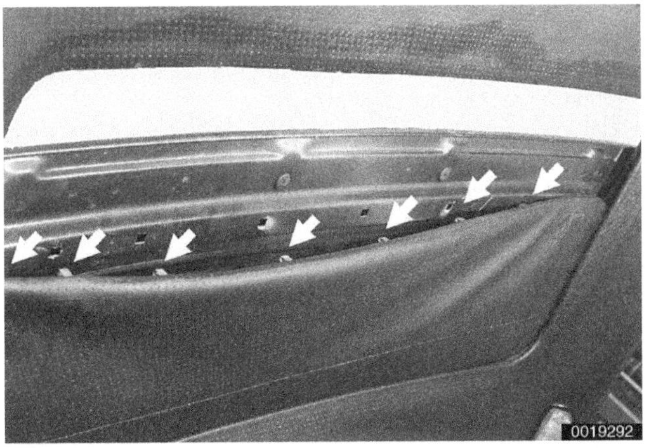

◄ Carefully pull down on front of trim panel to free securing clips (**arrows**).

– Slide headliner panel back into roof cavity.

– Close sunroof.

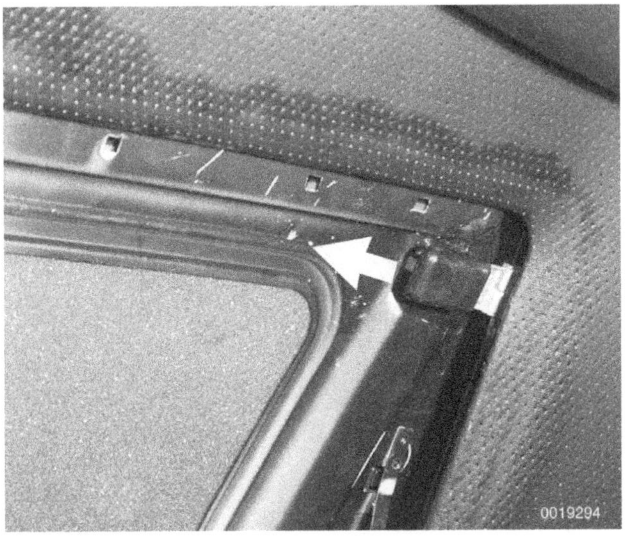

◄ Remove plastic caps covering front guide screws on left and right sides. Left side is shown (**arrow**).

To adjust front height of sunroof panel:

- Remove front guides.
- Turn knurled nuts to adjust height (**arrow**).
- Reinstall guides and check roof adjustment.

NOTE —
Adjust front height so that panel is on same plane as roof or just slightly lower.

To adjust rear height of sunroof panel:

- Loosen slotted height adjustment screws. Left side shown (**arrow**).
- Adjust ratchet (**inset**) so that sunroof panel is on same plane as roof or just slightly higher.
- Tighten screws.

– Open sunroof approximately 100 mm (4").

– Slide headliner trim panel forward and snap clips into place.

– Check operation of sunroof.

Sunroof panel, removing and installing

◄ Open sunroof approximately 100mm (4").

– Carefully pull down on front of trim panel to free securing clips.

– Slide sunroof headliner panel back into roof.

– Close sun roof.

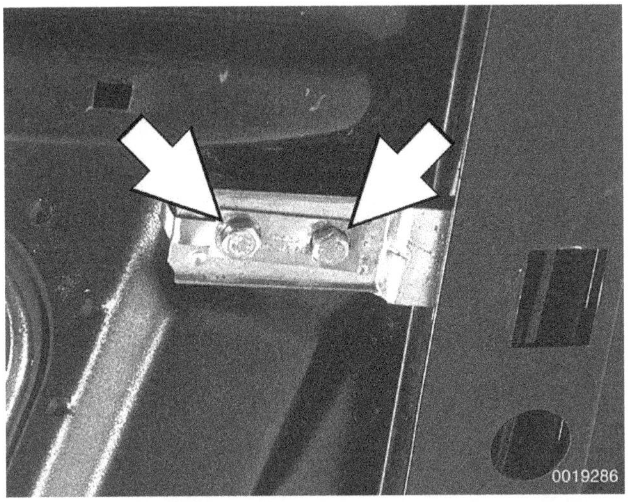

◄ Remove guide bracket covers from front guide brackets (left and right sides). Remove front guide bracket mounting bolts (**arrows**).

NOTE —
Place a strip of masking tape across front of sunroof opening to protect the paint when removing sunroof parts.

◄ Working at rear of panel, lift trim panel leaf spring off pin and turn to side (left and right sides).

– Remove screws from rear guide brackets (left and right sides).

– Carefully push front of sunroof panel up and slide forward and out of body.

– Installation is reverse of removal. Check sunroof adjustment as described earlier.

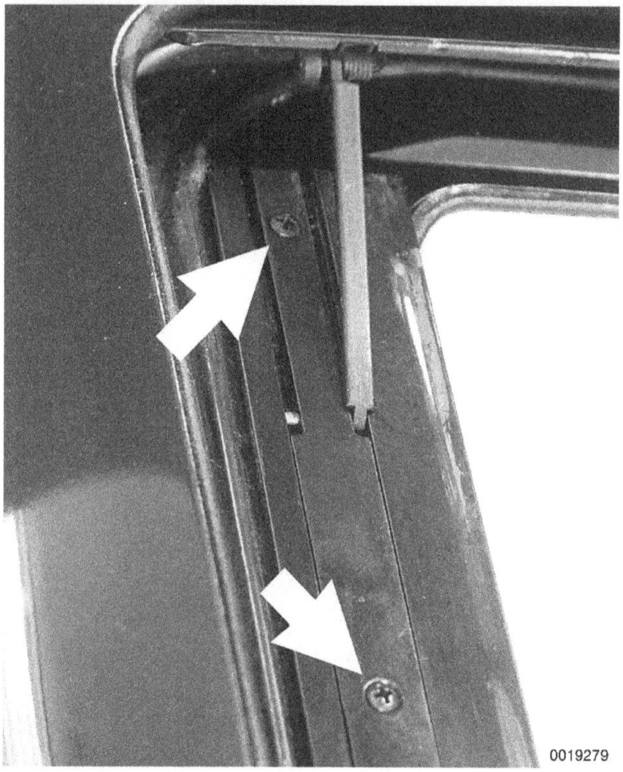

Wind deflector, removing and installing

– Open sunroof.

< Remove intermediate cover screws (**arrows**) and remove covers (left and right sides).

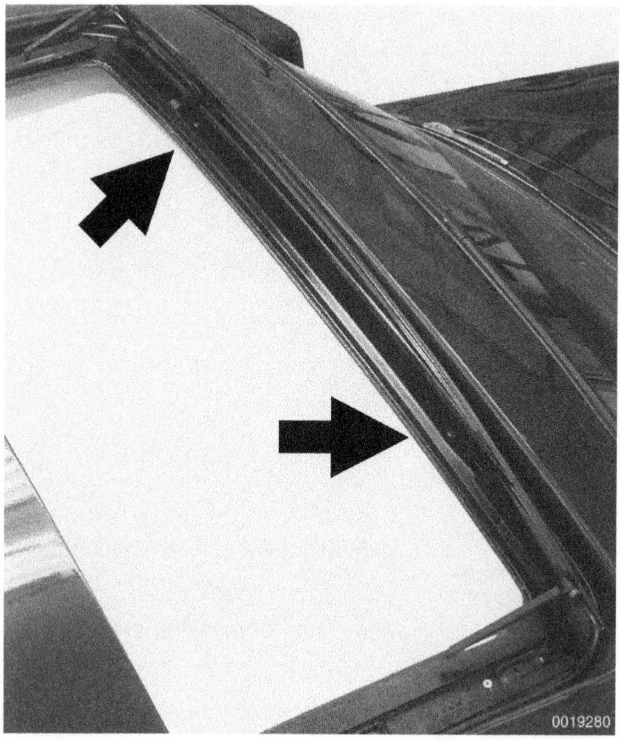

< Remove mounting screws from trim cover (**arrows**).

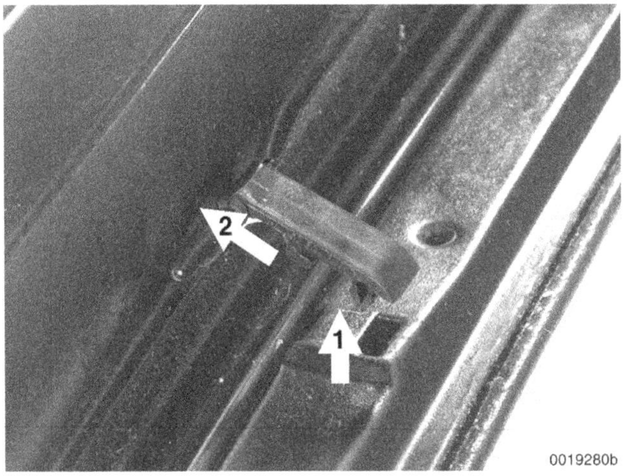

While lifting up cover, carefully pry up on rear of deflector clip (**1**) and slide forward clip off cover (**2**). Remove wind deflector with clips.

– Installation is reverse of removal.

Sunroof cables, replacing

– Remove sunroof panel, described later.

– Remove sunroof drive motor, described earlier.

– Remove wind deflector, described earlier.

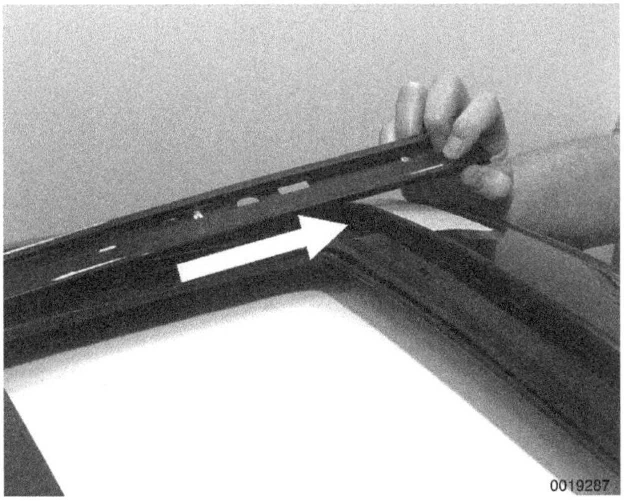

Pull sunroof guide rails straight forward and out of sunroof opening to remove.

– Pull sunroof cables from guides.

– Clean guide rails with hot soapy water and wipe dry.

NOTE—
Do not use a lubricant on sunroof cables.

– Installation is reverse of removal
- Inspect all parts for wear and replace as necessary.
- Adjust sunroof height as described earlier.

Sunroof motor and drive unit, removing and installing

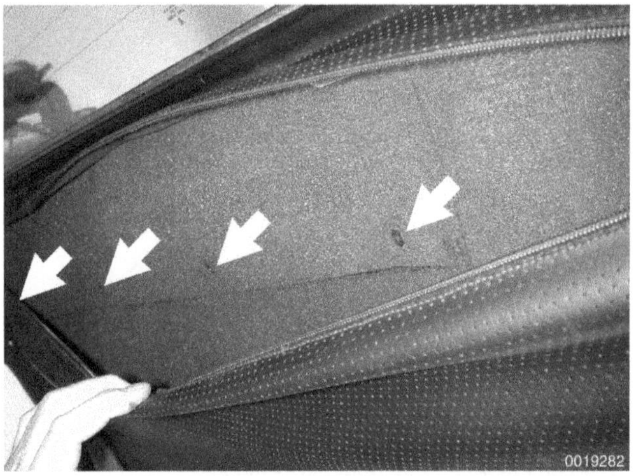

- Working in rear headliner, just forward of top of rear window, open zipper in headliner.

- Remove sunroof drive assembly access panel screws (**arrows**). Remove panel.

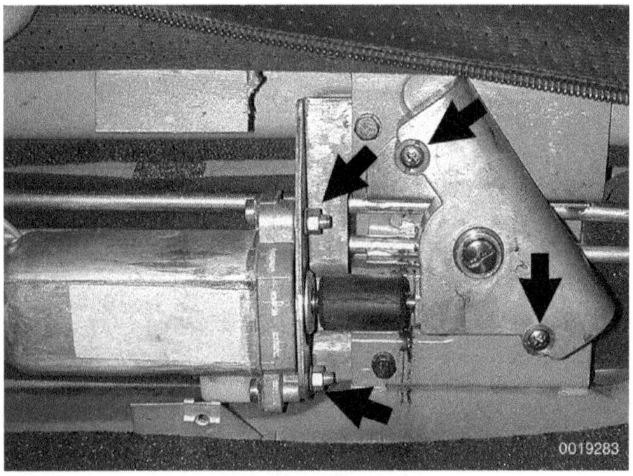

- Disconnect electrical harness connector. Remove sunroof drive assembly mounting screws (**arrows**). Lower drive assembly with motor.

- Installation is reverse of removal.

Sunroof guide rails, removing and installing

- Remove sunroof panel, sunroof drive unit and wind deflector, as described earlier

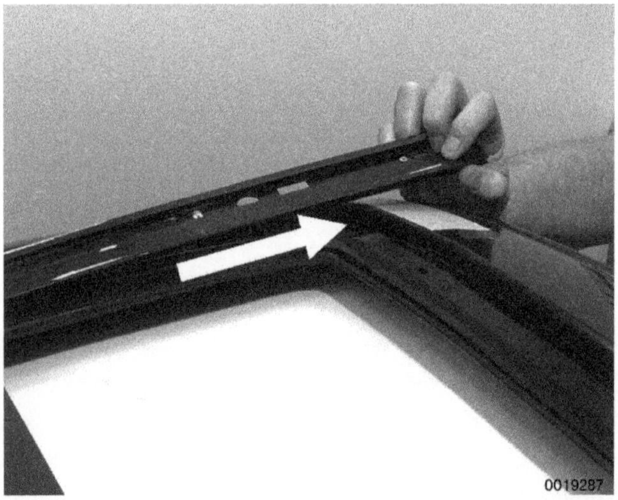

- Pull sunroof guide rails straight forward out of sunroof opening to remove.

- Installation is reverse of removal..

612 Targa Top

GENERAL

The targa top is secured to the roof bar by means of three pins, and to the windshield frame by two locks, facilitating easy removal and installation.

Targa top assembly

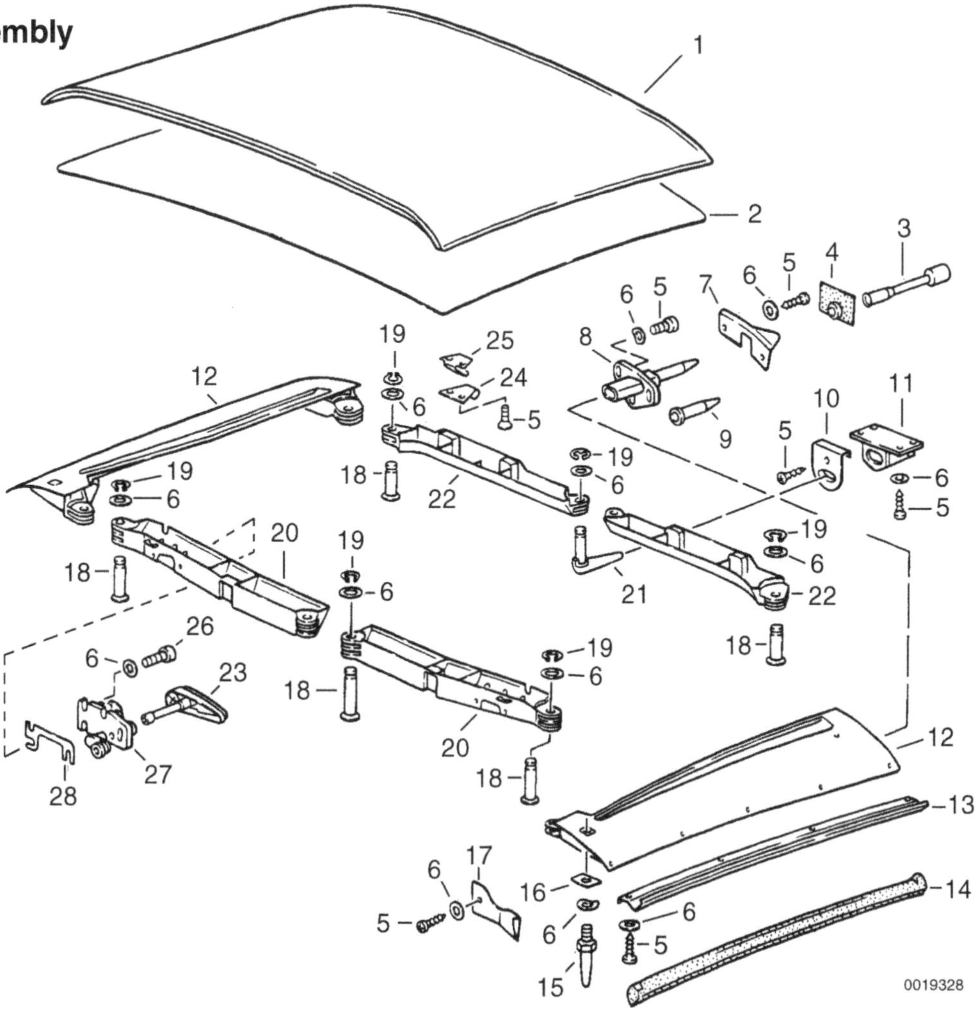

1. Outer lining
2. Inner lining
3. Silent block
4. Sealing
5. Screw
6. Washer
7. Trim
8. Silent block
9. Covering
10. Cover plate
11. Mounting
12. Threaded plate
13. Water strip
14. Sealing
15. Fixing bolt
16. Spring washer
17. Trim
18. Bolt
19. Lock ring
20. Roof frame, front
21. Locking bolt
22. Roof frame, rear
23. Handle
24. Latch
25. Clamp
26. Screw
27. Catch
28. Shim

TARGA TOP

Prior to removal, clean the targa roof panel to prevent road dust or dirt from soiling clothes or the luggage compartment carpet. When storing the top in the luggage compartment, protect it from objects that could damage the panel surfaces.

NOTE —
If the top is to be left removed from the vehicle for an extended period, store it away from the vehicle in an open, not folded, position.

Targa top, removal and installation

Special targa top removal handles are provided in the tool kit.

- Working inside car in front seat, insert both release handles in lock receptacles in front roof support bow of targa top (above sun visors). Start with handles pointed left.
- Turn handles counterclockwise until they are at the 4 o'clock position.
- Continue turning left handle while gently lifting left side of top until it snaps out of lock.
- Perform same operation on opposite side of top.
- Till front of top up and slide forward to remove from car.
- Installation is reverse of removal.

Targa top, folding and unfolding

Once it is removed from the car, the targa top must be folded in order to be stored in the luggage compartment. To fold or unfold, support the rear edge of the roof on a soft surface.

NOTE —
The toes of your shoes work well in this operation.

- **Folding:** Apply pressure simultaneously to middle of both hinged bows. Finish folding by applying light pressure to sides.
- **Unfolding:** Simultaneously push both hinged bows outward to locked position.

Targa top latches, adjusting

The front latch assemblies can be adjusted in or out by means of shims. Remove targa top and plastic latch covers. Loosen latch mounting screws and add or remove shims as necessary.

630 Bumpers

GENERAL

This repair group covers removal and installation of the front and rear bumper assemblies and bumper impact absorbers.

FRONT BUMPER

Front bumper assembly

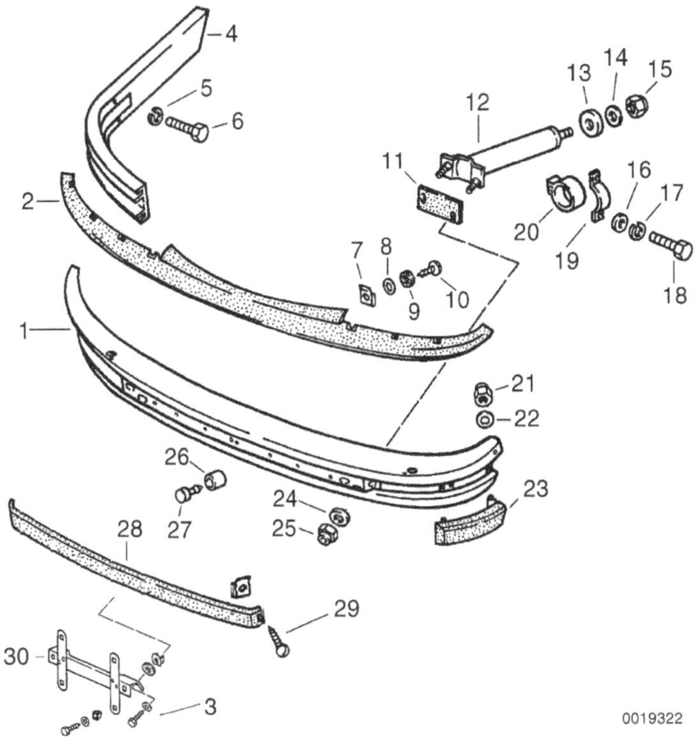

1. Bumper frame
2. Lip seal
3. License plate mounting hardware (kit)
4. Reinforcing bracket
5. Washer
6. Bolt
7. Clip
8. Washer
9. Washer
10. Bolt
11. Gasket
12. Impact absorber
13. Fender washer
14. Washer
15. Nut
16. Washer
17. Lock washer
18. Bolt
19. Clamping bracket
20. Insulator
21. Nut
22. Washer
23. Protective strip
24. Washer
25. Nut
26. Insert
27. Clip
28. Protective strip
29. Screw
30. License plate frame

0019269

Front bumper, removing and installing

- Remove parking light lens.

◄ Remove screw securing reflectors to parking light lens assembly. Disconnect wiring harness from reflectors and remove reflectors from left and right sides.

- Remove fog lights. See **940 Exterior Lighting**.

- Working through foglight openings, withdraw parking light wiring harnesses from rear of parking light housings.

0019270

◄ Remove protective bumper strip securing screws. Left side shown (**arrow**).

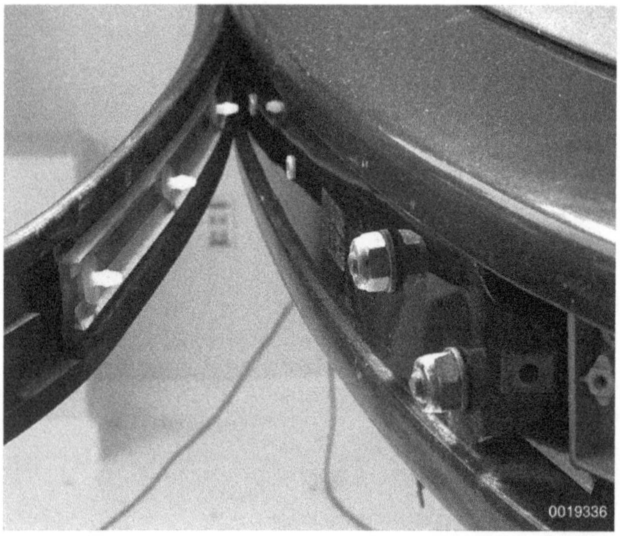

0019336

◄ Remove protective strip by carefully prying out clips on rear of strip.

NOTE —

Be sure to have trim clips and plugs in case they break while being removed.

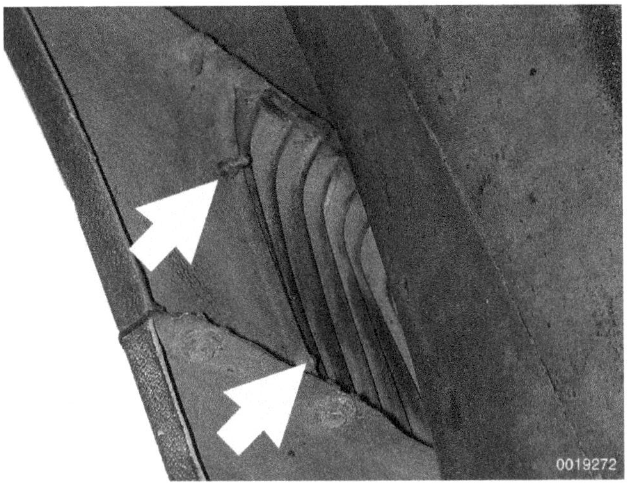

◀ Working below bumper, remove nuts securing bumper bellows to fenders on left and right sides.

- Pull bumper bellows out so that studs clear fender. Wrap studs with tape to prevent damaging painted surfaces.

◀ Remove bumper securing nuts and washers from left and right sides. Left side shown (**arrows**).

- Remove bumper by sliding straight off.

- Installation is reverse of removal.

Front bumper impact absorber, replacing

- Remove front bumper as described earlier.

- Left side: Remove washer fluid tank.
  Right side: Remove horn mounting bracket with horns.

◀ Remove mounting nut (**1**) from rear of impact absorber .

- Remove side clamping bracket bolts (**2**).

- Remove impact absorber unit.

- Installation is reverse of removal.

Rear bumper assembly

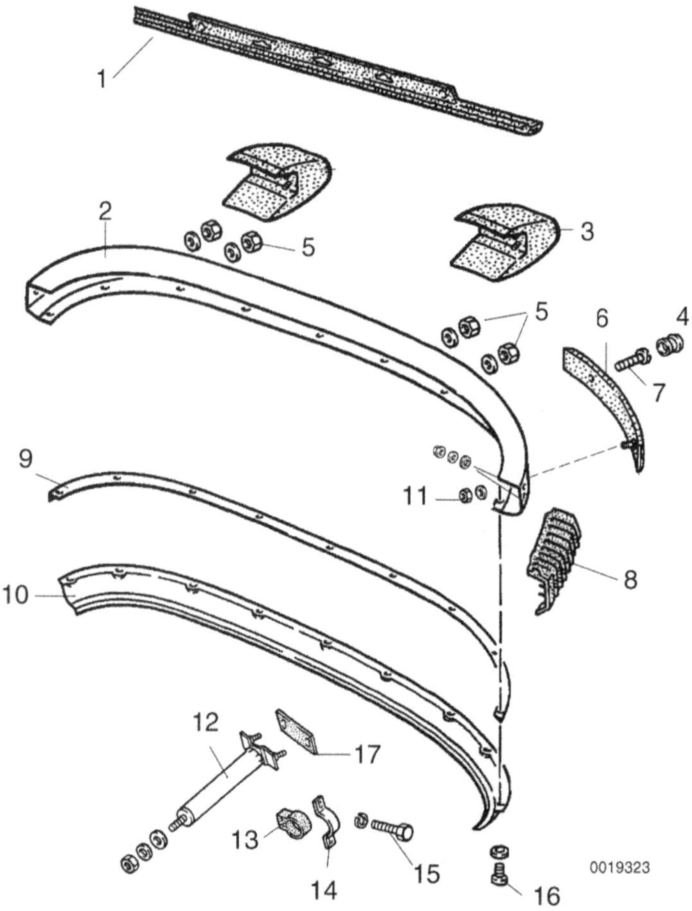

1. Trim seal
2. Bumper frame
3. Buffer
4. Concealing cap
5. Nut and washer
6. Protective strip
7. Screw
8. Bellows
9. Valence sealing strip
10. Valence
11. Nut and washer
12. Impact absorber
13. Insulator
14. Clamping bracket
15. Bolt
16. Screw
17. Gasket

REAR BUMPER

Rear bumper, removing and installing

- Working below bumper, remove bellows fasteners (left and right sides of bumper).
- Carefully pull bellows away from bumper and body.
- Remove license plate light assemblies from bumper buffers and disconnect wiring harnesses.
- Working below bumper, remove bumper buffer securing fasteners. Pull buffers off bumper on left and right sides.

< Remove concealing cap from bumper protective strip.

- Remove screw behind cap.
- Pull back protective strip.

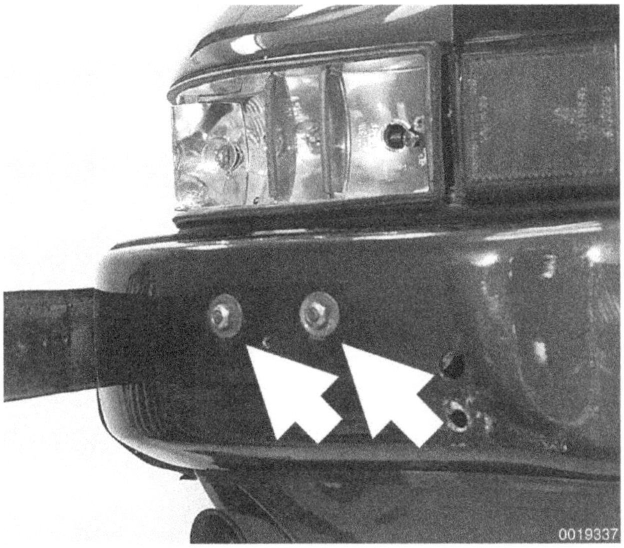

< Remove bumper securing nuts and washers from left and right sides. Left side shown (**arrows**).

– Remove bumper by sliding it straight off body.

– Installation is reverse of removal.

Rear bumper impact absorber, replacing

– Remove rear bumper as described earlier.

< Remove mounting nut from rear of impact absorber (**1**).

– Remove bolts holding side clamping bracket in place (**2**).

– Remove impact absorber unit.

– Installation is reverse of removal.

640 Windows

GENERAL

This repair group covers replacement of the vent and door glass. Also covered here is the replacement of the power window regulator and power window motor. For removal and installation of the doors, door panels, and window frame, see **570 Doors**.

WINDOW SERVICE

Vent window, removing and installing

– Remove door trim panel. See **570 Doors**.

– Remove window frame from door. See **570 Doors.**

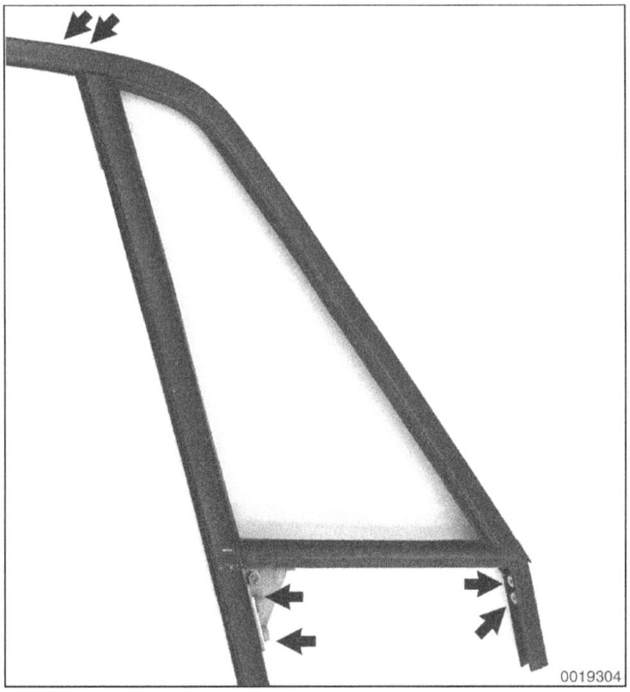

Remove screws (**arrows**) from frame assembly.

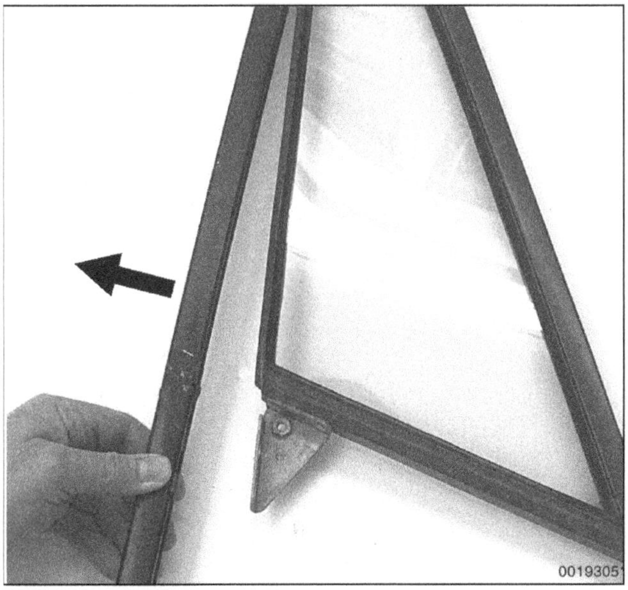

Carefully separate center window guide rail from window frame.

– Remove vent window with rubber seal from window frame.

– Remove rubber seal from vent window.

– Installation is reverse of removal.

640

Door window, removing and installing

– Position window about halfway down in door.

– Remove door trim panel. See **570 Doors**.

– Remove window frame from door. See **570 Doors**.

NOTE —

- *When removing the window frame, the window glass stays in the door.*
- *Be careful not to drop two bolts located at the top front of the door. If bolts do drop into the channel, use a magnet to fish them out through the drain hole in the bottom of the door.*

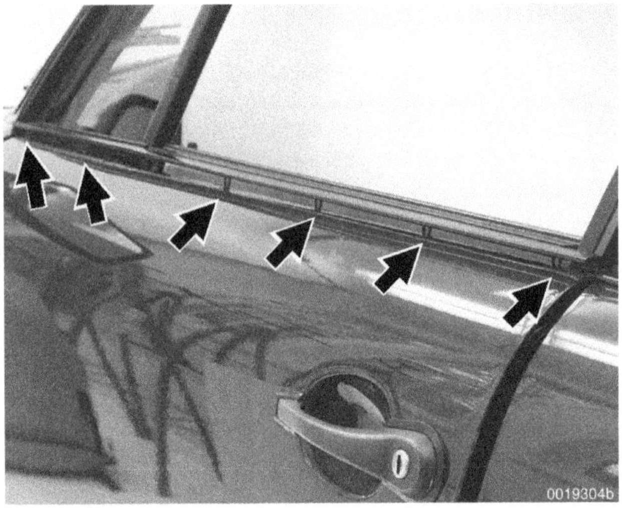

◄ Carefully pry outer window weather-strip from top of door at retaining clips (**arrows**).

– Carefully peel off inner weather strip.

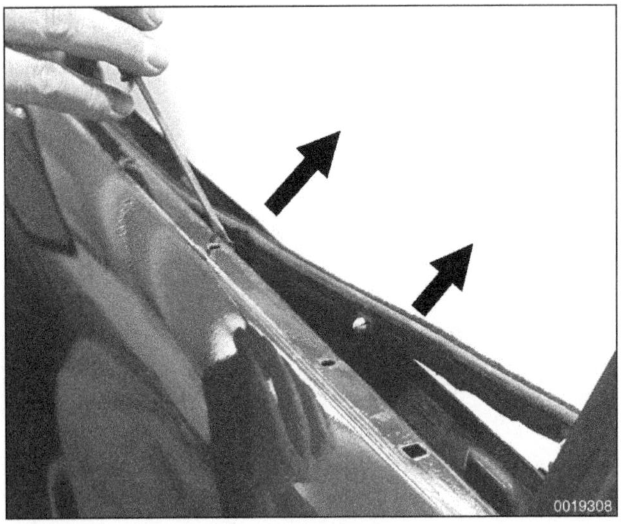

◄ Carefully pry out felt weather-strip from inside of door.

◄ Remove window regulator arm mounting bolts (**arrows**).

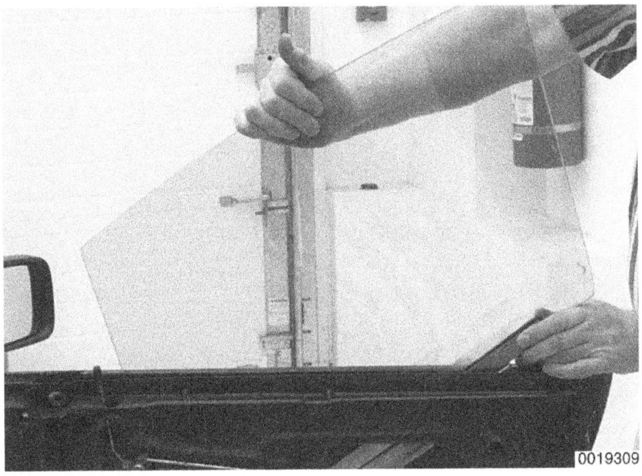

◄ Tilt window up at rear and pull out of door while sliding both guide wheels out of front of window guide channel.

NOTE —
The guide wheels will not come out through the rear of the window guide rail.

– Installation is reverse of removal.

CAUTION —
If replacing a broken glass be very sure to clean all the glass from window channels.

Door window regulator, removing and installing

- Position window about halfway down.
- Remove door trim panel. See **570 Doors**.

- Remove window regulator arm mounting bolts (**arrows**).
- Disconnect electrical harness connector from window motor.

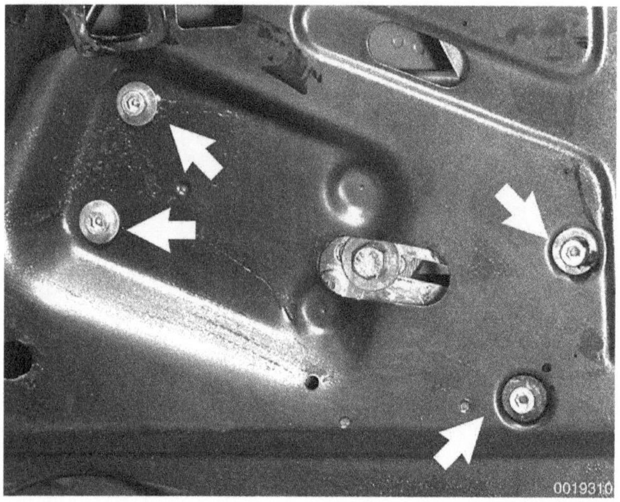

- Remove window regulator mounting bolts (**arrows**).
- Slide both regulator guide wheels out of front window guide rail.

NOTE —

The guide wheels will not come out through the rear of the window guide rail.

Push window into closed position and tape in place.

> ***CAUTION —***
> *Be sure tape holds well. If the window drops into the door unexpectedly, it will most likely shatter.*

Lower regulator into door to access window motor. Remove window motor mounting bolts (**arrows**) from regulator and set motor inside door.

Remove regulator through hole in door.

Door window, adjusting (Coupe models)

Whenever the window regulator is removed, window adjustment should be checked. The glass should contact the top of the window frame squarely and should seat against the window seal uniformly. If the window stop is not adjusted correctly the door may become hard to close or wind noise may be a problem.

The window regulator has two adjustments, one is for in-out tilt and the other is for front/rear height parallelism.

– Remove door trim panel. See **570 Doors**.

◄ Loosen window regulator arm mounting bolts (**arrows**).

– Move rear of arm up or down in elongated slot (**small arrow**) to change window parallelism.

◄ Loosen height adjusting bolt (**arrow**).

– Reposition window stop left or right in elongated slot (**small arrow**) and check window height adjustment.

- Move stop to rear to decrease amount of window travel.
- Move stop to front to increase amount of window travel.
- Tighten bolt when adjustment is correct.

– Install door trim panel.

Door window, adjusting (Targa and Cabriolet models)

Whenever the window regulator is removed, the window adjustment should be checked. The glass should contact the top of the window frame squarely and should seat against the window seal uniformly. If the window stop is not adjusted correctly the door may become hard to close or wind noise may be a problem.

The window regulator used in the targa and convertible uses stops for setting window travel.

– Remove door trim panel. See **570 Doors**.

< Lower window until stop bracket is aligned with access hole in door (**arrow**).

– Loosen bolt securing stop to regulator.
  - Move stop forward to raise window height.
  - Move stop backward to lower window height.
  - Tighten bolt when adjustment is correct.

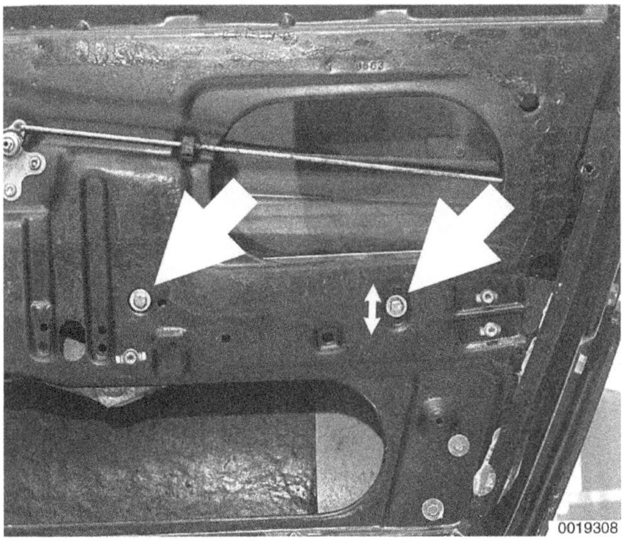

< Loosen window regulator arm mounting bolts (**arrows**).
  - Move rear of arm up or down in elongated slot (**small arrow**) to change window parallelism.
  - Tighten bolts when adjustment is correct.

– Install door trim panel.

660 Exterior Trim

Special Tools

Special tools

A Baum 9700/9701/9702. . . . . Wiring connector removal tool (source: Baum Tools Unlimited)

GENERAL

This repair group covers removal and installation of the outside rear view mirrors and the exterior trim.

For front and rear bumper removal, see **630 Bumpers**.

Special tools

< Some of the procedures require special tools.

OUTSIDE REAR VIEW MIRROR

Outside mirror housing, removing and installing

– Remove mirror glass and motor (if applicable).

< Use wiring connector removal tool to push wire terminals out of mirror motor harness connector.

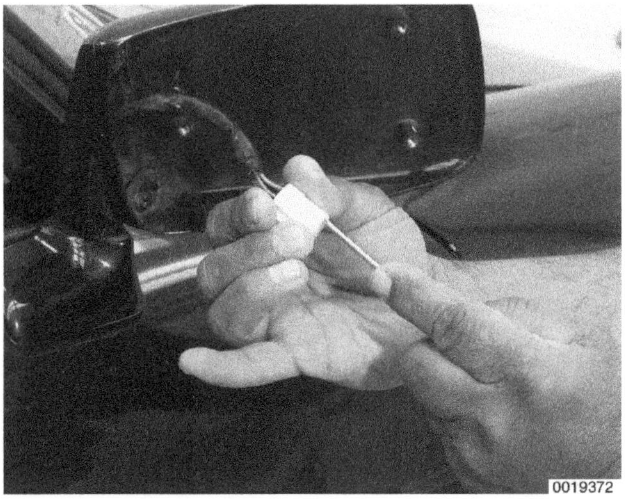

Remove allen mounting bolt (**arrow**) at base of mirror housing.

- Remove mirror while feeding wiring harness through mirror base.

– Installation is reverse of removal. Wiring colors are shown in **Table a**.

Table a. Outside mirror drive motor wire colors

Terminal	Left	Right
1	white/green	white/green
2	white/brown	white/blue
3	black	black/blue
4	green/white/green	green/white/green
5	grey/green	grey/green
6	brown	brown

EXTERIOR TRIM

Rocker panel, removing and installing

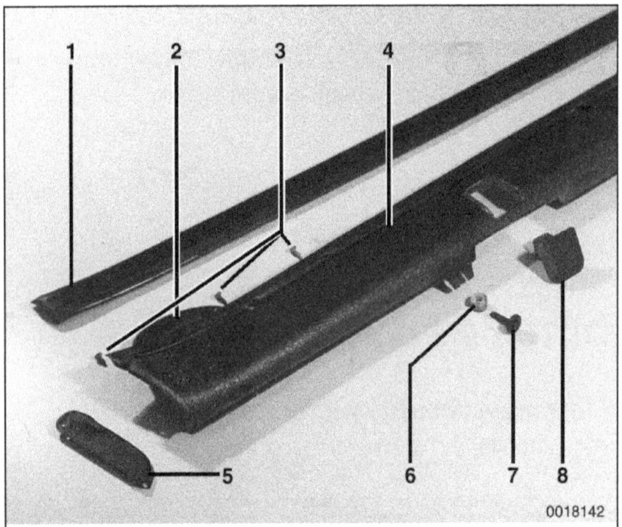

1. Rubber protector
2. Rear torsion bar end cover
3. Rocker panel mounting screws
4. Rocker panel
5. Rocker panel end cover
6. Plastic spacer
7. Rocker panel bottom mounting screw
8. Jack hole cover

Pull off rubber rocker panel protector.

- Remove top and bottom mounting screws and remove rocker panel.

NOTE —
*The rear torsion bar end cover (**2**) is one piece with the rocker cover.*

– Remove screws securing end cover and remove covers.

– Installation is reverse of removal.

NOTE —
To reinstall the rubber protective strip use a rubber lubricant and slide the strip in from one end.

0019371

Rear reflector panel, removing and installing

– Open engine compartment lid.

◄ Lift up seal in back of engine compartment to expose reflector panel plastic fasteners.

- Release fasteners by turning each a half turn.

NOTE —
When the fastener is released it will pop-up.

– Tilt reflector panel out at top and lift out of retaining clips at bottom.

– Installation is reverse of removal.

0019370

Porsche emblem

◄ Two speed nuts (**arrows**) are used to mount the Porsche emblem to the luggage compartment hood.

– Use a deep 8 mm socket to remove nuts. Use new nuts when installing emblem.

680 Interior Trim

GENERAL

This repair group covers interior trim removal and installation procedures.

NOTE —
- *For removal and installation of center console mounted switches, see* **Center Console** *below.*
- *For removal and installation of interior switches, including interior lighting equipment, see* **960 Electrical Switches, Interior Lighting**.

CENTER CONSOLE

Access to the A/C controls, as well as removal of the shifter boot, makes it necessary to partially or completely disassemble the center console.

Center console, removing and installing

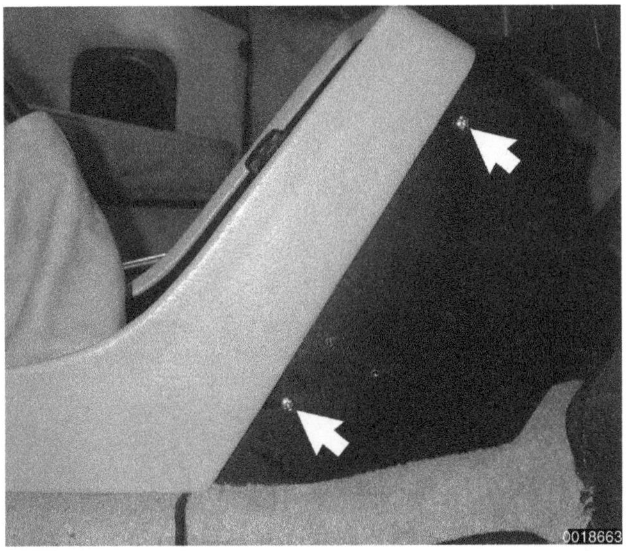

Remove left and right side screws (**arrows**) securing front cover to center console. (Right side shown; left is similar.) Slide cover forward (toward front of car) off center console.

◄ Remove center console mounting screws.

- Screw on left shown (**A**). Right side is similar.
- Pull back rug slightly to reveal screw in middle of rear (**B**).

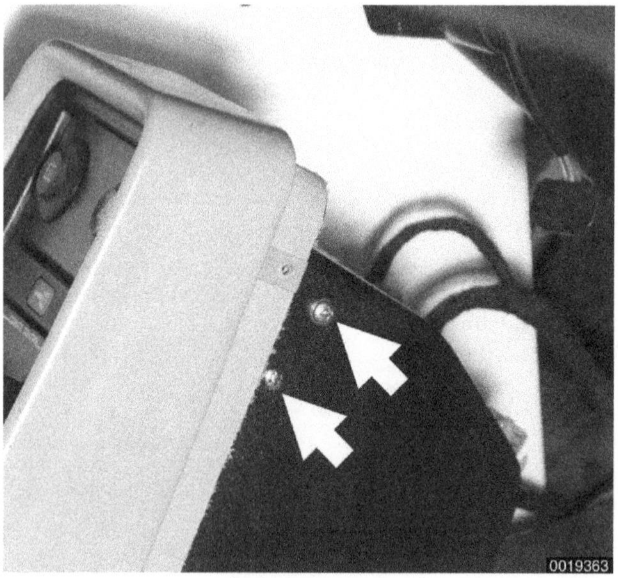

◄ Remove screws (**arrows**) securing switch supports to center console. (Right side shown; left is similar.)

– Pull off A/C temperature control knob, then remove slotted nut in back and push A/C switch forward out of support.

– Pull switch supports and switches out of console, carefully feeding harnesses into front of console housing.

> ***CAUTION —***
> *Do not pull on the A/C temperature control switch. This can damage the temperature sensor conduit.*

NOTE —
The A/C temperature control switch is attached to the temperature sensor conduit which is inserted into the evaporator in the luggage compartment. The A/C switch must be removed through the front of the console housing.

– Remove switch support retaining screw on left side.

– Detach wiring harness connectors from blower switch. Remove switch support assembly from console

– Lift console over shifter and remove.

– Installation is reverse of removal.

DASHBOARD

Most gauges and instruments are removed from the dashboard cover by gentle prying. See **901 Instruments**.

A cracked or damaged dashboard cover can be replaced using the procedure given below.

Dashboard cover, removing and installing

– Remove windshield.

– Use a sharp blade to slice off the lip of the windshield gasket inside the car. Then gently tip windshield forward and out.

0019417

◄ Remove five screws (**arrows**) securing defroster outlet duct to dashboard cover. Remove duct.

– Working from inside luggage compartment locate and remove seven lock nuts securing dashboard cover to dashboard.

– Carefully lift dashboard cover to release snap clips located above gauges.

– Carefully work dashboard cover away from dashboard.

– Installation is reverse of removal.

Knee bar, removing and installing

0019134

◄ Working under dashboard, remove knee bar screw (**arrow**). (Left side shown. Right is similar.)

0019133

Remove outer A/C vent from dash.

- Remove screw (**A**) from trim panel on side of dash.
- While pulling on A/C vent, release locking tabs (**arrows**) at top and bottom of vent with a small screwdriver.

0019108

Remove knee bar:

- Working through A/C vent opening, remove knee bar mounting nut and washer from stud (**A**).
- Working just below heater controls or radio in center of dashboard, remove knee bar mounting nut and washer from stud (**B**).
- Right side: Remove radio as described in **910 Radio**. Remove knee bar mounting nut and washer through radio opening.

– Installation is reverse of removal.

GLOVE COMPARTMENT

Glove compartment, removing and installing

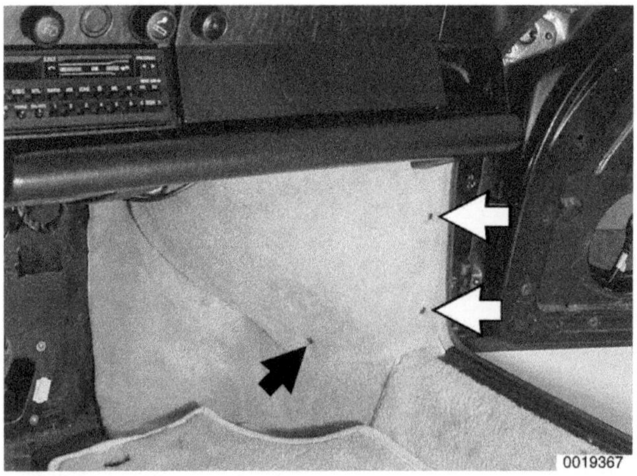
0019367

Remove right side kickpanel mounting screws (**arrows**) and remove panel.

Remove screw from glove compartment mounting strap (**arrow**).

- Remove glove compartment.
- Installation is reverse of removal.

Glove compartment door, removing and installing

- Remove right knee bar as described earlier.
- Open glove compartment door.

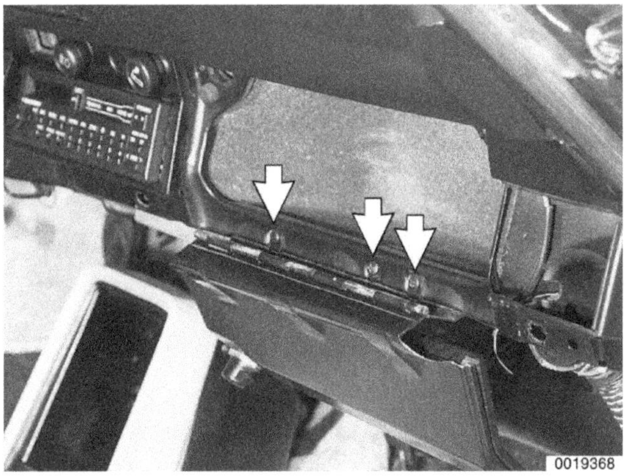

Remove hinge screws (**arrows**) and remove door.

- Installation is reverse of removal.

SUN VISORS

Visor tension, adjusting

Visor tension can be changed by loosening or tightening the adjusting screw (**arrow**).

Visor, removing and installing

- Snap visor out of center support and turn 90° to side.
- Remove two screws securing visor to headliner and remove visor.
- Installation is reverse of removal.

720 Seats

GENERAL

This repair group covers removal and installation of the front seats.

A variety of seat styles including leatherette, leatherette/leather and all-leather versions are installed in the models covered by this manual.

Upholstery repairs are beyond the scope of this manual.

SEAT REMOVAL

Front seat, removing and installing

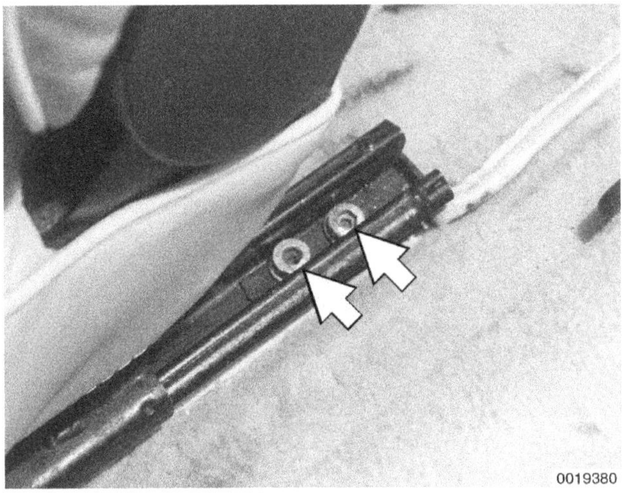
0019380

< Move seat to full forward position. Remove mounting bolts (**arrows**) at rear of each seat rail. (Left rail shown. Right is similar.)

0019381

Slide seat back to rear position. Remove mounting bolts (**arrow**) at front of seat rails.

– Cut wire ties as necessary.

– Remove seat from car.

NOTE —
Use a blanket to protect door sill from scuffing during seat removal.

– Installation is reverse of removal.
  - Use wire ties or equivalent means to keep seat harness wiring from damage.
  - Install all seat mounting bolts finger-tight before torquing.

Tightening torque
- Seat to seat rail. 23 Nm (17 ft-lb)

800 Heating, Ventilation, Air Conditioning–General

GENERAL

This section covers general information for the heating, ventilation and air conditioning (A/C) systems. Diagnostic and troubleshooting information for these systems is at the end of the section.

The Porsche flow-through heating and ventilation system provides draft-free air-flow while driving. Outside air flows through the grille in front of the windshield and into the passenger compartment via the windshield defroster, dashboard and footwell vents. The flow-through air is heated or cooled according to settings on the heating and A/C controls.

Heating,ventilation, A/C system components

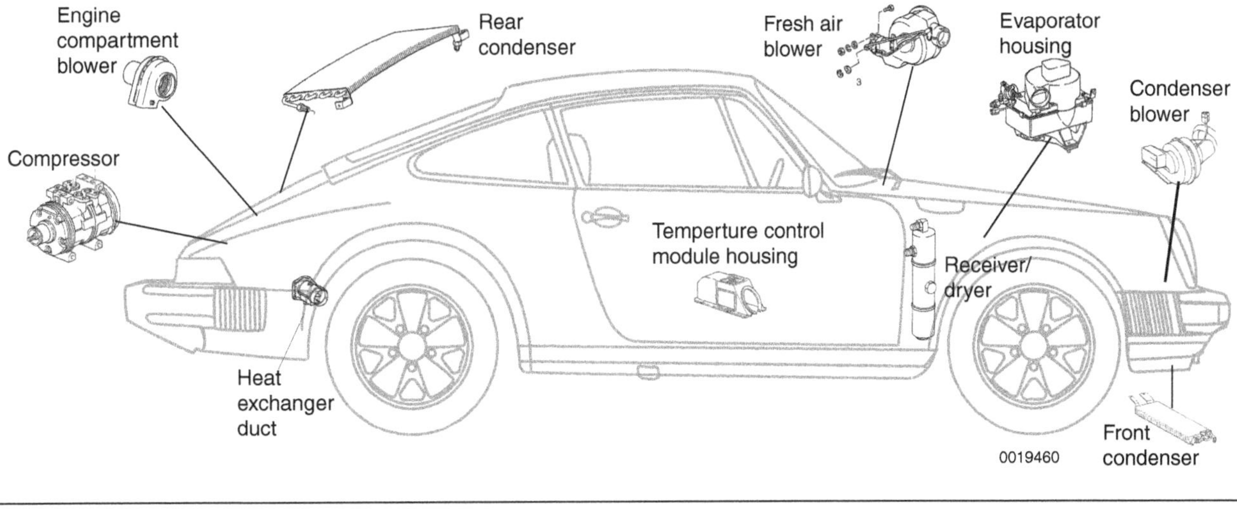

Fresh and heated air distribution

0019813

The air distribution levers control fresh air and heated air to the footwells and the defroster outlets.

NOTE —
The slider controls have no control over air distribution to the center dash vent or the side dash vents. This center vent is used for A/C only. The left and right side vents are for A/C or heated air and are turned on and off by the vertical knurled control knobs.

- **Top lever(s)**: Controls fresh air intake from outside the vehicle. When the lever(s) is in the extreme left position, air intake from the outside is closed off. When the lever(s) is in the extreme right position, maximum ventilation of the car interior is obtained.
- **Middle lever**: Fresh air footwell/defrost blend door. This lever directs fresh air to the footwells (extreme left position) or the defroster (extreme right position) or a blend of the two.
- **Bottom lever**: Heated air footwell/defrost blend door. This lever directs heated air to the footwells (extreme left position) or the defroster (extreme right position) or a blend of the two. On cars with automatic heating, heated air temperature is controlled by the temperature control knob between the seats.

NOTE —
- *For information on heating and defrosting component replacement, see* **801 Heating**.
- *For information on ventilation system component replacement, see* **850 Ventilation**.

Automatic heating system

The automatic heating system controls the passenger compartment temperature based on the setting of the temperature control knob.

Heater control module assembly

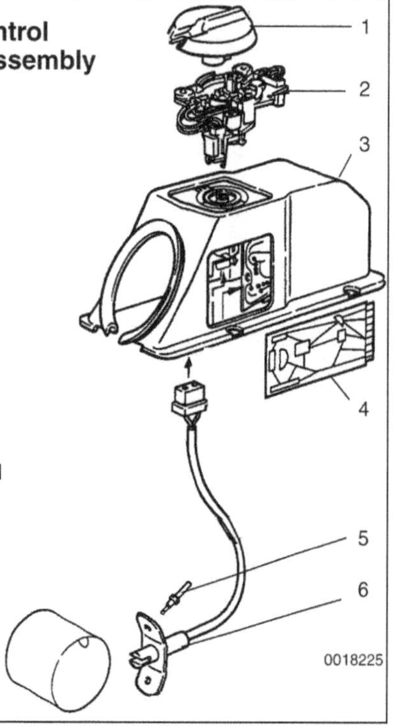

1. Knob
2. Control module with servo motor
3. Housing
4. Printed circuit board
5. Rivet
6. Heat exchanger temperature sensor

0018225

By comparing the passenger compartment temperature to the setting at the temperature control knob, the control module servo motor automatically opens or closes the heater control box valves to maintain the desired temperature.

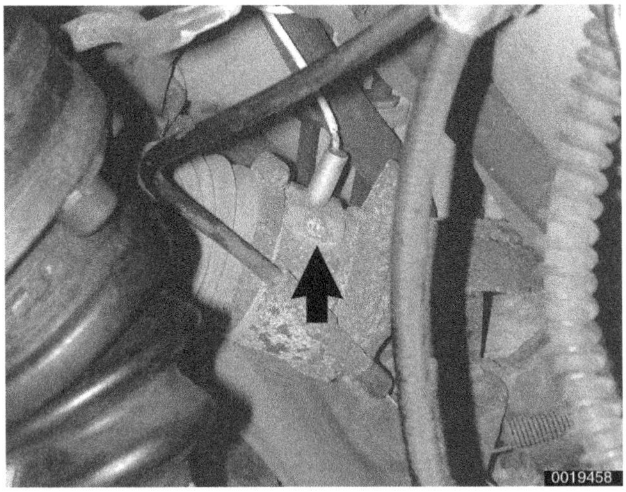

The **heat exchanger temperature sensor** (**arrow**) is used as part of the automatic heater system to automatically control heat distribution into the passenger compartment. The sensor is installed inside the left heater duct next to the transmission and prevents operation of the footwell blower motors when the engine is cold.

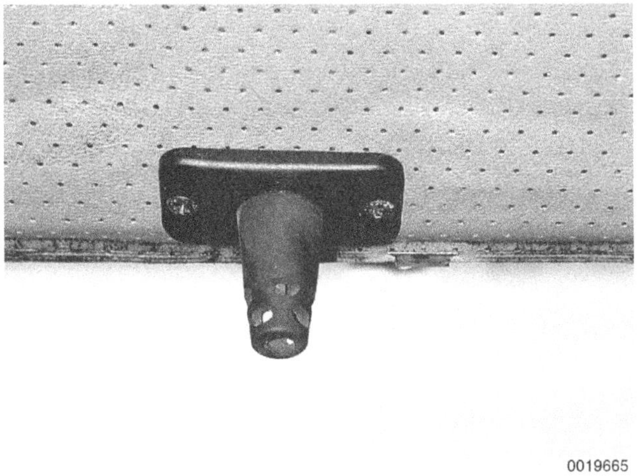

The **passenger compartment temperature sensor** monitors the passenger compartment temperature.The sensor is located behind a trim cover in the headliner between the sun visors.

The **temperature regulating control module** is operated through the temperature control knob (**arrow)** and regulates the opening and closing of the heater control box valves via an integrated servo motor. The control module is located below the handbrake console.

The **engine compartment blower** forces air through the heat exchangers and into the passenger compartment when heat is called for at the temperature control knob. The blower is mounted on the top of the engine.

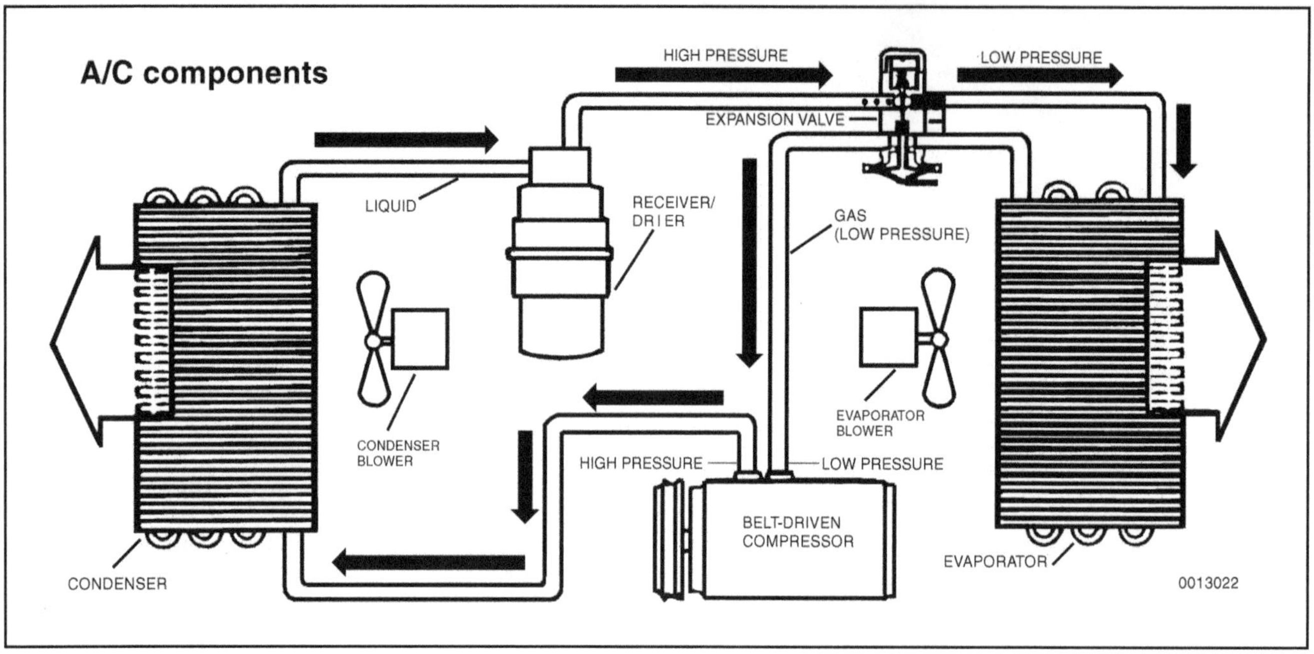

Air Conditioning

With the A/C on, the air is cooled and dehumidified by the evaporator to maintain the desired temperature.

Cooling efficiency of the air conditioner system is dependent on the engine rpm. The greater the rpm the higher the output from the air conditioner. The air conditioner works only with the engine running. The rear lid must be closed for the A/C system to operate properly.

The schematic view above is a typical A/C system. In the Porsche system there are two condensers in the vehicle, one in front, cooled by a blower, and another in the rear.

The A/C system removes heat and moisture from the passenger compartment. It accomplishes this through the application of four physical principles: materials absorb heat as they change from a liquid to a gas (evaporate); materials give off heat as they change from a gas to a liquid (condense); the boiling point of a liquid varies with its pressure; and, heat always flows from hot to cold. For example, the first principle is demonstrated by wetting your hand and then blowing on it. As the water evaporates, it takes some heat with it and your hand feels cooler.

In the A/C system, the heat from the passenger compartment boils (evaporates) the refrigerant (R-12) in the **evaporator**, causing heat to be absorbed by the R-12. This heat is then released into the atmosphere when the R-12 is cooled and condensed into a liquid in the condensers.

Moisture is removed at the evaporator in the same way that water drops form on a cold glass. The moisture drips beneath the evaporator and onto the ground. This is the reason a water puddle may often be seen under the car when the A/C is operating.

The **A/C compressor** forces refrigerant through the system and at the same time pressurizes it, raising the refrigerant boiling point to force it to condense more readily. The compressor is engaged by an electro-magnetic clutch that is actuated when the A/C is turned on. The compressor clutch disengages when the temperature in the passenger compartment reaches the level set on the operating controls.

The **condensers**, which look like small radiators, are located in front and rear of the car. The front condenser is located below the front of the car, protected by a perforated pan. It has a separate blower, accessible through the front of the luggage compartment. The rear condenser is integral with the engine cover.

The **receiver/drier**, located in the left front wheel housing, removes small amounts of moisture and dirt from the system.

0019552

◀ Two switches in the center console operate the A/C system.

The **temperature control knob** (left hand switch) regulates the temperature of cooling air coming from the A/C vents in the dash.

- Fully left. . . . . low cooling
- Fully right . . . . full cooling

The **blower switch** (right hand switch) also turns on the compressor clutch.

- Position I . . . . no cooling, blower off
- Position II to IIII . . . . A/C on, increasing blower speed

Vents. Air is drawn in from the passenger footwell and blown through the air conditioner evaporator where it is cooled. Air conditioned air is then emitted from the center and side dash vents.

NOTE —
For maximum cooling, the console A/C control switches should be turned to the extreme right.

For information on A/C system component replacement, see **870 Air Conditioner**.

AUTOMATIC HEATING TROUBLESHOOTING

For all practical purposes, the heating system is separate from the A/C system. Therefore, problems should fall within one of two categories: heating faults or air conditioning faults. Each system has its own causes and corrective actions.

Troubleshooting of the heater electronic controls begins by removing the heater control module housing as described in **801 Heater**. Heater control module troubleshooting is summarized in **Table a**. Use the photos that follow for connector and test terminals referred to in **Table a**.

Table a. Temperature control troubleshooting

Symptom	Probable cause	Corrective action (refer to photos)
Temperature control knob in position 1, control servo moves to maximum heating	Open circuit in passenger compartment temperature sensor	Test sensor resistance at connector **B**. Passenger compartment temperature sensor resistance specification: 1.9 kΩ at 20°C (68°F).
	Open circuit in temperature control potentiometer	Test resistance at temperature control potentiometer terminals with wires connected and switch in position 5. Potentiometer resistance specification: 950 ± 20 Ω
Control module regulates to maximum heating in defrost switch position only	Short circuit in passenger compartment temperature sensor	Test sensor resistance at connector **B**. Passenger compartment temperature sensor resistance specification: 1.9 kΩ at 20°C (68°F).
	Short circuit in temperature control potentiometer	Test resistance at temperature control potentiometer terminals with wires connected and switch in position 5. Potentiometer resistance specification: 950 ± 20 Ω
Too much heat, does not regulate correctly	Open circuit in heat exchanger temperature sensor	Test sensor resistance at connector **C**. Heat exchanger temperature sensor resistance specification: 1.7 kΩ at 20°C (68°F).
	Temperature control potentiometer set incorrectly	Set potentiometer switch to position 5 and remove knob. Connect ohmmeter to potentiometer. Hold shaft of switch with pliers and adjust potentiometer with a screwdriver to specification: 950 ± 20 Ω
Too little heat, does not regulate correctly	Short circuit in heat exchanger temperature sensor	Test sensor resistance at connector **C**. Heat exchanger temperature sensor resistance specification: 1.7 kΩ at 20°C (68°F).
	Temperature control potentiometer set incorrectly	Set potentiometer switch to position 5 and remove knob. Connect ohmmeter to potentiometer. Hold shaft of switch with pliers and adjust potentiometer with a screwdriver to specification: 950 ± 20 Ω
Control module not regulating	Open circuit in power supply	Check for battery voltage at connector **A** terminal 15. Check for continuity to ground at connector **A** terminal 31.
	Printed circuit board defective	Replace circuit board.
	Control module defective	Replace control module.

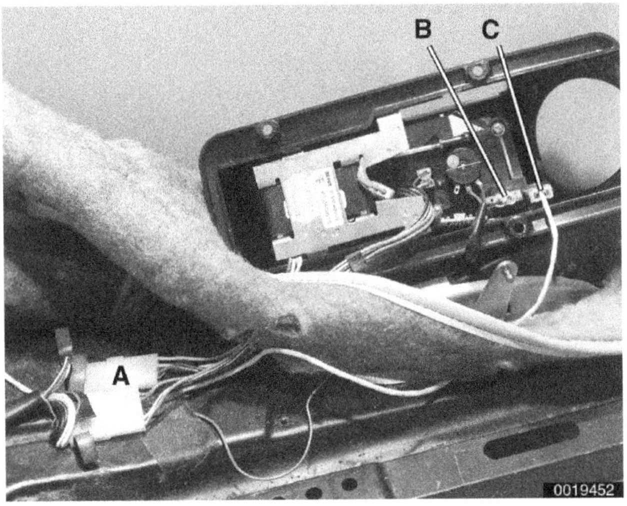

0019452

< Test connections at temperature control module:

- Harness connector **A** (power supply).
- Harness connector **B** (passenger compartment temperature sensor).
- Harness connector **C** (heat exchanger temperature sensor).

0019349

< Test connections (**arrows**) at temperature control potentiometer.

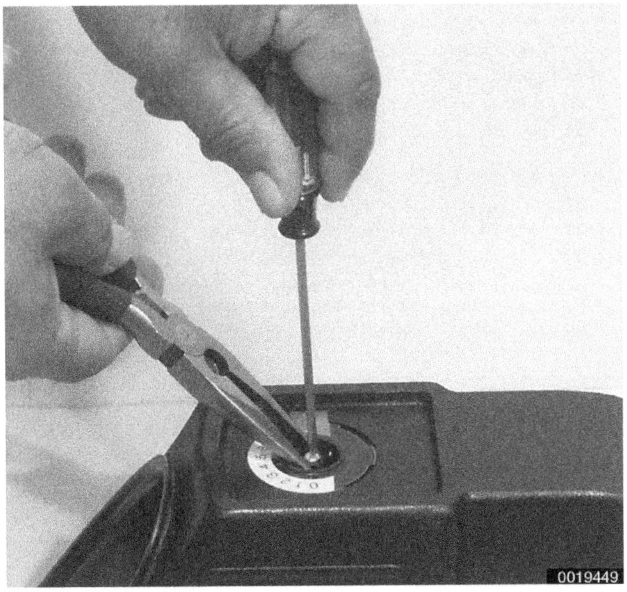
0019449

< Adjust temperature sensor potentiometer:

– Set potentiometer switch to position 5 and remove knob. Connect ohmmeter to potentiometer (shown above). Hold shaft of switch with pliers and adjust potentiometer with a screwdriver to specification: 950 ± 20 Ω

A/C TROUBLESHOOTING

Although detailed A/C diagnosis is beyond the scope of this manual, this troubleshooting section may help to pinpoint common air conditioning problems.

Before starting to troubleshoot the A/C system, check the following:

- Evaporator blower motor: do all three speeds work?
- Is the A/C compressor drive belt tight?
- Is the A/C compressor clutch engaging when switched on?

A/C warnings and cautions

Heeding the warnings and cautions will help prevent personal injury or damage to the system.

WARNING —

- *Work in a well ventilated area. Avoid breathing refrigerant vapors. Exposure may irritate eyes, nose, and throat. Switch on exhaust/ventilation systems when working on the refrigerant system.*

- *Always be careful that refrigerant does not come in contact with your skin. Always wear hand and eye protection (gloves and goggles) when working around the A/C system. If refrigerant does come in contact with your skin or eyes:*
  -Do not rub skin or eyes
  -Immediately flush skin or eyes with cool water for 15 minutes
  -Rush to a doctor or hospital.
  -Do not attempt to treat yourself.

- *Refrigerant gases are heavier than air, displace oxygen and may cause suffocation in areas of poor circulation, for example under a car.*

- *Do not discharge/charge the A/C system without proper equipment and training. Damage to the vehicle and personal injury may result.*

- *The A/C system is filled with refrigerant gas which is under pressure. Pressurized refrigerant in the presence of oxygen may form a combustible mixture. Never introduce compressed air into any container (full or empty), capped off A/C component, or piece of service equipment.*

- *Do not expose any component of the A/C system to high temperatures (Above 80°C/176°F) or open flames. Excessive heat will cause a pressure increase which could burst the system.*

- *Keep refrigerant away from open flames. Poisonous gas will be produced if it burns. Do not smoke near refrigerant gases for the same reason.*

CAUTION—
- *Immediately plug open connections on A/C components to prevent dirt and moisture contamination.*
- *Do not steam clean condensers or evaporators. Use only cold water or compressed air.*
- *It is recommended that all service to the A/C system refrigerant components be left to an authorized Porsche dealer or other qualified A/C service facility. If any of the refrigerant components are disconnected or removed, special equipment, supplies, and new parts will be needed to remove moisture from the system and recharge the refrigerant before it is closed back up.*
- *As of January 1, 1992, any person who services a motor vehicle air conditioner must, by law, be properly trained and certified, and use approved refrigerant recycling equipment. Technicians must complete an EPA-approved recycling course to be certified*
- *Do not top-off a partially charged refrigerant system. Discharge system, evacuate and then recharge system.*
- *State and local governments may have additional requirements regarding air conditioning servicing. Always comply with state and local laws.*

A/C inspection and tests

Periodic inspections will help keep the A/C operating at its peak. There are almost always small leaks in the system that will require that it eventually be recharged with R-12. The V-belt that drives the compressor is subject to wear. The condenser fins can become bent or covered with debris, reducing air flow and raising system pressure to damaging levels.

NOTE—
Run the air conditioning system for a few minutes every few weeks to keep the seals lubricated.

Clean any debris or bugs from the front of the condenser. Straighten any bent fins using a fin comb. Check the fresh air intake for obstructions. Inspect the compressor, the hoses, and all visible components for any oil leaks. These leaks are often seen at the bottom side of the fittings and components. Inspect the wiring to the compressor clutch. Check the compressor mountings for tightness, and check the condition and tension of the V-belt as described later.

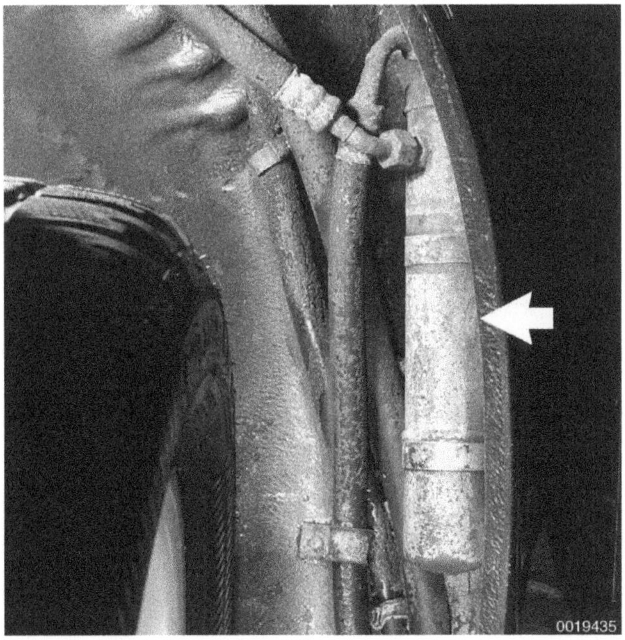

Refrigerant charge, checking

The refrigerant (R-12) charge is quickly checked by looking at the sight glass on the receiver/dryer (**arrow**), located in the left front wheel housing.

NOTE—
It may be necessary to loosen the receiver/dryer mounting clamps in order to rotate the sight glass into view.

– Start engine and turn A/C on to its maximum setting. With compressor running (clutch cycled on) view sight glass. There should be few or no bubbles visible in the glass. A constant foaming indicates that system charge is low. Streaks on interior of glass indicate that system is discharged.

If using an aftermarket refrigerant product to recharge the system, follow the manufacturer's directions closely. To prevent component damage, a totally discharged system must be evacuated (sometimes called pulling a vacuum) using special equipment before recharging. This removes any harmful moisture from lines and components.

A/C system pressure, testing

Test system pressure at an engine speed of 2,000 rpm and an outside temperature of approximately 77°F (25° C). Keep engine lid closed.

A/C system pressure
- Low side . . . . 0.81 - 2.0 bar (12 - 30 psi)
- High side. . . . . 7.93 - 14.13 bar (115 - 205 psi)

WARNING —
Wear eye protection when testing the system. A/C refrigerant at normal atmospheric pressures can evaporate and freeze anything it contacts.

Table b lists A/C refrigerant system capacities.

Table b. A/C system capacities

Compontent	Oil capacity
Refrigerant (R12): Total system capacity	 1350 gr. (47 oz.)
A/C lubricating oil: Compressor Evaporator Condensers Receiver/dryer	 120 gr. (5.0 oz) 20 gr. (0.8 oz.) 20 gr. (0.8 oz.) 10 gr. (0.4 oz.)

A/C evaporator blower motor, testing

– To test evaporator blower, first remove luggage compartment floor mat.

◀ Open access door in floor behind fuel tank.

Detach connector (**arrow**) for A/C evaporator blower motor and turn ignition and A/C blower switch on (position IIII, maximum setting).

- Check for battery voltage at connector red/white wire.
- If voltage is present and blower does not operate, blower motor is faulty and should be replaced.
- If voltage is not present, check for a faulty fuse (luggage compartment fuse box 3, fuse 2), damaged wires or connectors, or a faulty A/C blower switch.

A/C blower resistor pack, testing

The A/C blower resistor pack, located in a compartment behind the passenger side footrest, is also referred to as the bimetallic switch.

– If the evaporator blower works at the maximum setting, but not in A/C blower switch positions II and III, test resistor pack.

- Detach harness connector at resistor pack.
- With ignition switch on and blower switch in position II, there should be battery voltage at harness connector red/green wire.
- With ignition on and blower switch in position III there should be battery voltage at harness connector red/black wire.
- If voltage is present and blower does not operate in the lower speeds, blower resistor pack is faulty and should be replaced.
- If voltage is not present in either switch setting, check for damaged wires or connectors or a faulty A/C blower switch.

A/C compressor clutch, testing

To quickly check the operation of the compressor clutch, turn the ignition on and turn the temperature regulating knob to its full cold position. Turn the A/C switch on and listen for the compressor clutch to click on.

If no click is heard, the compressor clutch may not be receiving voltage or the clutch itself may be faulty.

- Check for battery voltage at single-wire connector (**arrow**) leading out of compressor.
- If voltage is present, compressor clutch is faulty and should be replaced.
- If no voltage is present, check for a faulty fuse (luggage compartment fuse box 3, fuse 2), or damaged wiring.

Table c suggests some corrective actions for additional A/C fault symptoms. Most other testing of the A/C system requires specialized knowledge and equipment and is beyond the scope of this manual. Servicing by an authorized Porsche dealer or other qualified air conditioning shop is highly recommended.

Table c. A/C troubleshooting

Symptom	Probable cause	Corrective action
Cooling has failed completely	Burst seal on receiver/drier due to A/C system overheating	Check front condenser blower fuse, relay and motor. See **971 Electrical Component Locations**. Replace receiver drier. See **870 Air Conditioning**.
Low pressure in both high and low sides	Leak in system	Check all hoses, lines and the compressor seal for leaks. Replace any seal found leaking.
Low side pressure to high High side pressure to low	Faulty compressor	Replace A/C compressor. See **870 Air Conditioning**.
Low side pressure normal High side pressure too high	Refrigerant overfilled	Discharge system and recharge.
Low side pressure too low High side pressure too low	Refrigerant low	Charge system. Check for leaks.
Low side pressure too high High side pressure normal	Expansion valve faulty	Replace expansion valve. See **870 Air Conditioning**.

801 Heater

GENERAL

This group covers automatic heating system component replacement. Refer to the following groups for further information:

- **800 Heating, Ventilation, Air Conditioning–General**
- **850 Ventilation**
- **870 Air Conditioner**

NOTE—
- *Some of the cars covered by this manual may be fitted with a manual heating control system and not the optional automatic system (M424 option code). For information on the ducting and ventilation controls for these cars, see* **850 Ventilation***.*
- *Automatic heating system troubleshooting is covered in* **800 Heating, Ventilation, Air Conditioning–General***.*

Special tools

Some of the procedures require special tools.

Special Tools

Special tools
A Baum 9700/9701/9702. . . . . Wiring connector removal tool (source: Baum Tools Unlimited)

Automatic heating system

The automatic heating system controls the passenger compartment temperature based on the setting of the temperature control knob (**arrow**). By comparing the passenger compartment temperature to the setting at the temperature control knob, the control module servo motor automatically opens or closes the heater control box valves to maintain the desired temperature. For a description of the automatic heating system, see **800 Heating, Ventilation, Air Conditioning–General**.

Heated air is blown through the exhaust system heat exchangers by an engine-mounted electric fan.

801

Heating system components

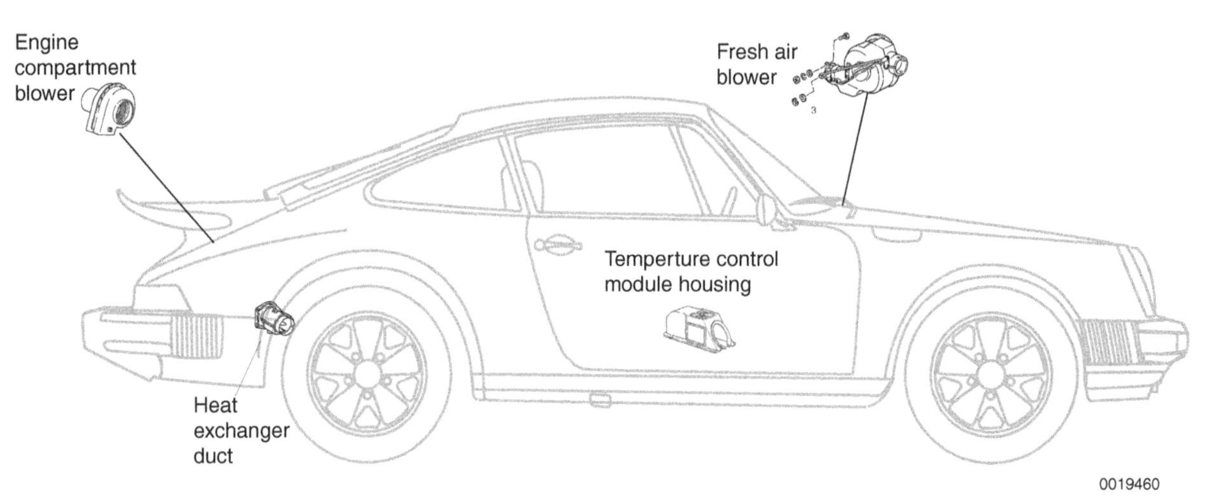

Temperature is controlled by automatic regulation of the flaps on the heat exchanger valve housings. Control of the heat exchanger valves is via the servo motor in the control module housing.

NOTE —
Heat exchanger replacement is covered in **260 Exhaust System**.

AUTOMATIC HEATER COMPONENT REPLACEMENT

Control module housing, removing and installing

- Remove passenger seat to gain access to heater controls and parking brake handle assembly.
- Remove access panel from right side of heater control housing.

- ◄ Disconnect heater valve servo control rod (**arrow**) from lever.
- Remove heater control housing trim screws from floor.
- Remove rubber sleeve covering parking brake lever.

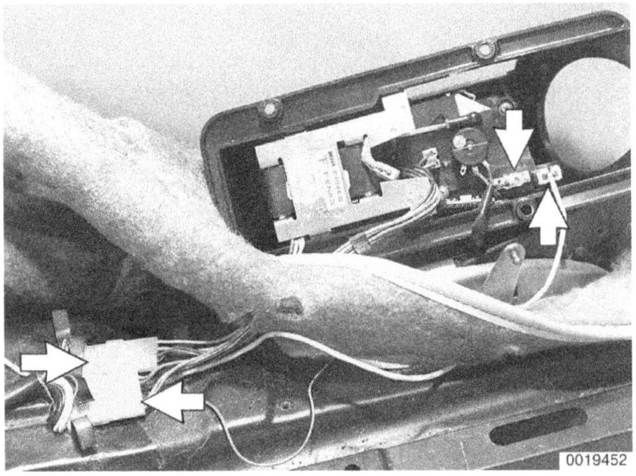
0019452

Lift control module housing up and turn over.

- Lift rug behind and under housing.
- Disconnect electrical harness connectors (**arrows**).

– With all wiring harnesses disconnected remove control module housing.

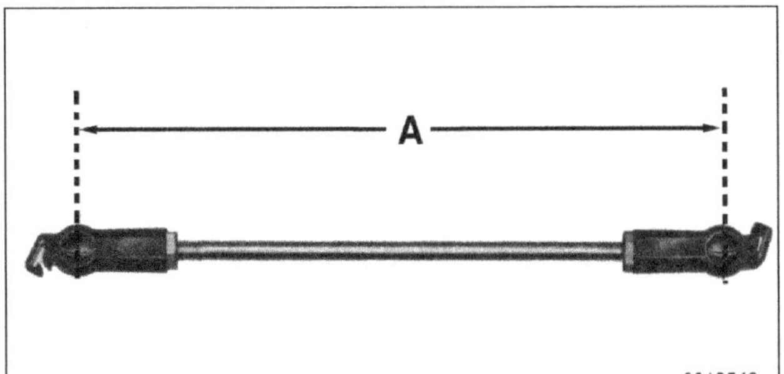

0019549

If replacing servo control rod, set length (**A**) to specification below.

Servo control rod

- Length **A** . . . . 124 ± 2 mm (4.646 ± 0.079 in.)

– Installation is reverse of removal.

Passenger compartment temperature sensor, removing and installing

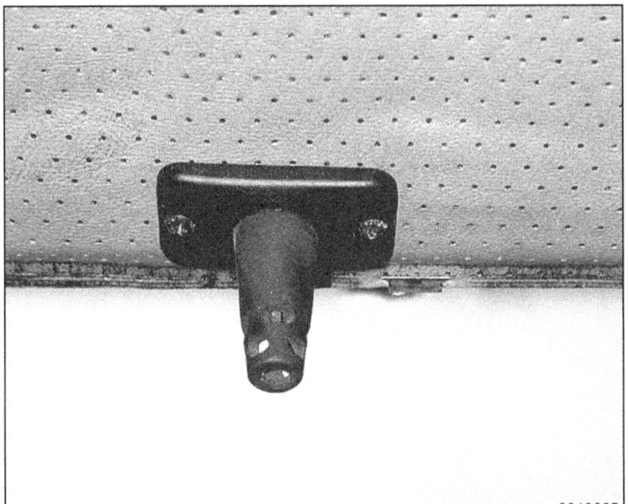
0019665

The passenger compartment temperature sensor monitors the passenger compartment temperature. The sensor is located behind a trim cover in the headliner between the sun visors.

– Remove screws holding sensor trim cover to headliner and remove cover.

– Pull sensor out of trim cover and unplug.

– Installation is reverse of removal.

Heat exchanger temperature sensor, removing and installing

– Lift temperature control module housing as described earlier.

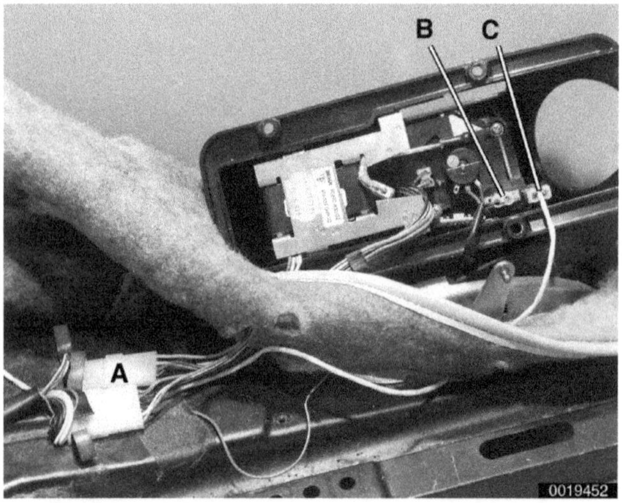

∢ Disconnect harness connector **C** at front of module.

- Use special wire connector tool to push wire ends out of plastic connector.

NOTE —

The plastic connector will not fit through the hole in the rear bulkhead and must first be removed from the harness.

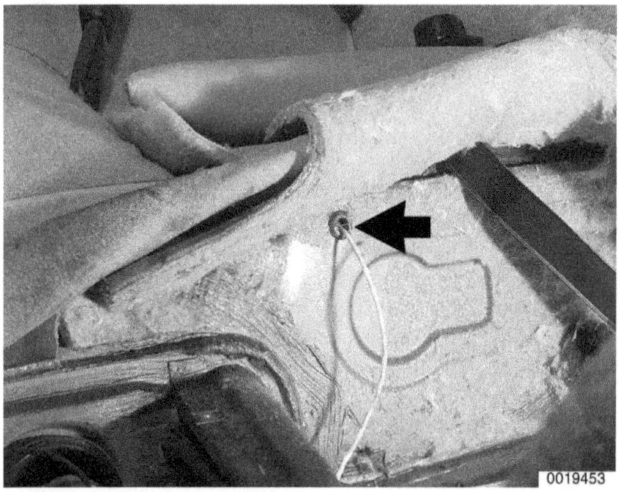

∢ Follow white wiring harness back to bulkhead (behind driver's seat) and push rubber grommet (**arrow**) and harness through bulkhead.

– Remove heater hose from left heat exchanger valve housing.

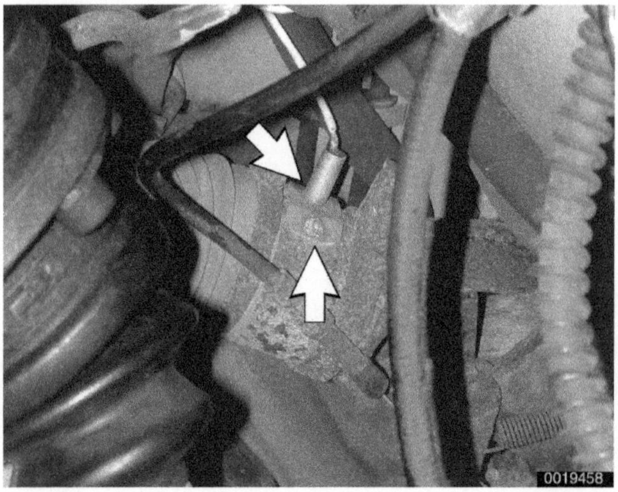

∢ Loosen clamp bolt and remove heat sensor sleeve.

- Drill out rivets (**arrows**) and remove sensor from sleeve.

– Rivet new sensor to sensor sleeve. The remainder of installation is reverse of removal.

Heater blower motor, removing and installing

The heater blower motor is located on top of the engine on the left side.

◄ Disconnect electrical harness connector (**A**) from blower.

- Detach blower air intake duct.
- Loosen hose clamp and detach elbow hose.

– Remove blower motor mounting clamp (**B**). Remove blower motor.

– Installation is reverse of removal.

Heat exchanger control cable, replacing and adjusting

The heat exchanger valve on each side is held open by a spring. Closing of the valves is controlled by a cable attached to the temperature control module or control lever.

– Raise rear of car and support it securely.

> **WARNING —**
> *Make sure the car is stable and well supported at all times. Use a professional automotive lift or jack stands designed for the purpose. A floor jack is not adequate support.*

◄ Working underneath car, disconnect cable ends (**arrow**) from heat exchanger valves. (Left side shown. Right is similar.)

NOTE —
One cable is looped and used to connect the control module to both heat exchanger valves.

– Working inside passenger compartment, remove heater control module housing as described earlier.

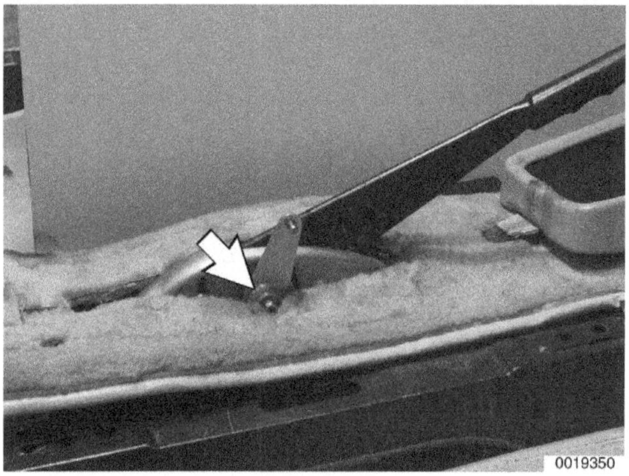
0019350

◀ Remove lock nut (**arrow**) from heater lever mounting nut at base of parking brake lever.

- Remove heater lever and pull heater cable out of guide tubes.
- Install new cable on heater control lever.
- Install cable into right and left guide tubes.
- Reinstall heater control lever and nut.
- Push heater control lever forward to stop.

0019547

◀ Install cable ends into heat exchanger valve attachment pins (**arrow**) and tighten set screws on left and right sides.

- Reinstall heater control module housing and attach servo control rod.
- Check heater operation.

MANUAL HEATER CONTROLS

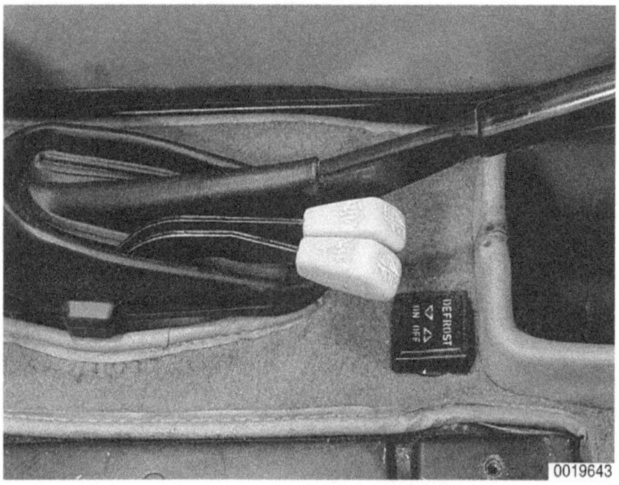
0019643

◀ Some SC models were equipped with manual heater valve control. This is a simple system with levers and a cable operating the heater valve doors. The levers are mounted to the side of the parking brake lever.

Heater cable, replacement

– Raise car and support safely.

> ***WARNING —***
> *Make sure the car is stable and well supported at all times. Use a professional automotive lift or jack stands designed for the purpose. A floor jack is not adequate support.*

◀ Working under engine near clutch bellhousing, loosen cable clamps on heater control valves and pull cables out of clamps.

– Remove parking brake lever rubber boot.

– Remove parking brake assembly mounting bolts

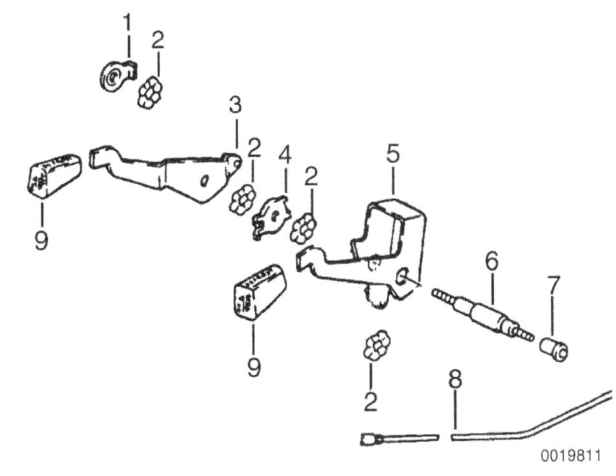

1. Guide washer
2. Spacers
3. Lever
4. Guide clip
5. Lever
6. Mounting pin
7. Lock nut
8. Cable
9. Button

◀ Remove lock nuts and washers from heater lever support.

– Slide heater control levers off support and pull levers and cables out of guide tubes.

– Install new cables on levers.

– Guide new cables through opening in parking brake support and into guide tubes.

– Reinstall heater levers, washers and lock nut and push levers down into off position.

– Reinstall parking brake mounting bolts.

– Install heater cables in heater valve clamps and tighten with valve ports open.

– Check operation of heater and adjust lever lock nut so lever does not move from set position.

– Install rubber boot and carpet.

850 Ventilation

GENERAL

This repair group covers component replacement and adjustment procedures for the ventilation system. Refer to the following repair groups for further information:

- **800 Heating, Ventilation, Air Conditioning–General**
- **801 Heater**
- **870 Air Conditioner**

Fresh and heated air distribution

0019813

The air distribution levers use cables to control fresh air and heated air to the footwells and the defroster outlets. A fresh air blower is located in the luggage compartment below the windshield.

NOTE —

The ventilation controls have no control over air distribution to the center dash vent or the side dash vents. This center vent is used for A/C only. The left and right side vents are for A/C or heated air and can are turned on and off by the vertical knurled control knobs.

- **Top lever(s)**: Controls fresh air intake from outside the vehicle. When the lever(s) is in the extreme left position, air intake from the outside is closed off. When the lever(s) is in the extreme right position, maximum ventilation of the car interior is obtained.
- **Middle lever**: Fresh air footwell/defrost blend door. This lever directs fresh air to the footwells (extreme left position) or the defroster (extreme right position) or a blend of the two.
- **Bottom lever**: Heated air footwell/defrost blend door. This lever directs heated air to the footwells (extreme left position) or the defroster (extreme right position) or a blend of the two. On cars with automatic heating, heated air temperature is controlled by the temperature control knob between the seats.

NOTE —

The HVAC system uses two different blowers for heating and A/C air distribution:

- *Heater blower motor, in engine compartment. See* **801 Heater**.
- *A/C evaporator blower, in luggage compartment. See* **870 Air Conditioner**.

FRESH AIR BLOWER

The fresh air blower is located in the luggage compartment below the windshield. It circulates fresh air to the footwell vents and the windshield defroster outlets.

Fresh air blower, removing and installing

– Disconnect negative (–) battery cable.

– Unsnap luggage compartment floor carpet and lift off.

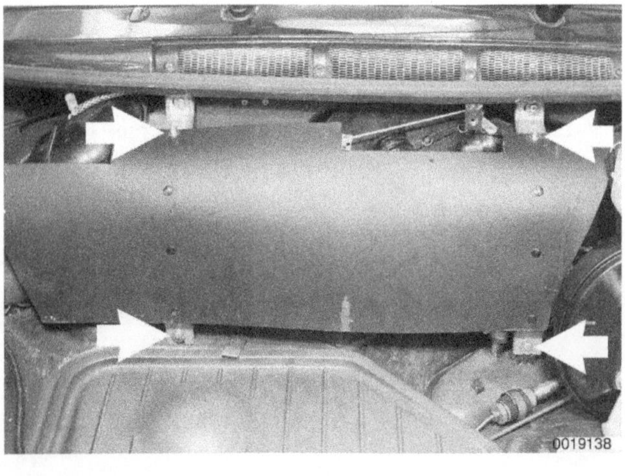

Remove fresh air plenum cover retaining fasteners (**arrows**). Remove cover.

FRESH AIR BLOWER

< Remove fresh air vent trim screws (**arrows**) in cowl.

– Remove vent trim and screen.

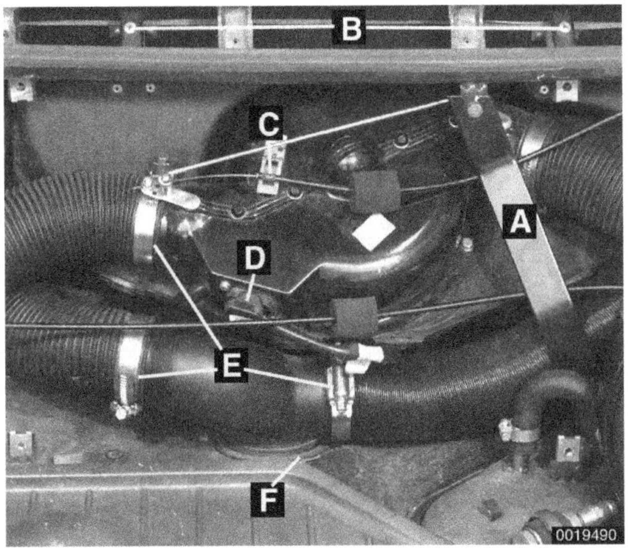

< Working inside luggage compartment, detach fresh air plenum and blower connections:

- Remove diagonal brace (**A**).
- Remove screws (**B**) holding fresh air blower housing in place.
- Disconnect heater/defroster cable (**C**) from housing.
- Disconnect electrical harness connector (**D**).
- Disconnect fresh air hoses (**E**) from housing.
- Remove retaining screw (**F**) and pull off large plastic air hose T-fitting.
- Remove blower.

– Installation is reverse of removal.

VENTILATION CONTROLS

< In order to access the ventilation controls and operating cables, first remove fresh air plenum cover retaining fasteners (**arrows**) inside luggage compartment. Remove cover.

Ventilation controls, removing and installing

– Remove fresh air blower assembly as described earlier.

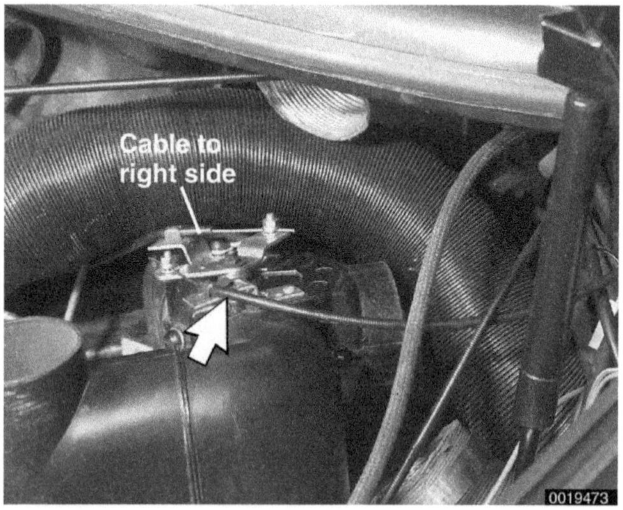

∢ Working behind brake master cylinder inside luggage compartment, detach control cables (**arrow**) from left fresh air distribution valve housing. (Heated air cable shown. Fresh air cable is similar.)

NOTE —

- *If necessary, lift housing off lower duct to access heated air cable.*
- *On each side of left distribution housing, one cable is attached to the controls while the other is attached to the right side air distribution housing.*

– Working under dashboard, just ahead of ashtray:

- Remove bolt securing ventilation control assembly to firewall.
- Disconnect ground wire to left of ashtray.

– Remove ventilation controls (with control cables attached) through dashboard opening, detaching harness connector.

– Installation is reverse of removal.

Fresh air valve cable, adjusting

– Access fresh air blower as described earlier.

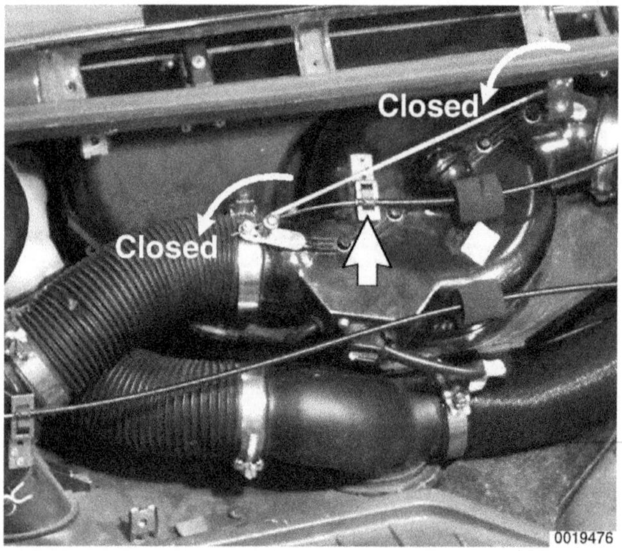

∢ Remove bowden cable retaining spring clip (**arrow**).

– Close fresh air flaps.

– Working at dashboard ventilation controls, slide fresh air lever to extreme left.

– Reinstall retaining spring clip. Check that flap operates smoothly and correctly.

Fresh air distribution cable, adjusting

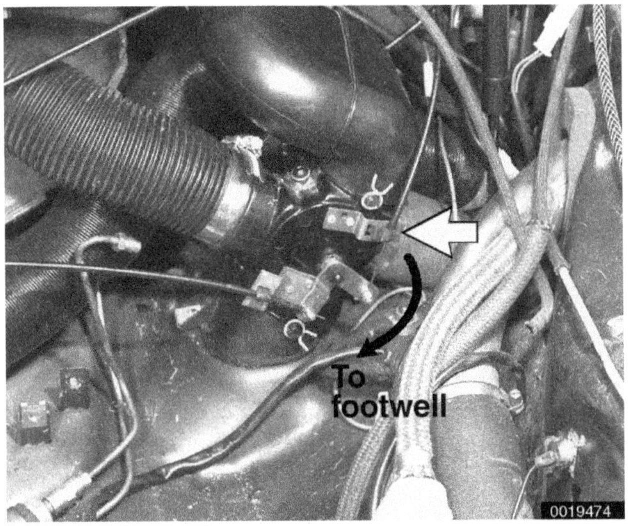

< Working in luggage compartment behind brake master cylinder, remove bowden cable retaining spring clip (**arrow**) at left fresh air distribution valve housing.

- Set valve on distribution box so air flows to footwell.

NOTE —
Brake master cylinder and booster removed for clarity.

– Working at dashboard ventilation controls, slide fresh air lever to left.

– Reinstall retaining spring clip.

Heated air distribution cable

The procedure for adjusting the heated air distribution cable is similar to the fresh air cable adjustment. However, the linkage for the cable is hidden between the valve housing and the dashboard bulkhead. Lift off the valve housing to access and adjust the cable.

870 Air Conditioner

TABLES

GENERAL

This section covers removal and installation of air conditioning components. A/C refrigerant discharge, evacuation and recharge methods, techniques and equipment are not covered here.

> ***WARNING —***
> *When discharging and charging R-12 refrigerant, refer to* **A/C warnings and cautions** *later in this group.*

A/C system components

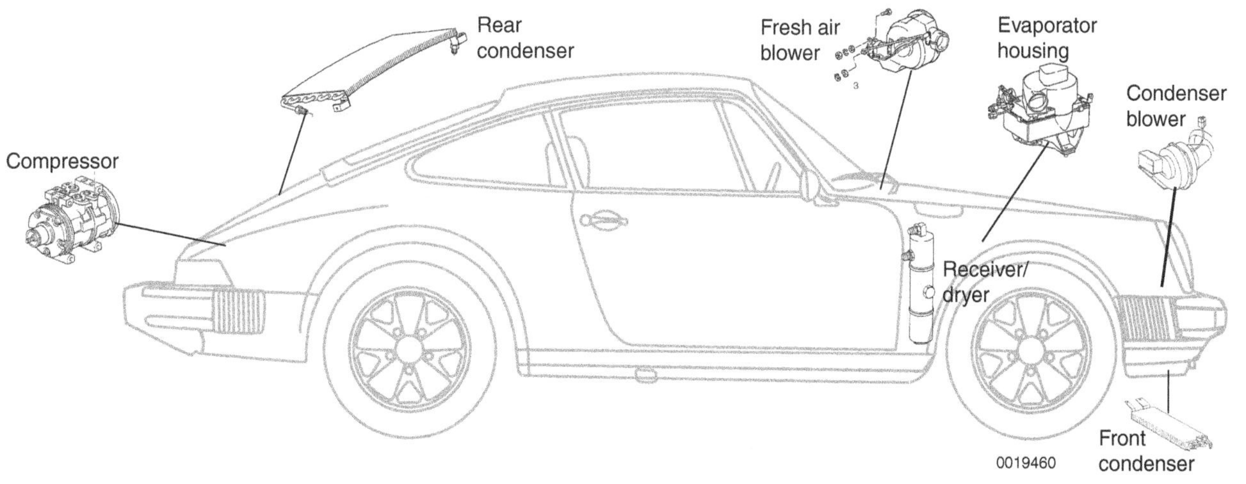

870

A/C service

NOTE —

- *Removal and installation of A/C console mounted switches is covered in* **680 Interior Trim**.
- *For further information on the ventilation system, refer to the following groups:* **801 Heater** *and* **850 Ventilation**.
- *The A/C system is separate from the heating and ventilation system and uses its own blower motor.*

A/C service and repair requires special equipment and knowledge. Incorrect procedures may not only damage the system, but also may be hazardous. Pressures in excess of 300 psi are created in the system when it is operating. The refrigerant used (R-12) is not poisonous, but in its vapor form it can accumulate in areas with poor ventilation and cause suffocation. Also, in vapor or liquid form, R-12 can immediately freeze anything it contacts, including eyes and skin.

Although tests can determine A/C efficiency, it is recommended that all service to the system be left to an authorized Porsche dealer or other qualified repair shop. Some diagnostic and troubleshooting information for A/C components is provided in **800 Heating, Ventilation, Air Conditioning–General**.

Dismounting of the A/C compressor and the condenser, without disconnecting the hoses, is covered as part of the engine removal procedure in **101 Engine Removal and Installation**. If any of the hoses or components are disconnected and the system is opened, special equipment will be needed to remove moisture from the system before it can be closed up and refilled with refrigerant and refrigerant oil.

A/C warnings and cautions

Heeding the warning and cautions on the following pages will help prevent personal injury or damage to the system.

WARNING —

- *Work in a well ventilated area. Avoid breathing refrigerant vapors. Exposure may irritate eyes, nose, and throat. Switch on exhaust/ventilation systems when working on the refrigerant system.*

- *Always be careful that refrigerant does not come in contact with your skin. Always wear hand and eye protection (gloves and goggles) when working around the A/C system. If refrigerant does come in contact with your skin or eyes:*
  -Do not rub skin or eyes;
  -Immediately flush skin or eyes with cool water for 15 minutes;
  -Immediately go to a doctor or hospital;
  -Do not attempt to treat yourself.

- *Refrigerant gases are heavier than air, displace oxygen and may cause suffocation in areas of poor circulation, for example under a car.*

- *Do not discharge/charge the A/C system without proper equipment and training. Damage to the vehicle and personal injury may result.*

- *The A/C system is filled with refrigerant gas which is under pressure. Pressurized refrigerant in the presence of oxygen may form a combustible mixture. Never introduce compressed air into any container (full or empty), capped off A/C component, or piece of service equipment.*

- *Do not expose any component of the A/C system to high temperatures (above 80°C/176°F) or open flames. Excessive heat will cause a pressure increase which could burst the system.*

- *Keep refrigerant away from open flames. Poisonous gas will be produced if it burns. Do not smoke near refrigerant gases for the same reason.*

CAUTION —

- *Immediately plug open connections on A/C components to prevent dirt and moisture contamination.*

- *Do not steam clean condensers or evaporators. Use only cold water or compressed air.*

- *It is recommended that all service to the A/C system refrigerant components be left to an authorized Porsche dealer or other qualified A/C service facility. If any of the refrigerant components are disconnected or removed, special equipment, supplies, and new parts will be needed to remove moisture from the system and recharge the refrigerant before it is closed back up.*

- *As of January 1, 1992, any person who services a motor vehicle air conditioner must, by law, be properly trained and certified, and use approved refrigerant recycling equipment. Technicians must complete an EPA-approved recycling course to be certified.*

- *State and local governments may have additional requirements regarding air conditioning servicing. Always comply with state and local laws.*

- *Do not top off a partially charged refrigerant system. Remove refrigerant, evacuate and then recharge system.*

A/C COMPRESSOR AND RECEIVER/DRYER

A/C compressor, replacing

– Following manufacturer's instructions, connect an approved refrigerant recovery/recycling/recharging unit to A/C system and remove refrigerant.

– Disconnect negative (–) battery cable.

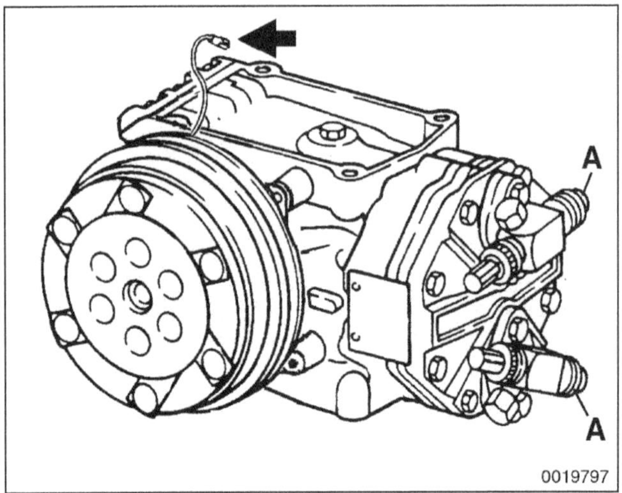

< Disconnect compressor clutch electrical connection (**arrow**).

– Disconnect A/C refrigerant lines (**A**) and plug openings.

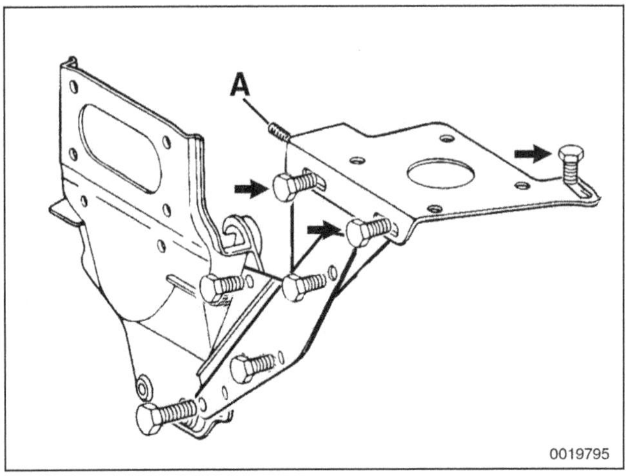

< Loosen compressor mounting plate fasteners (**arrows**). Do not remove fasteners. Loosen V-belt adjuster lock nut and screw (**A**). Slide compressor to left and remove V-belt.

– Remove compressor mounting bolts and remove compressor from adjusting plate.

– Installation is reverse of removal.
  - Be sure there is sufficient oil in compressor. See **Table a**.
  - Before recharging, evacuate system for at least 30 minutes.
  - Recharge system following equipment manufacturer's instructions.

> ***WARNING —***
> *A new Porsche A/C compressor is shipped pressurized with refrigerant. Use care when removing sealing caps and follow all manufacturer's instructions.*

> ***CAUTION —***
> *Always replace O-rings when reconnecting refrigerant lines.*

Table a. A/C system capacities

Component	Oil capacity
Refrigerant (R-12):	
Total system capacity	1350 gr. (47 oz.)
A/C lubricating oil:	
Compressor	120 gr. (2.0 oz)
Evaporator	20 gr. (1.6 oz.)
Condensers	20 gr. (0.8 oz.)
Receiver/dryer	10 gr. (0.4 oz.)

Tightening torques

- A/C compressor adjusting plate to mount . . . . 22 Nm (17 ft-lb)
- A/C compressor to adjusting plate . . . . . 22 Nm (17 ft-lb)
- Refrigerant lines to A/C compressor
  - 3/4 in. fitting . . . . 20 - 28 Nm (14 - 20 ft-lb)
  - 7/8 in. fitting . . . . 29 - 37 Nm (21 - 27 ft-lb)

A/C receiver/drier, replacing

The A/C receiver/drier is mounted in the rear of the left front wheel housing.

> ***CAUTION —***
> *The receiver/drier must be replaced when;*
> *-there is dirt in the A/C system;*
> *-the compressor has seized or has been replaced for any reason;*
> *-the condenser or evaporator are replaced for any reason;*
> *-the A/C system is leaking and there is no more refrigerant;*
> *-the A/C system was open for 24 hours or more.*

– Following manufacturer's instructions, connect an approved refrigerant recovery/recycling/recharging unit to A/C system and remove refrigerant.

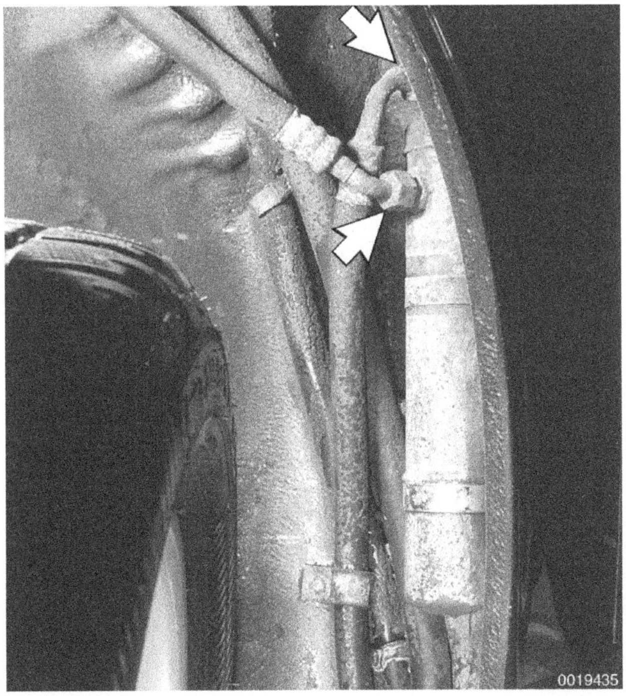

◄ Disconnect and plug refrigerant lines (**arrows**).

– Remove hose clamps used to mount receiver/drier to inner fender panel.

– Remove receiver/drier.

– Installation is reverse of removal.
- Be sure there is sufficient oil in receiver/dryer. See table below.
- Before recharging, evacuate system for at least 30 minutes.
- Recharge system following equipment manufacturer's instructions.

> ***CAUTION —***
> *Always replace O-rings when reconnecting refrigerant lines.*

A/C system capacities

Component	Oil capacity
Refrigerant (R-12): Total system capacity	1350 gr. (47 oz.)
A/C lubricating oil: Compressor Evaporator Condensers Receiver/dryer	 120 gr. (2.0 oz) 20 gr. (1.6 oz.) 20 gr. (0.8 oz.) 10 gr. (0.4 oz.)

Tightening torque

- Refrigerant line to receiver/dryer:
  5/8 in. fitting . . . . 14 - 20 Nm (10 - 14 ft-lb)

A/C CONDENSERS

A/C condenser (front), removing and installing

– Following manufacturer's instructions, connect an approved refrigerant recovery/recycling/recharging unit to A/C system and remove refrigerant.

> ***WARNING —***
> *Do not discharge/charge the A/C system without proper equipment and training. Damage to the vehicle and personal injury may result. Read warnings and cautions given earlier.*

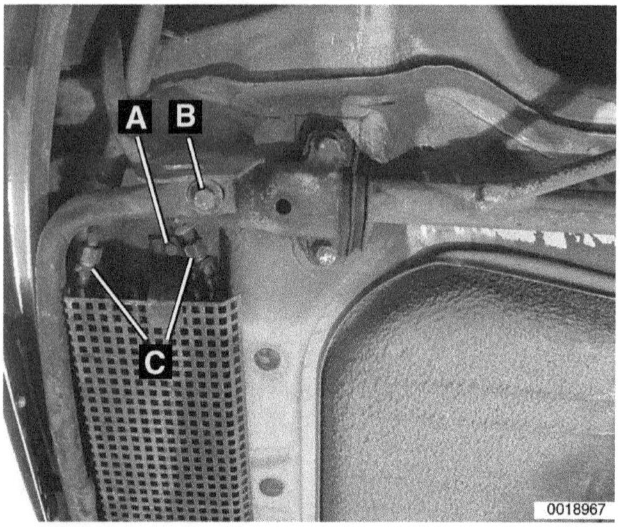

◄ Remove front condenser protective bar and pan mounting bolts (**A** and **B**). (Left side shown. Right is similar.)

– Disconnect refrigerant lines (**C**) from left side of condenser. Plug lines and fittings immediately.

– Remove A/C condenser and protective pan.

– Installation is reverse of removal.
  - Be sure there is sufficient oil in condenser. See table below.
  - Before recharging, evacuate system for at least 30 minutes.
  - Recharge system following equipment manufacturer's instructions.

> ***CAUTION —***
> - *Always replace O-rings when reconnecting refrigerant lines.*
> - *When installing a new condenser always replace the receiver/drier unit.*

Tightening Torque

- Refrigerant lines to front condenser
  5/8 in. fitting . . . . 14 - 20 Nm (10 - 14 ft-lb)

A/C system capacities

Component	Oil capacity
Refrigerant (R-12):	
Total system capacity	1350 gr. (47 oz.)
A/C lubricating oil:	
Compressor	120 gr. (2.0 oz)
Evaporator	20 gr. (1.6 oz.)
Condensers	20 gr. (0.8 oz.)
Receiver/dryer	10 gr. (0.4 oz.)

A/C condenser (rear), removing and installing

The rear A/C condenser is mounted to the underside of the engine lid.

– Following manufacturer's instructions, connect an approved refrigerant recovery/recycling/recharging unit to A/C system and remove refrigerant.

> ***WARNING —***
> *Do not discharge/charge the A/C system without proper equipment and training. Damage to the vehicle and personal injury may result. Read warnings and cautions given earlier.*

◀ Working at underside of engine lid, disconnect refrigerant lines (**arrows**) from sides of condenser. Plug lines and fittings immediately.

– Remove condenser mount (**A**) from center of engine lid.

– Remove refrigerant line support clamps on hinges.

– Installation is reverse of removal.

- Be sure there is sufficient oil in condenser. See table below
- Before recharging, evacuate system for at least 30 minutes.
- Recharge system following equipment manufacturer's instructions.

> ***CAUTION —***
> *Always replace O-rings when reconnecting refrigerant lines.*

A/C system capacities

Component	Oil capacity
Refrigerant (R-12):	
Total system capacity	1350 gr. (47 oz.)
A/C lubricating oil:	
Compressor	120 gr. (2.0 oz)
Evaporator	20 gr. (1.6 oz.)
Condensers	20 gr. (0.8 oz.)
Receiver/dryer	10 gr. (0.4 oz.)

Tightening torque

- Refrigerant lines to rear condenser
  3/4 in. fittings. . . . . 20 - 28 Nm (14 - 20 ft-lb)

A/C condenser blower (front), removing and installing

– Disconnect negative (–) battery cable. See **030 Maintenance**

– Remove luggage compartment floor mat.

– Remove battery.

– Remove spare tire.

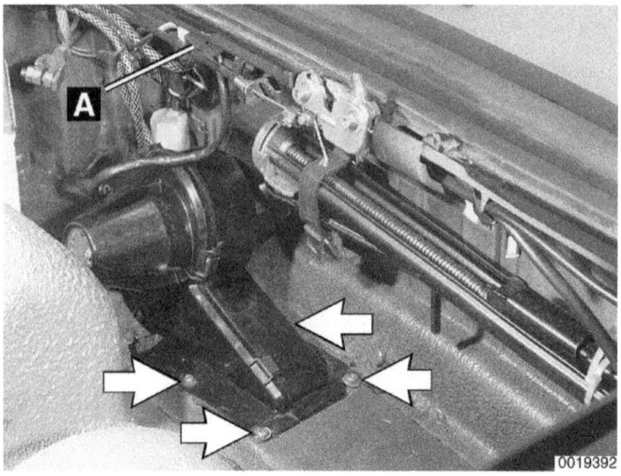

< Disconnect blower electrical harness connector (**A**).

– Remove screws (**arrows**) securing blower to floor.

– Slide blower back and lift out.

– Installation is reverse of removal.

A/C EVAPORATOR ASSEMBLY

The A/C evaporator assembly includes the following:

- **Evaporator** and **evaporator blower**
- **Expansion valve** attached to the outside of the evaporator housing
- **A/C temperature control switch** in the center console, connected to the **evaporator temperature probe** in the evaporator housing.
- **Evaporator blower resistor** in the evaporator blower air intake port in the passenger compartment

NOTE —
The blower resistor is also referred to as the bimetallic switch.

Expansion valve, replacing

– Following manufacturer's instructions, connect an approved refrigerant recovery/recycling/recharging unit to A/C system and remove refrigerant.

> ***WARNING —***
> *Do not discharge/charge the A/C system without proper equipment and training. Damage to the vehicle and personal injury may result. Read warnings and cautions given earlier.*

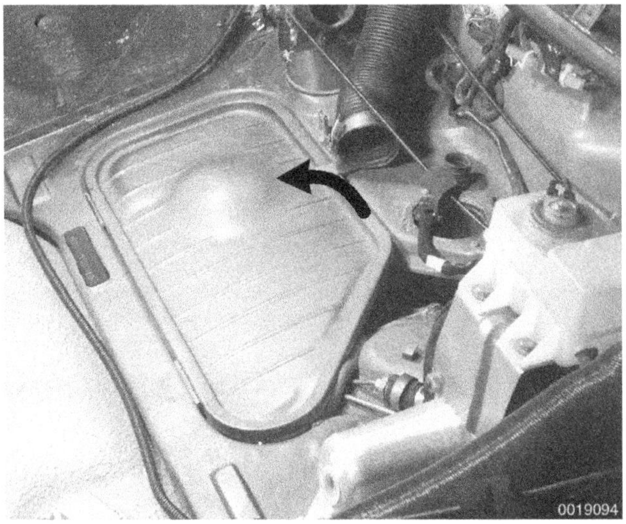

◄ Remove luggage compartment floor mat and open access panel in compartment floor.

◄ Remove evaporator temperature sensor duct support from expansion valve and gently push duct away from valve.

– Remove insulating tape (**A**) from expansion valve.

– Disconnect refrigerant connections (**B**) from expansion valve and remove valve.

– Installation is reverse of removal.
  - Be sure there is sufficient oil in evaporator. See table below.
  - Before recharging, evacuate system for at least 30 minutes
  - Recharge system following equipment manufacturer's instructions.

> ***CAUTION—***
> *Always replace O-rings when reconnecting refrigerant lines.*

Tightening torques

- Refrigerant lines to expansion valve or evaporator
  - 5/8 in. fitting .... 14 - 20 Nm (10 - 14 ft-lb)
  - 3/4 in. fitting .... 20 - 28 Nm (14 - 20 ft-lb)
  - 7/8 in. fitting .... 29 - 37 Nm (21 - 27 ft-lb)

A/C system capacities

Component	Oil capacity
Refrigerant (R-12):	
Total system capacity	1350 gr. (47 oz.)
A/C lubricating oil:	
Compressor	120 gr. (2.0 oz)
Evaporator	20 gr. (1.6 oz.)
Condensers	20 gr. (0.8 oz.)
Receiver/dryer	10 gr. (0.4 oz.)

A/C evaporator housing, removing and disassembling

– Following manufacturer's instructions, connect an approved refrigerant recovery/recycling/recharging unit to A/C system and remove refrigerant.

> ***WARNING —***
> *Do not discharge/charge the A/C system without proper equipment and training. Damage to the vehicle and personal injury may result. Read warnings and cautions given earlier.*

– Disconnect negative (–) battery cable. See **030 Maintenance**.

– Remove expansion valve as described earlier.

– Loosen hose clamp and detach large air duct at top of evaporator housing.

◄ Disconnect evaporator blower harness connector (**1**).

– Remove both support brackets (**2**).

– Gently pull evaporator temperature probe (**3**) out of base of evaporator and lay aside without damaging duct.

– Remove evaporator assembly from well.

– Remove spring clips securing lower evaporator housing to upper housing. Remove upper housing.

– Remove, service and replace blower motor or evaporator, as needed.

– Installation is reverse of removal.

- Be sure there is sufficient oil in evaporator. See table below.
- Before recharging, evacuate system for at least 30 minutes.
- Recharge system following equipment manufacturer's instructions.

> ***CAUTION —***
> *Always replace O-rings when reconnecting refrigerant lines.*

A/C system capacities

Component	Oil capacity
Refrigerant (R-12):	
Total system capacity	1350 gr. (47 oz.)
A/C lubricating oil:	
Compressor	120 gr. (2.0 oz)
Evaporator	20 gr. (1.6 oz.)
Condensers	20 gr. (0.8 oz.)
Receiver/dryer	10 gr. (0.4 oz.)

Tightening torques

- Refrigerant lines to expansion valve or evaporator
  - 5/8 in. fitting . . . . 14 - 20 Nm (10 - 14 ft-lb)
  - 3/4 in. fitting . . . . 20 - 28 Nm (14 - 20 ft-lb)
  - 7/8 in. fitting . . . . 29 - 37 Nm (21 - 27 ft-lb)

A/C temperature control switch, replacing

The A/C temperature switch (**arrow**) in the center console is connected via a capillary tube to the temperature probe in the evaporator housing.

– Access evaporator in compartment below luggage compartment.

Remove temperature probe from hole in evaporator housing.

– Remove temperature control switch from center console. See **680 Interior Trim**.

– Carefully feed temperature sensor tube and probe through opening in bulkhead into passenger compartment.

– Installation is reverse of removal.

Evaporator blower resistor, replacing

The evaporator blower resistor, also referred to as the bi-metallic switch, is located in the intake port of the evaporator, in front of the passenger foot-rest.

- Detach resistor harness connector (**arrow**).
- Remove intake port mounting screws (**A**) and remove resistor and port.

– Installation is reverse of removal.

900 Electrical System–General

TABLES

Special Tools

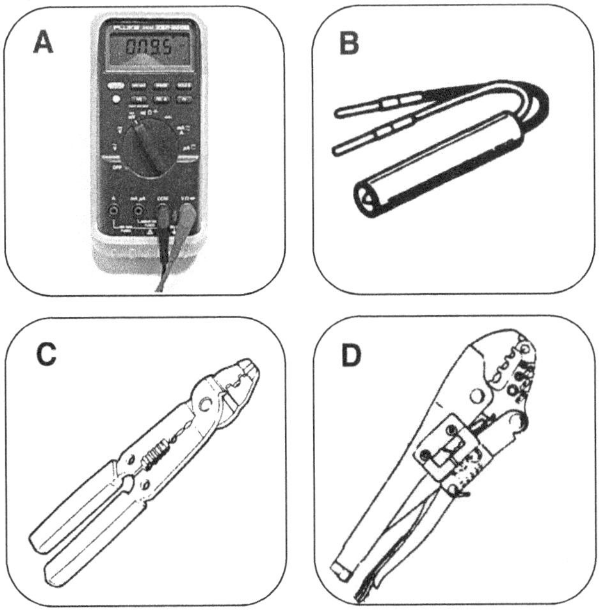

Special tool

A Fluke 87 . . . . Automotive digital multimeter
B Baum 1115 . . . . LED tester with thin spade probes (source: Baum Tools Unlimited)
C T746 . . . . Wire stripper/crimper (source: Baum Tools Unlimited)
D T748 . . . . One-step crimping pliers (source: Baum Tools Unlimited)

GENERAL

A brief description of the principal parts of the electrical system is presented here. Also covered here are basic electrical system troubleshooting tips.

Special tools

Many procedures and tests in this partition of the manual require the use of some electronic test equipment. Read the procedure through before starting work.

Electrical system safety precautions

Please read the following warnings and cautions before doing any work on your electrical system.

> ***WARNING —***
> - *The ignition system of the car operates at lethal voltages. People with pacemakers or weak hearts should not expose themselves to the ignition system. Extra caution must be taken when working on the ignition system or when servicing the engine while it is running or the key is on. See* **280 Ignition System** *for additional ignition system warnings and cautions.*
> - *Before operating the starter without starting the engine (as when making a compression test), disable the ignition system by removing the harness connector from the CDI control unit. See* **280 Ignition System**.

> ***CAUTION —***
> - *Always turn off the engine and disconnect the negative (–) cable from the battery before removing any electrical components.*
> - *Connect and disconnect ignition system wires, multiple connectors, and ignition test equipment leads only while the ignition is off.*
> - *Do not disconnect the battery with engine running.*
> - *Do not quick-charge the battery (for boost starting) for longer than one minute, and do not exceed 16.5 volts at the battery with the boosting cables attached. Wait at least one minute before boosting the battery a second time.*
> - *Do not use a test lamp that has a normal incandescent bulb to test circuits containing electronic components. The high electrical consumption of these test lamps may damage the components.*
> - *Do not use an analog meter. Use only a digital multimeter.*
> - *Solid-state modules are static-sensitive. Static discharge will permanently damage them. Always handle the modules using proper static prevention equipment and techniques.*
> - *To avoid damaging harness connectors or relay panel sockets, use jumper wires with flat-blade connectors that are the same size as the connector or relay terminals.*
> - *Always switch a digital multimeter to the appropriate function and range before making test connections.*
> - *Do not try to start the engine of a car which has been heated above 176°F (80°C), (for example, in a paint drying booth). Allow it to cool to normal temperature.*
> - *Disconnect the battery before doing any electric welding on the car.*
> - *Do not wash the engine while it is running or any time the ignition is switched on.*

Voltage and polarity

The vehicle electrical system is a 12-volt direct current (DC) negative-ground system. A voltage regulator controls system voltage at 12.6 - 14.5 volts. All circuits are grounded by direct or indirect connection to the negative (–) terminal of the battery. A number of ground connections throughout the car connect the wiring harness to chassis ground. These circuits are completed by the battery cable or ground strap between the body and the battery negative (–) terminal.

Wiring, fuses and relays

Nearly all parts of the wiring harness connect to components of the electrical system with keyed, push-on connectors that lock into place. Notable exceptions are the heavy battery cables and the starter wiring. The wiring is color-coded for circuit identification.

With the exception of the battery charging system, most electrical power is routed from the ignition switch or the battery through the main fuse/relay panel, located in the left rear corner of the luggage compartment. Fuses are color coded to indicate current capacities.

The relays and control units/modules are mounted in various places throughout the vehicle. See **971 Electrical Component Locations**.

Electrical test equipment

Many of the electrical tests described in this manual call for measuring voltage, current or resistance using a digital multimeter (DMM). Digital meters are preferred for precise measurements and for electronics work because they are generally more accurate than analog meters. The numerical display is also less likely to be misread, since there is no needle position to be misinterpreted by reading at an angle.

An LED test light is a safe, inexpensive tool that can be used to perform many simple electrical tests that would otherwise require a digital multimeter. The LED indicates when voltage is present between any two test-points in a circuit.

CAUTION —

- *Use a digital multimeter with at least 10 megaohm input impedance, or an LED test light. An analog meter (swing-needle) or a test light with a normal incandescent bulb may draw enough current to damage sensitive electronic components.*
- *An ohmmeter must not be used to measure resistance on solid state components such as control units or time delay relays.*
- *Disconnect battery before making resistance (ohm) measurements on the circuit.*

WIRING DIAGRAMS

The wiring diagrams shown in **970 Electrical Wiring Diagrams** have been specially designed to enable quick and efficient diagnosis and troubleshooting of electrical malfunctions.

Wiring codes and abbreviations

A lot of information is included in each wiring diagram if you know how to read them. Wire colors in the diagrams are abbreviated. Combined color codes indicate a multi-colored wire. For example the code BLU/RED indicates a blue wire with a red stripe. See **Table a**.

Many electrical components, connectors, fuses, and ground locations are identified using a unique number. Each of these numbers correspond to a particular part in the circuit commonly found in **970 Electrical Wiring Diagrams**.

NOTE —
Sometimes the color of an installed wire may be different than the one on the wiring diagram. Do not be concerned. Just be sure to confirm that the wire connects to the proper terminals.

Table a. Wire color codes

BLU	Blue
BRN	Brown
YEL	Yellow
GRN	Green
GRY	Gray
ORG	Orange
RED	Red
BLK	Black
VIO	Violet
PNK	Pink
WHT	White

Most terminals are identified by numbers on the components and harness connectors. The terminal numbers for major electrical connections are shown in the diagrams. Though many terminal numbers appear only once, several other numbers appear in numerous places throughout the electrical system and identify certain types of circuits. Some of the most common circuit numbers are listed below in **Table b**.

Table b. Terminal and circuit numbers

Number	Circuit description
1	Low voltage switched terminal of coil
4	High voltage center terminal of coil
X or +X	Originates at ignition switch. Supplies power when the ignition switch is in the PARK, RUN or START position
15	Originates at ignition switch. Supplies power when ignition switch is in RUN or START position
30	Battery positive (+) voltage. Supplies power whenever battery is connected. (Not dependent on ignition switch position, unfused)
31	Ground, battery negative (–) terminal
50	Supplies power from battery to starter solenoid when ignition switch is in START position only
+54	Originates at ignition switch. Supplies power when ignition switch is in the RUN position only
85	Ground side (–) of relay coil
86	Power-in side (+) of relay coil
87	Relay actuated contact
D	Alternator warning light and field energizing circuit

Additional abbreviations shown in the wiring diagrams are given below.

Abbreviations
- A/C . . . . air conditioning
- CDI . . . . Capacitor Discharge Ignition
- CIS . . . . Continuous Injection System

ELECTRICAL TROUBLESHOOTING

Four things are required for current to flow in any electrical circuit: a voltage source, wires or connections to transport the voltage, a load or device that uses the electricity and a connection to ground. Most problems can be found using a digital multimeter (volt/ohm/amp meter) to check for voltage supply, for breaks in the wiring (infinite resistance/no continuity), or for a path to ground that completes the circuit.

Electric current is logical in its flow, always moving from the voltage source toward ground. Electrical faults can usually be located through a process of elimination. When troubleshooting a complex circuit, separate the circuit into smaller parts. The general tests outlined below may be helpful in finding electrical problems. The information is most helpful when used with the wiring diagrams.

Be sure to analyze the problem. Use the wiring diagrams to determine the most likely cause. Get an understanding of how the circuit works by following the circuit from ground back to the power source.

When making test connections at connectors and components, use care to avoid spreading or damaging the connectors or terminals. Some tests may require jumper wires to bypass components or connections in the wiring harness. When connecting jumper wires, use blade connectors at the wire ends that match the size of the terminal being tested. The small internal contacts are easily spread apart, and this can cause intermittent or faulty connections that can lead to more problems.

Voltage and voltage drop

The wires, connectors, and switches that carry current are designed with very low resistance so that current flows with a minimum loss of voltage. A voltage drop is caused by higher than normal resistance in a circuit. This additional resistance actually decreases or stops the flow of current. A voltage drop can be noticed by problems ranging from dim headlights to sluggish wipers. Some common sources of voltage drops are corroded or dirty switches, dirty or corroded connections or contacts and loose or corroded ground wires and ground connections.

A voltage drop test is a good test to make if current is flowing through the circuit, but the circuit is not operating correctly. A voltage drop test will help to pinpoint a corroded ground strap or a faulty switch. Normally, there should be less than 1 volt drop across most wires or closed switches. A voltage drop across a connector or short cable should not exceed 0.5 volts.

NOTE —

- *A voltage drop test is generally more accurate than a simple resistance check because the resistances involved are often too small to measure with most ohmmeters. For example, a resistance as small as 0.02 ohms would results in a 3 volt drop in a typical 150 amp starter circuit. (150 amps x 0.02 ohms =3 volts).*
- *Keep in mind that voltage with the key on and voltage with the engine running are not the same. With the ignition on and the engine off (battery voltage), voltage should be approximately 12.6 volts. With the engine running (charging voltage), voltage should be approximately 14.0 volts. Measure voltage at the battery with the ignition on and then with the engine running to get exact measurements.*

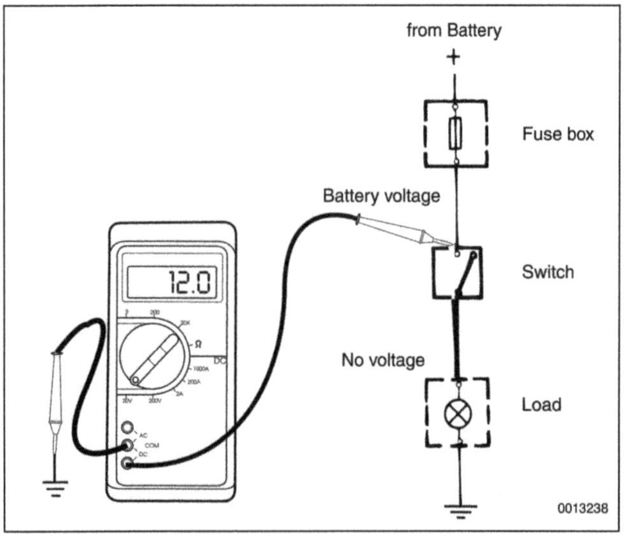

Voltage, measuring

– Connect digital multimeter negative lead to a reliable ground point on car.

NOTE —
The negative (–) battery terminal is always a good ground point.

< Connect digital multimeter positive lead to point in circuit you wish to measure. If a reading is obtained, current is flowing through circuit.

NOTE —
The voltage reading should not deviate more than 1 volt from the voltage at the battery. If the voltage drop is more than this, check for a corroded connector or loose ground wire.

Voltage drop, testing

Voltage drop can only be checked when current is running through the circuit, such as by operating the starter motor or turning on the headlights. A digital multimeter should be used to ensure accurate readings.

– Connect digital multimeter positive lead to positive end of cable or switch being tested.

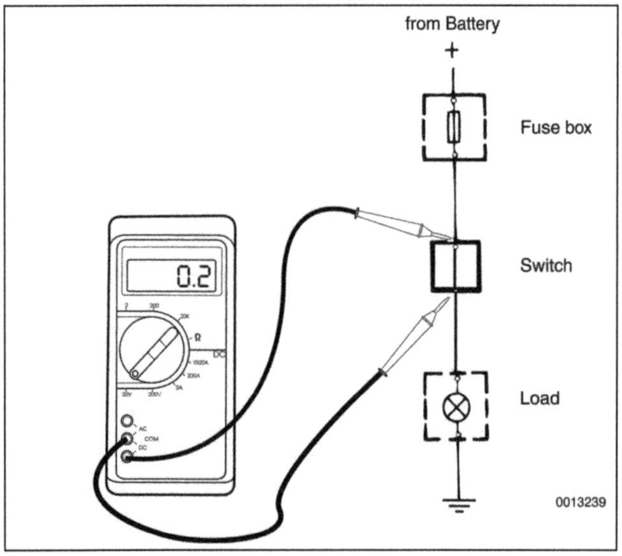

< Connect digital multimeter negative lead to other end of cable or switch being tested.

– With power on and circuit working, meter shows voltage drop (difference between two points). This value should not exceed 1 volt.

NOTE —
The maximum voltage drop in an automotive circuit, as recommended by the Society of Automotive Engineers (SAE), is as follows: 0 volts for small wire connections; 0.1 Volts for high current connections; 0.2 volts for high current cables; and 0.3 volts for switch or solenoid contacts. On longer wires or cables, the drop may be slightly higher. In any case, a voltage drop of more than 1.0 volt usually indicates a problem.

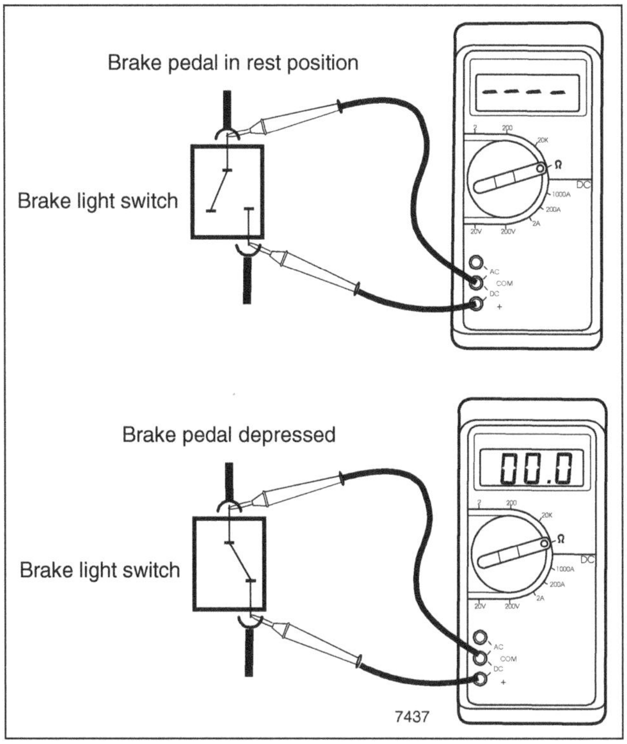

Continuity, checking

The continuity test can be used to check a circuit or switch. Because most automotive circuits are designed to have little or no resistance, a circuit or part of a circuit can be easily checked for faults using an ohmmeter. An open circuit or a circuit with high resistance will not allow current to flow. A circuit with little or no resistance allows current to flow easily.

> ***CAUTION—***
> *Use only a high quality digital ohmmeter having high input impedance when checking electronic components. The internal power source used in most analog (swing-needle) meters can damage solid state components.*

◀ Brake light switch being tested for continuity.

- With brake pedal in rest position (switch open) there is no continuity (infinite ohms). With pedal depressed (switch closed) there is continuity (zero ohms).
- When checking continuity, the ignition should be off. On circuits that are powered at all times, the battery should be disconnected. Using the appropriate wiring diagram, a circuit can be easily tested for faulty connections, wires, switches, relays, and engine sensors by checking for continuity.

Short circuits

A short circuit is exactly what the name implies. The circuit takes a shorter path than it was designed to take. The most common short that causes problems is a short to ground where the insulation on a positive (+) wire wears away and the metal wire is exposed. When the wire rubs against a metal part of the car or other ground source, the circuit is shorted to ground. If the exposed wire is live (positive battery voltage), a fuse will blow and the circuit may be damaged.

Shorts to ground can be located with a digital multimeter. Short circuits are often difficult to locate and may vary in nature. Short circuits can be found using a logical approach based on the current path.

> ***CAUTION—***
> - *On circuits protected with high rating fuses (25 amp and greater), the wires or circuit components may be damaged before the fuse blows. Always check for damage before replacing fuses of this rating.*
> - *When replacing blown fuses, use only fuses having the correct rating. Always confirm the correct fuse rating printed on the fuse/relay panel cover.*

Short circuit, testing with ohmmeter

– Remove blown fuse from circuit and disconnect cables from battery.

– Disconnect harness connector from circuit's load or component.

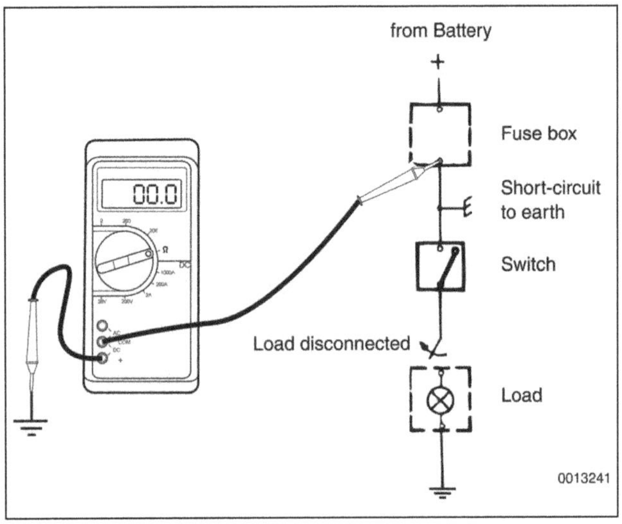

◄ Using an ohmmeter, connect one test lead to load side of fuse terminal (terminal leading to circuit) and the other test lead to ground.

– If there is continuity to ground, there is a short to ground.

– If there is no continuity, work from wire harness nearest to fuse/relay panel and move or wiggle wires while observing meter. Continue to move down harness until meter displays a reading. This is the location of short to ground.

Visually inspect the wire harness at this point for any faults. If no faults are visible, carefully slice open the harness cover or the wire insulation for further inspection. Repair any faults found.

Short circuit, testing with voltmeter

– Remove blown fuse from circuit.

– Disconnect harness connector from circuit's load or component.

NOTE —
Most fuses power more than one component. Be sure all components are disconnected when checking for a short circuit.

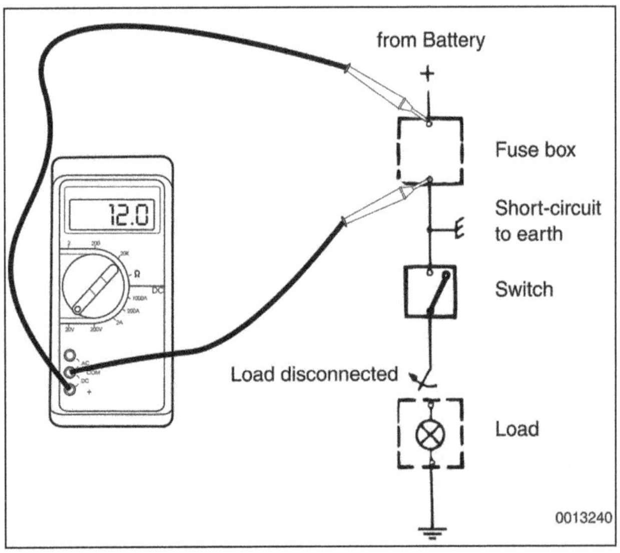

◄ Using a digital multimeter, connect test leads across fuse terminals. Make sure power is present in circuit. If necessary turn key on.

– If voltage is present at voltmeter, there is a short to ground.

– If voltage is not present, work from wire harness nearest to fuse/relay panel and move or wiggle wires while observing meter. Continue to move down harness until meter displays a reading. This is the location of short to ground.

– Visually inspect wire harness at this point for any faults. If no faults are visible, carefully slice open harness cover or wire insulation for further inspection. Repair any faults found.

901 Instruments

Special Tools

A

B

Special tools

A Fluke 87 . . . . . Automotive digital multimeter
B 1115 . . . . . LED tester with thin spade probes
(source: Baum Tools Unlimited)

GENERAL

This repair group covers the removal and installation of the gauges and warning lights. Bulb replacement information for the gauges is also included.

Special tools

Many procedures and tests in this partition of the manual require the use of some electronic test equipment. Read the procedure through before starting work.

INSTRUMENTS

The instrument gauges are mounted in rubber retaining rings in the dashboard on the driver's side. Five gauges and several warning lights are used. All gauges are electronic.

When working on the instruments or warning lights, disconnect the negative (–) battery cable and cover battery terminal to prevent cable from accidentally contacting terminal.

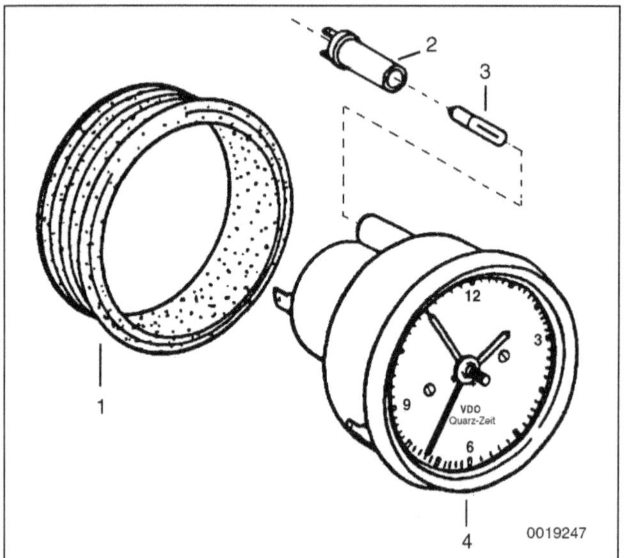

1. Rubber ring
2. Bulb holder
3. Bulb
4. Clock

Instruments, removing and installing

- Using a plastic pry tool or other blunt tipped object to pry the gauge out of its rubber retaining ring.

> **CAUTION—**
> *Use care not to damage the dashboard or gauge.*

- Label and disconnect wires.
- Installation is reverse of removal.

Instrument bulbs, replacing

- Remove instrument as described earlier.
- Pull bulb holder from back of instrument.
- Replace bulb.
- Installation is reverse of removal.

Seatbelt/brake warning lamp, removing and installing

The seatbelt/brake warning lamp is contained in a plastic housing with a control unit and circuit board. The bulbs are soldered directly to the circuit board and are not replaceable without special tools.

- Remove clock. See **Instruments, removing and installing**.
- Disconnect harness connector from back of seatbelt/brake warning lamp (**A**).
- Press warning lamp out from behind dashboard (**B**).

> **CAUTION—**
> *Do not pry on front of seatbelt/brake warning lamp housing. The lens is brittle and may break.*

- Installation is the reverse of removal.

Heating and ventilation control lamps, replacing

– Remove heating and ventilation control mounting bolt as described in **801 Heater**.

– Slide heating and ventilation control out of dash to access bulb sockets.

– Remove socket from control and replace bulb.

> ***CAUTION —***
> *Replace bulbs with proper bulbs only. See Porsche parts for the latest information.*

– Installation is reverse of removal.

HORNS

The components of the horn system are the dual horns, the horn relay, and the horn buttons on the steering wheel. A brush and slip ring assembly makes the electrical contact between the rotating steering wheel and the steering column. The horns are located below the right headlight and are accessible by removing the front bumper.

The horns are connected to ground. Depressing the horn button activates the horn relay to supply the horns with positive voltage.

Horns, removing and installing

– Remove front bumper. See **630 Bumpers**.

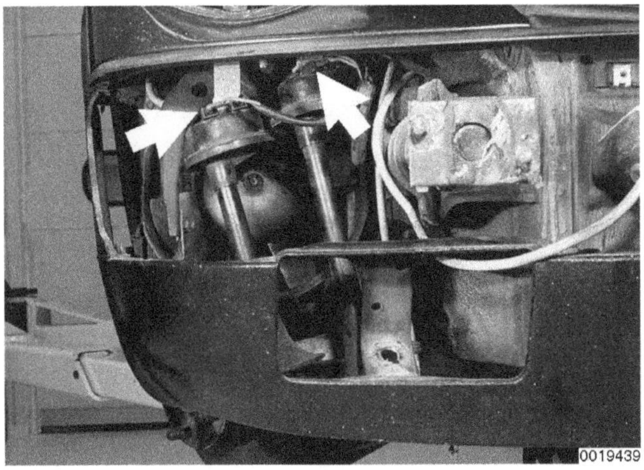
0019439

◀ Unplug wiring harnesses from horns (**arrows**).

NOTE —
Fog light is shown removed to illustrate horn removal.

– Remove mounting bracket nut and both horns with mounting bracket.

– Installation is reverse of removal.

SPEEDOMETER

The electronic speedometer uses the signal from the speedometer sensor to display the road speed. The speedometer sensor is mounted in the right side of the final drive above the drive axle.

Speedometer sensor, removing and installing

- Working inside passenger compartment, remove floor mats at rear of center tunnel.

NOTE —
Slide driver's seat forward to gain access to rear center tunnel.

- Remove four securing screws and tunnel access cover.

0018824

< Disconnect sensor wires connector (**arrow**) in tunnel.

- Raise vehicle to allow access to transmission.

> ***WARNING —***
> *Be sure vehicle is stable and fully supported on the manufacturer's lift points as shown in* ***010 General****. Use proper jack stands to support the vehicle at all four corners.*

0019635

< Remove retaining bolt (**arrow**) and sensor holder from transmission.

- Remove wires from retaining clips on transmission and pull wires from tunnel.
- Remove sensor from transmission.
- Installation is reverse of removal.

MILEAGE COUNTER

A mileage counter is installed to monitor operating miles. It turns on the OXS light in the dashboard every 30,000 miles. Once the light has been turned on, the mileage counter must be reset to turn it off.

Resetting OXS lamp

- Disconnect cable from negative battery terminal.
- Remove speedometer from dash as described in **Instruments, removing and installing**.
- Using a small punch, press white reset button on back of speedometer housing until mileage counter clicks.
- Reinstall speedometer and reconnect battery.

909 Alarm System

Special Tools

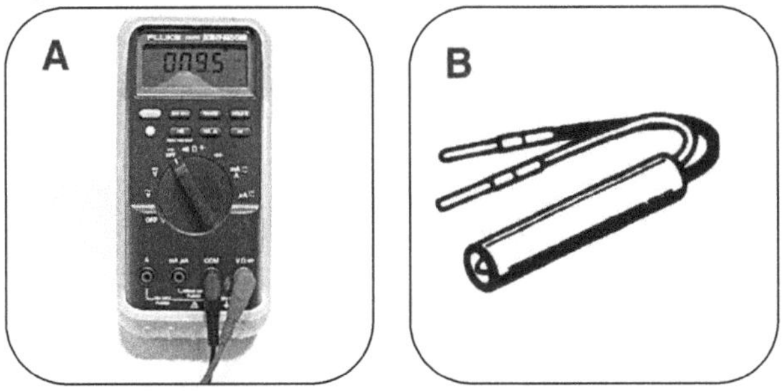

Special tools

A Fluke 87 . . . . . Automotive digital multimeter
B 1115. . . . . LED tester with thin spade probes (source: Baum Tools Unlimited)

GENERAL

This repair group covers factory installed alarm. Aftermarket anti-theft systems are not covered.

Special tools

Many procedures and tests in this partition of the manual require the use of some electronic test equipment. Read the procedure through before starting work.

Alarm system

The alarm system uses its own key for activation. Keys are not available separately. If the alarm key is lost the alarm switch must be replaced.

When it is set, if the alarm is tripped it will sound a horn and interrupt power to the fuel pump. Even with the alarm not set, a faulty alarm control module may interrupt power to the fuel pump.

The switch for setting the alarm (**arrow**) is located in the driver's door behind the door lock. The door must be partially opened to access the switch. To set the alarm, insert key and turn 90° clockwise. To deactivate the alarm insert key and turn 90° counter-clockwise.

> ***CAUTION—***
> *Always remove alarm key from door before closing.*

When set, the alarm will be tripped by opening the driver's door, passenger door or front hood, or by turning on the ignition switch.

TROUBLESHOOTING

Alarm system troubleshooting requires a digital multimeter, 12V (max. 3W) test lamp and the appropriate wiring diagrams. See **970 Electrical Wiring Diagrams**.

Alarm system switch, testing

– Make sure alarm switch and ignition switch are OFF.

– Remove alarm switch as described later. Detach electrical harness connectors.

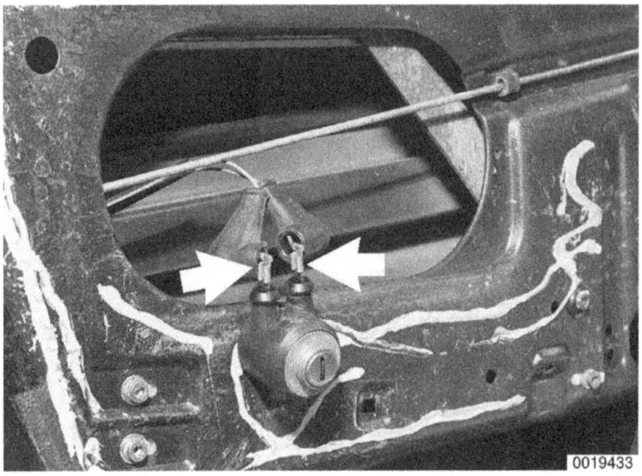
0019433

◀ Measure resistance between terminals (**arrows**) and compare to specifications given below.

Alarm switch resistance specifications

- Alarm switch ON, (key horizontal). . . . . 2 - 3 kΩ
- Alarm switch OFF, (key vertical) . . . . 4 - 5.5 kΩ

– Replace switch if resistance is not as specified.

Alarm control module, accessing

For the electrical tests that follow, it is necessary to access the alarm control module inside the dashboard.

– Working inside luggage compartment, remove fresh air plenum cover from rear of compartment. Then remove fresh air plenum and blower. See **850 Ventilation**.

Alarm circuit, testing

> ***CAUTION —***
> *The alarm circuit is tested with the battery connected. Use care when connecting and disconnecting plugs. DO NOT allow any wires to touch the vehicle body.*

0019391

◀ Access alarm module as described above. Detach harness connectors from module (**arrow**).

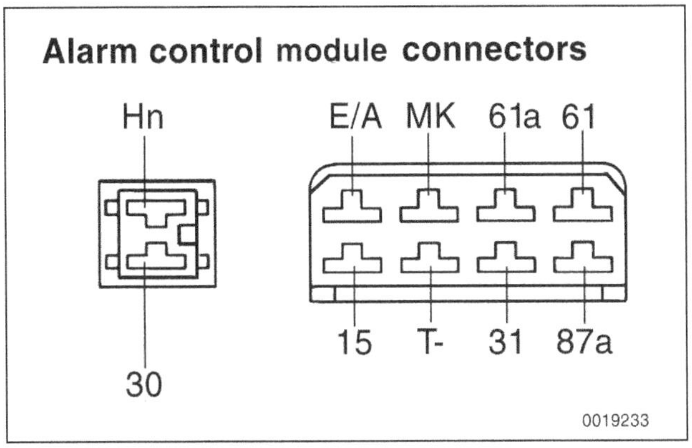

Test harness connectors with a test lamp and digital multimeter. Tests are detailed in **Table a**.

Table a. Alarm control unit connector tests

Test terminals	Result
30 and 87a	Test lamp ON.
Hn and 30	Test lamp ON.
E/A and 31	Alarm switch ON (key horizontal): 2 - 3 kΩ Alarm switch OFF (key vertical): 4 - 5.5 kΩ
MK and 30	Test lamp ON. Press trunk light contact switch: lamp OFF.
61 and 30	Test lamp ON. Turn ignition switch ON: lamp OFF.
15 and 31	Turn ignition switch ON: lamp OFF.
T- and 30	Open door (driver or passenger): test lamp ON. Close door: lamp OFF.
31 and 30	Test lamp ON.

Alarm control module, testing

Testing the alarm control module requires a 2-wire jumper harness and 12V test light (max. 3w).

- Access alarm module as described earlier.
- Unplug 2-pin connector from module. 8-pin connector remains connected.
- Depress contact switch for trunk and secure with tape or clamp.

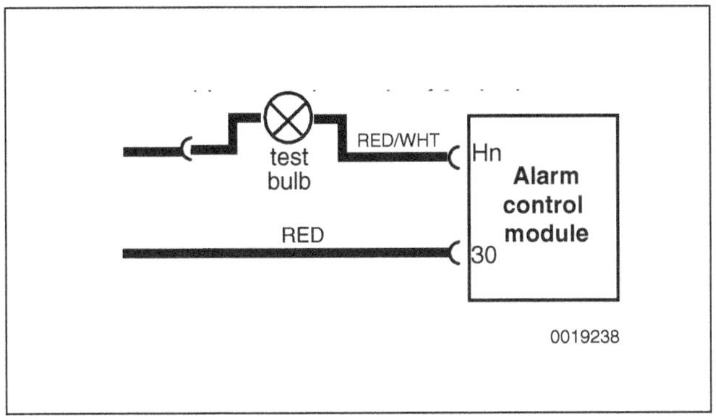

Install 2-wire jumper harness with test bulb as shown.

- Test alarm module as follows:
  - Turn alarm ON. Open driver side door fully, then close: Light should flash for approx. 30 seconds.
  - After light stops flashing, turn ignition switch ON: Light should flash.
  - Attempt to start engine: Engine should not start.
  - Open passenger door fully: Light should flash.
  - Remove tape or clamp from trunk contact switch: Light should flash.
- Disconnect jumper harness and reconnect 2-pin plug.
- Installation is reverse of removal.

COMPONENT REPLACEMENT

This section covers removal and installation of alarm system components. Always disconnect battery before connecting or disconnecting electrical components.

Alarm switch, removing and installing

- Raise driver's side window.
- Disconnect negative (–) battery cable and cover battery terminal to keep cable from accidentally contacting battery terminal.
- Remove inner door panel from driver's door. See **570 Doors**.

- Working at outer edge of driver's door, remove plastic nut cover and alarm switch mounting nut (**arrow**).
- Remove switch from inside door and unplug wires.
- Installation is reverse of removal.

Alarm control module, removing and installing

- Disconnect negative (–) battery cable and cover battery terminal to keep cable from accidentally contacting terminal.
- Working inside luggage compartment, remove fresh air plenum cover from rear of compartment. Then remove fresh air plenum and blower. See **850 Ventilation**.

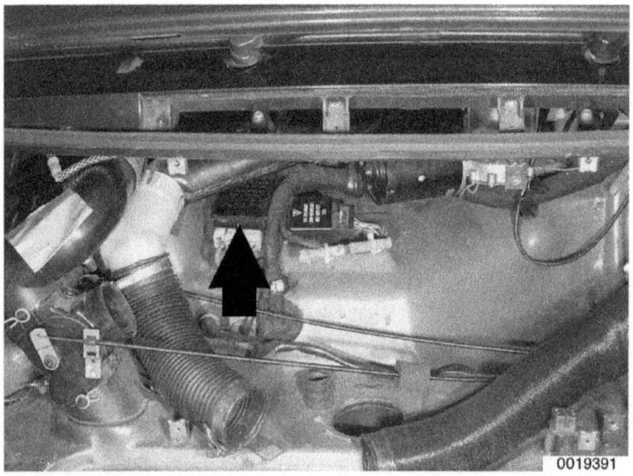

- Push alarm control module (**arrow**) back towards dashboard until clip is free of bracket.
- Disconnect multi pin connectors and remove control unit.
- Installation is reverse of removal.

Alarm system horn, removing and installing

The alarm horn is bolted to front suspension cross-member, behind fuel pump.

– Raise car and support in a safe manner.

> **CAUTION—**
> *Make sure the car is stable and well supported at all times. Use a professional automotive lift or jack stands designed for the purpose. A floor jack is not adequate support.*

◀ Remove stone guard fasteners (**arrows**) from front suspension. Remove stone guard.

◀ Remove alarm horn mounting bolt (**arrow**) and disconnect wires.

– Installation is reverse of removal.

910 Radio

Special Tools

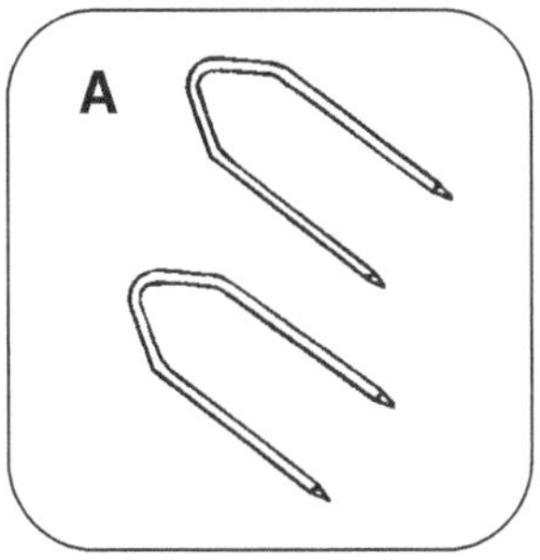

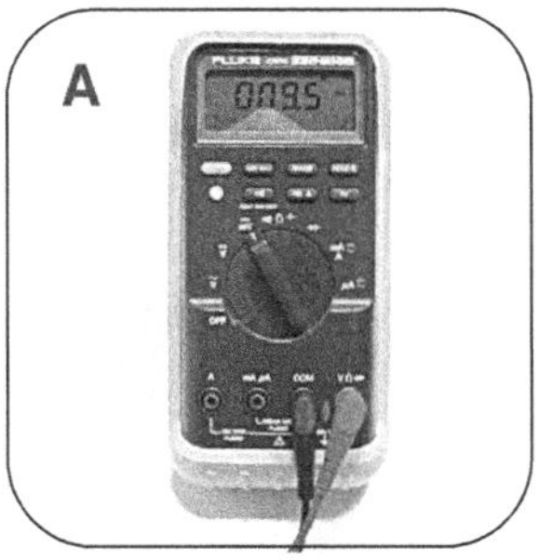

Special tool

A PNA 721 021 .... Radio removal tool (1983 models only)
B Fluke 87 .... Automotive digital multimeter

GENERAL

This section covers the Porsche factory installed sound system. Antenna and speaker removal are covered. Servicing of components is beyond the scope of this manual and should be referred to an authorized agent of the equipment in question.

Aftermarket radios, amplifiers, speakers, etc. are beyond the scope of this manual. Servicing of aftermarket sound equipment should be referred to an authorized agent of the equipment in question.

Special tools

Removal of some radio models requires the use of a special radio removal tool to release the retaining clips of the pull-out radio unit.

Sound system

The basic sound system uses a radio and four speakers. The audio option pack uses a radio, a separate power amplifier and may have additional speakers. The antenna is mounted on the right front fender.

CAUTION —
- *Before beginning work on the sound system verify that the sound system is original equipment and the wiring harness has not been modified.*
- *Do not remove the radio or disconnect the battery without obtaining the anti-theft code.*
- *The radio is held in with tamper resistant clips. Special tools are necessary to remove radio without damage.*

910

Radio, removing and installing

NOTE —

This removal method only applies to certain factory-installed radios. There are many varied brands and styles of radio installed in Porsche cars which are beyond the scope of this manual.

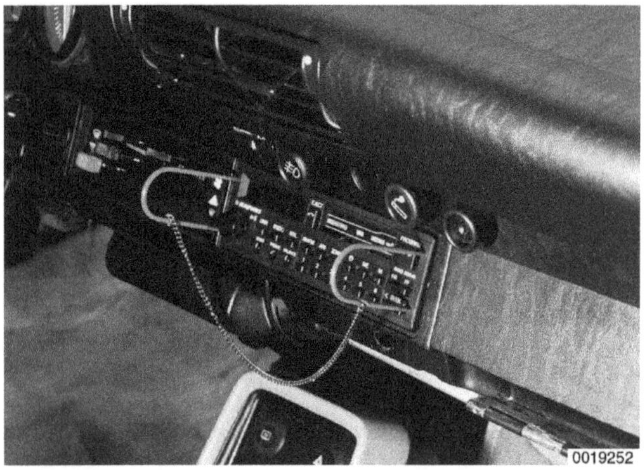

- Insert radio removal tools into holes in faceplate and gently pull radio out.

- Label and disconnect wires and antenna. Remove radio.

- Install radio by sliding into dash until retaining clips engage.

POWER ANTENNA

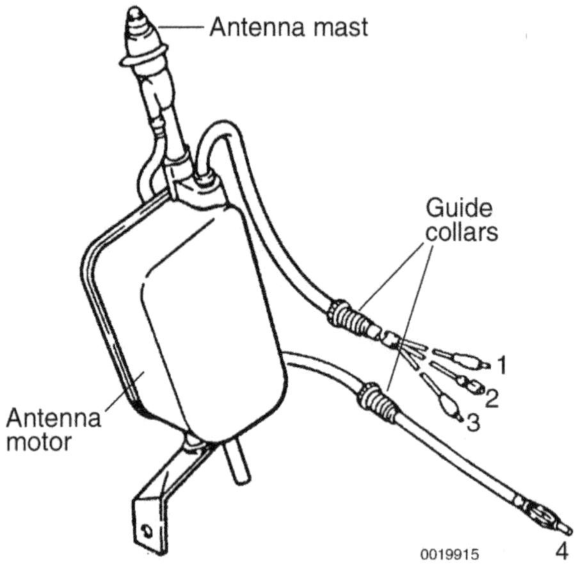

1. From head light switch terminal 30 (red wire)
2. Ground behind heater control (brown wire)
3. Antenna activation lead from radio (white wire)
4. Antenna lead

The power antenna is installed in one of the front fenders, usually on the right. An electric motor and drive unit are used to raise and lower the antenna mast.

Antenna troubleshooting

- Refer to illustration for numbered antenna wire leads.

- **1** Check for battery voltage at terminal 30 on headlight switch.

- **2** Check ground connection located behind heater controls in dash.

- 3 Check at antenna feed wire at back of radio.
  - With radio on: battery voltage.
  - With radio off: no voltage.

Antenna, removing and installing

- Disconnect battery negative (-) cable.

- Remove power supply wire from terminal 30 on headlight switch.

- Disconnect ground wire from central ground behind heater controls.

- Remove white antenna control wire and antenna lead from rear of radio.

- Turn antenna wiring harness guide collars counterclockwise to remove insulation with the harnesses from the inner fender.

– Remove lower antenna support bracket bolt.

– Loosen antenna mounting nut.

– Remove mounting nut while holding antenna. Guide antenna out of fender.

– Installation is reverse of removal.

FRONT SPEAKERS

The front speakers are mounted in the door panels. Sound systems without audio option pack use one full range speaker per side. Sound systems with audio option package may use separate high and low frequency speakers on each side.

Front speaker, removing and installing

Front speaker assembly

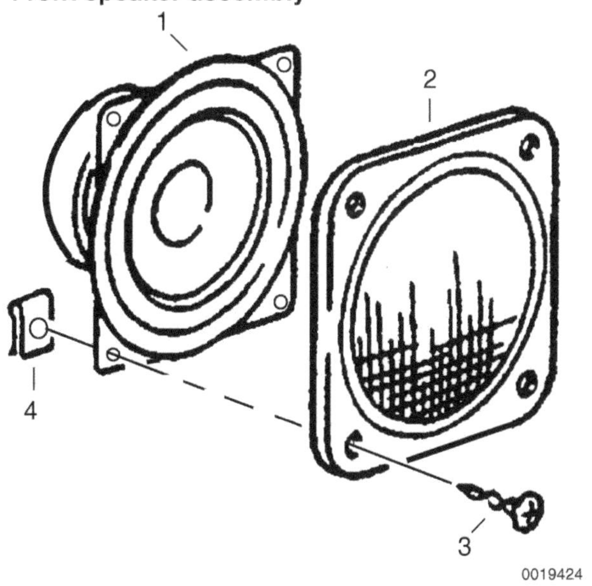

1. Front speaker
2. Grille
3. Mounting screw
4. Speed nut

< Remove speaker mounting screws.

NOTE —
If speaker mounting screws are not visible, carefully remove speaker grill.

– Remove speaker from door. Label and disconnect wires.

– Installation is reverse of removal.

Rear speaker assembly

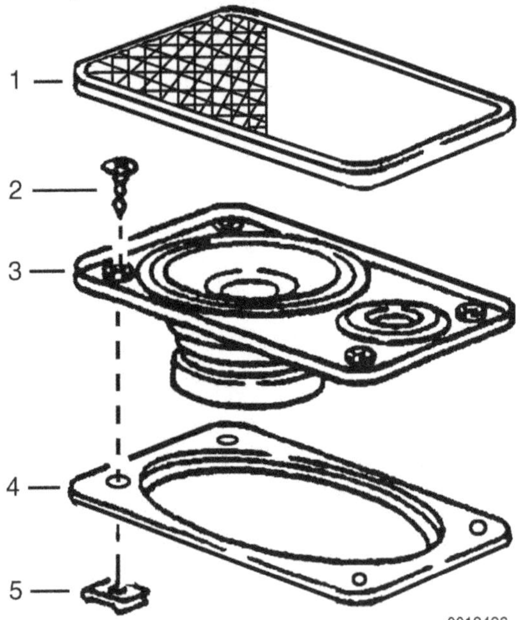

1. Grill
2. Mounting screw
3. Speaker
4. Frame
5. Speed nut

REAR SPEAKERS

Rear speakers are mounted in the shelf below the rear window on the left and right. Sound systems without audio option pack use one full range speaker per side. Sound systems with audio option package may use separate high and low frequency speakers on each side.

Rear speaker, removing and installing

- Remove speaker mounting screws.

NOTE —
If speaker mounting screws are not visible, carefully remove speaker grill.

- Remove speaker from shelf. Label and disconnect wires.
- Installation is reverse of removal.

920 Windshield Wipers and Washers

GENERAL

This repair group covers windshield wiper and washer system repair information. Removal and replacement of the windshield wiper/washer stalk switch assembly is covered in **960 Electrical Switches, Interior Lighting**.

Electrical wiring diagrams and relays for the wiper/washer are covered in **970 Electrical Wiring Diagrams**.

WIPER BLADES AND ARMS

Wiper blade symptoms and fixes

Common problems with the windshield wipers include streaking or sheeting, water drops after wiping, and blade chatter.

Streaking is usually caused when wiper blades are coated with road film or car wash wax. Clean the blades using soapy water. If cleaning does not cure the problem, the blades or inserts should be replaced. Insert replacement is the most economical method, although over time the wiper blade carrier itself will become worn.

Drops that remain behind after wiping are usually caused by oil, road film or diesel exhaust residue on the glass. Use a luke warm soap/water solution or a commercial window cleaning product. Sekuriflex windshields may be cleaned with normal drugstore alcohol if very dirty.

> ***CAUTION —***
> *Use only non-abrasive cleaners on windshield and windows.*

Wiper blade chatter may be caused by dirty or worn blades or by wiper arms that are out of alignment. Clean the blades and windshield as described above. Adjust the wiper arm so that there is even pressure along the blade, and so that the blade is perpendicular to the windshield at rest. If the problem persists, the blade carrier and wiper arms should be replaced.

Wiper blades, replacing

- Pivot wiper arm away from glass.

- Position wiper blade carrier approximately perpendicular to wiper arm.

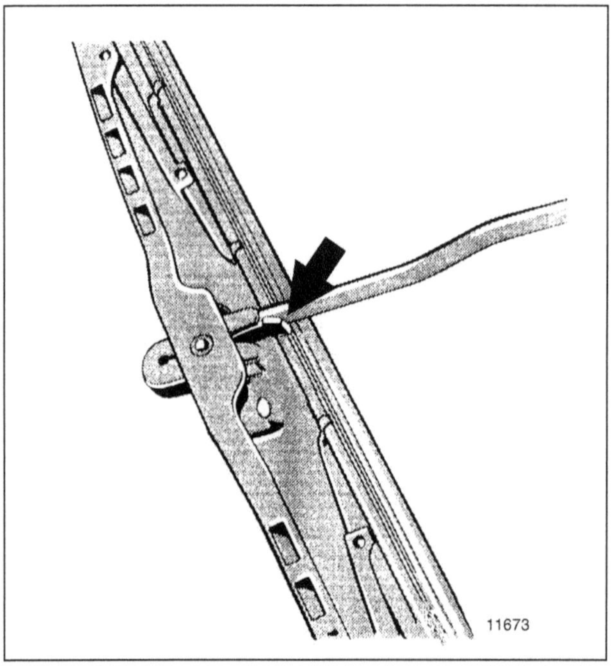

◄ Remove wiper blade carrier from wiper arm by depressing retaining tab (**arrow**) and sliding blade out of arm.

NOTE —
Some wiper blade versions may have two retaining tabs.

- Installation is reverse of removal. Install wiper blade carrier to wiper arm until it clicks into position.

Wiper blade inserts, replacing

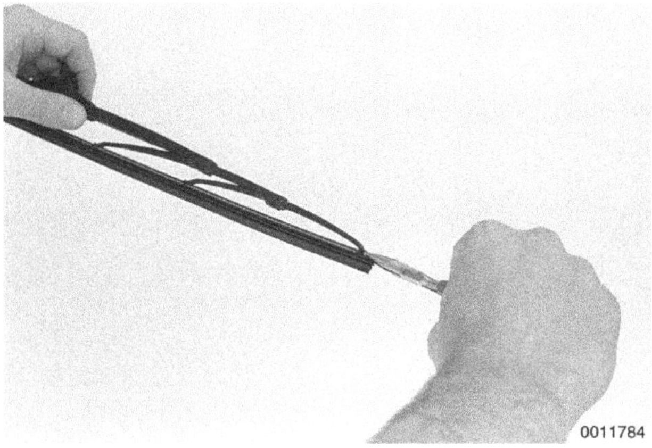
0011784

- Remove wiper blade as described earlier.
- ◄ Unhook end of wiper blade insert from wiper arm guides.

NOTE—
If necessary, spread guides slightly using needle nose pliers.

- Pull old insert from wiper arm guides.
- Remove metal support strips from old insert and install into slots in new insert, noting installation direction of cutouts in support strips.

NOTE—
The notched cutouts in the retaining strips should engage the molded tangs in the inserts.

- Slide new insert through wiper blade guides. Lock insert in place at end guides.

Wiper arms, removing and installing

0019494

- Swing hinged plastic cover upwards from base of wiper arm.
- ◄ Loosen, but do not remove wiper arm retaining nut (**arrow**).

NOTE—
Marking the position of the wiper arm on the shaft will aid installation.

- Pivot wiper arm up slightly to release from tapered mounting shaft.
- Remove mounting nut and take off wiper arm.
- Install driver's side wiper arm first. Wiper arm should not touch bottom part of windshield gasket.

NOTE—
When installing wiper arm, be sure to carefully align end of arm onto tapered shaft before tightening mounting nut.

- Install passenger wiper arm.

NOTE—
Passenger side wiper blade should be positioned approximately 25 mm (1 inch) above driver's side wiper blade when both arms are in park position.

Windshield wiper assembly

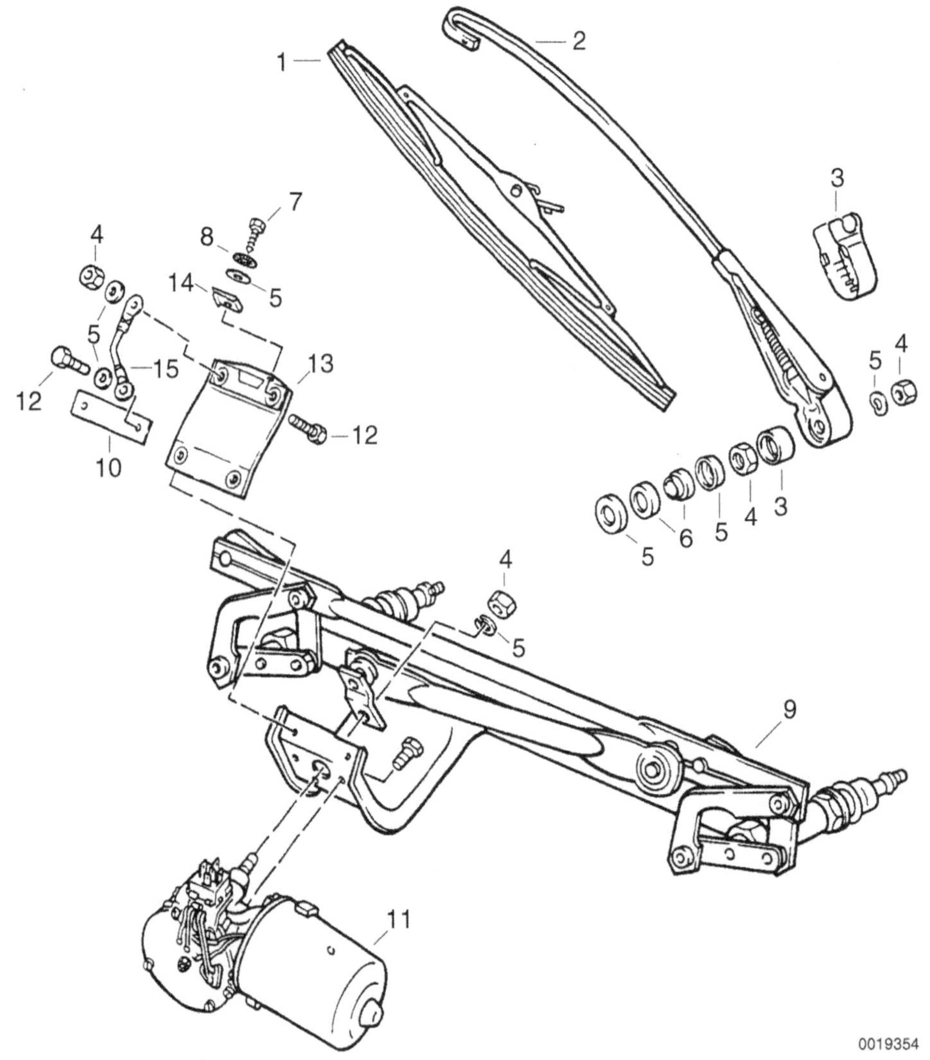

1. Wiper blade
2. Wiper arm
3. Cap
4. Nut
5. Washer
6. Rubber bushing
7. Tapping screw
8. Shake proof washer
9. Crank mechanism
10. Lock plate
11. Motor
12. Bolt
13. Fabric mounting
14. Speed nut
15. Ground cable

WINDSHIELD WIPER ASSEMBLY

The windshield wiper assembly (linkage and motor) is removed as a single unit. Once the assembly is removed, the wiper motor and other linkage parts can be repaired or replaced.

Windshield wiper assembly, removing and installing

NOTE —
The wiper arms should be in the parked position before removing the wiper assembly. To avoid damaging the wiper arms and pivots, do not manually slide or force the wiper arm across the windshield.

– Disconnect negative (–) battery cable.

– Unsnap luggage compartment carpet and lift out.

0019138

◄ Remove fresh air plenum cover retaining fasteners (**arrows**). Remove cover.

– Remove fresh air blower assembly. See **850 Ventilation**.

– Remove wiper arms as described earlier.

– Remove clock and speedometer from dashboard. See **901 Instruments**.

– Working though speedometer opening, unplug wiper motor and remove M6 wiper motor mounting fasteners.

– Support windshield wiper assembly and remove plastic caps beneath wiper arms. Remove retaining nuts.

– Lower windshield wiper assembly and remove.

– Installation is reverse of removal.

> ***CAUTION—***
> - *Be sure ground wire at fabric mount is securely fastened.*
> - *Run wipers and allow to return to parked position before installing wiper arms.*

– To remove wiper motor from crank mechanism, see windshield wiper assembly component illustration.

WASHER SYSTEM

The windshield washer system includes the windshield spray nozzles, the washer fluid pump and the washer fluid reservoir. A separate washer fluid tank, pump and nozzles are used for the intensive washer system (where applicable).

Cars with the headlight washer system are equipped with an additional fluid pump, hoses and nozzles on top of the front bumper.

Washer spray nozzles can be aimed by using a sewing needle or a similar diameter stiff piece of wire.

Windshield spray nozzle, removing and installing

◀ Working inside luggage compartment, press washer nozzle from the rear (**arrow**) and pivot nozzle up out of the cowl.

NOTE —
It may be necessary to slide washer nozzle slightly toward front of car while pushing from below.

- Remove clamp from nozzle and remove washer hose.
- Unplug 2-pin connector for heated nozzle (as applicable) and remove nozzle.

NOTE —
2-pin connector for heater washer nozzle is not located at the nozzle itself but in the plenum area.

- Installation is reverse of removal.

NOTE —
Be sure to replace washer nozzle gasket as necessary.

Windshield washer pump, removing and installing

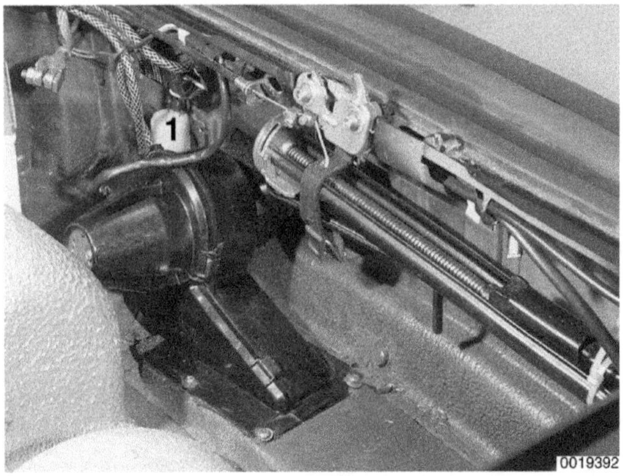

◀ Unplug wires from windshield washer pump (**1**).

- Remove clamps and fluid hoses.
- Remove pump bracket from rubber mounts and remove pump.
- Connectors on pump are labeled plus (+) and minus (-). Brown wire connects to minus (-) connector.
- Installation is reverse of removal.

Washer fluid reservoir, removing and installing

- Raise and safely support vehicle.

> ***WARNING —***
> *Make sure car is stable and well supported at all times. Use a professional automotive lift or jack stands designed for the purpose. A floor jack is not adequate support.*

- Remove left front wheel.

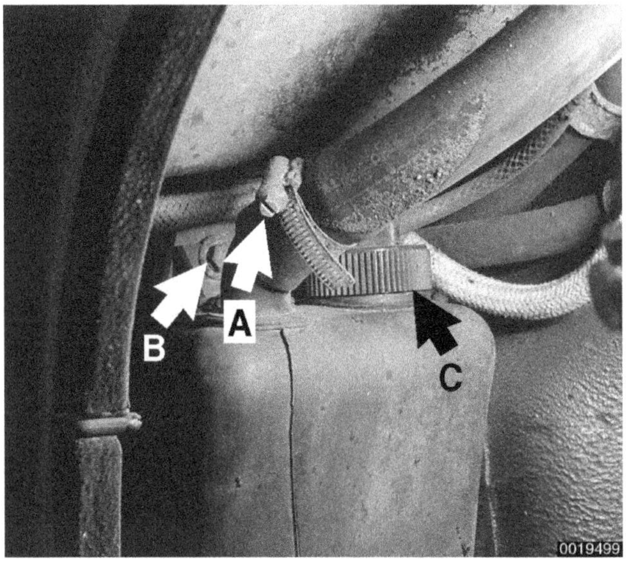

- Unscrew filler tube clamp (**A**) and remove tube from reservoir.
- Unscrew collar (**C**) for washer pump feed and return lines and set aside.
- Remove lower mounting bolt and retaining strap from underneath reservoir.
- Remove upper reservoir mounting bolt (**B**) and lower reservoir.
- Remove washer pump feed and return lines from reservoir. Remove reservoir.

NOTE —
Feed line should be equipped with a small filter at the end.

- Installation is reverse of removal.

Windshield wiper/washer switches

The windshield wiper/washer switch is mounted on the right side of the steering column. The wipers have three speeds and an OFF position. Pushing up on the wiper lever increases the speed. Removal of the wiper lever/switch is covered in **960 Electrical Switches, Interior Lighting**.

The windshield washer system is operated by pulling the lever towards the steering wheel.

Intermittent wiper action switch, removing and installing

The intermittent wiper rheostat is located on the dashboard between the speedometer and the clock. The wiper switch on the steering column must be in the OFF position (all the way down) to actuate intermittent operation.

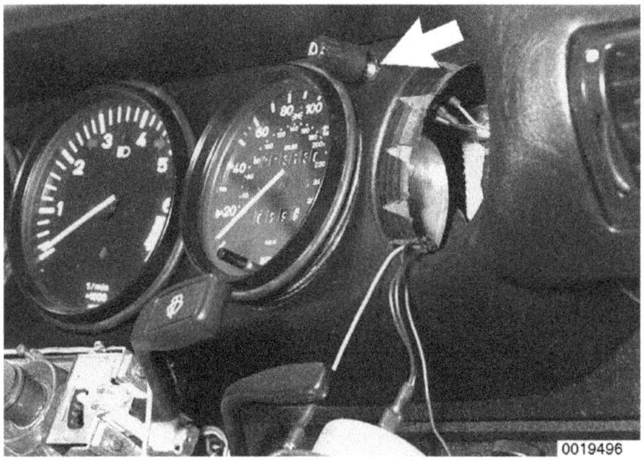

- Disconnect negative (–) battery cable. See **030 Maintenance**.
- Remove clock from the dashboard. See **901 instruments**.
- Pull off intermittent switch knob and remove mounting nut (**arrow**).
- Slide switch back into dash and remove through clock opening.
- Label and disconnect wires.
- Installation is reverse of removal.

940 Exterior Lighting

Special Tools

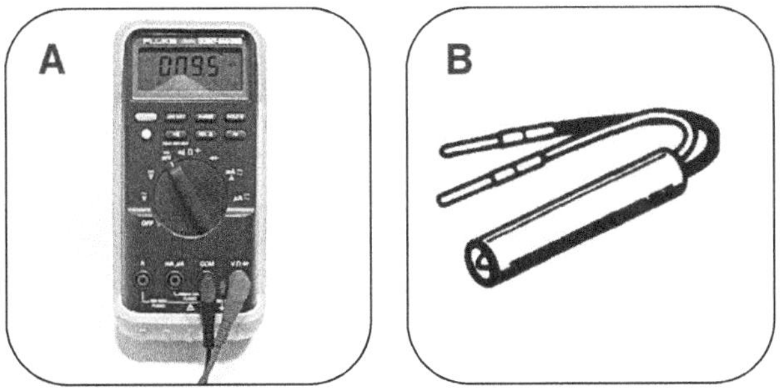

Special tool

A Fluke 87 . . . . Automotive digital multimeter
A Baum 1115 . . . . LED tester with thin spade probes (source: Baum Tools Unlimited)

GENERAL

This section covers exterior light troubleshooting and component replacement.

See **960 Electrical Switches, Interior Lighting** for exterior light switch replacement procedures.

See **970 Electrical Wiring Diagrams** for information on individual circuits.

Test equipment

Electrical and electronic automotive components require the use of low impedance test equipment.

Troubleshooting

Many lighting problems are due to dirty or corroded sockets, or poorly fitting bulb contacts or connectors. Check that voltage is reaching the bulb and that the socket has a good connection to ground.

Most exterior lighting involves pairs of lights, and this fact is an aid to troubleshooting. If only one of a pair of lights is out then the problem is most likely due to a failed bulb or some other problem with that particular light socket or its wiring. A simple test is to exchange the failed bulb with one that is known to be good.

If a pair of bulbs are both out then the problem is most likely in some part of the system common to both lights. Begin by checking for a failed fuse. Test switches using simple continuity checks made with a digital ohmmeter. Check the switch connectors for voltage and continuity to ground. Always check the bulbs as they are the most common cause of lighting problems.

For a general description of electrical test procedures, see **900 Electrical System–General**.

NOTE —
Replacement exterior bulbs are listed in **Table a**.

> ***CAUTION —***
> *When working on electrical systems, detach negative (–) battery cable. See* **030 Maintenance**.

Table a. Bulb specifications

Location	Rating	DIN (SAE) number
Back-up light	32 cp	(1073)
Brake/taillight	32/3 cp	(1034)
Front fog light H3	55 W	PK 22 S
Headlight Sealed beam H4	 60/50 W 60/55 W	 (H6014) P43t-38
License plate	2 cp	(1895)
Side marker	2 cp	(1895)
Sign light	2 cp	2.9 x 4.6
Turn signal/side marker	32/3 cp	(1034)

NOTE —
In the table above, "cp" stands for candle power.

FRONT LIGHTING

> ***CAUTION —***
> *Do not touch halogen bulbs with your bare hands. Dirt and skin oils may cause premature bulb failure. If necessary clean bulb using a clean cloth dampened with rubbing alcohol.*

Sealed beam headlights are standard in SC cars. Halogen headlights were available as an update.

Headlights, aiming

> ***CAUTION —***
> *Headlights must be aimed using a headlight aiming tool. DO NOT aim headlights by eye.*

NOTE —
Prior to adjusting headlights, make sure the car has a full fuel tank, correct tire pressures and 75 kg (165 lbs.) in driver's seat.

– Remove rubber plugs from headlight trim ring.

NOTE —
If necessary, remove trim ring to expose headlight adjusting screws.

– Mount headlight aiming tool.

◄ Working through holes in headlight trim ring (**arrows**), adjust aiming screws.

– Reinstall rubber plugs.

Sealed beam headlight assembly

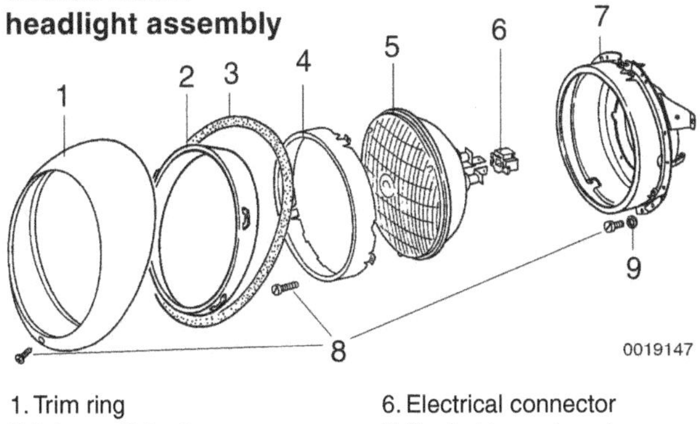

1. Trim ring
2. Intermediate ring
3. Sealing ring
4. Sealed beam retaining ring
5. Sealed beam
6. Electrical connector
7. Sealed beam housing
8. retaining crews
9. Washer

Headlight sealed beam, replacing

– Remove retaining screw and headlight trim.

◄ Remove three screws and sealed beam retaining ring.

– Remove sealed beam and unplug electrical connector.

– Installation is reverse of removal.

Halogen headlight assembly

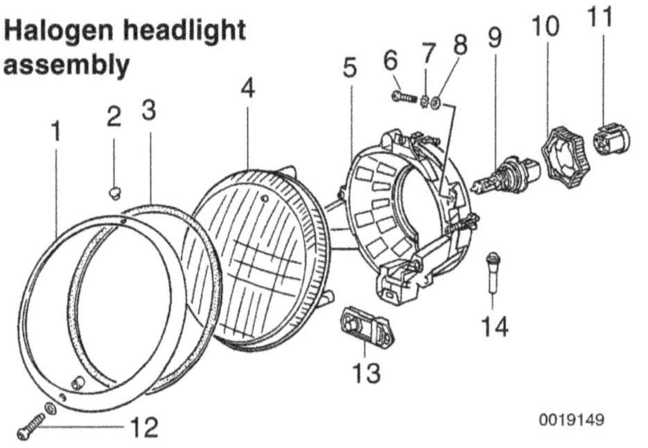

1. Trim ring
2. Rubber plug
3. Sealing ring
4. Reflector housing
5. Headlight housing
6. Aligning screw
7. Lock washer
8. Washer
9. Halogen bulb
10. Plastic retaining nut
11. Electrical connector
12. Retaining screw
13. Humidity absorber
14. Drain plug

Headlight halogen H-4 bulb, replacing

- Remove retaining screw and headlight trim.
- Remove four screws and pull headlight assembly out of fender.
- Unplug harness connector.
- Remove large plastic retaining nut and remove bulb.

> ***CAUTION—***
> - *Bulb has O-ring seal and may be difficult to remove. DO NOT apply excessive force as bulb may shatter.*
> - *Do not touch halogen bulbs with your bare hands. Dirt and skin oils may cause premature bulb failure. If necessary clean bulb using a clean cloth dampened with rubbing alcohol.*

- Installation is reverse of removal.
  - Align grooves in bulb with tabs in lens.

Front fog light assembly, removing and installing

The following procedure includes fog light lens replacement and fog light bulb replacement.

NOTE—
There are several different fog lights. Procedures may vary slightly.

Front fog light assembly

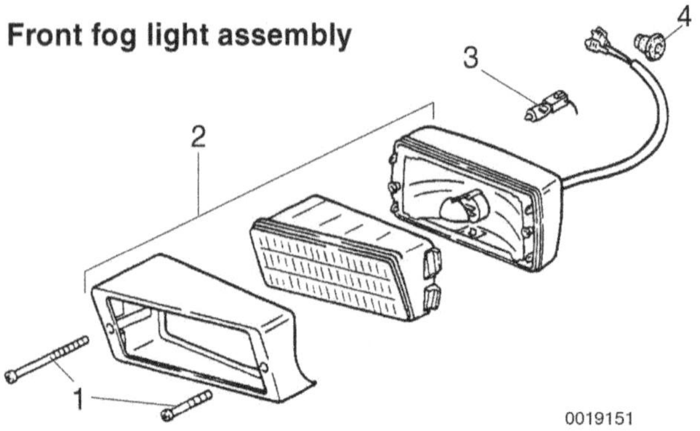

1. Retaining screw
2. Fog light assembly housing
3. Halogen bulb H3
4. Grommet

- Remove fog light mounting screws.
- Remove fog light lens assembly.
- Disconnect electrical harness connector from fog light bulb. Remove bulb from rear of fog light assembly.
- Installation is reverse of removal.
  - Check electrical connector for corrosion.
  - Spray with contact cleaner, if necessary.

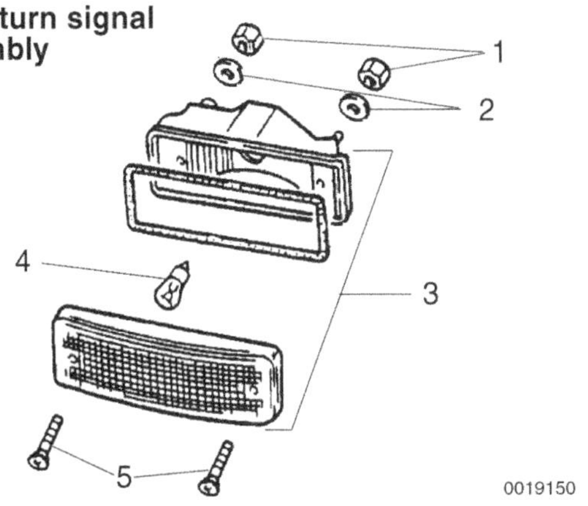

1. Housing retaining nut
2. Washer
3. Turn signal assembly housing
4. Bulb
5. Lens screw

Front turn signal bulb, replacing

- Remove front turn signal lens screws. Remove lens.
- Remove bulb.
- Inspect socket for corrosion.
- Installation is reverse of removal.

Front turn signal assembly, removing and installing

- Remove front bumper. See **630 Bumpers**.
- Remove turn signal mounting nuts.
- Unplug wires and remove turn signal assembly.
- Installation is reverse of removal.

REAR LIGHTING

Taillight bulb, replacing

- Remove taillight lens mounting screws and remove lens.
- To remove bulb from bulb socket, push bulb in gently and rotate counterclockwise. Pull bulb from socket.
- Inspect socket for corrosion. Clean or replace as required.
- Installation is reverse of removal.

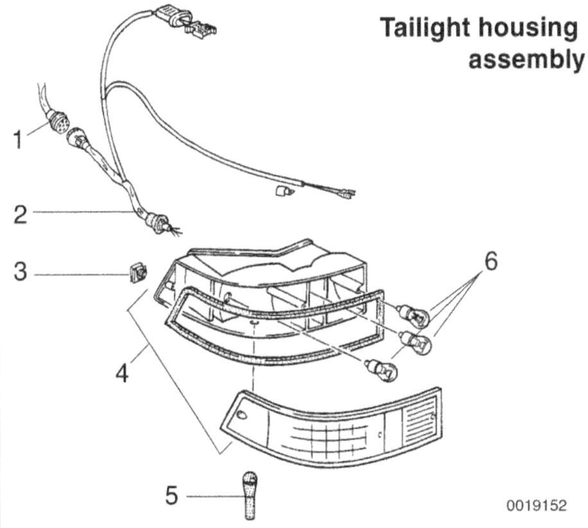

1. Harness connector (7-pin)
2. Harness
3. Nut
4. Tailight housing
5. Drain plug
6. Bulb

Taillight assembly, removing and installing

- Remove taillight lens mounting screws and remove lens.
- Remove taillight drain plug.
- Remove taillight housing mounting screws.
- Pull taillight housing from car.
- Disconnect multipin connector, in engine compartment, and remove taillight housing.
- Installation is reverse of removal.

Center brake light bulb, replacing

Center brake light assembly

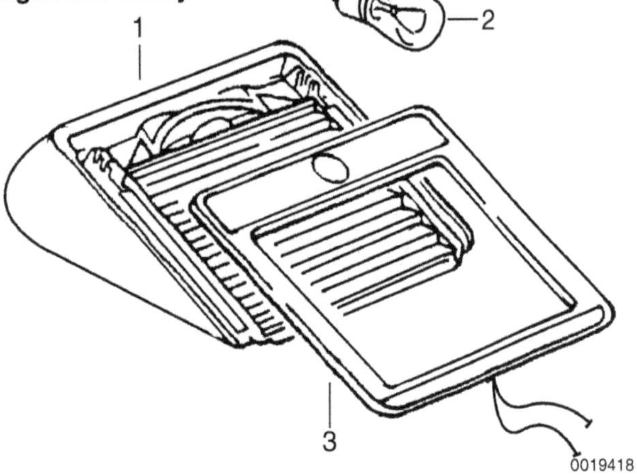

1. Housing
2. Bulb
3. Frame

- Working at utility shelf below rear window, depress locking tab in center of center brake light plastic housing and lower housing.
- Pull housing straight out from rear window.
- Press two locking tabs at top of red lens and lift lens off.
- To remove bulb from bulb socket, push bulb in gently and rotate counterclockwise. Pull bulb from socket.
- Installation is reverse of removal.

License plate light, removing and installing

License plate light assembly

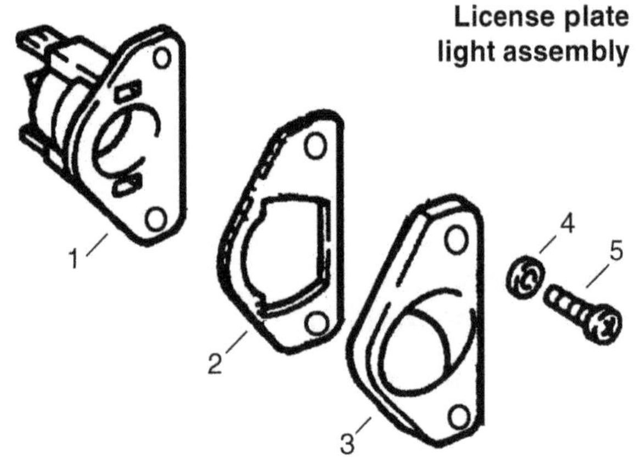

1. Light socket
2. Gasket
3. Lens
4. Washer
5. Screw

- Remove screws retaining license plate light assembly to rear bumper pad.
- Pull license plate light assembly from bumper pad and remove lens.
- To remove bulb from bulb socket, push bulb in gently and rotate counterclockwise. Pull bulb from socket.
- Installation is reverse of removal.

960 Electrical Switches, Interior Lighting, Cruise Control

TABLES

GENERAL

This section covers replacement of electrical switches and interior lights. The cruise control system is also covered here.

See **901 Instruments** for dash gauge and instrument repair information.

See **940 Exterior Lighting** for exterior light repair information.

See **970 Electrical Wiring Diagrams** for information on individual circuits.

Special Tools

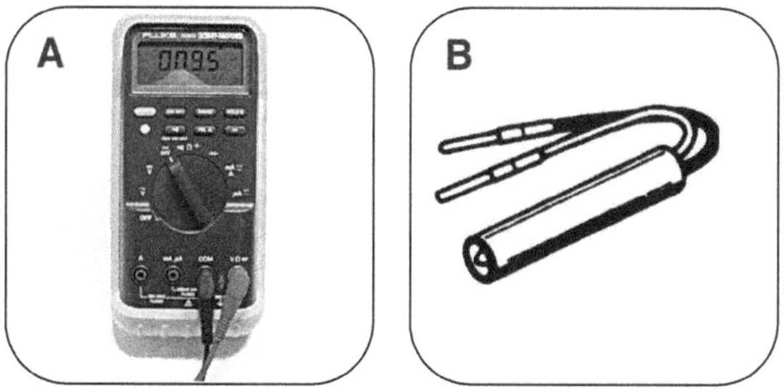

Special tool

A Fluke 87 . . . . Automotive digital multimeter
A Baum 1115 . . . . LED tester with thin spade probes (source: Baum Tools Unlimited)

Test equipment

Electrical and electronic automotive components require the use of low impedance test equipment.

> ***CAUTION —***
> - *When working on electrical systems, detach negative (–) battery cable. See* **030 Maintenance**.
> - *Prior to disconnecting the battery, read the battery disconnection cautions given at the front of this manual on page viii.*

ELECTRICAL SWITCHES

Console mounted switches

The switches mounted in the center console are covered in **680 Interior Trim.** This may include the following switches:

- A/C blower switch
- A/C temperature switch
- Hazard warning switch
- Rear window defogger switch

Door contact switch, removing and installing

The door contact switches are located in each front door post.

– Disconnect negative (–) battery cable.

< Remove rubber cap from switch.

Door contact switch assembly

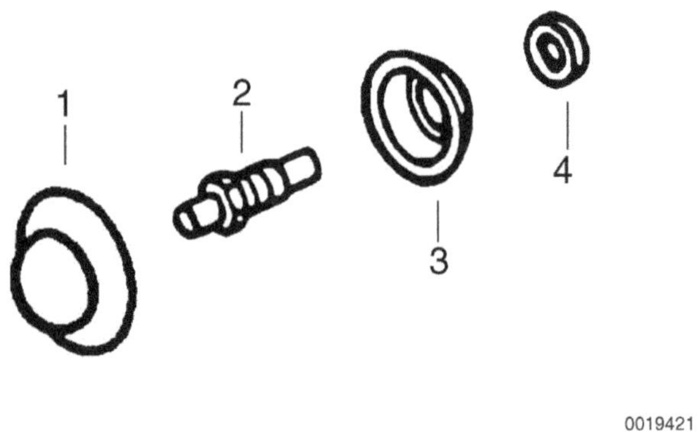

1. Rubber cap
2. Door contact switch
3. Concave washer
4. Washer

– Using an 11 mm socket, unscrew switch and disconnect wire.

– Installation is reverse of removal.

> ***CAUTION —***
> *DO NOT allow wire to fall into door pillar.*

Fog light switch, removing and installing

– Disconnect negative (–) battery cable.

– Working inside passenger compartment at dashboard, unscrew fog light switch knob in counter-clockwise direction.

– Use two-pin spanner wrench to remove switch retaining collar from switch.

– Working inside luggage compartment, remove fresh air blower cover and plenum. See **850 Ventilation**.

– Remove switch from inside luggage compartment.

– Label and disconnect wires.

– Installation is reverse of removal.

Headlight switch, removing and installing

- Disconnect negative (–) battery cable.
- Working at dashboard, pry out plastic center from rubber headlight switch knob.
- Unscrew nut and remove headlight switch knob.

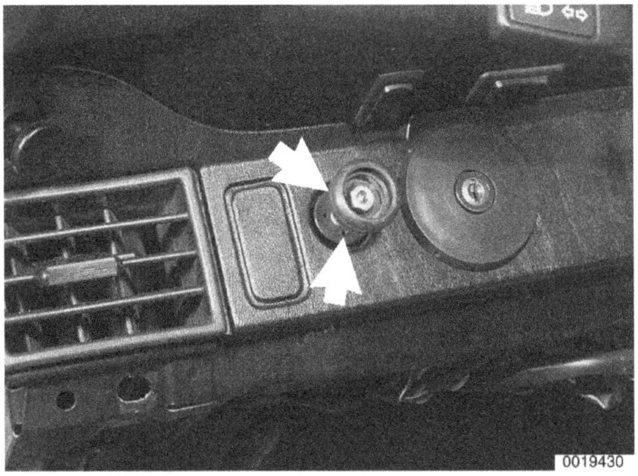

- Use two-pin spanner wrench to remove switch retaining collar from switch.

NOTE —
*Note pin-holes (**arrows**) for spanner wrench.*

- Push switch through dashboard and remove from below.
- Label and disconnect wires.
- Installation is reverse of removal.

Horn contact ring, removing and installing

- Disconnect negative (–) battery cable.
- Remove steering wheel, see **480 Steering**.

- Remove horn contact ring mounting screws (**arrows**) and contact ring.
- Installation is reverse of removal.

Ignition switch, removing and installing

- Remove left side knee protection trim from dashboard.
- Working under dashboard, remove electrical wiring harness from ignition switch.
- Remove ignition switch mounting screws and remove from ignition/steering lock mechanism.

See **480 Steering** for additional information on the steering lock.

Luggage compartment light/alarm switch, removing and installing

The luggage compartment light switch is located on the right side hinge, between the hinge bolts. A copper connector is crimped to the wire and to the switch.

- Disconnect negative (–) battery cable.

- Unscrew switch retaining nut (**arrow**) from hinge.
- Remove copper connector wire from switch and remove switch.

> ***CAUTION —***
> *DO NOT allow wire to fall into hood.*

- Installation is reverse of removal.

Power mirror selector switch, removing and installing

The power mirror selector switch is located above the ignition switch to the right of the sunroof switch (if applicable). This switch selects which mirror (driver or passenger side) is selected for adjustment with the mirror adjustment switch.

- Disconnect negative (–) battery cable.
- Use a plastic pry tool or other blunt tipped object to pry temperature/oil pressure gauge out of dashboard.

- Working in luggage compartment, squeeze switch retaining tabs (**arrows**) while pushing switch down out of dashboard.
- Disconnect harness connector.
- Install switch by pushing back up into dashboard.

Power mirror adjustment switch, removing and installing

– Disconnect negative (–) battery cable.

– Remove mirror adjustment switch knob by pulling upward.

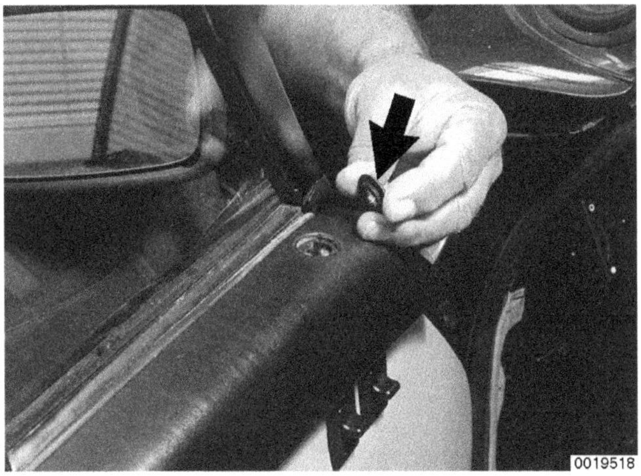
0019518

◄ Unscrew switch retaining ring (**arrow**).

– Remove upper door panel trim.

– Disconnect harness connector and remove mirror adjustment switch.

– Installation is reverse of removal.

Power window switch, removing and installing

– Disconnect negative (–) battery cable.

– Loosen and/or remove enough door panel mounting hardware to pull front part of panel away from door. See **570 Doors** for more information.

0019517

◄ Push out power window switch from behind door panel while depressing spring tabs on top and bottom of switch (**arrows**).

– Pull rubber boot back and disconnect wires from switch.

– Installation is reverse of removal.

Table a. Window switch wiring connections

Switch terminal	Left window driver's door switch	Right window driver's door switch	Right window passenger door switch
1	Blue	Red / white	Blue
2	Blue / black	Red / black	Blue / black
3	Single red / blue	Single red / blue	Red / white
4	Brown	Brown	Brown
5	Double red / blue	Double red / blue	Red / black

Steering column switches, removing and installing

The turn signal/headlight dimmer switch, windshield wiper/washer switch and cruise control switch are all mounted on the steering column.

- Disconnect negative (–) battery cable.
- Working inside passenger compartment, remove steering wheel as described in **480 Steering**.

◀ Remove steering column cover outer mounting screws (**arrows**) located on sides of cover.

NOTE —
Screws are located behind switch arms.

◀ Remove horn contact plate mounting screws (**arrows**).

- Disconnect ground wire from back of plate.
- Remove plate.

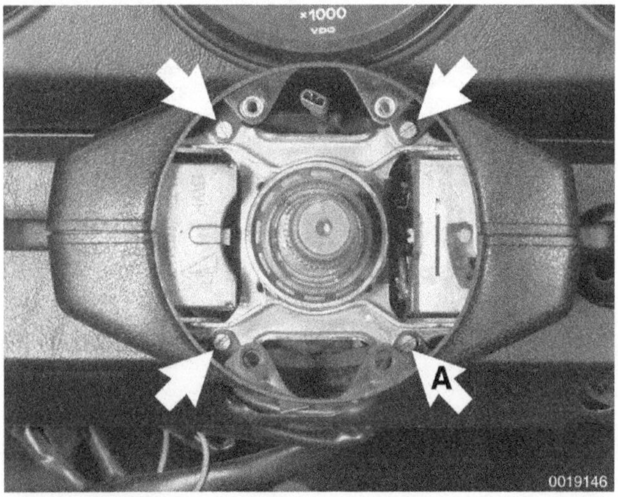

◀ Remove steering column cover inner mounting screws (**arrows**). Remove covers.

NOTE —
*Ground wire connection is on lower right screw (**A**).*

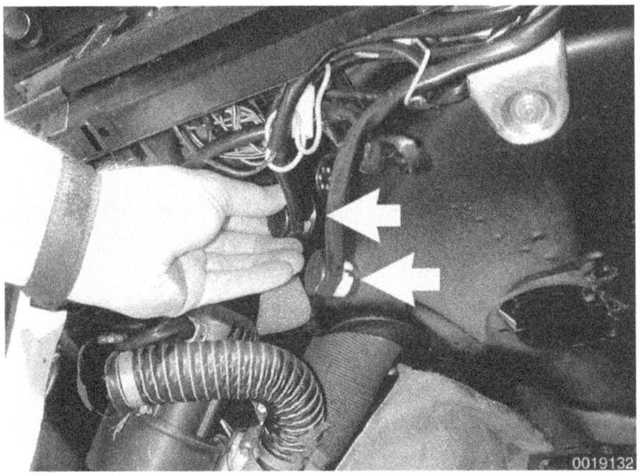

◄ Unplug switch wiring harness (**arrows**) for appropriate switch from underneath dashboard.

NOTE —

Harness connector for cruise control switch is not shown in illustration. Follow wires down to right of steering column to locate harness connector.

◄ To remove turn signal switch from column:

- Cut wire ties.
- Remove small mounting screws (**arrows on left side**).
- Remove switch.

– To remove windshield wiper/washer switch or cruise control switch from column:

- Cut wire ties.
- Remove small mounting screws (**arrows on right side**).
- Remove switch.

– Installation is reverse of removal.

- Tighten mounting screws, then lock in place with lacquer or paint.
- Reattach harnesses with new wire tires.
- When installing steering column covers, be sure to anchor ground wire under lower right cover screw.

Tightening torque

- Steering wheel to steering column . . . . . 50 Nm (36 ft-lb)

Sunroof switch, removing and installing

– Disconnect negative (–) battery cable.

– Use a blunt tipped screwdriver to pry temperature/oil pressure gauge out of dashboard.

◄ Reach through gauge opening and push sunroof switch (**arrow**) downward out of dashboard.

– Disconnect harness connector.

– Install switch by pushing back up into dashboard.

INTERIOR LIGHTS

The dome lights for the coupe are located over each door and can be operated by simply tilting the lens face. The glove compartment light is located in the dash above the glove compartment door. A switch in the dash is released when the glove compartment door is opened that turns on the light. Both the dome and glove compartment light can be set to operate by the door contact switches.

The targa and cabriolet use one interior light located above the windshield between the sun visors. The light has a built-in three position rocker switch which is operated by pressing on the lens.

The various interior light bulb specifications are provided in **Table b**.

Table b. Bulb specifications

Location	Rating	DIN (SAE) number
Passenger compartment and trunk	10 W	BA 15s
Instruments, ashtray	2 W	BA 7
Glass base indicator lamp	1.2 W	W2 x 4.6d
Sign light (USA only)	1.2 W	2.9 x 4.6

NOTE —
Use only original equipment replacement bulbs.

Engine compartment bulb, replacing

The engine compartment light is located in the center of the engine compartment lid. The sliding bulb cover is also an on/off switch. The engine compartment light operates only when the headlights are on.

- Disconnect negative (–) battery cable.

0019413

- Pull housing cover off light assembly at arrows.

NOTE —
It may be necessary to pry off the housing cover with a small screw-driver.

- Rotate bulb cover/switch to expose bulb.
- Remove bulb.
- Installation is reverse of removal.

Glove compartment bulb, replacing

The glove compartment light is located in the dashboard, directly above the glove compartment. The light has a built-in three position rocker switch which is operated by pressing on the lens.

– Disconnect negative (–) battery cable.

– Insert screwdriver in cut-out on light housing and carefully pry out light assembly.

– Remove bulb from between contacts.

– Install new bulb and press light into place.

Interior light bulb, replacing

– Disconnect negative (–) battery cable.

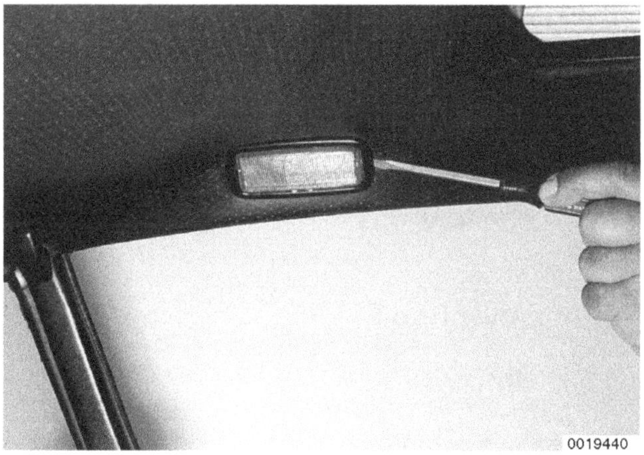
0019440

◄ Insert screwdriver into front of light bulb housing and carefully pry out interior light assembly.

– Remove bulb from between contacts.

– Install new bulb and press interior light into place.

Luggage compartment bulb, replacing

The luggage compartment light is installed in the luggage compartment lid near the windshield, on the right.

0019400

◄ Remove two mounting screws (**arrows**) and remove light from lid.

– Slide bulb holder out of the lens.

– Remove bulb from socket.

– Installation is reverse of removal.

> ***CAUTION —***
> *Align plastic gasket before installing light.*

CRUISE CONTROL

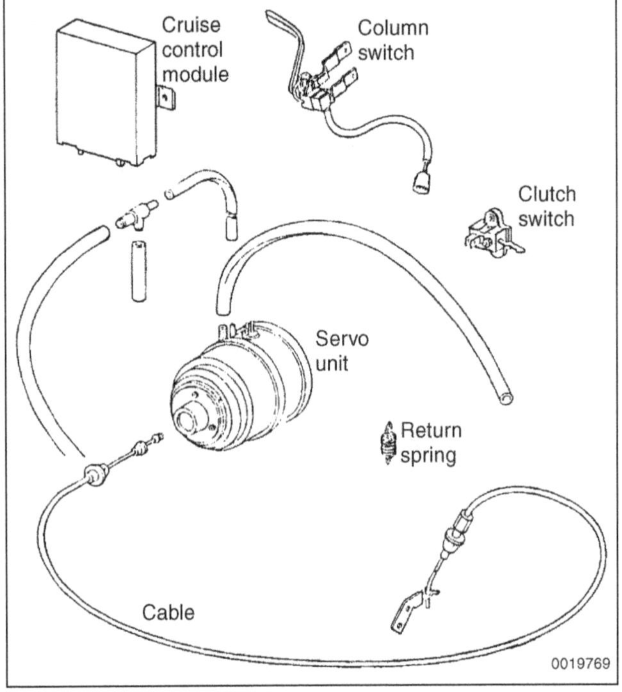

The cruise control system consists of an electronic control unit, a cruise control servo, a control switch on the steering column, and switches at the brake and clutch pedals.

The control module uses a signal from the speedometer to compare the car's true speed to the speed selected by the driver via the column switch. The control unit then signals the servo to open or close the throttle via a cable.

The system will maintain the set speed while going up and down hills within the engine's power range. The set speed can be increased at any time by depressing the accelerator pedal. When the accelerator pedal is released the vehicle speed will return to the previously set speed. If either the clutch or brake pedal are depressed the system will be disengaged. If the vehicle speed drops below 15 mph at any time the system will automatically disengage.

To disengage cruise control when the clutch pedal is depressed, a clutch pedal switch is used. To disengage cruise control when the brakes are used an extra connection at one of the brake light switches is used.

NOTE —
Cruise control switch removal and installation is covered earlier in this group under **Steering column switches, removing and installing**.

Troubleshooting

– When troubleshooting cruise control system, first check for a faulty fuse 6 in the fuse/relay panel in luggage compartment.

– To begin troubleshooting cruise control components and electrical circuit, access cruise control module as described later.

- Remove fresh air plenum in luggage compartment.
- Detach electrical harness connector at cruise control module.
- Proceed to tests in **Table c**.

– If no faults are found with cruise control components or electrical circuits, replace cruise control module.

Table c. Cruise control troubleshooting

Test (terminals at cruise control module harness connector)	Results	Probable cause
Connect ohmmeter between terminal 3 and ground	Meter reads infinite Ω	None
Depress clutch pedal	Meter reads approx. 10 Ω	None
	Meter reads infinite Ω	Clutch pedal switch or servo unit faulty

Table c. Cruise control troubleshooting

Test (terminals at cruise control module harness connector)	Results	Probable cause
Connect ohmmeter between terminals 3 and 7	Meter reads 11 -17 Ω	None
	Meter reads <11 Ω or >17 Ω	Servo unit faulty
Connect ohmmeter between terminal 12 and ground	Meter reads 0 Ω	See wiring diagram in **970 Electrical Wiring Diagrams**
Connect ohmmeter between terminal 11 and ground. Turn rear wheels	Meter deflects	None
	Meter does not deflect; speedometer functions	Wiring between cruise control module and speedometer faulty
	Meter does not deflect; speedometer does not function	Speedometer sensor at transmission faulty or wiring from speedometer sensor faulty
Connect ohmmeter between terminal 6 and ground	Meter reads infinite Ω	None
	Meter reads 0 Ω	Brake light switch faulty
Connect voltmeter between terminal 6 and ground, turn ignition on Depress brake pedal	Meter reads battery voltage	None
	Battery voltage not present	Brake light switch faulty
Connect voltmeter between terminal 9 and ground Push up on column switch	Meter reads battery voltage	None
	Battery voltage not present	Fuse 6 faulty
Connect voltmeter between terminal 8 and ground Pull column switch toward steering wheel	Battery voltage not present	None
	Meter reads battery voltage	Cruise control column switch faulty
Connect voltmeter between terminal 10 and ground Push column switch down	Meter reads battery voltage	None
	Battery voltage not present	Cruise control column switch faulty

Cruise control module, removing and installing

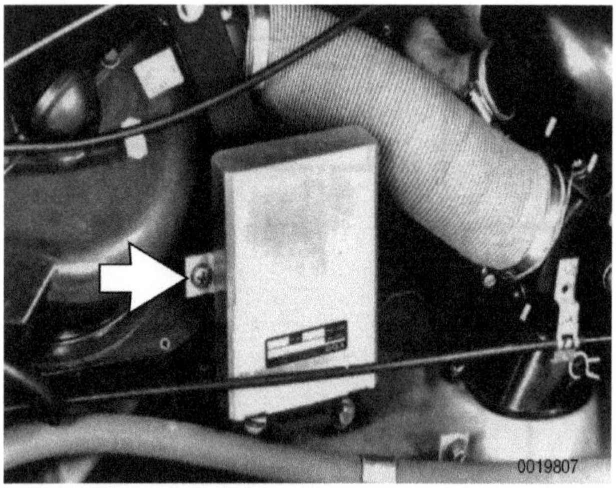

- Working in luggage compartment, remove fresh air plenum cover in luggage compartment.
- Pull electrical harness connectors off module.
- Remove phillips head screw (**arrow**) and remove module.
- Installation is reverse of removal.

Clutch pedal switch, removing and installing

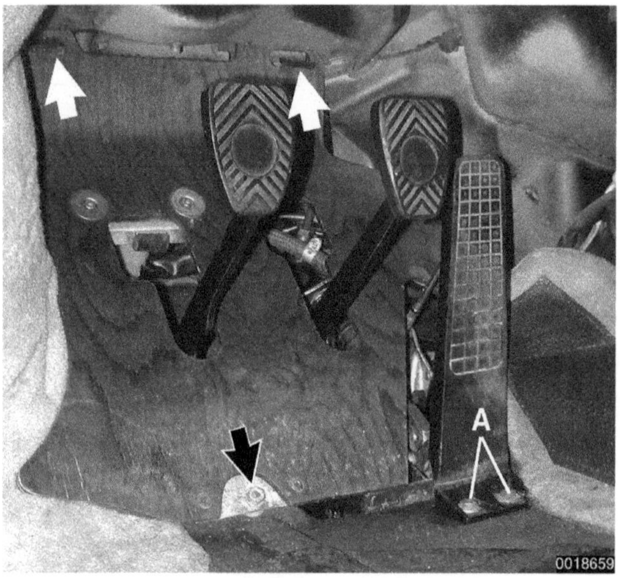

- Working inside passenger compartment at the pedals, remove accelerator pedal mounting screws (**A**) and remove pedal.
- Remove pedal cluster cover after removing carpet securing straps at top and fastener at bottom (**arrows**).

- Remove clutch pedal switch bracket fasteners, disconnect wires and remove bracket with switch.
- Installation is reverse of removal.
  - Make sure switch is adjusted correctly for clutch pedal. Use adjusting slots so that switch is not damaged when pulling back clutch pedal to check clutch play.

Cruise control servo unit, removing and installing

- Working inside engine compartment on left side of engine, remove hose between heater blower motor and heater distribution pipe.
- Disconnect cruise control servo electrical harness connector (**A**).
- Detach vacuum line (**B**) on servo.
- Working at throttle assembly:
  - Detach cruise control cable attachment bracket from throttle lever.
  - Slide plastic cable housing end out of holder bracket.
- Remove servo mounting screws. Remove servo with cable.
- Installation is reverse of removal.

Cruise control cable, removing and installing

- Working at throttle assembly:
  - Detach cruise control cable attachment bracket from throttle lever.
  - Slide plastic cable housing end out of holder bracket.
- Working inside engine compartment on left side of engine, remove hose between heater blower motor and heater distribution pipe.
- Working at cruise control servo, remove screws holding cable end flange to vacuum diaphragm at cruise control servo.
- To disconnect cable from servo, squeeze plastic retainer clip and push forward to release cable.
- Remove plastic clip from cable. Remove flange from cable.
- Installation is reverse of removal.
  - Be sure there are no kinks in cable.
  - Make sure there is as little slack as possible in cable at throttle end.
  - With engine off, there should be no tension on cable.
  - Use plastic hexagon nut at throttle housing end of cable to adjust cable tension.

970 Electrical Wiring Diagrams

ELECTRICAL COMPONENT LOCATIONS

Relay, fuse and control module positions, ground locations . . . . see **Repair Group 971**

GENERAL

This section contains wiring diagrams for 1978 to 1983 Porsche 911 SC vehicles.

WIRING COLOR CODE		
BLK	-	black
BRN	-	brown
RED	-	red
YEL	-	yellow
GRN	-	green
BLU	-	blue
VIO	-	violet
GRY	-	grey
WHT	-	white
PNK	-	pink

The various wires used in the wiring diagrams use a three digit color coding system.

CAUTION —
- *Always switch the ignition off and disconnect the negative (–) battery cable before removing any electrical components. Always make sure ignition is off before disconnecting battery.*
- *Connect and disconnect ignition system wires, multiple connectors and ignition test equipment leads only while ignition is switched off.*
- *Always switch a test meter to the appropriate function and range before making test connections.*

NOTE —
- *The alarm system is also referred to as the anti-theft system.*

970

Power Distribution (1 of 3)

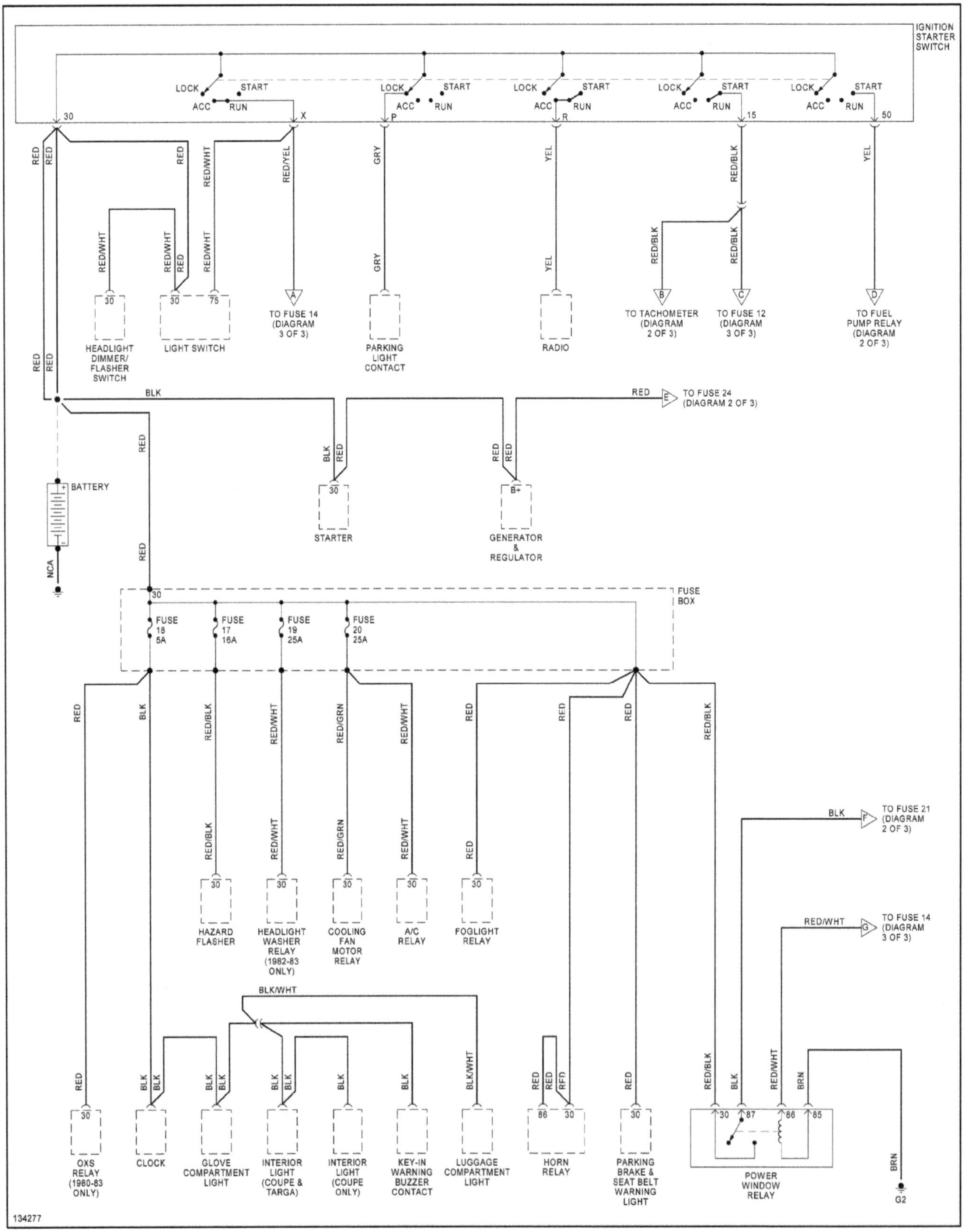

Power Distribution
(2 of 3)

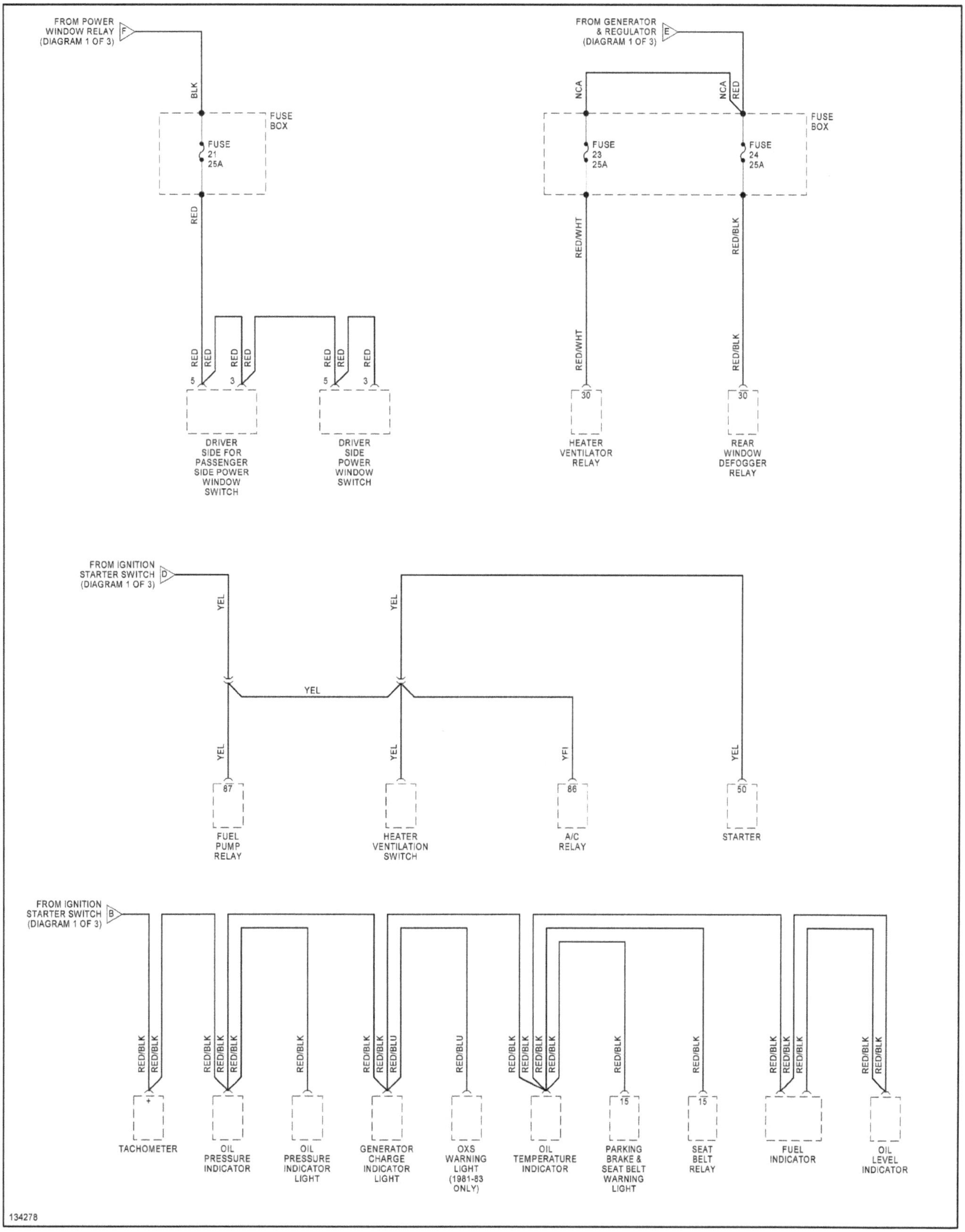

Power Distribution
(3 of 3)

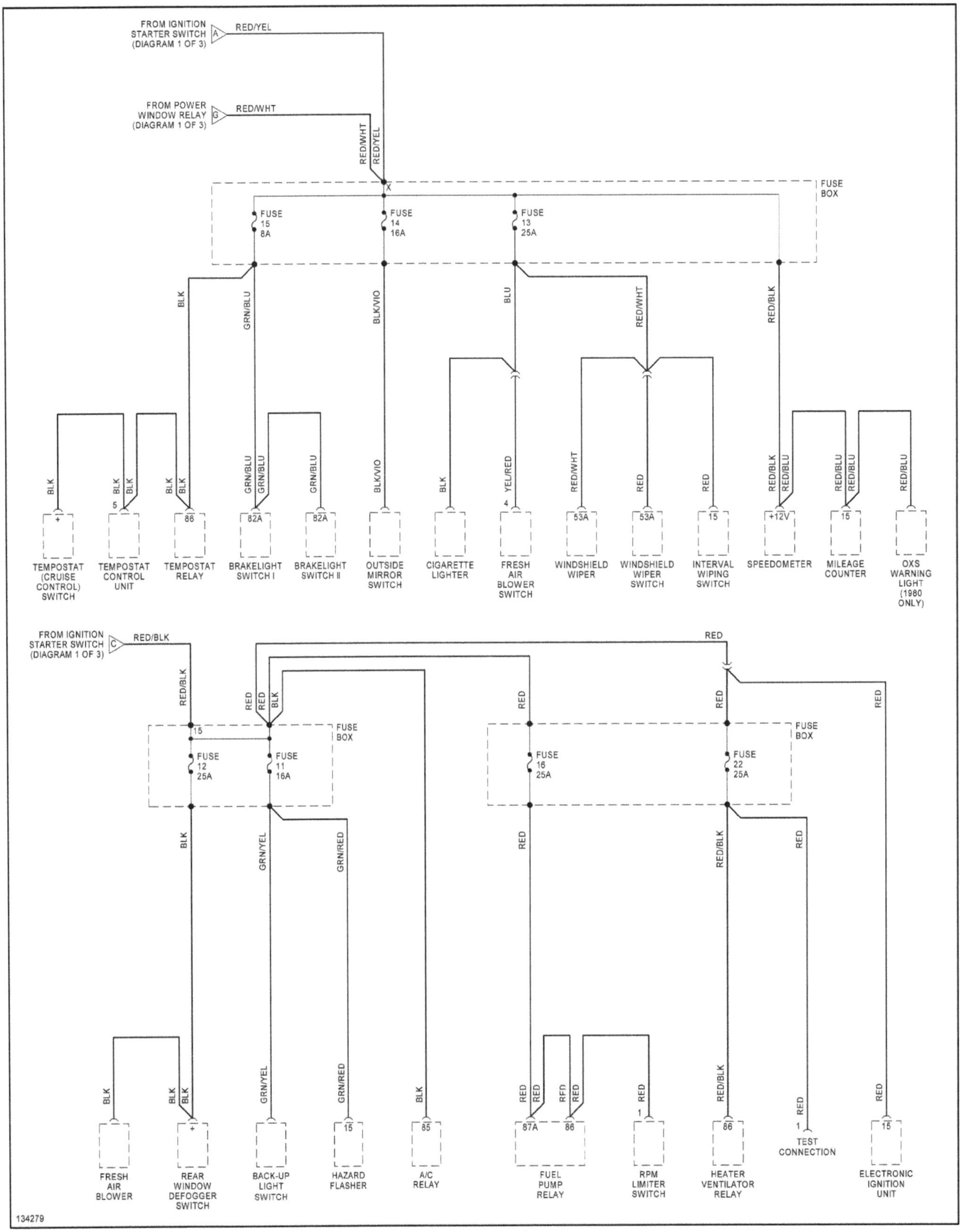

Ground Distribution

(1 of 3)

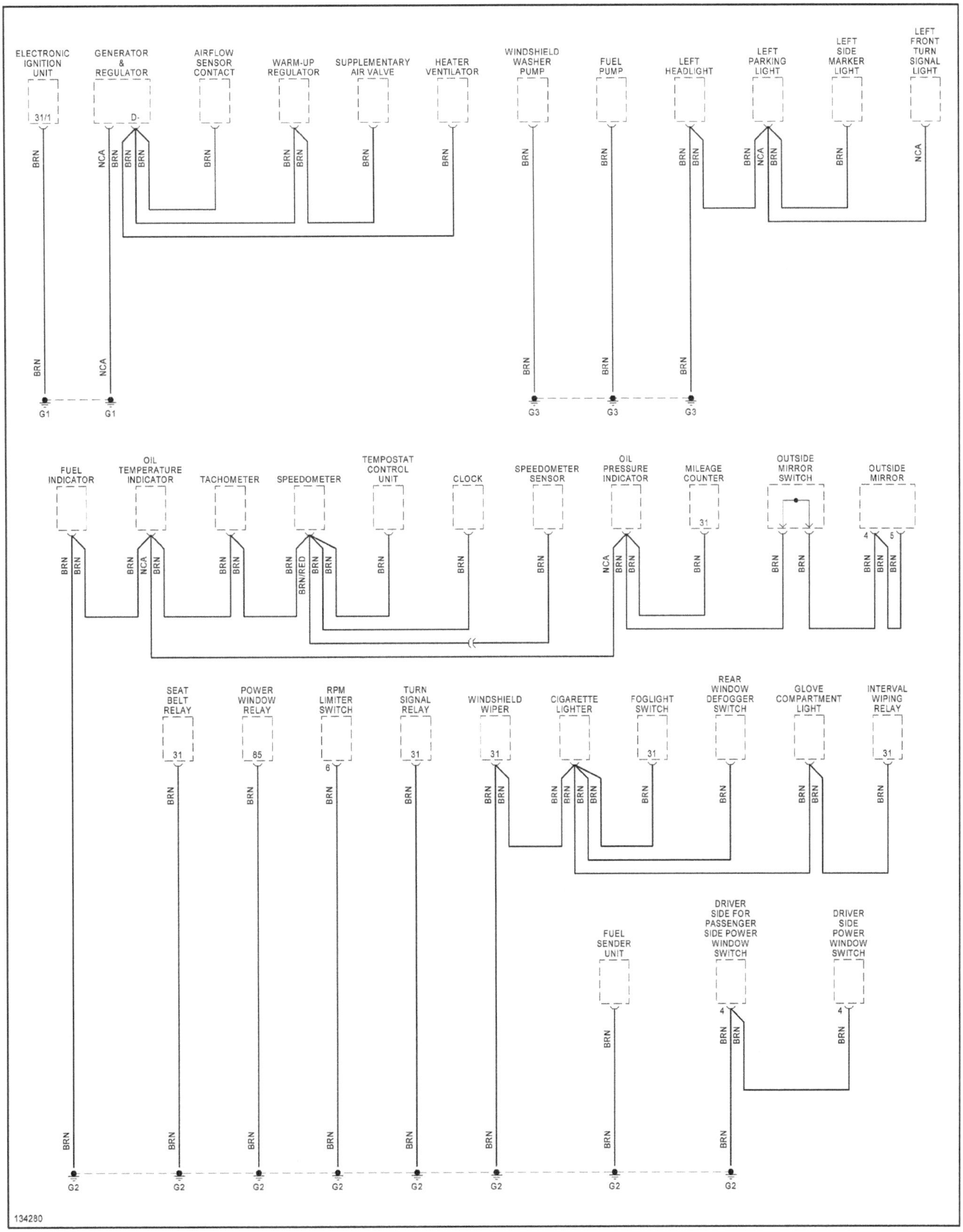

Ground Distribution

(2 of 3)

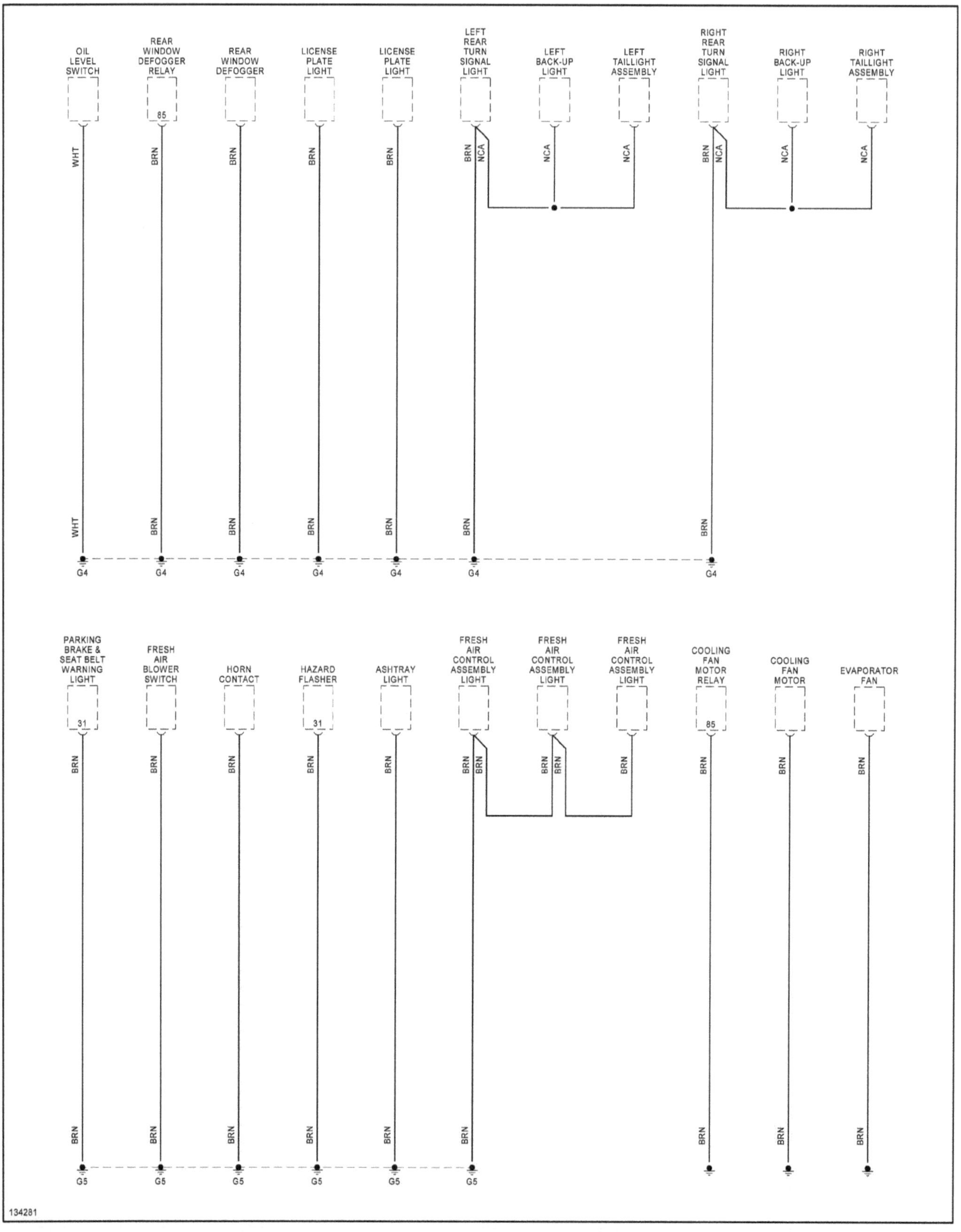

Ground Distribution
(3 of 3)

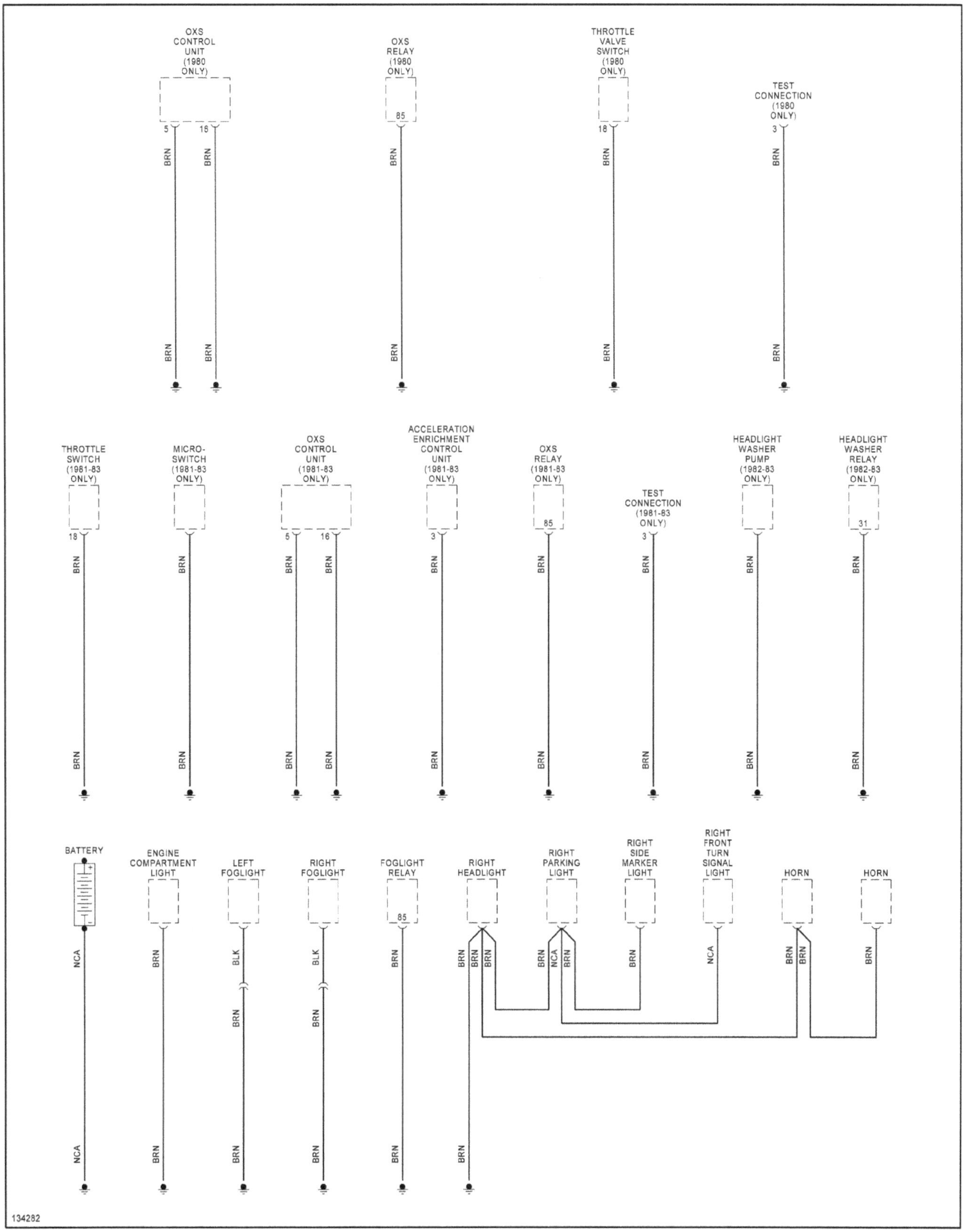

Fuel Injection, Oxygen Sensor System, Ignition System

(1 of 2)

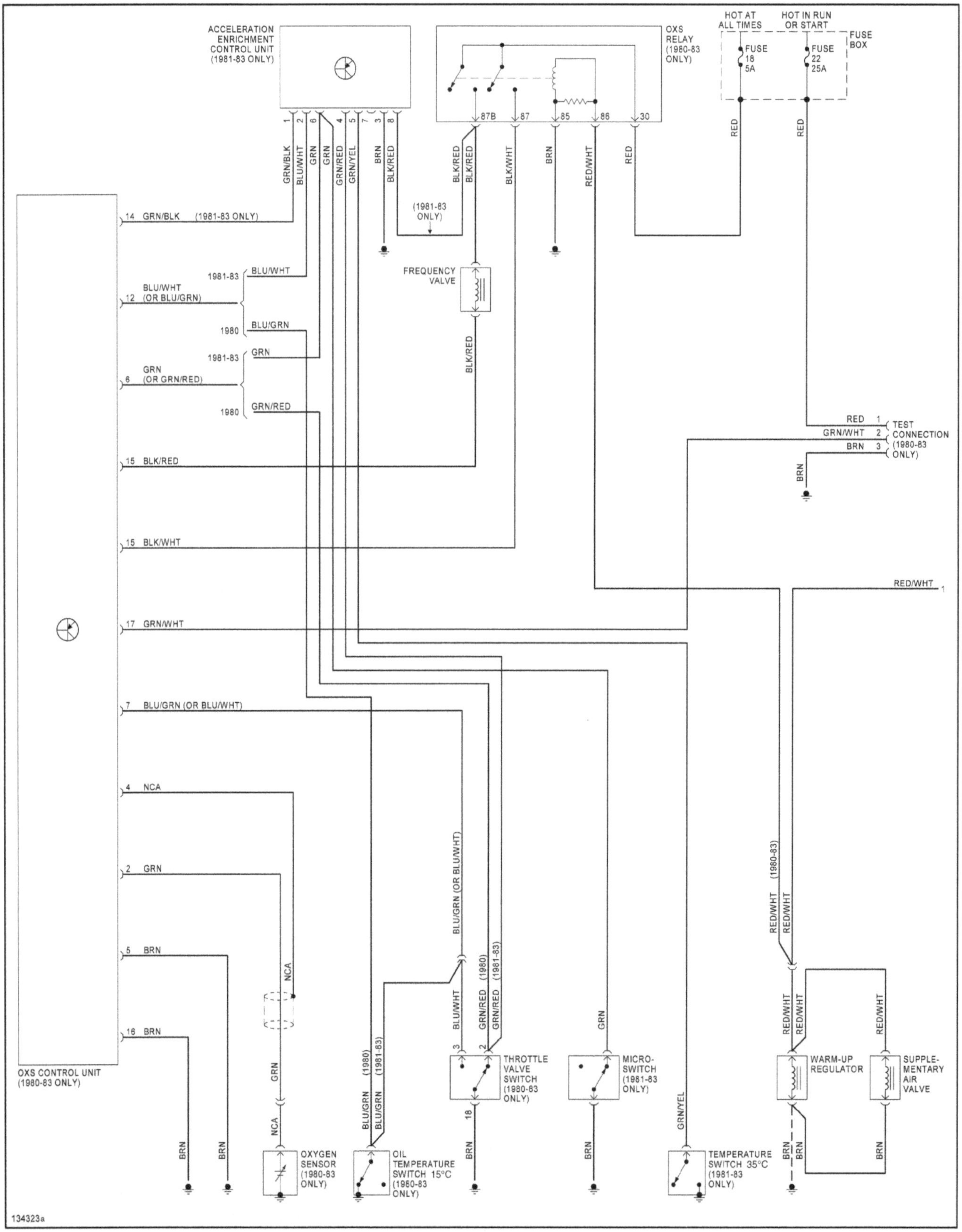

Fuel Injection, Oxygen Sensor System, Ignition System
(2 of 2)

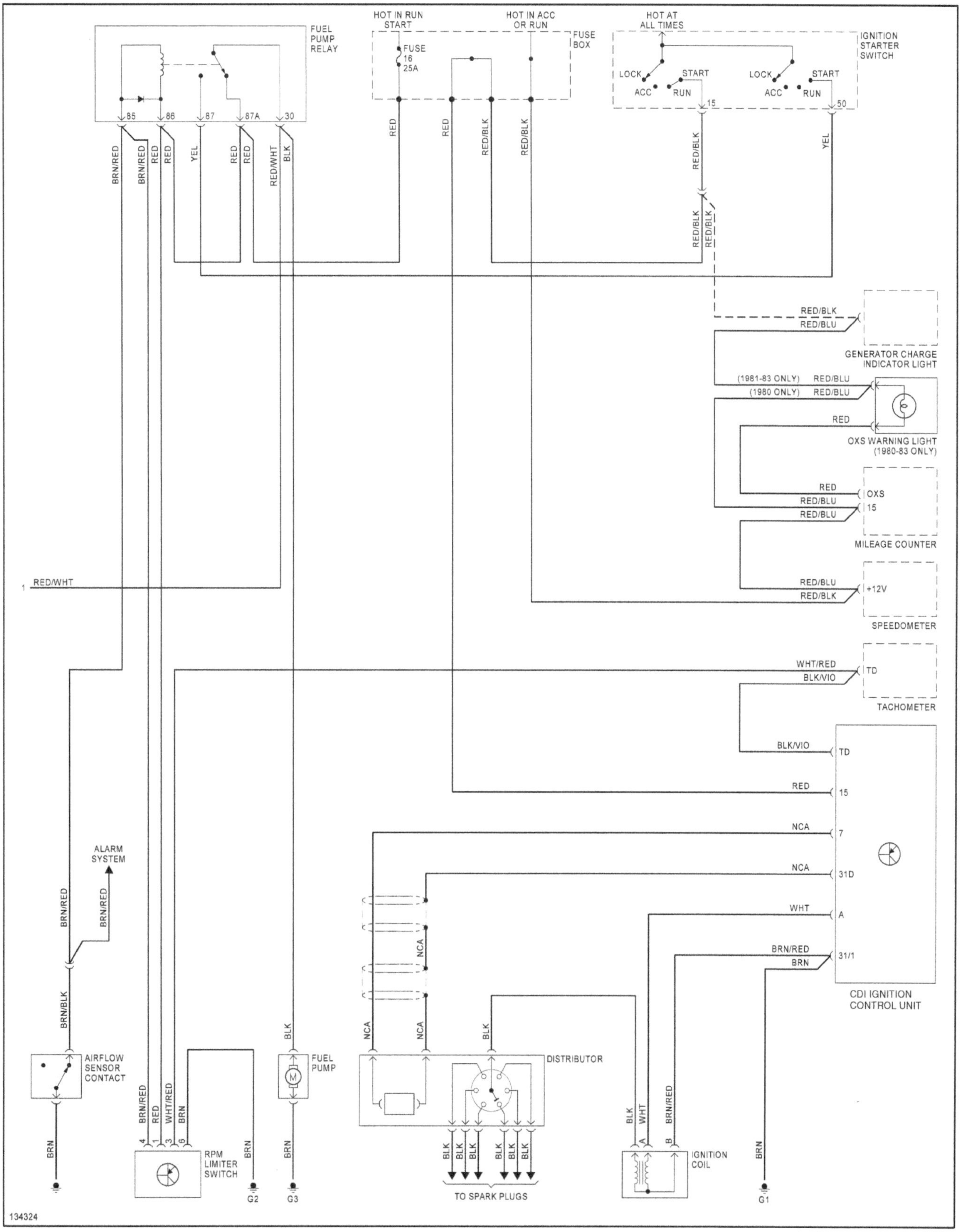

Charging

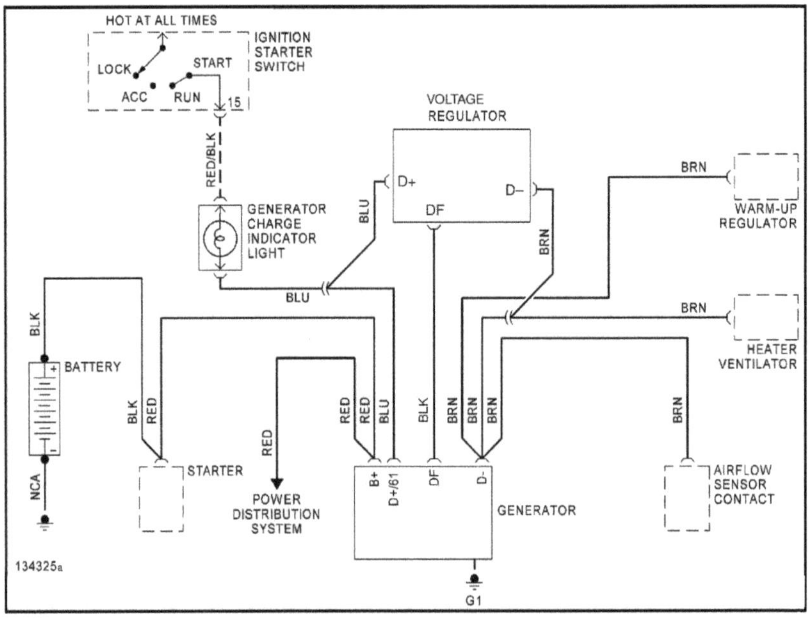

1978-1981 Charging

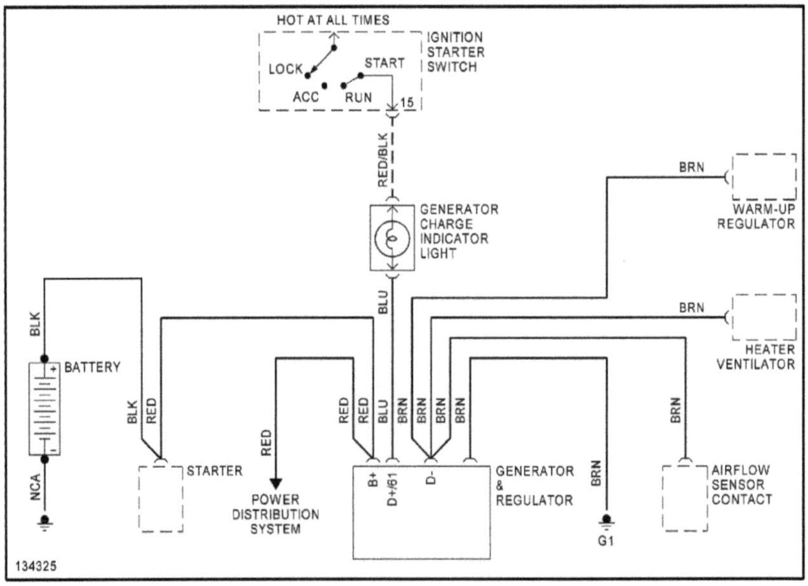

1982-1983 Charging

Starting/Cold Start Valve

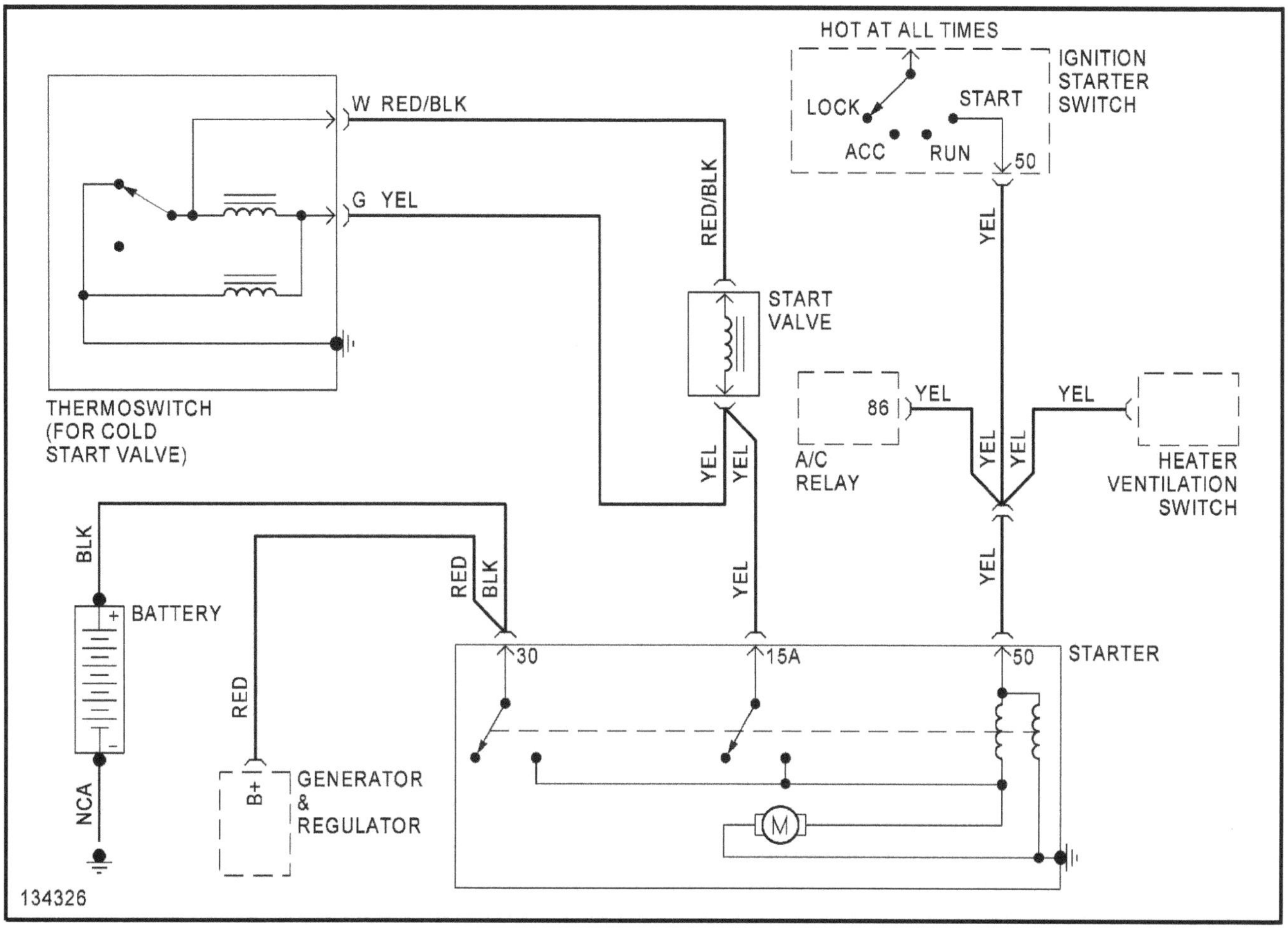

Air Conditioning

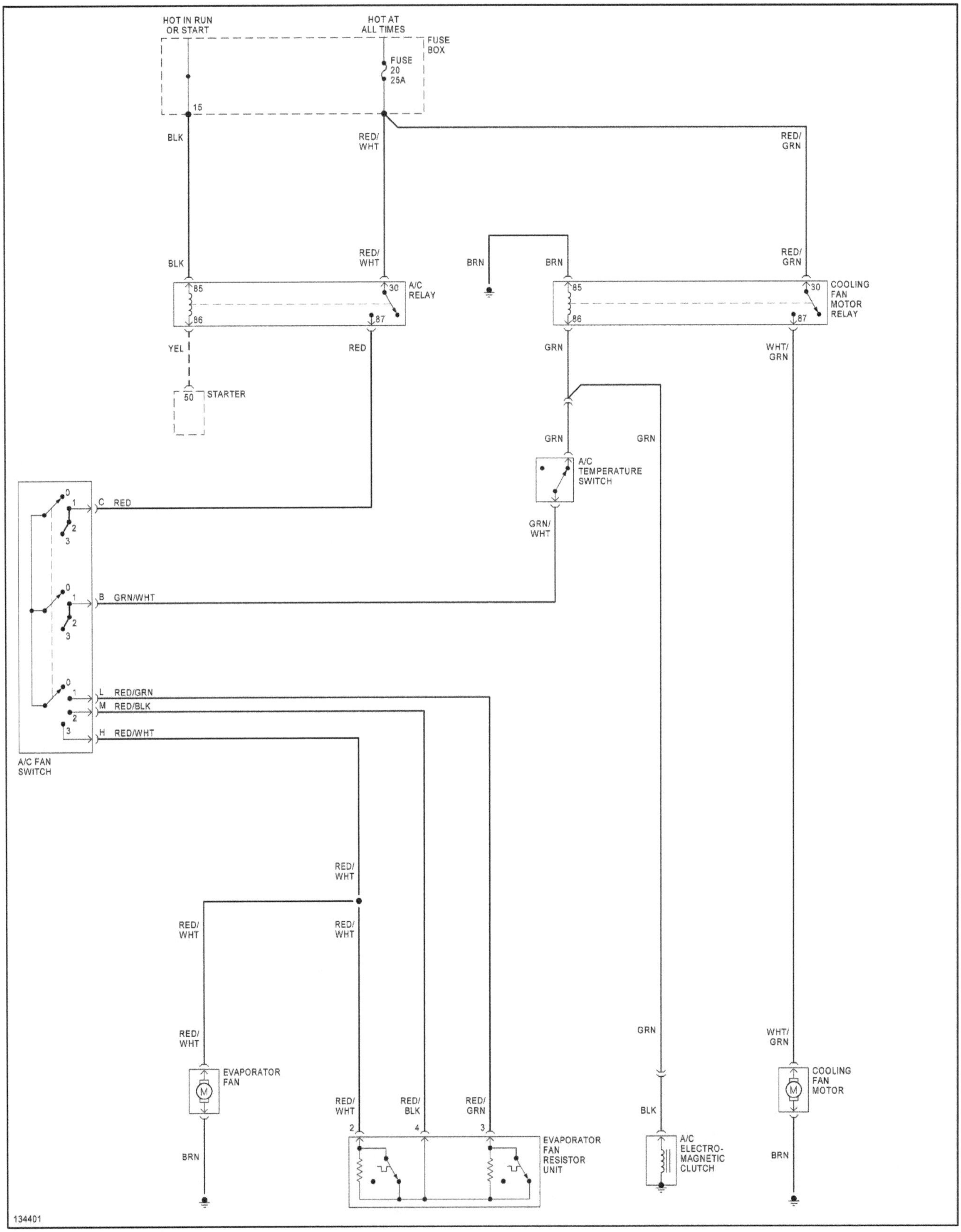

Anti-theft

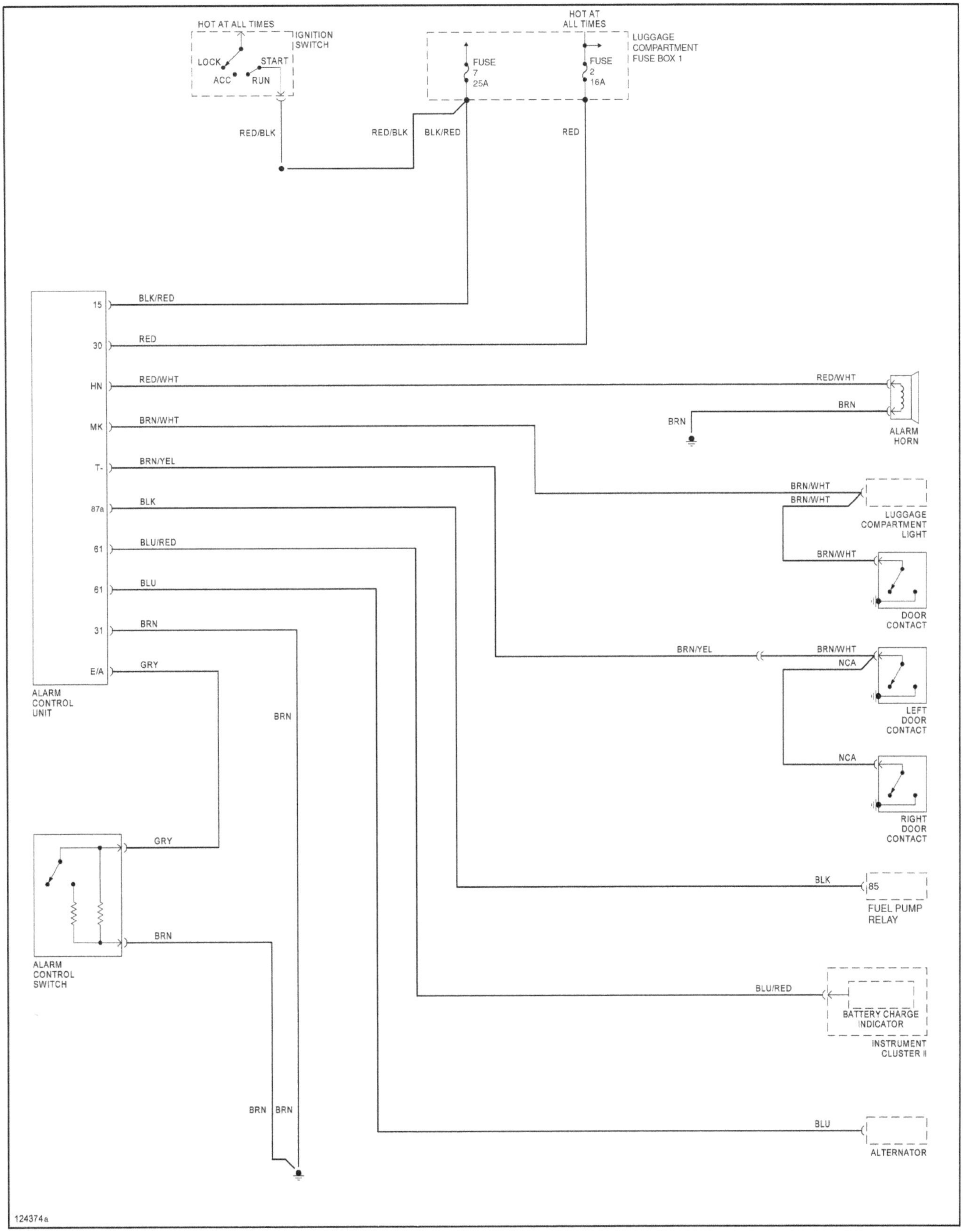
HOT AT ALL TIMES
IGNITION SWITCH
LOCK
START
ACC
RUN
HOT AT ALL TIMES
LUGGAGE COMPARTMENT FUSE BOX 1
FUSE 7 25A
FUSE 2 16A
RED/BLK
RED/BLK
BLK/RED
RED
15
BLK/RED
30
RED
HN
RED/WHT
MK
BRN/WHT
T-
BRN/YEL
87a
BLK
61
BLU/RED
61
BLU
31
BRN
E/A
GRY
ALARM CONTROL UNIT
RED/WHT
BRN
BRN
ALARM HORN
BRN/WHT
BRN/WHT
LUGGAGE COMPARTMENT LIGHT
BRN/WHT
DOOR CONTACT
BRN/YEL
BRN/WHT
NCA
LEFT DOOR CONTACT
NCA
RIGHT DOOR CONTACT
BRN
GRY
BRN
ALARM CONTROL SWITCH
BLK
85
FUEL PUMP RELAY
BLU/RED
BATTERY CHARGE INDICATOR
INSTRUMENT CLUSTER II
BRN
BRN
BLU
ALTERNATOR
124374a

Back-up Lights

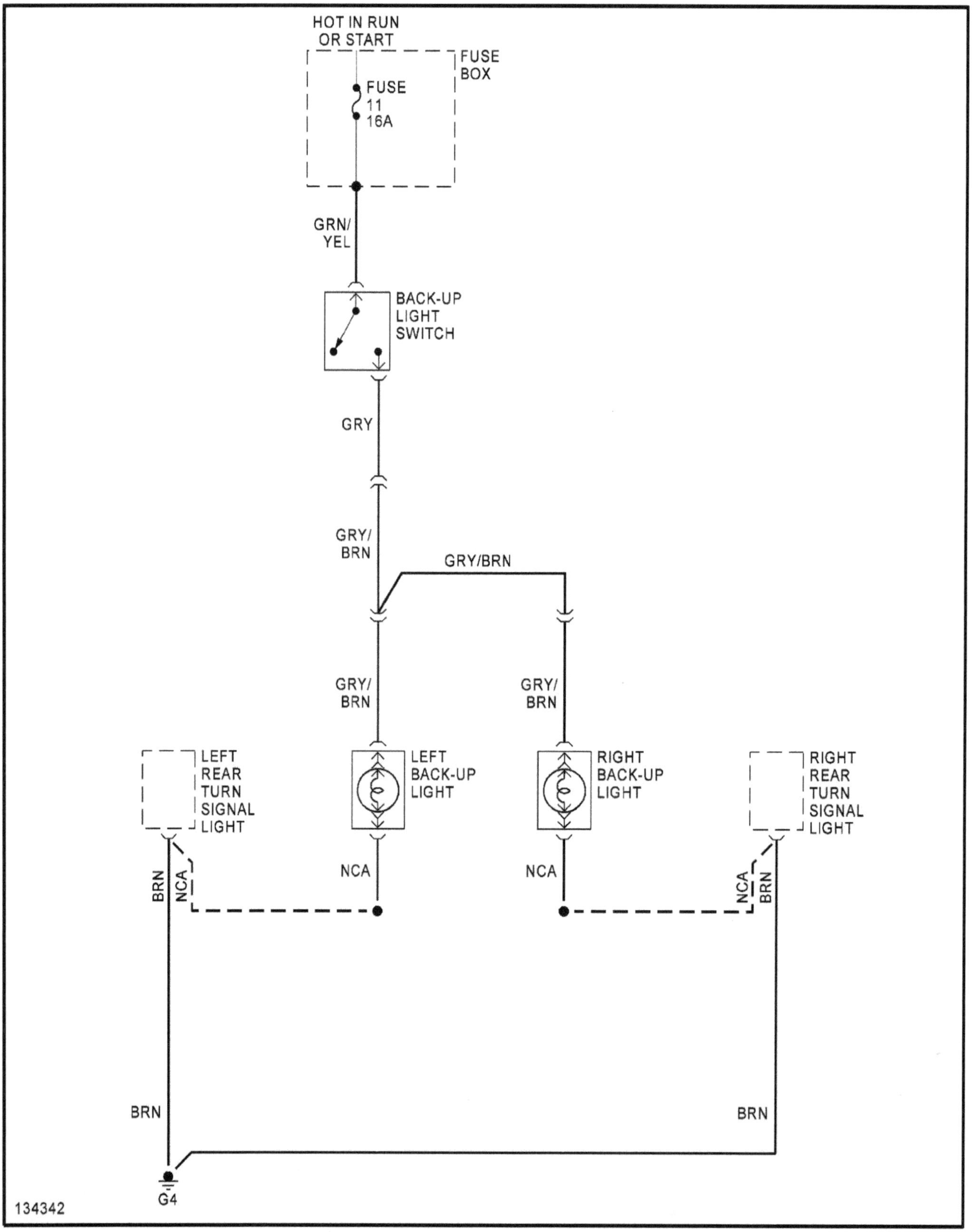

Cruise Control

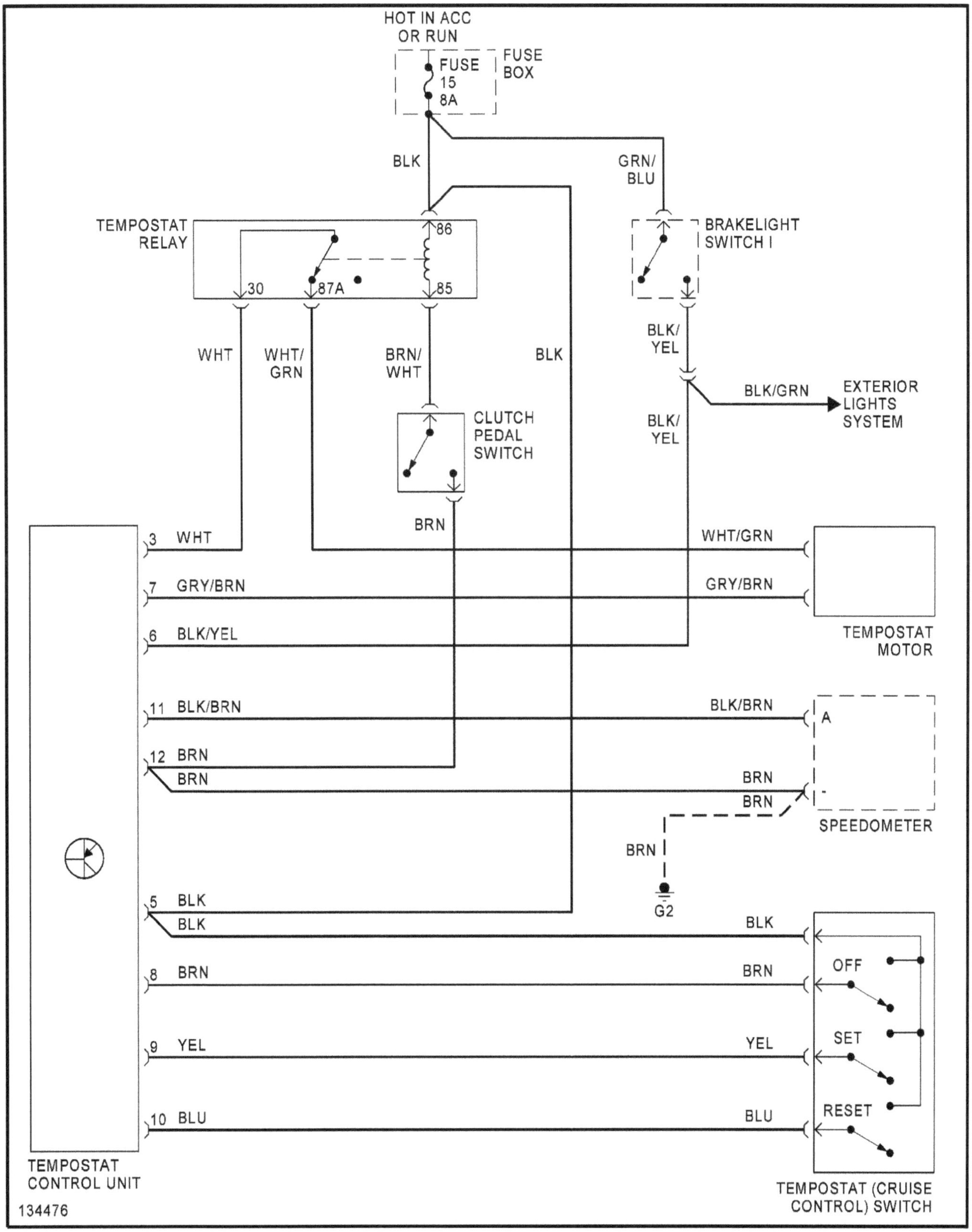

Defogger

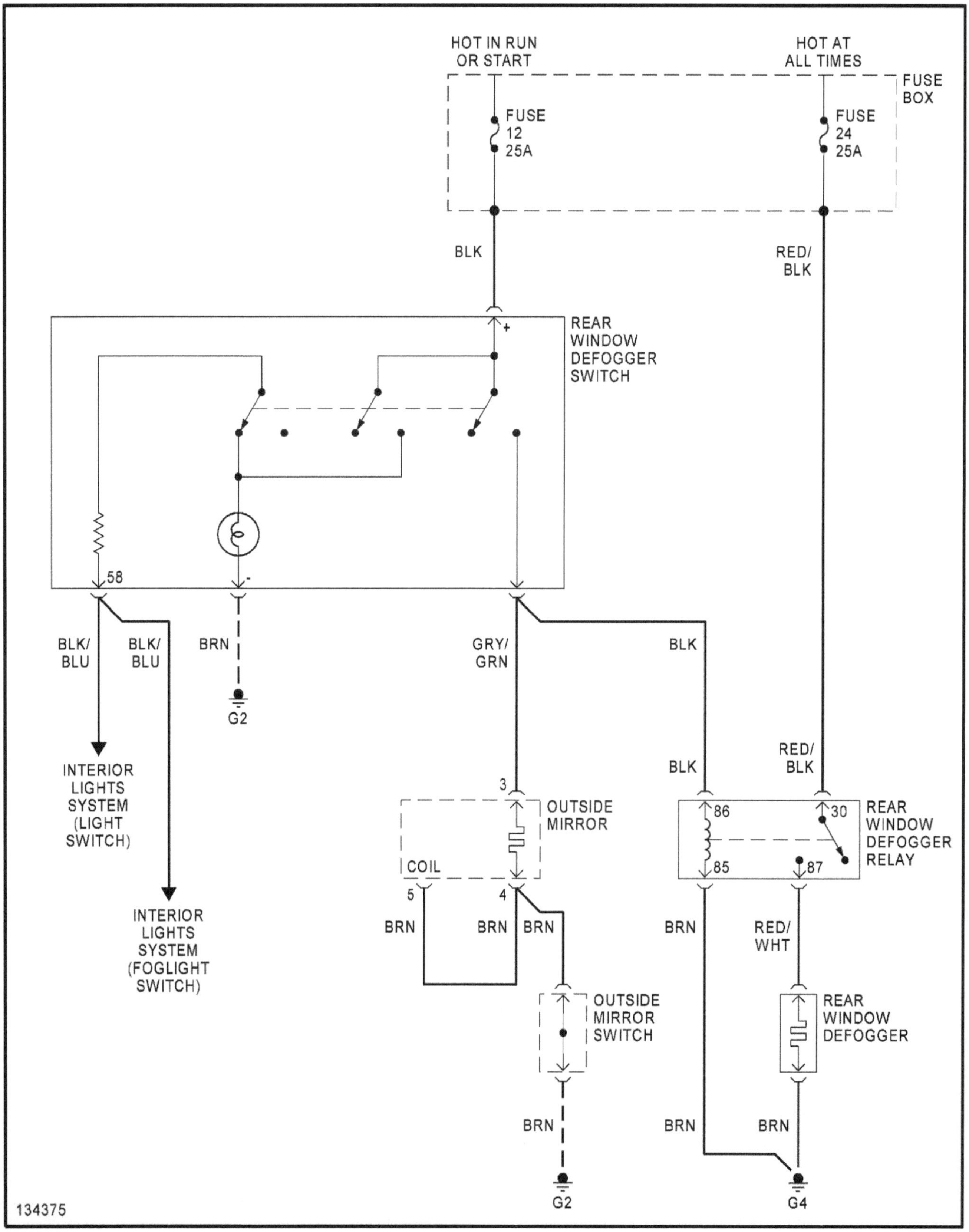

HOT IN RUN OR START
HOT AT ALL TIMES
FUSE BOX
FUSE 12 25A
FUSE 24 25A
BLK
RED/ BLK
REAR WINDOW DEFOGGER SWITCH
+
58
-
BLK/ BLU
BLK/ BLU
BRN
G2
GRY/ GRN
BLK
INTERIOR LIGHTS SYSTEM (LIGHT SWITCH)
INTERIOR LIGHTS SYSTEM (FOGLIGHT SWITCH)
BLK
RED/ BLK
3
OUTSIDE MIRROR
COIL
5
4
86
30
85
87
REAR WINDOW DEFOGGER RELAY
BRN
BRN
BRN
BRN
RED/ WHT
OUTSIDE MIRROR SWITCH
REAR WINDOW DEFOGGER
BRN
BRN
BRN
G2
G4
134375

Exterior Lights

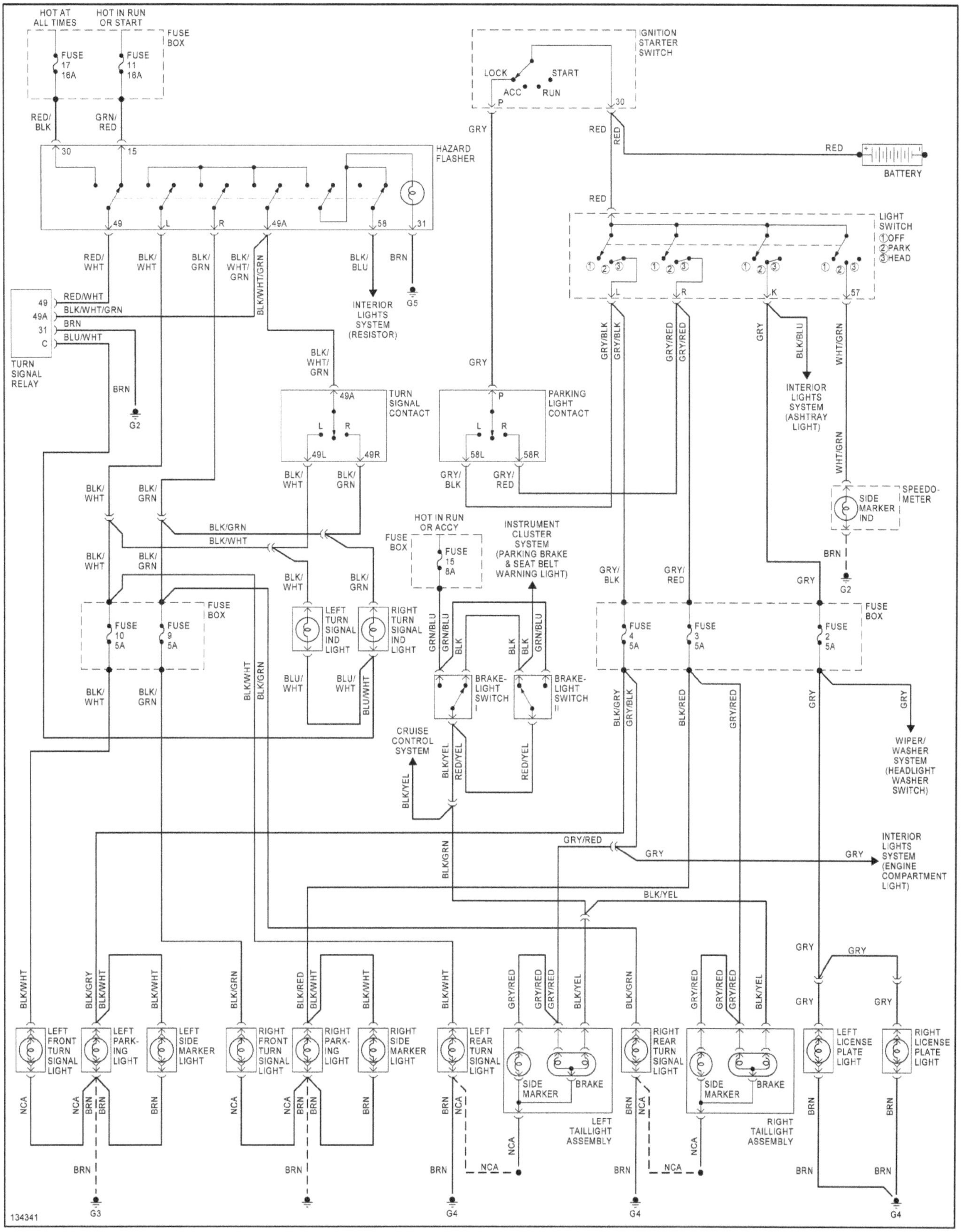

Fog Lights

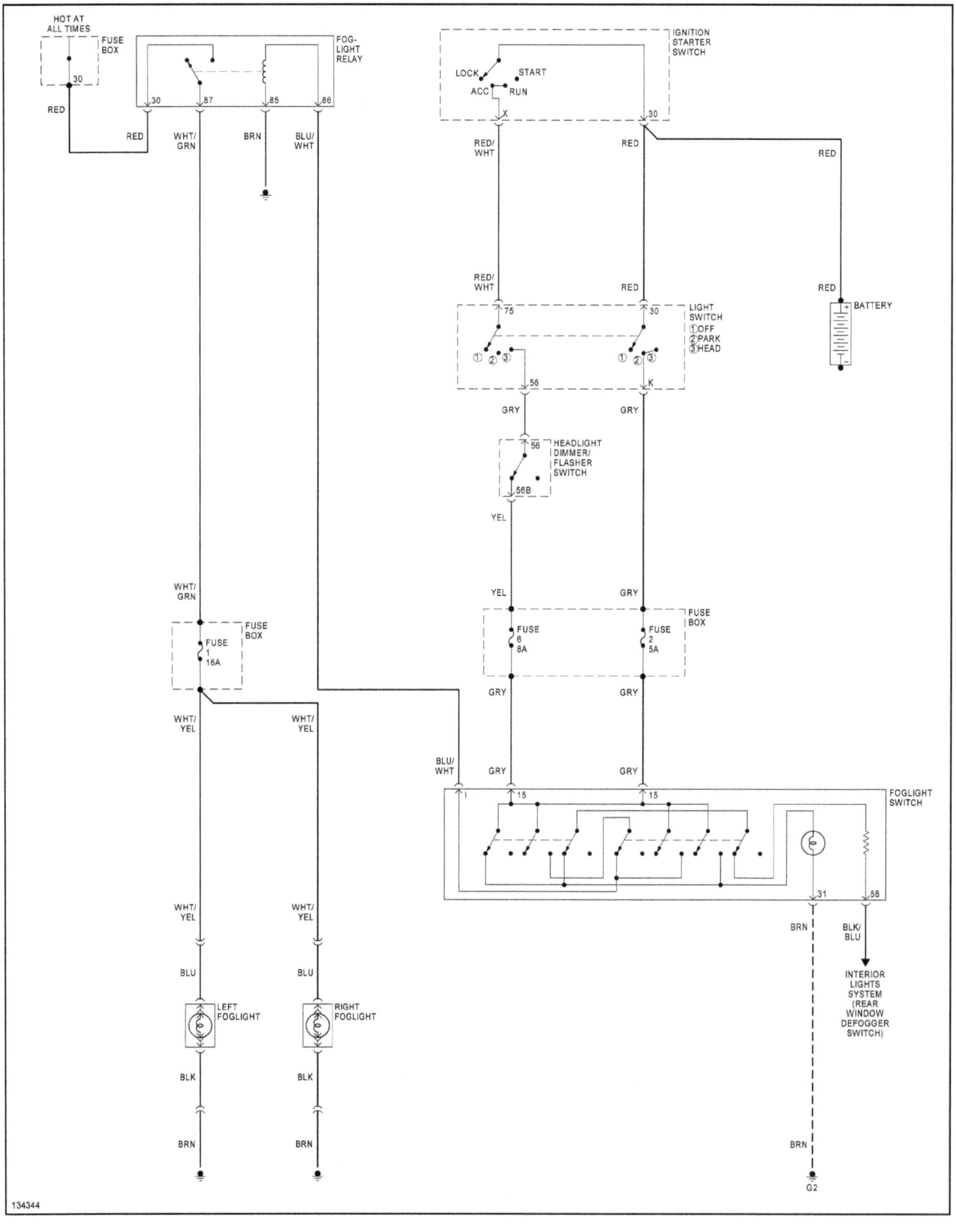
HOT AT ALL TIMES
FUSE BOX
30
RED
FOG-LIGHT RELAY
30
87
85
86
RED
WHT/GRN
BRN
BLU/WHT
IGNITION STARTER SWITCH
LOCK
START
ACC
RUN
X
30
RED/WHT
RED
RED
RED/WHT
RED
RED
BATTERY
75
30
LIGHT SWITCH
①OFF
②PARK
③HEAD
56
K
GRY
GRY
56
HEADLIGHT DIMMER/FLASHER SWITCH
56B
YEL
YEL
GRY
FUSE BOX
FUSE 6 8A
FUSE 2 5A
GRY
GRY
WHT/GRN
FUSE BOX
FUSE 1 16A
WHT/YEL
WHT/YEL
BLU/WHT
GRY
GRY
15
15
FOGLIGHT SWITCH
31
56
BRN
BLK/BLU
INTERIOR LIGHTS SYSTEM (REAR WINDOW DEFOGGER SWITCH)
WHT/YEL
WHT/YEL
BLU
BLU
LEFT FOGLIGHT
RIGHT FOGLIGHT
BLK
BLK
BRN
BRN
BRN
G2
134344

Headlights

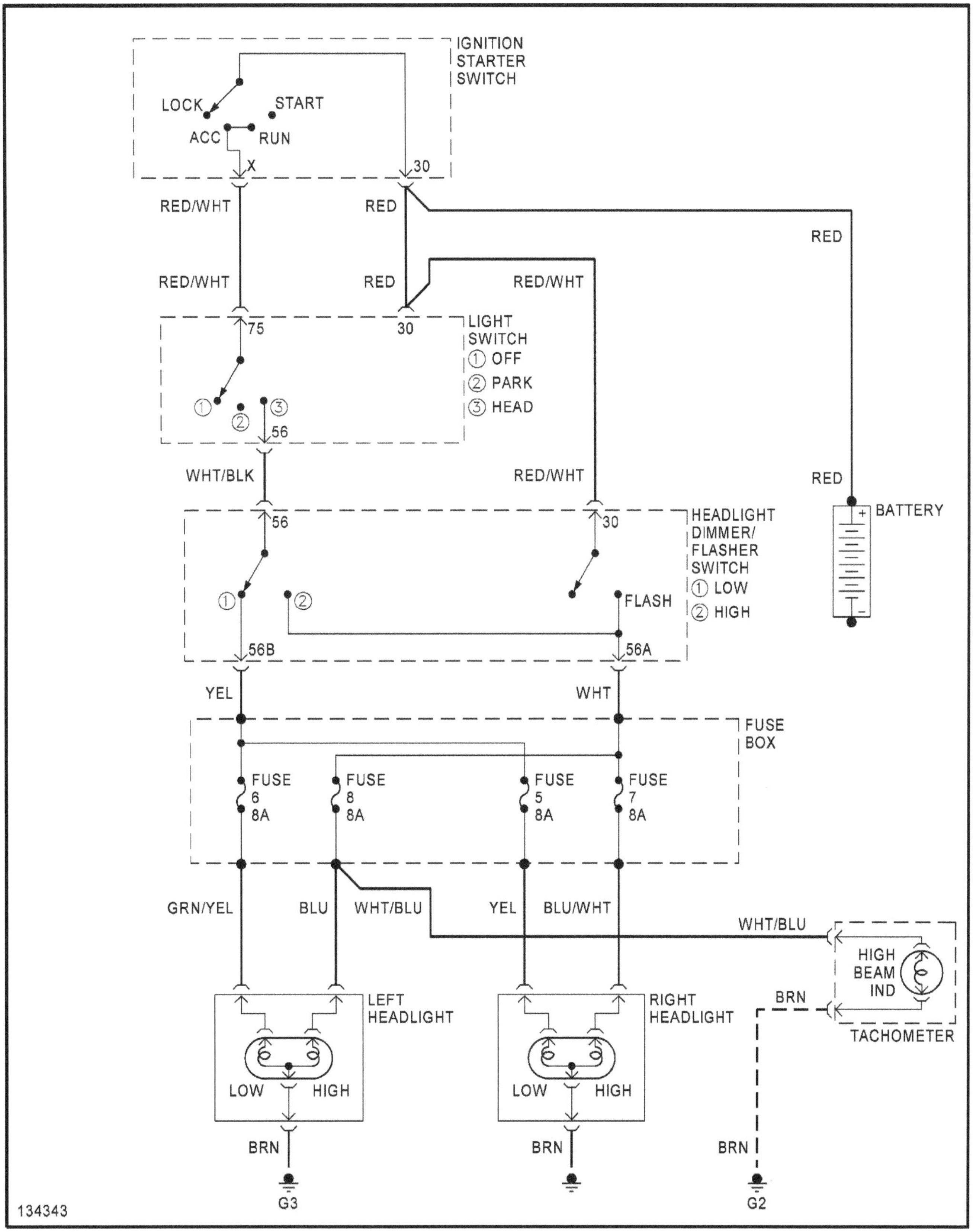

IGNITION
STARTER
SWITCH
LOCK
START
ACC
RUN
X
30
RED/WHT
RED
RED
RED/WHT
RED
RED/WHT
75
30
LIGHT
SWITCH
① OFF
② PARK
③ HEAD
56
WHT/BLK
RED/WHT
RED
56
30
HEADLIGHT
DIMMER/
FLASHER
SWITCH
① LOW
② HIGH
FLASH
56B
56A
BATTERY
YEL
WHT
FUSE
BOX
FUSE
6
8A
FUSE
8
8A
FUSE
5
8A
FUSE
7
8A
GRN/YEL
BLU
WHT/BLU
YEL
BLU/WHT
WHT/BLU
HIGH
BEAM
IND
TACHOMETER
BRN
LEFT
HEADLIGHT
RIGHT
HEADLIGHT
LOW
HIGH
LOW
HIGH
BRN
BRN
BRN
G3
G2
134343

Heating

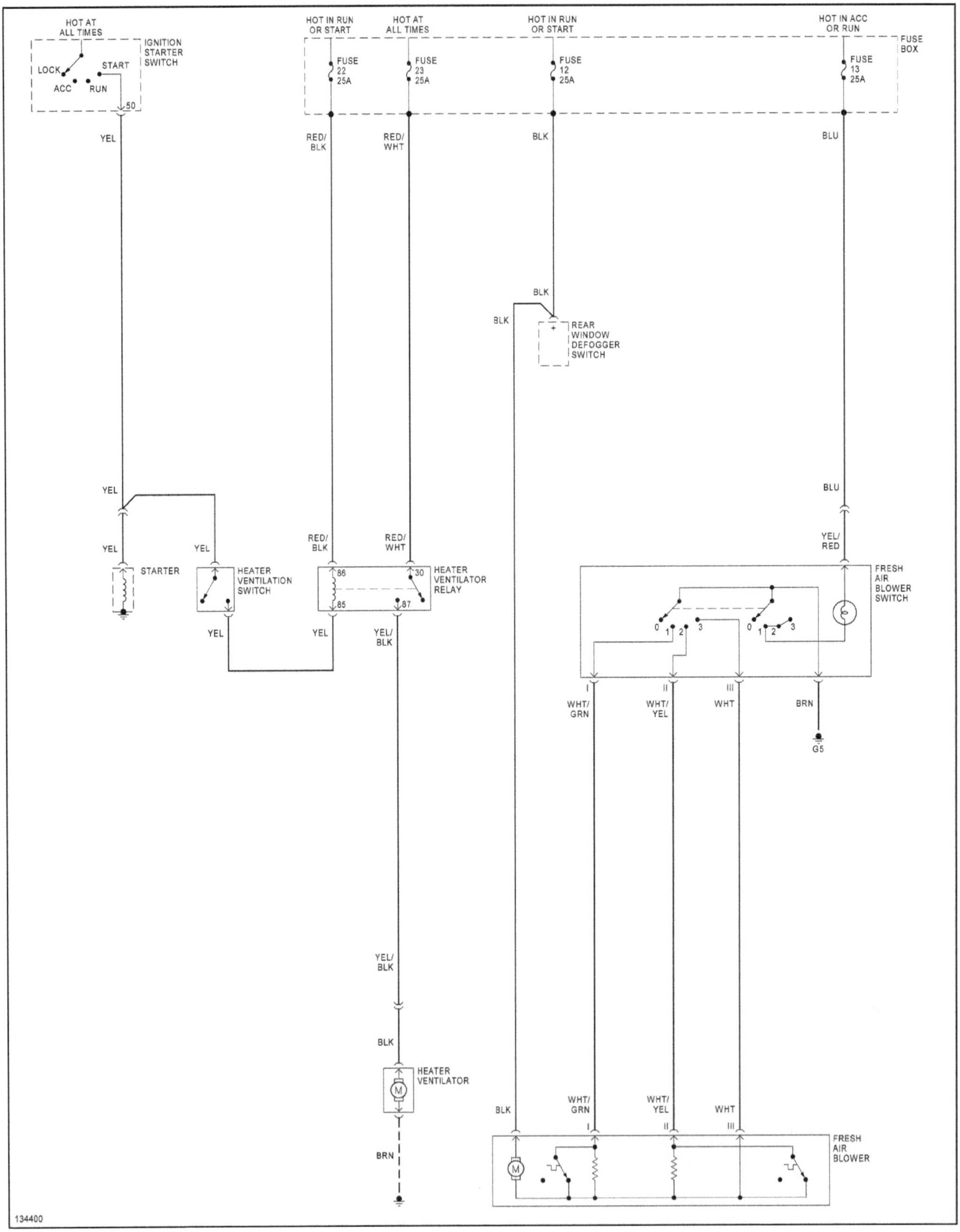
HOT AT ALL TIMES
IGNITION STARTER SWITCH
LOCK
START
ACC
RUN
50
HOT IN RUN OR START
HOT AT ALL TIMES
HOT IN RUN OR START
HOT IN ACC OR RUN
FUSE BOX
FUSE 22 25A
FUSE 23 25A
FUSE 12 25A
FUSE 13 25A
YEL
RED/ BLK
RED/ WHT
BLK
BLU
BLK
BLK
REAR WINDOW DEFOGGER SWITCH
YEL
YEL
YEL
STARTER
HEATER VENTILATION SWITCH
RED/ BLK
RED/ WHT
86
30
85
87
HEATER VENTILATOR RELAY
YEL
YEL
YEL/ BLK
BLU
YEL/ RED
FRESH AIR BLOWER SWITCH
0 1 2 3
0 1 2 3
I
II
III
WHT/ GRN
WHT/ YEL
WHT
BRN
G5
YEL/ BLK
BLK
HEATER VENTILATOR
M
BRN
BLK
WHT/ GRN
WHT/ YEL
WHT
I
II
III
FRESH AIR BLOWER
M
134400

Horns

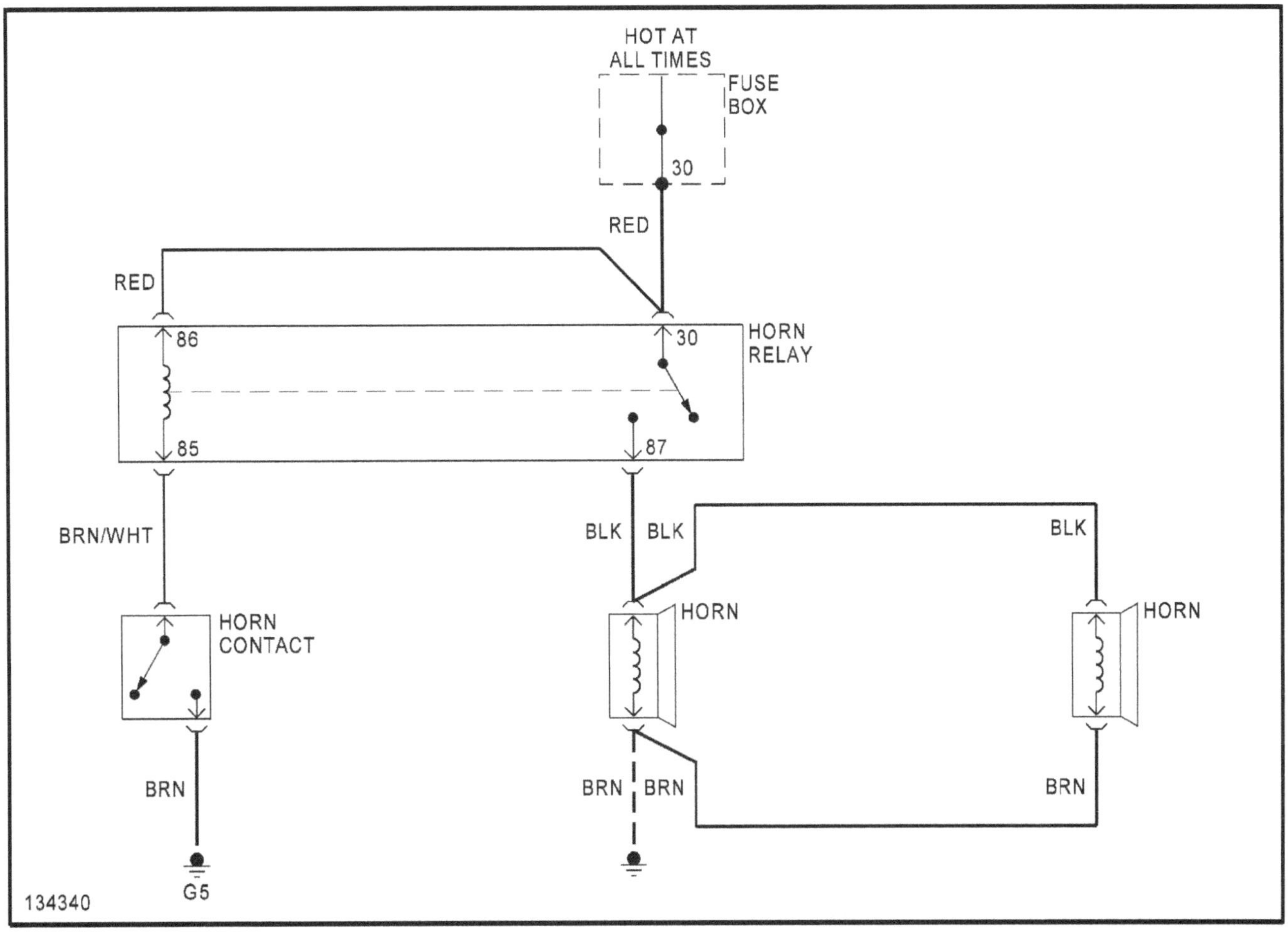

HOT AT
ALL TIMES
FUSE
BOX
30
RED
RED
86
30
HORN
RELAY
85
87
BRN/WHT
BLK
BLK
BLK
HORN
CONTACT
HORN
HORN
BRN
BRN
BRN
BRN
G5
134340

Instruments
(1 of 2)

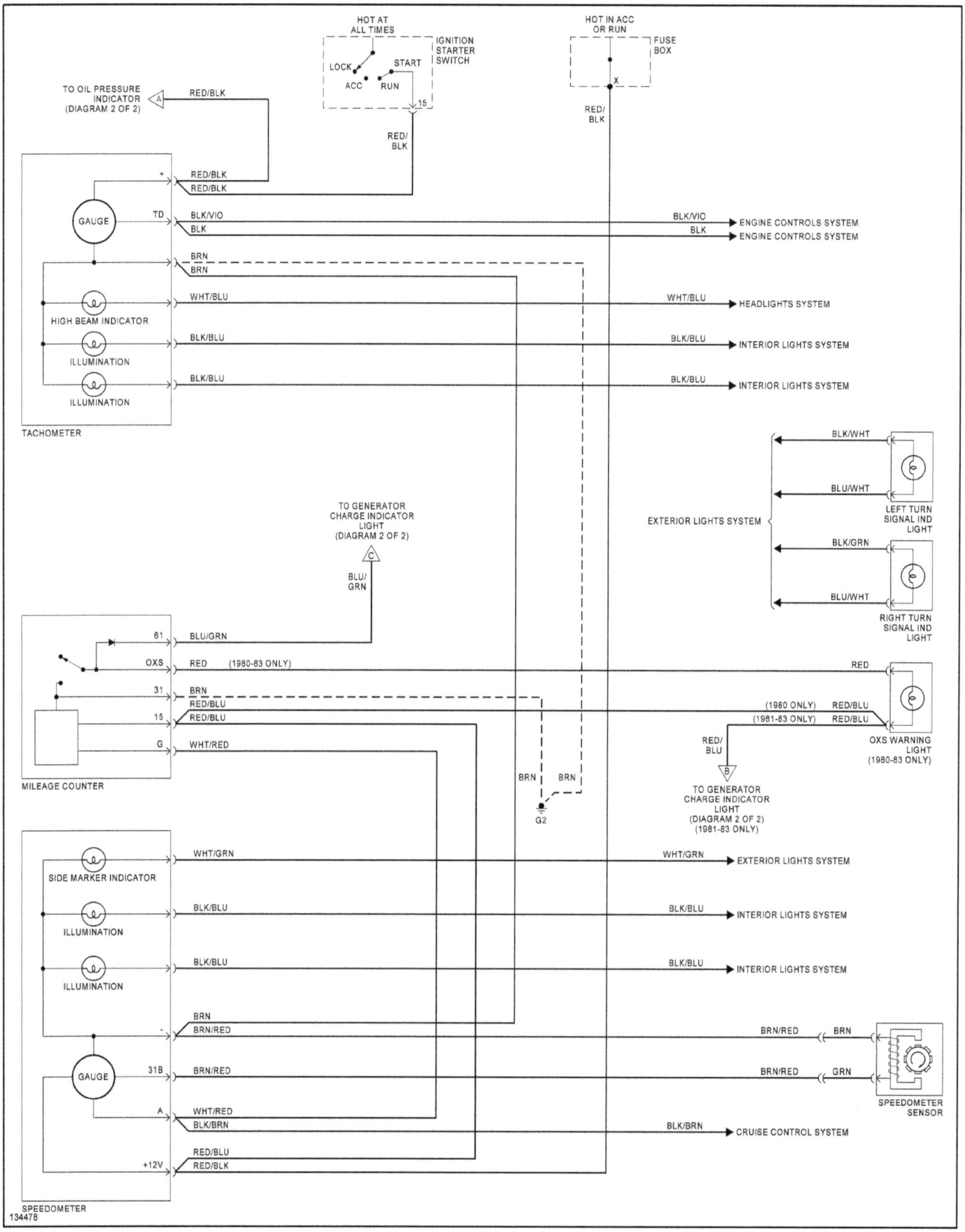

Instruments
(2 of 2)

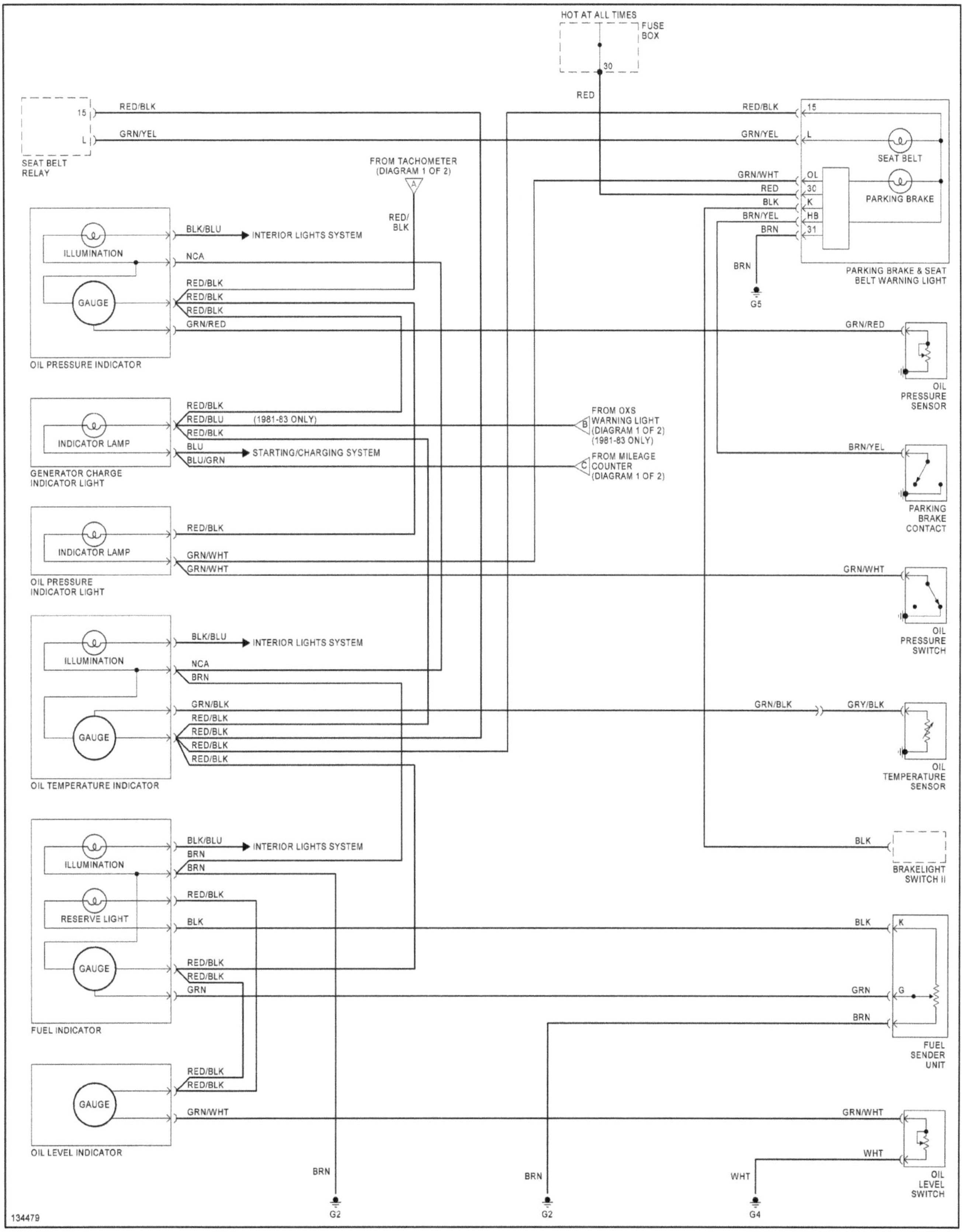

Interior Lights

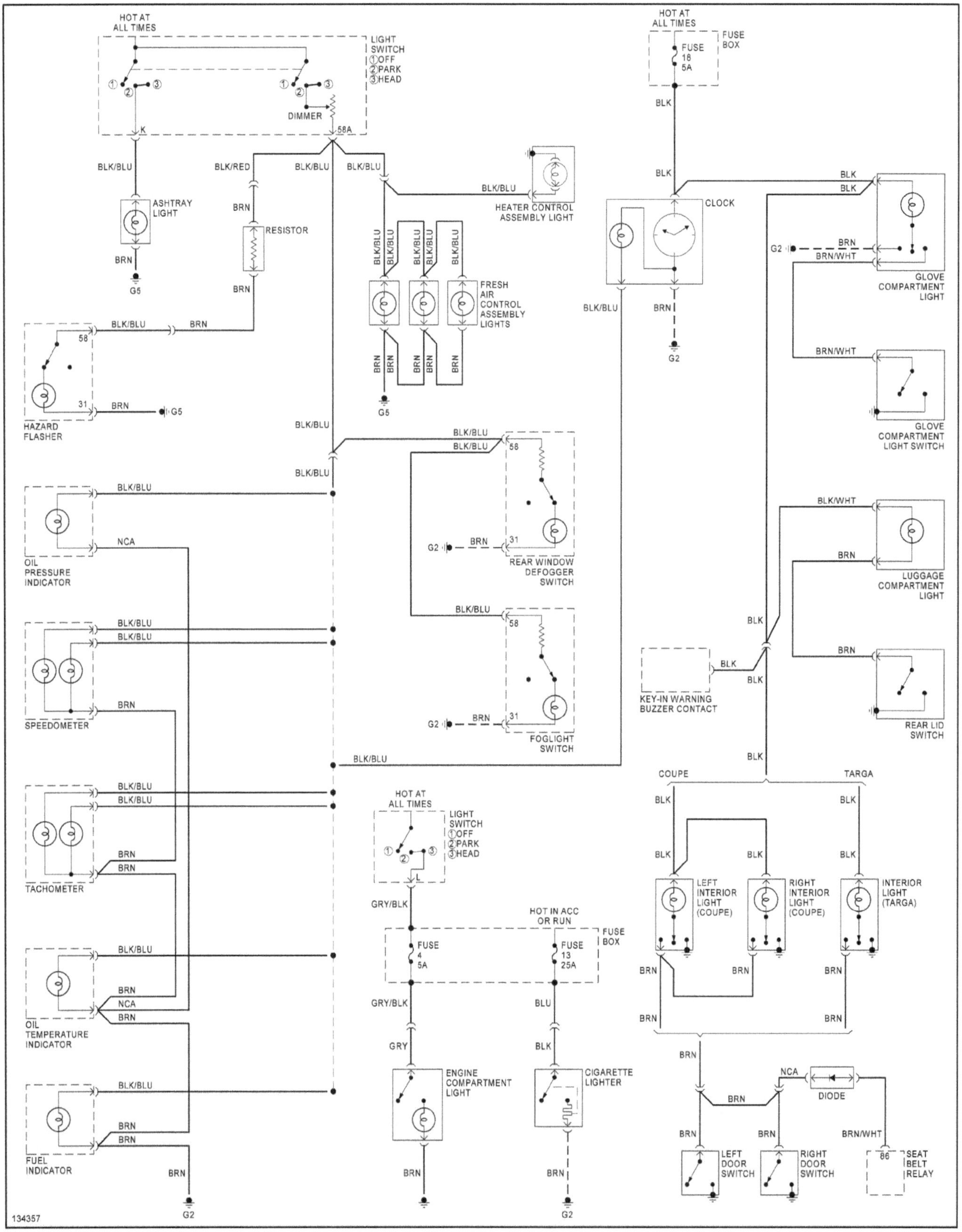

Mirrors

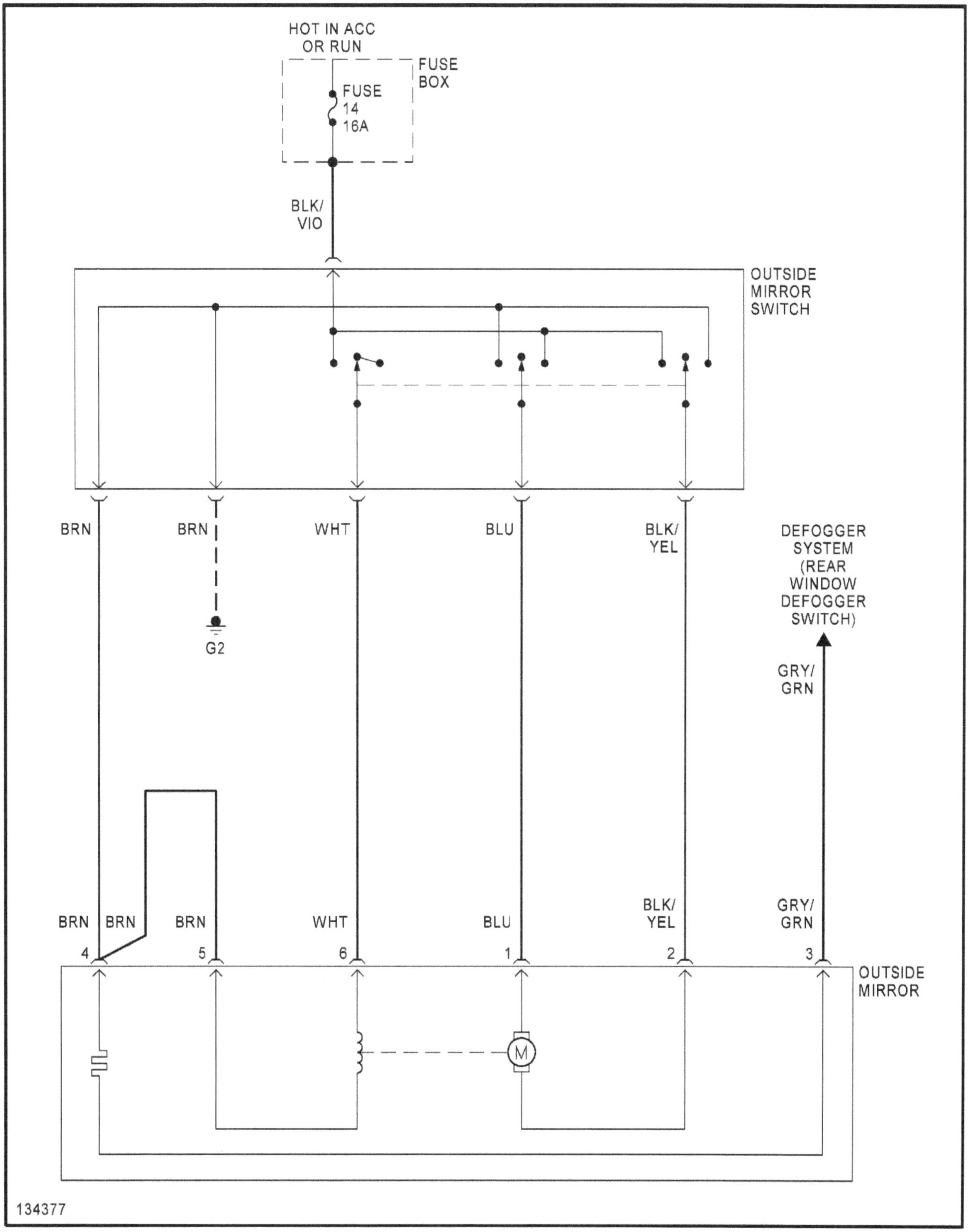

134377

Power Windows

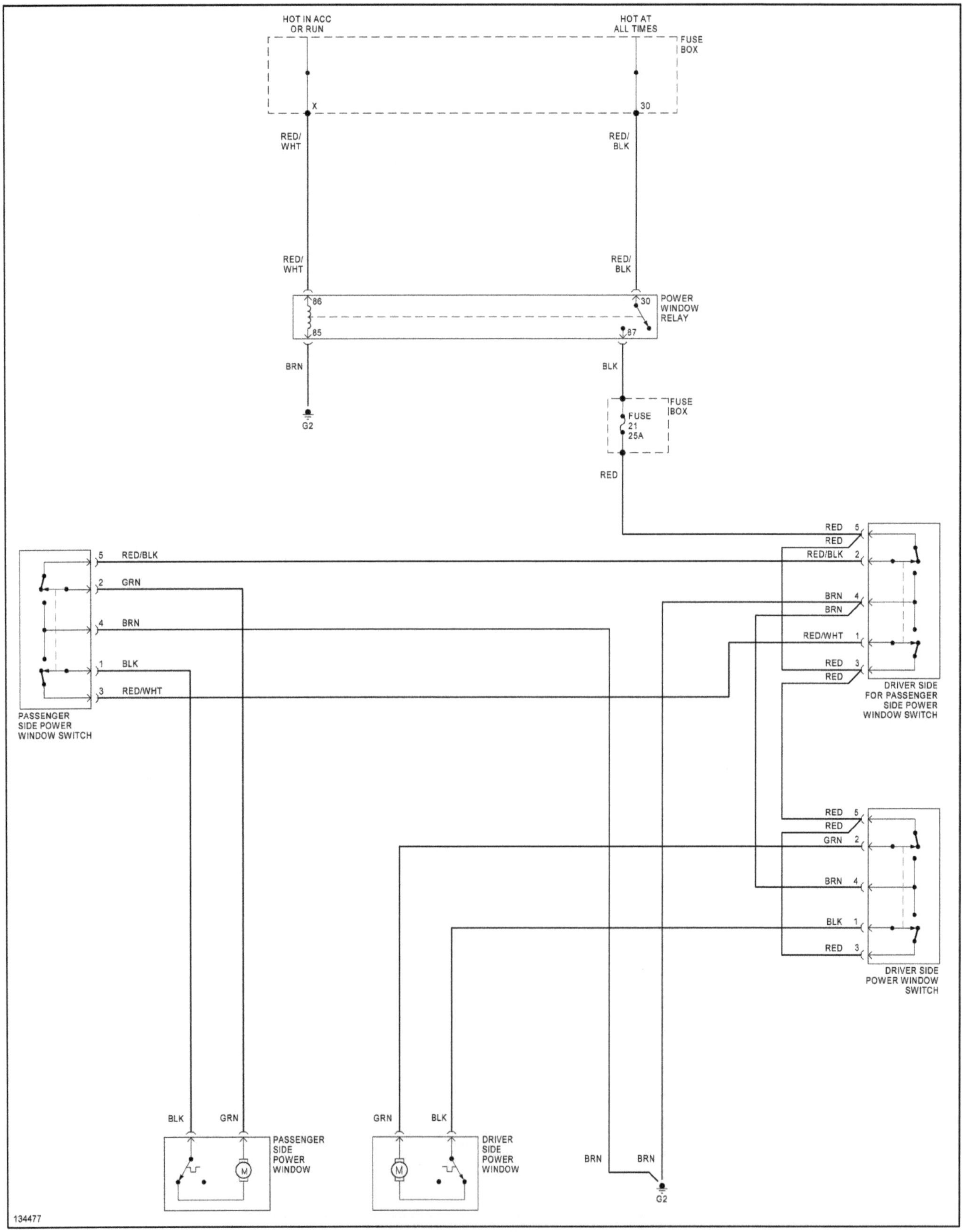

Sunroof

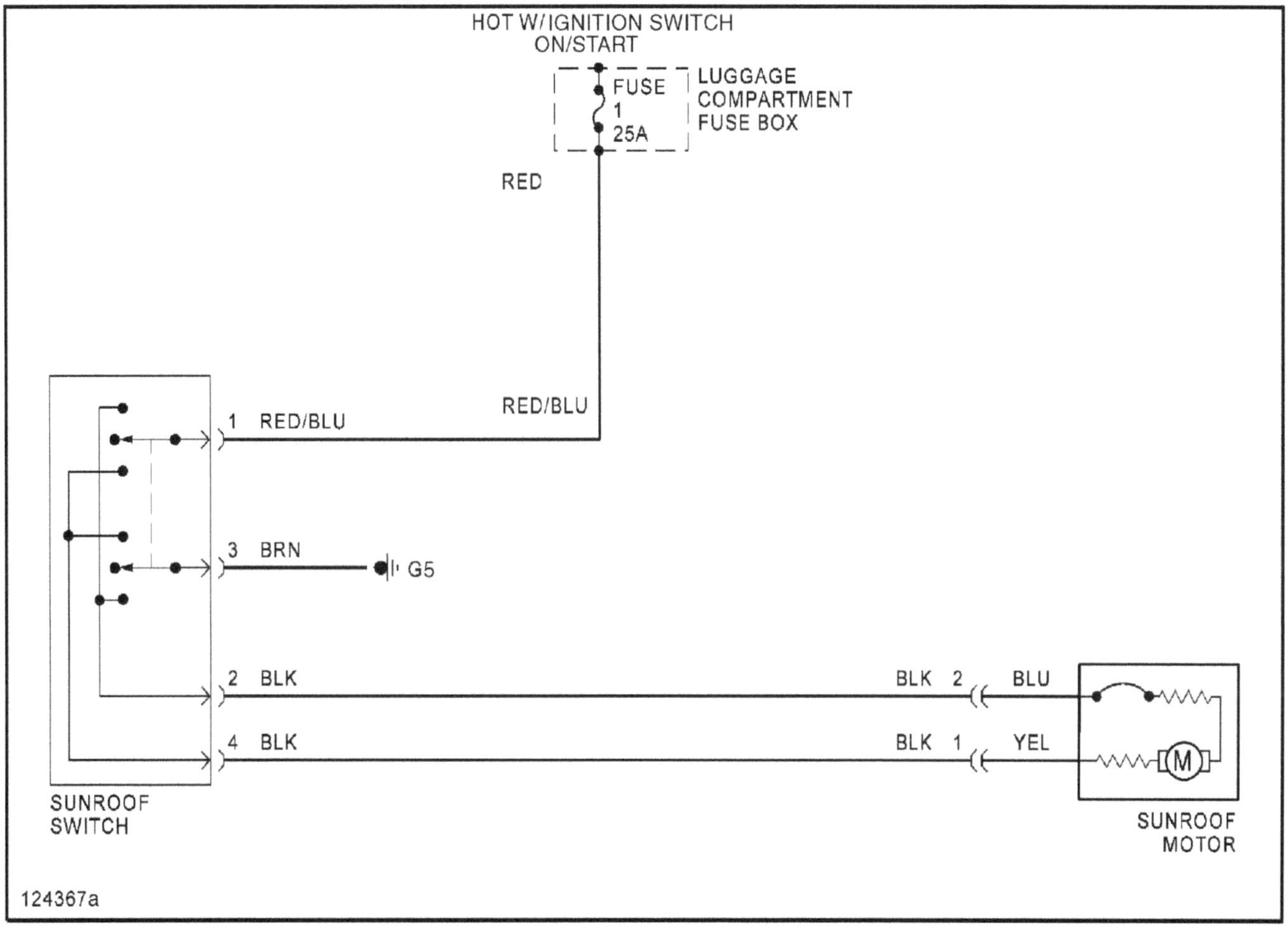
HOT W/IGNITION SWITCH
ON/START
FUSE
1
25A
LUGGAGE
COMPARTMENT
FUSE BOX
RED
RED/BLU
1 RED/BLU
3 BRN
G5
2 BLK
4 BLK
BLK 2
BLU
BLK 1
YEL
M
SUNROOF
SWITCH
SUNROOF
MOTOR
124367a

Warning System

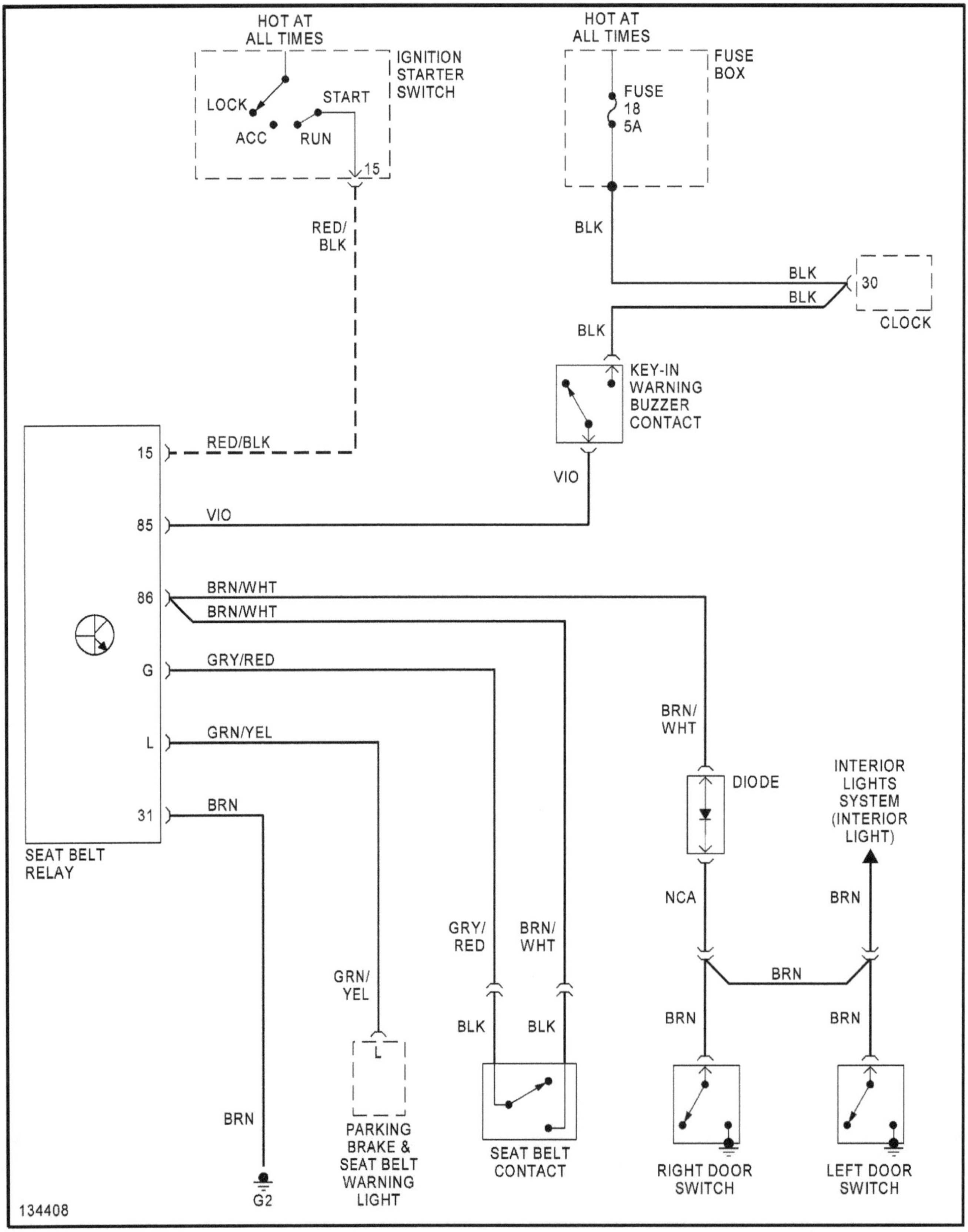

Wiper/Washer

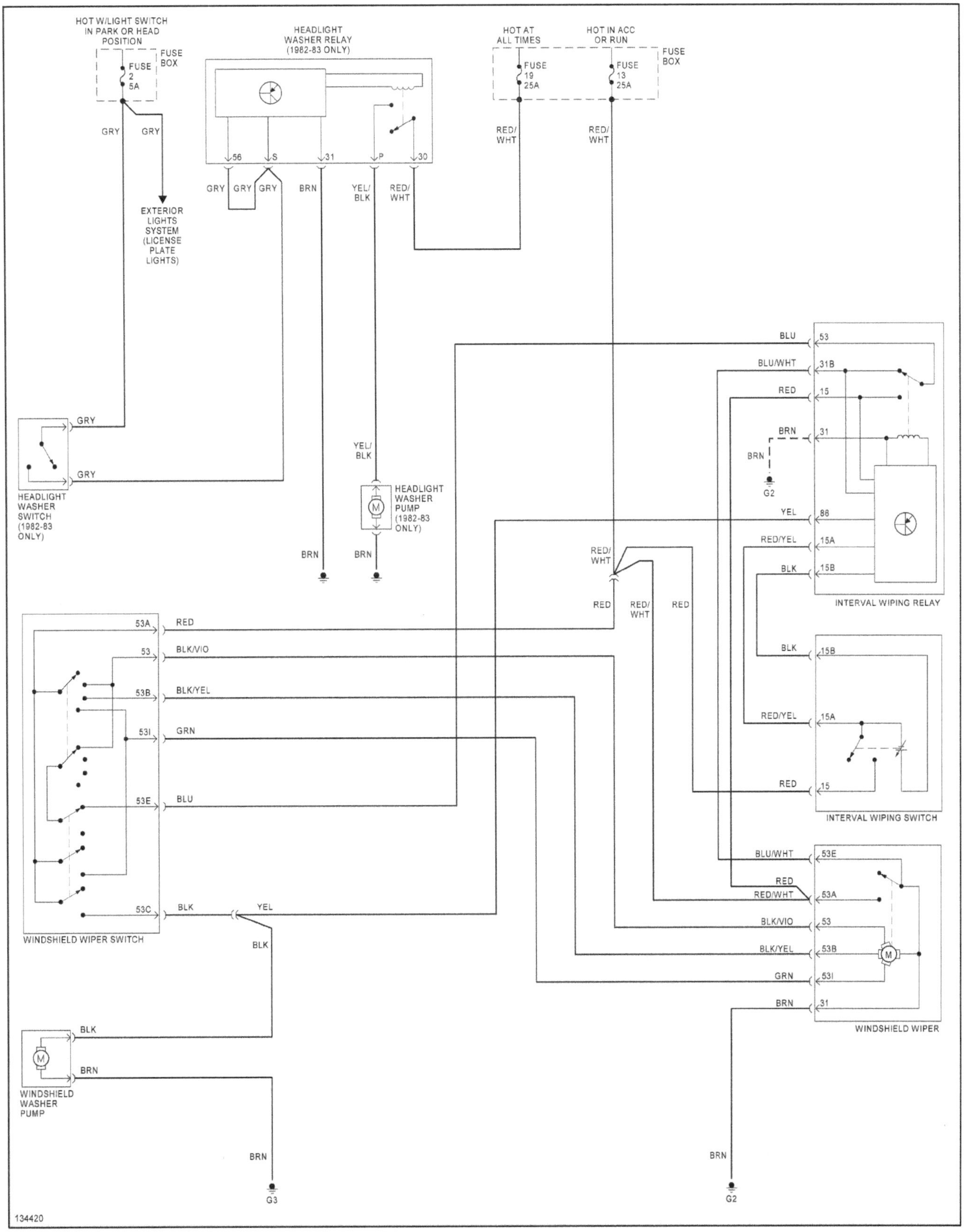

971 Electrical Component Locations

TABLES

GENERAL

This repair group covers fuse, relay, control unit and ground location information. Electrical equipment and accessories installed varies depending on model year.

> ***CAUTION —***
>
> - *Relay and fuse positions are subject to change and may vary from car to car. If questions arise, an authorized Porsche dealer is the best source for the most accurate and up-to-date information.*
> - *A good way to verify a relay position is to compare the wiring colors at the relay socket to the colors indicated on the wiring diagrams located at the rear of this manual. See* **970 Electrical wiring Diagrams**.
> - *Always switch the ignition off and remove the negative (–) battery cable before removing any electrical components.*
> - *Prior to disconnecting the battery, read the battery disconnection cautions given at the front of this manual on page viii.*
> - *Connect and disconnect ignition system wires, multiple connectors, and ignition test equipment leads only while the ignition is switched off.*
> - *Only use a digital multimeter for electrical tests.*

FUSES, RELAYS AND CONTROL UNITS

Fuses, relays and control units are located in various areas throughout the vehicle. Refer to the component location photos and tables in this repair group for electrical component locations.

Fuse positions

> ***CAUTION —***
> - *Replace fuses with those of the same rating. Installing a fuse with higher rating can lead to circuit failure and may also start a fire.*
> - *Fuse and relay designations and locations are subject to change and may vary from car to car, depending on options.*

Table a and **Table b** list fuse positions (1 - 21) in the front fuse/relay panel located in the left side of the luggage compartment and fuse positions (1 - 3) in the rear fuse/relay panel in the engine compartment.

NOTE —
In the wiring diagrams fuse positions 1 - 3 in the engine compartment fuse/relay panel may also be referred to as fuses 22 - 24.

Luggage compartment fuse/relay panel

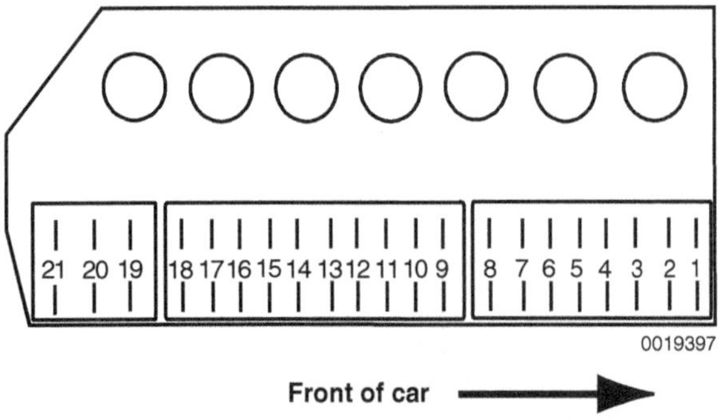

Table a. 1978 - 1980 fuse positions

Fuse	Rating	Protected circuit
Luggage compartment fuse/relay panel		
1	16	Fog lights
2	5	License plate lights Control light &relay for fog lights
3	5	Parking lights (right)
4	5	Parking lights (left) Engine compartment light
5	8	Low beam (right)
6	8	Low beam (left)
7	8	High beam (right)
8	8	High beam (left) High beam indicator
9	5	Front flasher light (right)
10	5	Front flasher light (left)
11	16	Brake lights, flasher relay, back-up lights, Automatic speed control
12	25	Rear window defogger relay, Fresh air blower, Control light for rear window defogger switch
13	25	Windshield wiper, windshield washers
14	25	Electric sunroof, rear window wiper, Outside mirrors
15	16	Cigarette lighter, control light for blower switch
16	25	Fuel pump
17	16	Emergency flashers
18	5	Interior lights, clock, glove compartment light, Luggage compartment light
19	16	Headlight washers
20	25	A/C blower fan
21	25	Power windows
Engine compartment fuse/relay panel		
1	5	Heater fan relay
2	25	Heater fan
3	25	Defogger Rear window wiper

Luggage compartment fuse/relay panel

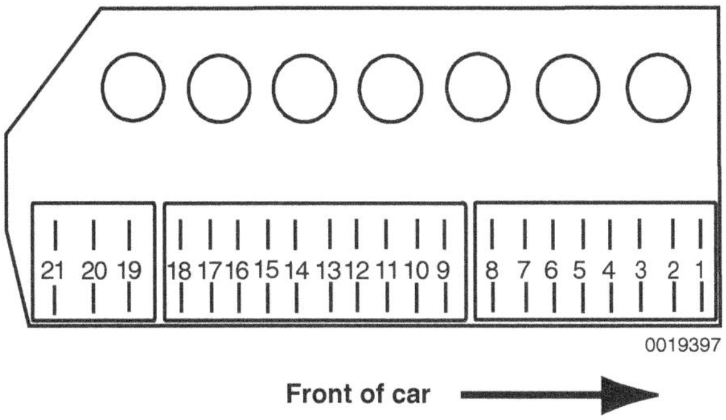

Table b. 1981 - 1983 fuse positions

Fuse	Rating	Protected circuit
Luggage compartment fuse/relay panel		
1	16	Fog lights
2	5	License plate lights
3	5	Parking lights (right)
4	5	Parking lights (left) Engine compartment light
5	8	Low beam (right)
6	8	Low beam (left)
7	8	High beam (right)
8	8	High beam (left) High beam indicator
9	5	Turn signal light (right)
10	5	Turn signal light (left)
11	16	Flasher relay, back-up lights,
12	25	Rear window defogger relay,Fresh air blower Control light for rear window defogger switch
13	25	Windshield wiper, windshield washers, cigarette lighter
14	25	Electric sunroof, rear window wiper, Outside mirrors
15	16	Brake lights, auto speed control, control light for blower switch
16	25	Fuel pump
17	16	Emergency flashers, wipers, defogger
18	5	Interior lights, clock, glove compartment light, Luggage compartment light
19	16	Headlight washers
20	25	A/C blower fan
21	25	Power windows
Engine compartment fuse/relay panel		
1	25	Rear window defogger and wiper
2	25	Heater blower
3	5	Fuse for fan relay

Relays and control units

The luggage compartment and the engine compartment fuse/relay panel are illustrated in Fig. 971-1 through Fig. 971-4.

Relay and control unit positions are included in the comprehensive list of electrical components in **Table d**. The last column of the table refers to applicable photos or illustrations shown in this section.

Front fuse and relay panel

The front fuse and relay panel is mounted in the left side of the luggage compartment.

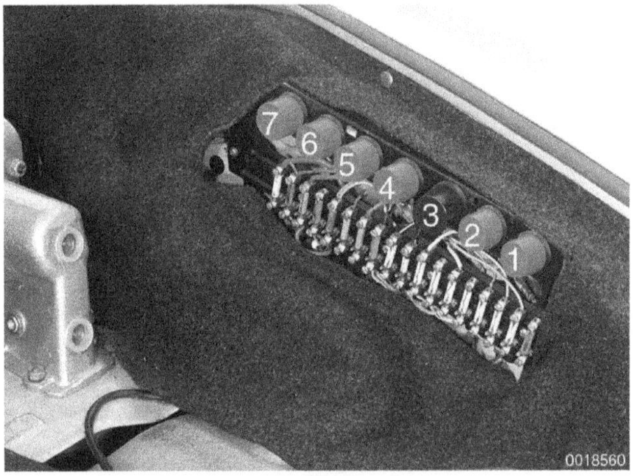

971-1 **Front fuse and relay panel 1978-1980**

Relay positions

1. Not used
2. Fog lights
3. Horn
4. Not used
5. Power windows
6. Not used
7. Fuel pump

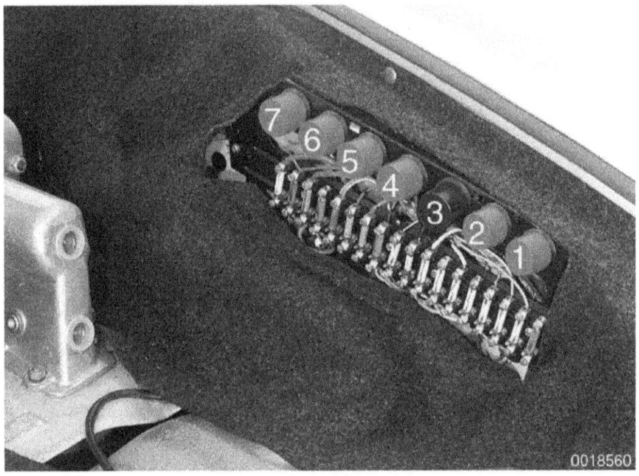

971-2 **Front fuse and relay panel 1981-1983**

Relay positions

1. Air conditioner
2. Fog lights
3. Horn
4. Automatic speed control
5. Power windows
6. Not used
7. Fuel pump

Rear fuse and relay panel

The rear fuse and relay panel is located to the left of the engine behind an access cover.

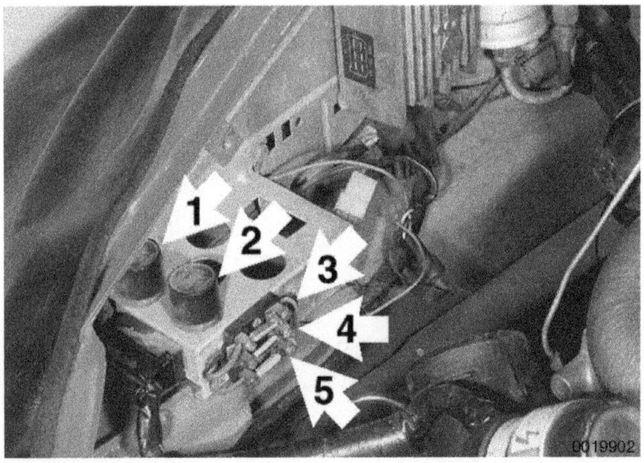

971-3 **Engine compartment fuse and relay 1978-1980**

1. Rear window defogger relay
2. Heater blower relay
3. Heater blower relay fuse
4. Heater blower fuse
5. Rear window defogger fuse

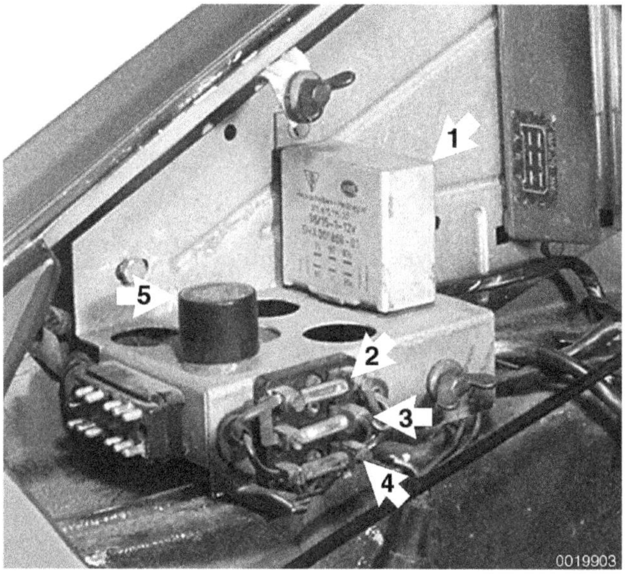

971-4 Engine compartment fuse and relay 1981-1983

1. Rear window defogger relay
2. Rear window defogger & rear wiper fuse
3. Heater fan fuse
4. Heater fan relay fuse
5. Heater fan relay

GROUND AND COMPONENT LOCATIONS

Ground locations

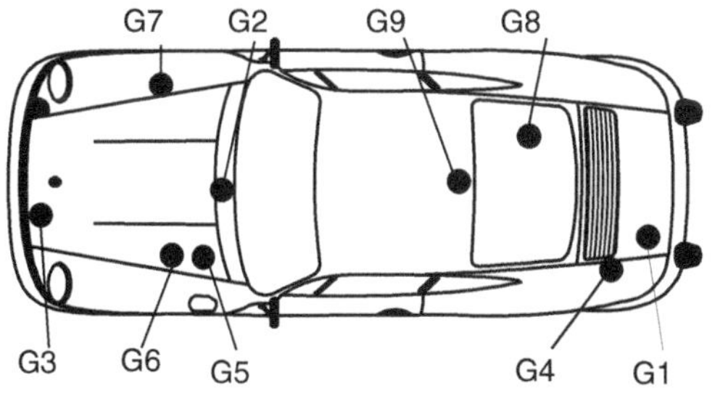

The grounds used in Porsche 911 SC models from 1978 to 1983 are shown in **Table c**. These ground points are also shown in **970 Electrical Wiring Diagrams**.

Table d lists electrical components and their locations. Numbers in the **Reference figure** column refer to photos or illustrations in this group.

Table c. Ground locations (quick reference)

Ground	Location	Reference figure
G1	Engine compartment, left side	
G2	Passenger compartment (behind heater/defroster controls)	
G3	Battery ground strap, luggage compartment (left front beside battery)	
G4	Engine compartment, left (behind fuel filter)	971-4
G5	Luggage compartment floor plate, left (below fuel filler tube)	971-8
G6	Luggage compartment, left (on inner fender panel at rear of fuse box)	971-7
G7	Electric antenna (right front fender inner panel)	
G8	Ground strap, engine/transmission/body (in front right transmission mount)	
G9	Roof, center, interior light and raised stop light ground (convertible only)	

Table d. Component locations

Component	Model Year	Location	Reference figure
A/C blower resistor	1978 - 1983	Under right front foot rest	971-13
A/C blower switch	1978 - 1983	Center console	
A/C clutch	1978 - 1983	Right side of engine mounted on A/C compressor	
A/C condenser blower	1978 - 1983	Luggage compartment, left front	971-5
A/C evaporator blower	1978 - 1983	Luggage compartment, center, under access panel	971-6
A/C evaporator blower relay	1978 - 1983	Luggage compartment, center, under access panel	971-6
A/C high pressure switch	1978 - 1983	Not used	
A/C low pressure switch	1978 - 1983	Not used	
A/C temperature switch	1978 - 1983	Center console	
Air flow meter	1978 - 1983	Top of engine	
Alarm horn	1980 -1983	Underneath car, behind fuel pump	971-23
Alarm system control unit	1980 -1983	Luggage compartment, under fresh air blower	971-10
Alarm system control switch	1980 -1983	Driver door edge, above latch	
Alternator	1978 - 1983	Rear of engine, behind cooling fan	
Ashtray light	1978 - 1983	Above ashtray	
Automatic heating control		See Heater blower switch	
Auxiliary air regulator	1978 - 1983	Right side of engine	971-19
Auxiliary fuse/relay panel	1984 - 1989	Engine compartment left side	971-3, 971-4
Back-up light switch	1978 - 1983	Next to gear selector rod at front of transmission	
Back-up lights	1978 - 1983	In taillight assemblies	
Battery	1978 - 1983	Left front luggage compartment	
Brake fluid level switch	1978 - 1983	In brake master cylinder cap	
Brake light switches	1978 - 1983	In luggage compartment mounted on brake master cylinder	971-9
Buzzer contact	1978 -1983	In ignition switch	
Capacitor discharge ignition (CDI) control unit	1978 - 1983	Engine compartment electrical panel	971-15
Charge indicator light (generator)	1978 -1983	In tachometer, dash	
Cigarette lighter	1978 - 1983	Next to glove box latch	
Clock	1978 - 1983	Dashboard, right side	
Cruise control actuator		See Cruise control servo	
Cruise control servo	1978 - 1983	Engine compartment, to left of rear shock mount	971-15
Cold acceleration control unit	1981- 1983	Under right side (passenger) seat	971-18
Cruise control unit	1978 - 1983	Below radio, under dash	
Cruise control switch	1978 - 1983	Right side steering column stalk	
Door contact switch, left	1978 - 1983	Door hinge post	
Door contact switch, right	1978 - 1983	Door hinge post	
Emergency/turn signal flasher	1978 - 1983	Behind dash, left side	971-12
Emergency flasher switch	1978 - 1983	In dash, next to radio	
Engine compartment fan		See Heater blower motor	
Fan temperature switch	1978- 1983	Top of engine, center, flywheel end	
Frequency valve		See Oxygen sensor frequency valve	
Fresh air blower	1978 - 1983	In luggage compartment, rear	971-11
Front fog light relay	1978 - 1983	In luggage compartment fuse/relay panel, relay # 2	971-1, 971-2
Front fog light switch	1978 - 1983	In dash, next to radio	

Table d. Component locations

Component	Model Year	Location	Reference figure
Front fog lights	1978 - 1983	Below front bumper	
Fuel injection harness connector	1978 - 1983	Rear of engine compartment, beside left rear shock tower	971-14
Fuel level gauge	1978 - 1983	Instrument cluster	
Fuel level sender	1978 - 1983	Top of fuel tank in luggage compartment	971-24
Fuel pump	1978 - 1983	Mounted to front suspension sub-frame	971-23
Fuel pump relay	1978 - 1983	Luggage compartment fuse/relay panel, relay 7	971-1, 971-2
Fuse/relay panel, front	1978 - 1983	Luggage compartment, left front fender inner panel	971-1, 971-2
Fuse/relay panel, rear	1978 - 1983	Engine compartment, left	971-3
Glove compartment light	1978 - 1983	In dash, above glove compartment door	
Ground points	1978 - 1983	Distributed throughout vehicle	Table c
Headlight cleaning control unit	1978 - 1983	Relay panel behind instrument gauges	
Headlight cleaning pump	1978 - 1983	Luggage compartment, in front of battery	
Headlight cleaning switch	1978 - 1983	Dashboard, between gauges	
Headlight dimmer switch	1978 - 1983	Right side steering column stalk	
Headlight switch	1978 - 1983	left side of dash, next to ignition switch	
Heater blower motor	1978 - 1983	Engine compartment, left	
Heater blower relay	1978 - 1983	Engine compartment fuse/relay panel	971-3, 971-4
Heater blower switch	1978 - 1983	Between front seats	
Heater control	1978 - 1983	In dash on right side of steering column	
Heater lever contacts	1978 - 1983	Between front seats	
Heater temperature sensor	1978 - 1983	Left heat exchanger	
High beam indicator light	1978 - 1983	In tachometer	
Horn contacts	1978 - 1983	In steering wheel pad	
Horn relay	1978 - 1983	In luggage compartment fuse/relay panel, relay 4	971-1, 971-2
Horns, high and low	1978 - 1983	Behind right headlight	
Ignition coil	1978 - 1983	Left side engine compartment mounted to inner panel	
Ignition control unit	1978 - 1983	Engine compartment, left side behind access cover	
Ignition distributor	1978 - 1983	Left side of crankcase, behind crankshaft pulley	
Instrument lamps	1978 - 1983	Dashboard gauges	
Interior light, left, right	1978 - 1983	Above doors	
Intermittent wiper control unit		See Windshield wiper intermittent relay	
Intermittent wiper switch		See Windshield wiper intermittent switch	
Light switch	1978 - 1983	Dashboard, next to ignition switch	
Luggage compartment light	1978 - 1983	Luggage compartment hood, right rear	
Mileage counter for OXS light	1980 - 1983	Behind speedometer in dash	971-17
Mirror control	1978 - 1983	Interior door panel below vent window	
Mirror heating switch		See Rear defogger switch	
Oil level transmitter	1978- 1983	In oil tank behind right rear wheel	
Oil pressure sending unit	1978 - 1983	Right side of crankshaft pulley	
Oil pressure switch	1978 - 1983	Top of crankcase on flywheel end	
Oil temperature sensor	1978 - 1983	Right side crankshaft pulley	
Oxygen sensor	1980 - 1983	Rear of catalytic converter	971-22
Oxygen sensor control unit	1981 - 1983	Under right front seat	971-18
Oxygen sensor frequency valve	1981 - 1983	Between engine and rear bulkhead	971-16

Table d. Component locations

Component	Model Year	Location	Reference figure
Oxygen sensor system power relay	1981 - 1983	Under right front seat	971-18
Oxygen sensor harness connector	1980 - 1983	Left side of engine compartment	
Parking brake contacts	1978 - 1983	Parking brake bracket, rear	
Power window motor	1978 - 1983	Inside door	
Power window relay	1978 - 1983	Relay panel, behind gauges	
Power window relay	1978 - 1983	Luggage compartment fuse/relay panel, relay 2	971-2
Power window switch	1978 - 1983	On door trim panel	
Pulse transmitter	1978 - 1983	See Windshield wiper intermittent relay	
Radio	1978 - 1983	Dashboard, center	
Radio speaker, front left, front right	1978 - 1983	On door trim panel, left, right	
Radio speakers rear	1978 - 1983	Utility shelf, below rear window	
Rear window defogger	1978 - 1983	In rear window	
Rear window defogger and mirror heating switch	1978 - 1983	Above radio	
Rear window defogger relay	1978 - 1983	Engine compartment fuse/relay panel	971-3, 971-4
Rear window wiper motor	1978 - 1983	Inside rear engine lid	
Rear window wiper switch	1978 - 1983	Dashboard, between gauges	
Reverse lights		See Back-up lights	
Seat belt contact	1978 - 1983	Seat belt latch	
Seat belt warning relay	1978 - 1983	In relay panel behind fuel gauge in dash	
Seat belt switch		See Seat belt contact	
Seat belt warning light	!978 - 1983	In dashboard	971-12
Speedometer	1978 - 1983	Instrument cluster	
Speedometer pick-up (speed transmitter)	1978 - 1983	Right side of transmission final drive, above output flange	
Starter	1978 - 1983	Right side transmission on bell housing	
Sunroof motor	1978 - 1983	Inside roof headliner between sunroof and rear window	
Sunroof switch	1978 - 1983	Above ignition switch	
Tachometer	1978 - 1983	Center gauge in instrument cluster	
Throttle switch (idle/full)	1980 - 1983	Right side of throttle body	
Throttle switch (15°)	1981-1983	Throttle housing, left side	
Turn signal flasher	1978 - 1983	In luggage compartment left rear, in front of power window relay	
Turn signal switch	1978 -1983	Left side steering column stalk	
Voltage regulator, external	1978 - 1980	Mounted in engine compartment electrical panel	971-28
Voltage regulator, internal	1981 - 1983	Mounted to rear of alternator	971-29
Warm-up (control pressure) regulator	1978 - 1983	Between intake runners 1 and 2	971-21
Windshield defroster blower		See Fresh air blower	
Windshield defroster switch	1978 - 1983	Beside radio	
Windshield washer pump	1978 - 1983	Luggage compartment, in front of battery	971-5
Windshield washer switch	1978 - 1983	See Windshield wiper switch	
Windshield wiper diode	1978 - 1983	Luggage compartment, under fresh air blower	
Windshield wiper intermittent relay	1978 - 1983	Luggage compartment, behind fresh air blower	
Windshield wiper intermittent switch	1978 - 1983	Dashboard, between speedometer and clock	
Windshield wiper motor	1978 - 1983	Behind speedometer and clock	

Table d. Component locations

Component	Model Year	Location	Reference figure
Windshield wiper pulse transmitter	1978 - 1983	See Windshield wiper intermittent relay	
Windshield wiper switch	1978 - 1983	Right side steering column	

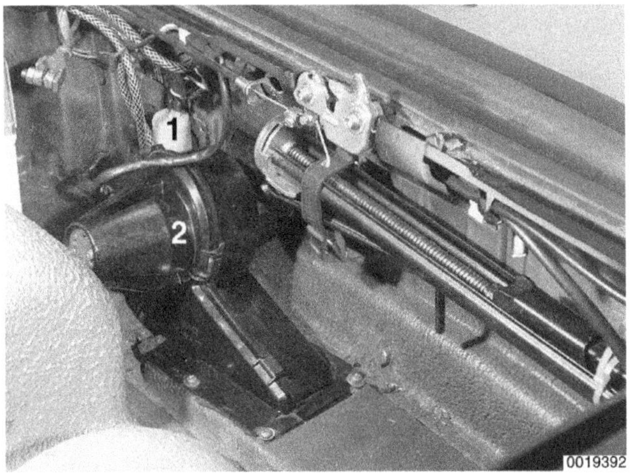

971-5 Luggage compartment, left front

1. Windshield washer pump
2. A/C condenser blower

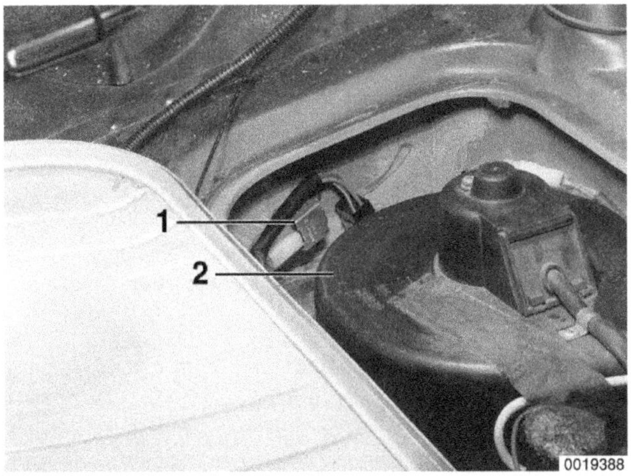

971-6 Luggage compartment, center, under hinged access panel

1. A/C evaporator blower relay
2. A/C evaporator blower

971-7 Luggage compartment, left, behind fuse/relay panel

1. Ground point G6

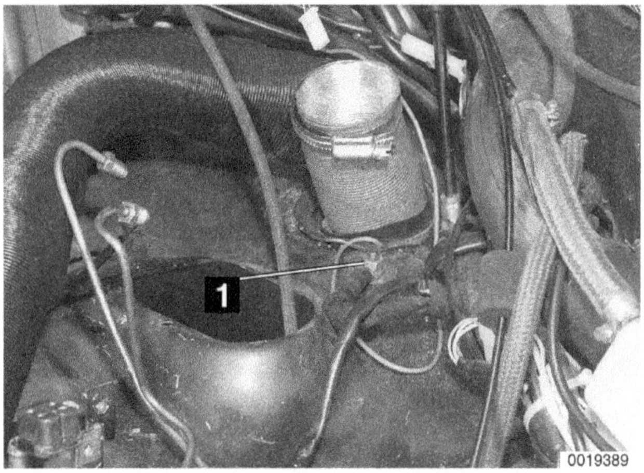

971-8 Luggage compartment, left rear (brake booster/master cylinder assembly removed)

1. Ground point G5

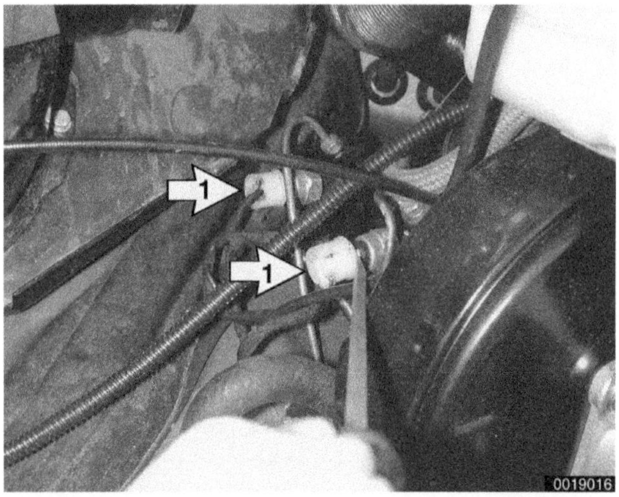

971-9 Luggage compartment, left rear, at brake master cylinder

1. Brake light switches

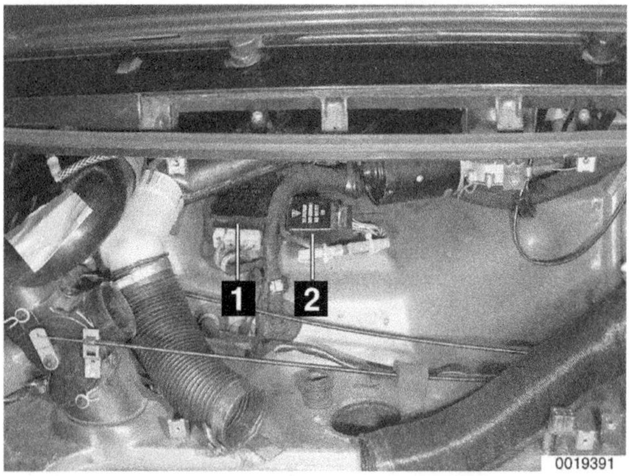

971-10 Luggage compartment, center rear (fresh air plenum removed)

1. Alarm system control unit
2. Wiper motor intermittent relay

971-11 Luggage compartment, center rear

1. Fresh air blower

971-12 Dashboard, left (left side instruments removed)

1. Seat belt warning relay
2. Windshield washer relay
3. Turn signal/emergency flasher relay (not shown)

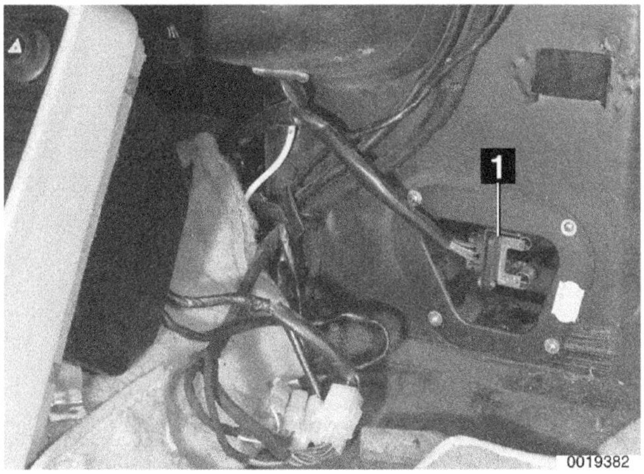

971-13 Front bulkhead, ahead of passenger seat

1. A/C blower resistor (bi-metallic switch)

971-14 Beside left rear shock tower

1. Engine fuel injection harness connection

971-15 Engine compartment, left front

1. Cruise control servo
2. Oxygen sensor plug connection
3. CDI ignition control unit
4. Fuel filter

971-16 Between engine and rear bulkhead

1. Oxygen sensor frequency valve

971-17 Behind speedometer in dash

1. OXS light mileage counter. Reset button shown at **arrow**.

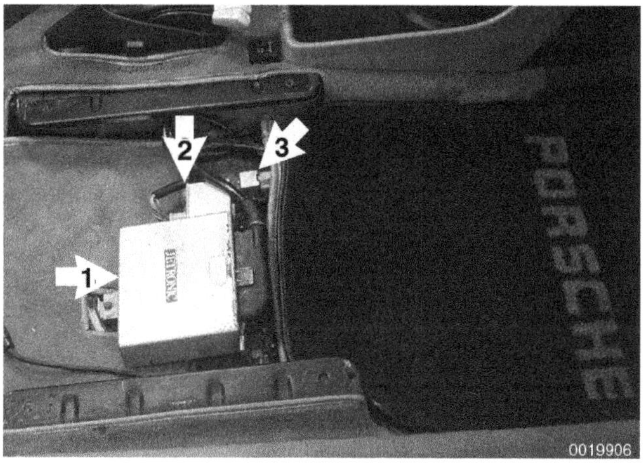

971-18 Under right side (passenger) seat

1. Oxygen sensor control unit (1980 and later) 2. Cold acceleration control unit (1981 and later) 3. Oxygen sensor control unit power relay (1980 and later)

971-19 Right side of engine

1. Auxiliary air regulator

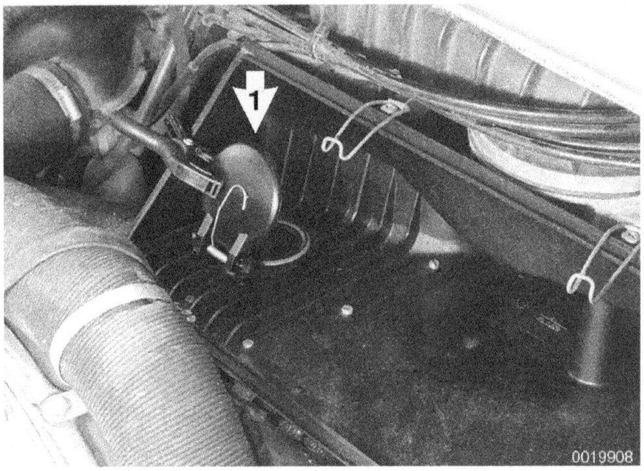

971-20 Inside air filter housing

1.	Pop-off valve (aftermarket add-on)

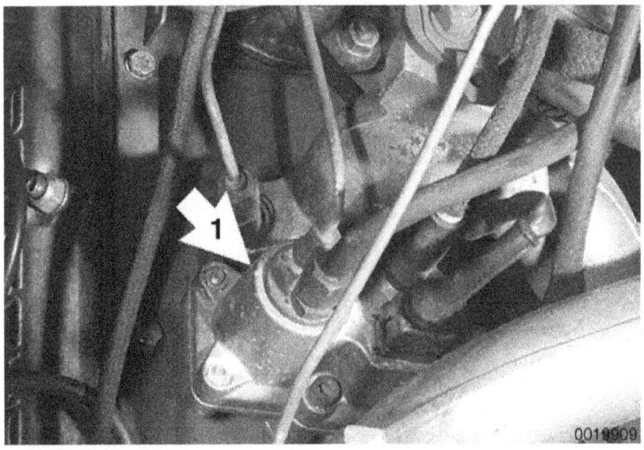

971-21 Between intake runners 1 and 2

1.	Warm-up regulator (control pressure regulator)

971-22 Catalytic converter

1.	Oxygen sensor

971-23 Behind front suspension cross member

1.	Fuel pump
2.	Alarm horn

971-24 Under luggage compartment mat

1.	Fuel gauge sending unit

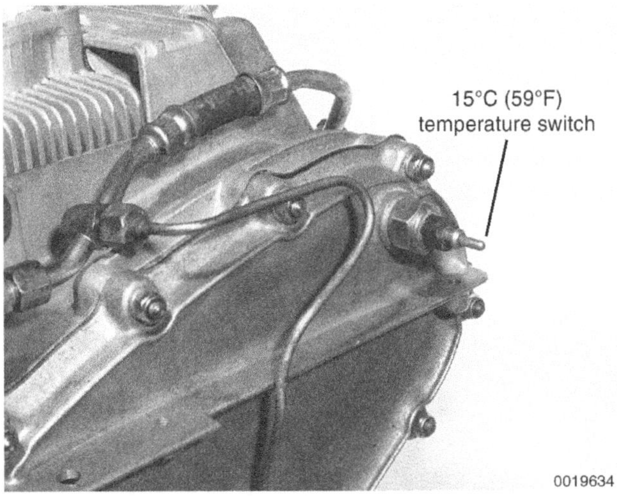

971-25 Right timing chain cover, front of engine

1.	Temperature switch 15°C (59°F)

971-26 Engine breather cover, rear of engine

1. Temperature switch 35°C/95°F (1981 and later)

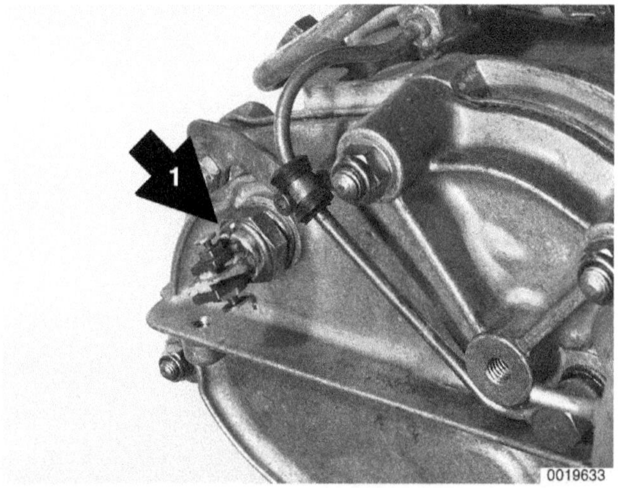

971-27 Left timing chain cover, front of engine

1. Thermo-time switch for cold start valve

971-28 Left side of engine compartment

1. Voltage regulator, external (1978 - 1981 models)

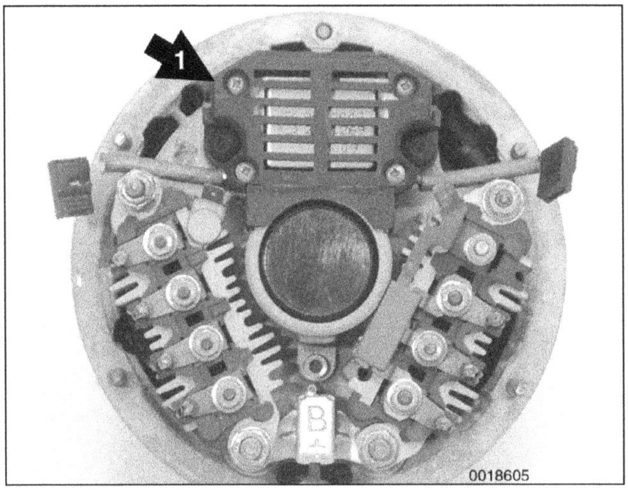

971-29 Alternator housing, rear

1. Voltage regulator, internal (1982 - 1983 models)

> **WARNING**
>
> Your common sense, good judgement, and general alertness are crucial to safe and successful service work. Before attempting any work on your Porsche, read the warnings and cautions on page vii and the copyright page at the front of the manual. Review these warnings and cautions each time you prepare to work on your Porsche. Please also read any warnings and cautions that accompany the procedures in the manual.

A

B

> **WARNING**
>
> Your common sense, good judgement, and general alertness are crucial to safe and successful service work. Before attempting any work on your Porsche, read the warnings and cautions on page vii and the copyright page at the front of the manual. Review these warnings and cautions each time you prepare to work on your Porsche. Please also read any warnings and cautions that accompany the procedures in the manual.

WARNING

Your common sense, good judgement, and general alertness are crucial to safe and successful service work. Before attempting any work on your Porsche, read the warnings and cautions on page vii and the copyright page at the front of the manual. Review these warnings and cautions each time you prepare to work on your Porsche. Please also read any warnings and cautions that accompany the procedures in the manual.

F

> **WARNING**
>
> Your common sense, good judgement, and general alertness are crucial to safe and successful service work. Before attempting any work on your Porsche, read the warnings and cautions on page vii and the copyright page at the front of the manual. Review these warnings and cautions each time you prepare to work on your Porsche. Please also read any warnings and cautions that accompany the procedures in the manual.

G

H

> **WARNING**
>
> Your common sense, good judgement, and general alertness are crucial to safe and successful service work. Before attempting any work on your Porsche, read the warnings and cautions on page vii and the copyright page at the front of the manual. Review these warnings and cautions each time you prepare to work on your Porsche. Please also read any warnings and cautions that accompany the procedures in the manual.

> ***WARNING***
>
> Your common sense, good judgement, and general alertness are crucial to safe and successful service work. Before attempting any work on your Porsche, read the warnings and cautions on page vii and the copyright page at the front of the manual. Review these warnings and cautions each time you prepare to work on your Porsche. Please also read any warnings and cautions that accompany the procedures in the manual.

Selected Books and Repair Information From Bentley Publishers

Motorsports

Alex Zanardi: My Sweetest Victory
Alex Zanardi and Gianluca Gasparini
ISBN 978-0-8376-1249-2

The Unfair Advantage *Mark Donohue and Paul van Valkenburgh*
ISBN 978-0-8376-0069-7

Driving Forces: The Grand Prix Racing World Caught in the Malestrom of the Third Reich
Peter Stevenson
ISBN 978-0-8376-0217-2

Equations of Motion - Adventure, Risk and Innovation *William F. Milliken*
ISBN 978-0-8376-1570-7

Engineering

The Hack Mechanic Guide to European Automotive Electrical Systems *Rob Siegel*
ISBN 978-0-8376-1751-0

Physics for Gearheads *Randy Beikmann*
ISBN 978-0-8376-1615-5

Bosch Automotive Handbook
Updated 9th Edition *Robert Bosch GmbH*
ISBN 978-0-8376-1732-9

Bosch Fuel Injection and Engine Management *Charles O. Probst, SAE*
ISBN 978-0-8376-0300-1

Maximum Boost: Designing, Testing, and Installing Turbocharger Systems
Corky Bell ISBN 978-0-8376-0160-1

Supercharged! Design, Testing and Installation of Supercharger Systems
Corky Bell ISBN 978-0-8376-0168-7

Race Car Aerodynamics *Joseph Katz*
ISBN 978-0-8376-0142-7

Audi Repair Manuals

Audi A4 Service Manual: 2002-2008, 1.8L Turbo, 2.0L Turbo, 3.0L, 3.2L
Bentley Publishers
ISBN 978-0-8376-1574-5

Audi A4 Service Manual: 1996-2001, 1.8L Turbo, 2.8L *Bentley Publishers*
ISBN 978-0-8376-1675-9

Audi TT Service Manual: 2000-2006, 1.8L turbo, 3.2 L *Bentley Publishers*
ISBN 978-0-8376-1625-4

Audi A6 (C5 platform) Service Manual: 1998-2004 *Bentley Publishers*
ISBN 978-0-8376-1670-4

BMW

BMW 3 Series (F30, F31, F34) Service Manual: 2012-2015 *Bentley Publishers*
ISBN 978-0-8376-1752-7

BMW 3 Series (E90, E91, E92, E93) Service Manual: 2006-2011 *Bentley Publishers* ISBN 978-0-8376-1723-7

BMW 3 Series (E36) Service Manual: 1992-1998 *Bentley Publishers*
ISBN 978-0-8376-1709-1

BMW 3 Series (E46) Service Manual: 1999-2005 *Bentley Publishers*
ISBN 978-0-8376-1657-5

BMW X3 (E83) Service Manual: 2004-2010 *Bentley Publishers*
ISBN 978-0-8376-1731-2

BMW X5 (E53) Service Manual: 2000-2006 *Bentley Publishers*
ISBN 978-0-8376-1643-8

BMW 5 Series (E60, E61) Service Manual: 2004-2010 *Bentley Publishers*
ISBN 978-0-8376-1689-6

BMW 5 Series (E39) Service Manual: 1997-2003 *Bentley Publishers*
ISBN 978-0-8376-1672-8

BMW Z3 Service Manual: 1996-2002
Bentley Publishers
ISBN 978-0-8376-1617-9

BMW 7 Series (E32) Service Manual: 1988-1994 *Bentley Publishers*
ISBN 978-0-8376-1619-3

Memoirs of a Hack Mechanic
Rob Siegel ISBN 978-0-8376-1720-6

Porsche

Porsche 911 (964): Enthusiast's Companion *Bentley Publishers*
ISBN 978-0-8376-0293-6

Porsche: Excellence Was Expected
Karl Ludvigsen ISBN 978-0-8376-0235-6

Ferdinand Porsche — Genesis of Genius
Karl Ludvigsen ISBN 978-0-8376-1557-8

Porsche – The Man and His Cars
Richard von Frankenberg
ISBN 978-0-8376-0329-2

Porsche – Origin of the Species
Karl Ludvigsen ISBN 978-0-8376-1331-4

Volkswagen

Volkswagen Jetta, Golf, GTI Service Manual: 1999-2005 *Bentley Publishers*
ISBN 978-0-8376-1678-0

Volkswagen Jetta Service Manual: 2005-2010 *Bentley Publishers*
ISBN 978-0-8376-1616-2

Volkswagen Passat Service Manual: 1998-2005 *Bentley Publishers*
ISBN 978-0-8376-1669-8

Volkswagen Jetta, Golf, GTI: 1993-1999, Cabrio: 1995-2002 Service Manual *Bentley Publishers*
ISBN 978-0-8376-1660-5

Volkswagen GTI, Golf, Jetta Service Manual: 1985-1992 *Bentley Publishers*
ISBN 978-0-8376-1637-7

Battle for the Beetle *Karl Ludvigsen*
ISBN 978-0-8376-0071-0

MINI Repair Manuals

MINI Cooper Service Manual: 2002-2006 *Bentley Publishers*
ISBN 978-0-8376-1639-1

MINI Cooper Service Manual: 2007-2013 *Bentley Publishers*
ISBN 978-0-8376-1730-5

Mercedes-Benz

Mercedes-Benz C-Class (W202) Service Manual 1994-2000
Bentley Publishers
ISBN 978-0-8376-1572-1

Mercedes Benz E-Class (W124) Owner's Bible: 1986-1995
Bentley Publishers
ISBN 978-0-8376-0230-1

Porsche Repair Manuals

Porsche 911 Carrera (Type 993) Service Manual: 1995-1998
Bentley Publishers
ISBN 978-0-8376-1719-0

Porsche 911 Carrera Service Manual: 1984-1989
Bentley Publishers
ISBN 978-0-8376-1696-4

Porsche 911 SC Service Manual: 1978-1983
Bentley Publishers
ISBN 978-0-8376-1705-3

Porsche 911 Carrera (Type 996) Service Manual: 1999-2005
Bentley Publishers
ISBN 978-0-8376-1710-7

BentleyPublishers®
Automotive Reference

Bentley Publishers has published service manuals and automobile books since 1950. For more information, please contact Bentley Publishers at 1734 Massachusetts Avenue, Cambridge, MA 02138 USA, or visit our web site at **BentleyPublishers.com**